Advanced Series in
FULLERENES
Vol. 1

PHYSICS & CHEMISTRY OF FULLERENES

A REPRINT COLLECTION

Advanced Series in
FULLERENES
Vol. 1

PHYSICS & CHEMISTRY OF FULLERENES

A REPRINT COLLECTION

edited by

Peter W. Stephens

*State University of New York
Stony Brook, NY
USA*

 **World Scientific**
Singapore • New Jersey • London • Hong Kong

Published by

World Scientific Publishing Co. Pte. Ltd.

P O Box 128, Farrer Road, Singapore 9128

USA office: Suite 1B, 1060 Main Street, River Edge, NJ 07661

UK office: 73 Lynton Mead, Totteridge, London N20 8DH

PHYSICS AND CHEMISTRY OF FULLERENES

ISBN 981-02-1116-3
ISBN 981-02-1117-1 (pbk)

The editor and publisher are grateful to the authors and the following publishers for their assistance and their permission to reproduce the articles found in this volume:

The American Association for the Advancement of Science (*Science*)
The American Chemical Society (*J. Phys. Chem. and J. Am. Chem. Soc.*)
The American Institute of Physics (*Appl. Phys. Lett.*)
The American Physical Society (*Phys. Rev. Lett.*)
Elsevier Science Publishers (*Chem. Phys. Lett.*)
Europhysics Physical Society (*Europhys. Lett.*)
Macmillan Magazines Limited (*Nature*)
Royal Society of Chemistry (*J. Chem. Soc. Chem. Comm.*)

Printed in Singapore.

"You are probably aware that fuller's-earth is a valuable product, ... quite as valuable as a gold-mine."

Sir Arthur Conan Doyle
"The Adventure of the Engineer's Thumb"

Contents

INTRODUCTION

Like so many important scientific discoveries, Buckminsterfullerene was found while alert individuals were looking for something else. In 1985, Kroto, Heath, O'Brien, Curl and Smalley were studying the formation of long-chain carbon molecules in circumstances analogous to interstellar space and stellar atmospheres. They observed that under certain conditions, clusters consisting of sixty carbon atoms were produced in unusual abundance. Their suggestion that this sixty-atom cluster might be a hollow sphere of carbon immediately captured the attention of scientists worldwide. However, for some time, experimental results came from only a few groups because of the elaborate laser vaporization supersonic cluster beam apparatus required to produce the species. Indeed, without bulk sample quantities available, the fullerene hypothesis was only indirectly testable. Nevertheless, quite precise predictions were made, many to be subsequently verified.

An explosion of fullerene research occurred with the invention of techniques for the production and isolation of bulk quantities of fullerenes in 1990. Due to the relative simplicity of sample preparation, many research groups have been able to bring the tools and insights of their specialty to the study of fullerenes, and we have seen extraordinary scientific progress.

This book is an attempt to bring together the most important research papers from these first few years of fullerene research. As such, it is a rather subjective collection, and some difficult choices had to be made (quickly). It is this editor's viewpoint that the field is growing, however rapidly, in however many directions, from a strong foundation in a certain number of excellent scientific papers. It is appropriate that researchers working in the field and entering the field should share this common canon of original literature. By and large, physicists do not read the *Journal of the American Chemical Society*, and chemists do not read *Physical Review Letters*. With this volume, we can hope to firm up the starting point for further progress.

I have broken down the general field of fullerenes into several chapters, and offer a brief discussion of the pathway that has been followed for each. There are a number of excellent review papers, which cover various topics in much greater detail and authority than is possible here. I have cited them in the appropriate chapter introductions. I have not presumed to discuss in depth those topics for which current reviews are available. The reader should not conclude that I do not find these areas significant, merely that I have little to add. However, I have taken the opportunity to discuss certain topics that have not been adequately covered by other review articles, or that have seen significant progress since the most recent review papers.

Not another word can be said about fullerenes without mentioning the bibliographic data base, started by R. Smalley at Rice University. It was upgraded to electronic mail by Jack Fischer and coworkers at the University of Pennsylvania, and is currently maintained by the fullerene research groups there and by Frank Tinker and Donald Huffman at Arizona State University. The system provides a comprehensive bibliography of published papers and preprints, updated approximately monthly, with various tools for online search, etc. In addition, one can subscribe to a frequent electronic mailing list of new research papers, conference announcements, job openings, etc. One can access and learn more about these services by sending an electronic mail message consisting of the single word, "INTRO" to the following address on Internet: BUCKY@SOL1.LRSM.UPENN.EDU. Further instructions will appear by return mail. This is a useful tool for anyone following the field, and has been invaluable to this editor.

Finally, I wish to acknowledge the help and cooperation of researchers too numerous to mention. I am especially pleased to thank those people with whom I have had particularly useful discussions and worked most directly on various fullerene projects: Yohji Achiba, Phil Allen, David Cox, Jack Fischer, Paul Heiney, Károly Holczer, Gerry Lavin, Peter Lee, Laszlo Mihaly, Rodney Ruoff, John Wiley, Fred Wudl and Taner Yilidrim. Last but not least, I am grateful to my wife Bonnie, for loving support and editorial assistance.

Peter Stephens
Stony Brook, New York
July 1993

THE C$_{60}$ MOLECULE

Our story opens in the very active field of research on clusters, exotic collections of atoms intermediate between molecules and condensed matter. Motivated by an interest in long-chain carbon molecules, Kroto and coworkers were using a powerful laser to vaporize graphite into a carrier gas of helium. Carbon species cooled and partially condensed in the helium, and then were expanded into a vacuum in a supersonic molecular beam. The carbon clusters were then ionized and weighed in a mass spectrometer. The startling discovery made was that under certain experimental conditions, the mass spectrometer signal from C$_{60}$ was many times stronger than any of its neighbors.[Kr85*] This implied that there must be a particularly stable structure for the C$_{60}$ cluster, relative to other carbon clusters of nearby sizes.

They suggested that C$_{60}$ has the soccer ball structure of a truncated icosahedron, a polyhedron having sixty vertices, ninety bonds, twelve pentagonal faces (five-membered rings) and twenty hexagonal faces (six-membered rings). This structure places each carbon atom in an identical molecular environment. Each atom is sp^2 bonded to three of its neighbors, leaving a sea of sixty π electrons to produce many resonance structures. The name buckminsterfullerene was chosen after the architect R. Buckminster Fuller, whose geodesic dome provided a guide to the hollow cage structural hypothesis. The general name, fullerene, has been adopted for all similar hollow shell carbon molecules C$_n$ closed by twelve five-membered rings. The term buckyball is often used more colloquially.

The general necessity for twelve pentagons, with a (nearly) arbitrary number of hexagons follows from Euler's theorem: a network of trivalent nodes comprising P pentagons and H hexagons contains $n = (6H + 5P)/3$ nodes, $b = (6H + 5P)/2$ bonds, and $a = H + P$ areas. In order to close the surface into a spheroid, the Euler characteristic $n - b + a$ must be 2, which implies $P = 12$, with no explicit constraint on H. (In fact, polyhedra exist for any value of H except 1.)

It is useful to consider this idea in the context of earlier experimental observations of carbon clusters.[Ro84] In a similar experiment, that group published, but did not remark on, a factor of two enhancement of the C$_{60}$ cluster relative to its immediate (even) neighbors. One should bear in mind that cluster abundance variations on that order are seen in many systems, but they are not thought to come from a uniquely stable cluster like the hypothesized buckminsterfullerene.

As beautiful as this suggestion might be, significantly more must be done to confirm it than to observe that the combination of sixty carbon atoms can be produced in greater abundance than other numbers. Unfortunately, the laser vaporization supersonic cluster beam technique does not produce enough material to perform direct structural techniques like x-ray crystallography or even indirect (but often decisive) experiments such as infrared or Raman spectroscopy, or nuclear magnetic resonance.

Focusing attention on what was learned from "first generation" cluster beam experiments, three observations are particularly noteworthy. First, mass spectra of carbon clusters heavier than about C$_{40}$ contained only even numbers of atoms.[Ro84, Kr85*] This is also the case if neutral or ionized C$_{60}$ is photofragmented by a high power laser.[O'B88, We88*] This observation supports the fullerene hypothesis, because a fullerene cage of 12 pentagons and m hexagons contains $2m + 20$ atoms, always an even number. Second, the C$_{60}$ signal was always accompanied by C$_{70}$, even when conditions were adjusted to reduce all other carbon clusters far below these two. As we shall see below, C$_{60}$ and C$_{70}$ are the two smallest fullerenes in which no two pentagons share an edge.[Kr87, Sc88, Sa93] Third, complexes of carbon with a single metal atom were produced by laser ablation from graphite disks impregnated with salts of La, K, and Cs.[He85] Their photophysical and chemical stability indicated that the reactive metal atom was trapped inside the fullerene cage. These endohedral fullerenes are discussed in Chapter V of the present volume.

Two excellent reviews during the first generation (cluster beam) of fullerene research are given by Curl and Smalley, and by Kroto.[Cu88, Kr88] The more comprehensive recent review by Kroto et al. also treats this period in detail.[Kr91] Mass spectrometric experiments are also reviewed by Smalley,[Sm92] and in detail by McElvany, Ross, and Callahan.[Mc92] The reader is encouraged to consult these papers for the full story of the important cluster-beam experimental work.

A useful theoretical analysis of the C_{60} buckminsterfullerene molecule (then just freshly hypothesized) was given by Haddon et al.[Ha86*] They worked out the Huckel molecular orbital levels, showing that there is a large gap between the highest occupied molecular orbital (HOMO) and the lowest unoccupied molecular orbital (LUMO), a sign of stability for the proposed structure. (This had been done by several earlier groups, as a purely theoretical exercise.[Bo73, Da81, Ha85]) Haddon et al. also showed that C_{60} had a relatively simple vibrational structure. Because of the high symmetry of the molecule, only four vibrations would be infrared active (observable as infrared absorption lines) and ten vibrations would be allowed in the Raman spectrum.

A number of increasingly sophisticated predictions of electronic structures and vibrational frequencies followed this work, but they are beyond the scope of the present review. However, one discussion of the molecular vibrations has particularly strong tutorial interest, containing stereoscopic pictures of the various vibrational modes.[We89]

On the theoretical front, it is clearly desirable to find more justification for the stability of icosahedral C_{60} than intuition. A first step to the comparative anatomy of fullerenes is an enumeration of all fullerene structures of a given size. Manolopoulos, May, and Down developed a systematic technique for doing this, although it is not proved to be mathematically rigorous.[Ma91*] They found at least 1760 distinct fullerene isomers for C_{60}, a number which has been subsequently improved to 1812.[Fo92] Only one of these is the icosahedrally symmetric structure hypothesized by Kroto et al.[Kr85*] with no pairs of adjacent pentagons. They also estimated the stability of every fullerene isomer up to C_{70}, using Huckel molecular orbital theory. They found that the gap between the highest occupied molecular orbital (HOMO) and the lowest unoccu-

pied molecular orbital (LUMO) was generally correlated very well with the experimental pattern of abundances under conditions of either mild clustering or strong fragmentation, in the mass range C_{42} to C_{70}. Extensions of this work to larger fullerenes is discussed in Chapter II of this volume.

Krätschmer, Fostiropoulos, and Huffman decisively ushered in the second generation of fullerene research with their invention of techniques to produce macroscopic quantities of C_{60} and other fullerenes.[Kr90a*] They produced carbon smoke by resistive heating of carbon rods in an inert gas atmosphere. The soot collected from this smoke had several sharp infrared (IR) absorption features, in contrast to soot collected in the absence of the inert quenching gas. When soot was generated from ^{13}C in place of ^{12}C starting material, the lines all shifted $3.75\% \pm 0.08\%$ lower, in agreement with the 3.9% shift expected for a pure carbon molecule. Four of the IR lines stood out above the rest, and they fell within the range of predictions of the four IR active modes for C_{60}. However, the material responsible for these IR lines was only a small fraction of the soot, and there was no direct evidence that it was C_{60}. (Soon thereafter, Meijer and Bethune used laser desorption mass spectroscopy to show that soot prepared by the Krätschmer et al. recipe did contain C_{60}.[Me90a, Me90b])

The same group (with the addition of L. D. Lamb) then demonstrated that the species responsible for the four IR lines was indeed C_{60}, by dissolving it out of the soot with benzene, and studying the crystals formed when the benzene evaporated. [Kr90b*] The resulting solid had a mass spectrum dominated by C_{60}, with notable C_{70} and fragmentation products, in agreement with the mass spectra of fullerene cluster beams. It had the same IR frequencies as the soot. Most compellingly, the powder x-ray diffraction pattern was that of a hexagonal close packed array of spheres 10.02 Å apart. (We shall have more to say about the x-ray diffraction data in Chapter III of this volume.) This is close to the size that one would have predicted for van der Waals packing of C_{60} molecules.[Kr90b*] The visible-ultraviolet absorption spectra of solid C_{60} was also reported in that work.

Bethune et al. used a different technique, fractional sublimation, to separate fullerene molecules from soot.[Be90*] They loaded soot into a cell, heated it in vacuum, and collected the material which evaporated. Mass spectra showed that it was C_{60} and C_{70}, and that the relative abundance of the two species could be controlled by changing the sublimation temperature. They took Raman spectra, and identified several vibrational modes for each molecule.

The final step to preparation of pure samples in macroscopic quantities was made by two groups, who described the chromatographic separation of C_{60} and C_{70} from soot.[Ta90*, Aj90*] The presence of C_{60} and C_{70} in the appropriate fractions was confirmed by mass spectroscopy of the precipitate. Both groups reported ^{13}C nuclear magnetic resonance (NMR) spectra in solution. ^{13}C carries a nuclear magnetic moment, which is sensitive to the local molecular environment. Observation of a single NMR line in C_{60} confirmed the highly symmetric fullerene structure, in which each atom had the same local environment. The five NMR lines observed in C_{70}, with intensities in the ratio 1:2:1:2:1, could be assigned to a plausible elongated structure with a five-fold axis (D_{5h} symmetry).

Various photophysical properties of C_{60} were given by the UCLA group. The longest wavelength absorption at 620 nm (2 eV) was shown ([Aj90*]) much more clearly than in the first work ([Kr90b*]). No fluorescence (emission of light from the decay of the singlet excited state) or phosphorescence (from the triplet state) was observed (in solution at room temperature), although it was observed that singlet C_{60} readily converted to triplet.[Ar91*] Subsequently, luminescence was observed from solid films of C_{60} at low temperature.[Re91] The triplet level lies only slightly below the lowest singlet excited state, and absorbs light more strongly than the ground state molecule.[Ar91*] The triplet state of C_{60} produces singlet oxygen very efficiently. One therefore expects to find significant amounts of this reactive molecule accompanying any C_{60} in room light. The optical spectrum of C_{70} was also reported.[Aj90*]

The molecular structure of the C_{60} molecule as a hollow sphere was confirmed by the first x-ray experiments.[Kr90b*] The most obvious Kekule structure has sixty single bonds on the twelve pentagons, and thirty double bonds, each shared by two hexagons. An electron diffraction measurement in the gas phase gave values of 1.458 Å and 1.401 Å for the single and double bonds, respectively.[He91] Nuclear magnetic resonance measurements of the solid at 77 K are in agreement, with slightly larger uncertainties.[Ya91] The most accurate measurements of bond lengths have been by x-ray crystallography in the solid state, and they are discussed later in this volume.

One of the most fundamental properties of any molecule is the strength with which it binds together its constituent atoms. Experimentally, this follows from a straightforward measurement of the heat of combustion, with analysis to reduce to a standard state. However, consensus has been elusive, with the following values for the standard heat of formation ΔH_f^0 (relative to graphite) in the literature:

545.0 ± 1.2 kcal/mole	Be92
578.9 ± 3.3	St92
524.2 ± 2.3	Sm93

These are quoted per mole of C_{60}. To compare with other forms of carbon, one must divide by 60, to get values in the neighborhood of 9.1 kcal/mole of carbon atoms. This may be compared with a value of 0.6 kcal/mole for diamond, or 172 kcal/mole for isolated carbon atoms.

A complimentary measure of an object's strength is to try to smash it. This has been done with fullerenes by three groups: two who directed ion beams of fullerenes against various solid surfaces and observed the recoil products,[Be91*, Mc92] and one who modeled the process by molecular dynamics computer simulations.[Mo91, Be91*] Experimentally, it is found that the fullerenes rebound from the surface intact, for C_{60}^+ energies up to 200 eV.[Be91*] This stands in contrast to the behavior of other molecules or clusters, which fragment extensively under comparable conditions. It was experimentally observed that a significant fraction of the cluster impact energy was converted to internal (vibrational) energy of the rebounding cluster. Because the rebounding cluster is detected within a fraction of a microsecond, it may not have time for the vibrational energy to dissociate the C_{60}^+ ion. Indeed, this is consistent with rate constants for dissociation.[S. L. Anderson, pers. commun.] There is also a more recent observation of fragmentation to C_{58}^+ and C_{56}^+ in higher energy collisions of C_{60}^{2+}.[Ca91]

Similar resilience was (experimentally) observed for C_{60}^-, C_{70}^+, and C_{84}^+.[Be91*] Molecular dynamics simulations (reprinted in this volume in Be91*) show that the fullerene ion is squashed flat at the moment of impact, but rebounds intact, even though its internal energy corresponds to a temperature of 3500 K. Like the experiments, the simulations do not continue for a sufficient length of time to determine whether the highly excited fullerene subsequently dissociates.

One of the greatest open questions in the field of fullerenes is the understanding of how such compact, closed structures form so efficiently from carbon vapor. Under appropriate conditions, as much as 20% of the carbon vaporized in an arc or a laser desorption source is incorporated into fullerenes; however, no laboratory chemical synthesis has succeeded! It is known from an isotope-scrambling experiment that the fullerenes are formed by the condensation of carbon atoms in the hot carbon vapor, as opposed, for example, to small sheets of graphite torn from the electrodes, and subsequently curling up.[Me90b]

Several pathways have been suggested. In Smalley's "pentagon road" [Sm92], it is hypothesized that, as atoms condense onto an open sheet of carbon hexagons and pentagons, the most favorable structures have the largest possible number of pentagons, while avoiding any adjacent pentagon pairs. This and other mechanisms are thoroughly reviewed by Curl.[Cu93] While there is no direct experimental evidence against any of the proposed mechanisms, there is not any strong evidence in their favor.

Recently, a group at Santa Barbara has developed a gas-phase ion chromatography technique which can give information about the shape of the cluster ions from a laser desorption source.[Vo91, Vo93*] Briefly, the drift time of an ion through a gas-filled cell depends on its collision cross section, allowing the separation of different isomers.[Bo93] That group finds that the population of ions of a given mass frequently contains several distinct isomers. Based on comparison of the mobilities of different cluster sizes and detailed modeling, they are able to distinguish linear chains, monocyclic rings, polycyclic rings, and fullerenes.[Vo91, Vo93*] In the paper reprinted in this volume, they inject a beam of ions, known to be principally mono-, bi-, and tricyclic planar rings, into the gas cell. If the injection energy is high enough, the rings are excited above the isomerization threshold by collisions with the gas atoms, and then anneal into fullerenes, with the loss of one or more carbon atoms.[Vo93*] Similar experiments have also been described by Hunter et al.[Hu93a, Hu93b] One can extend these results to a plausible picture of fullerene growth in a carbon arc reactor: the carbon vapor readily condenses into chains and large planar rings, which anneal into fullerenes under conditions of collisional heating.

The behavior of fullerenes in solution and in solids is strongly determined by its oxidation-reduction potentials, the energies to add or remove an electron. This information is directly accessed in electrochemical experiments, in which the flow of electrons onto a C_{60} molecule in solution is measured as a function of their electric potential. There have been several improvements in the choice of solvent and other experimental conditions, leading to an extension of the range of observable potentials, and therefore greater reduction. The paper reporting the current record of C_{60}^{6-} is reprinted in this volume.[Xi92*] This is the charge state of a fullerene with its six-fold degenerate LUMO fully occupied by electrons.

We continue this discussion of the C_{60} molecule with a few brief comments about other sources and observations of fullerenes. Between the discovery of fullerenes in laser ablation cluster beam experiments and their production in bulk from specially prepared soot, fullerene ions were seen in sooting hydrocarbon-oxygen flames.[Ge87] The C_{60}^{+} ion signal, and to a lesser extent, C_{50}^{+} and C_{70}^{+} stood out very strongly against other (even) carbon ions. Interestingly, when the experimental conditions were modified to collect negative ions, the C_{60}^{-} peak was not so distinctive, but C_{74}^{-} and C_{82}^{-} did stand out. Subsequently, after the procedures for the isolation of fullerenes from graphite arc-produced soot were developed, fullerenes C_{60} and C_{70} were purified from the soot derived from flames.[Ho91] Interestingly, it was found that the C_{70}/C_{60} ratio was much higher than that observed in graphite vaporization. The synthesis of fullerenes in combustion has been reviewed by Howard et al.[Ho92]

Returning to the question which motivated the experiments which in turn led to the discovery of the fullerenes: do such carbon clusters exist in nature? Astronomical searches for the distinctive fullerene signature of infrared absorption lines have been unsuccessful, and laboratory spectra of fullerenes have not shown any explicit connection to unsolved astrophysical problems such as the so-called diffuse interstellar bands or other unidentified spectral features.[Ha92]

Closer to home, C_{60} and C_{70} have been observed in geological samples of an unusual carbon-rich metamorphic rock known as shungite.[Bu92] The geological processes leading to its production are not known. Another geological sample containing fullerenes has been discovered in a fulgurite, the glassy material that is formed at the location of a lightning strike.[Da93]

We conclude this chapter with a few brief notes about the availability of fullerenes for research applications. The Bucky bibliography database, described in the introduction, lists several commercial suppliers of fullerenes. As of this writing, the quoted price is around $250 per gram of purified C_{60}. There are also several detailed descriptions of the construction and operation of fullerene soot reactors and solvent-reflux purification systems in the literature.[Ko91, Kh92].

Bibliography

References marked with * are reprinted in this volume.

Aj90*   H. Ajie et al., "Characterization of the Soluble All-Carbon Molecules C_{60} and C_{70}," J. Phys. Chem. 94, 8630–8633 (1990).

Ar91*   J. W. Arbogast et al., "Photophysical Properties of C_{60}," J. Phys. Chem. 95, 11–12, (1991).

Be90*   D. S. Bethune, G. Meijer, W. C. Tang and H. J. Rosen, "The Vibrational Raman Spectra of Purified Solid Films of C_{60} and C_{70}," Chem. Phys. Lett. 174, 219–222 (1990).

Be91*   R. D. Beck, P. St. John, M. M. Alvarez, F. Diederich and R. L. Whetten, "Resilience of All-Carbon Molecules C_{60}, C_{70}, and C_{84}: A Surface-Scattering Time-of-Flight Investigation," J. Phys. Chem. 95, 8402–8409 (1991).

Be92   H. D. Beckhaus, C. Rüchardt, M. Kao, F. Diederich and C. S. Foote, "The Stability of Buckminsterfullerine (C_{60}): Experimental Determination of the Heat of Formation," Angew. Chem. Int. Ed. Engl. 31, 63–64 (1992).

Bo73   D. A. Bochvar and E. G. Gal'pern, "Hypothetical Systems: Carbododecahedron, S-Icosahedrane, and Carbo-S-Icosahedron," Doklady Akademii Nauk SSSR, 209, 610–612 (1973) [Doklady Chemistry, 209, 239–241 (1973)].

Bo93   M. T. Bowers, P. R. Kemper, G. von Helden and P. A. M van Koopen, "Gas-Phase Ion Chromatography: Transition Metal State Selection and Carbon Cluster Formation," Science 260, 1446–1451 (1993).

Bu92   P. R. Buseck, S. J. Tsipursky and R. Hettich, "Fullerenes from the Geological Environ-ment," Science 257, 215–217 (1992). See also the exchange of letters, ibid. 258, 1718–1719 (1992).

Ca91   J. H. Callahan, M. M. Ross, V. H. Wysocki, J. Jones and M. Ding, "Fragmentation Yields in Surface-Induced Dissociation Tandem Mass Spectrometry," in Proceedings of the 39th ASMS Conference on Mass Spectrometry and Allied Topics, Nashville, TN May 19–24, 1991, pp. 829–830 (1991).

Cu88   R. F. Curl and R. E. Smalley, "Probing C_{60}," Science 242, 1017–1022 (1988).

Cu93   R. F. Curl, "On the Formation of the Fullerenes," Phil. Trans. R. Soc. Lond. A 343, 19–32 (1993).

Da81   R. A. Davidson, "Spectral Analysis of Graphs by Cyclic Automorphism Subgroups," Theoret. Chim. Acta (Berlin) 58, 193–231 (1981).

Da93   T. K. Daly, P. R. Buseck, P. Williams and C. F. Lewis, "Fullerenes from a Fulgurite," Science 259, 1599–1601 (1993).

Fo92   P. A. Fowler, "Solved and Unsolved Problems in Fullerene Systematics," in Fullerenes: Status and Perspectives, edited by C. Taliani, G. Ruani and R. Zamboni (World Scientific, Singapore, 1992).

Ge87   Ph. Gerhardt, S. Löffler and K. H. Homann, "Polyhedral Carbon Ions in Hydrocarbon Flames," Chem. Phys. Lett. 137, 306–310 (1987).

Ha85   A. D. J. Haymet, "C_{120} and C_{60}: Archimedean Solids Constructed from sp^2 Hybridized Carbon Atoms," Chem. Phys. Lett. 122, 421–424 (1985).

Ha86*   R. C. Haddon, L. E. Brus and K. Raghavachari, "Electronic Structure and Bonding in Icosahedral C_{60}," Chem. Phys. Lett. 125, 459–464 (1986).

Ha92   J. P. Hare and H. W. Kroto, "A Postbuckminsterfullerene View of Carbon in the Galaxy," Acc. Chem. Res. 25, 106–112 (1992).

He85   J. R. Heath et al., "Lanthanum Complexes of Spheroidal Carbon Shells," J. Am. Chem. Soc. 107, 7779–7780 (1985).

He91   K. Hedberg et al., "Bond Lengths in Free Molecules of Buckminsterfullerene, C_{60}, from Gas-Phase Electron Diffraction," Science 254, 410–412 (1991).

Ho91   J. B. Howard et al., "Fullerenes C_{60} and C_{70} in Flames," Nature 352, 139–141 (1991).

Ho92   J. B. Howard et al., "Fullerenes Synthesis in Combustion," Carbon 30, 1183–1201 (1992).

Hu93a   J. Hunter, J. Fye and M. F. Jarrold, "Carbon Rings," J. Phys. Chem. 97, 3460–3462 (1993).

Hu93b J. Hunter, J. Fye and M. F. Jarrold, "Annealing C_{60}^+: Synthesis of Fullerenes and Large Carbon Rings," *Science* **260**, 784–786 (1993).

Kh92 K. C. Khemani, M. Prato and F. Wudl, "A Simple Soxhlet Chromatographic Method for the Isolation of Pure C_{60} and C_{70}," *J. Org. Chem.* **57**, 3254–3256 (1992).

Ko91 A. S. Koch, K. C. Khemani and F. Wudl, "Preparation of Fullerenes with a Simple Benchtop Reactor," *J. Org. Chem.* **56**, 4543–4545 (1991).

Kr85* H. W. Kroto, J. R. Heath, S. C. O'Brien, R. F. Curl and R. E. Smalley, "C_{60}: Buckminsterfullerene," *Nature* **318**, 162–163 (1985).

Kr87 H. W. Kroto, "The Stability of the Fullerenes C_n, with $n = 24, 28, 32, 36, 50, 60$, and 70," *Nature* **329**, 529–531 (1987).

Kr88 H. Kroto, "Space, Stars, C_{60}, and Soot," *Science* **242**, 1139–1145 (1988).

Kr90a* W. Krätschmer, K. Fostiropoulos and D. R. Huffman, "The Infrared and Ultraviolet Absorption Spectra of Laboratory-Produced Carbon Dust: Evidence for the Presence of the C_{60} Molecule," *Chem. Phys. Lett.* **170**, 167–170 (1990).

Kr90b* W. Krätschmer, L. D. Lamb, K. Fostiropoulos and D. R. Huffman, "Solid C_{60}: A New Form of Carbon," *Nature* **347**, 354–358 (1990).

Kr91 H. W. Kroto, A. W. Allaf and S. P. Balm, "C_{60} : Buckminsterfullerene," *Chem. Rev.* **91**, 1213–1235 (1991).

Ma91* D. E. Manolopoulos, J. C. May and S. E. Down, "Theoretical Studies of the Fullerenes: C_{34} to C_{70}," *Chem. Phys. Lett.* **181**, 105–111 (1991).

Mc92 S. W. McElvany, M. M. Ross and J. H. Callahan, "Characterization of Fullerenes by Mass Spectrometry," *Acc. Chem. Res.* **25**, 162–168 (1992).

Me90a G. Meijer and D. S. Bethune, "Mass Spectrometric Confirmation of the Presence of C_{60} in Laboratory-Produced Carbon Dust," *Chem. Phys. Lett.* **175**, 1–2 (1990).

Me90b G. Meijer and D. S. Bethune, "Laser Depositions of Carbon Clusters on Surfaces: A New Approach to the Study of Fullerenes," *J. Chem. Phys.* **93**, 7800–7802 (1990).

Mo91 R. C. Mowrey, D. W. Brenner, B. I. Dunlap, J. W. Mintmire and C. T. White, "Simulations of C_{60} Collisions with a Hydrogen-Terminated Diamond {111} Surface," *J. Phys. Chem.* **95**, 7138–7142 (1991).

O'B88 S. C. O'Brien, J. R. Heath, R. F. Curl and R. E. Smalley, "Photophysics of Buckminsterfullerene and Other Carbon Cluster Ions," *J. Chem. Phys.* **88**, 220–230 (1988).

Re91 C. Reber *et al.*, "Luminescence and Absorption Spectra of C_{60} Films," *J. Phys. Chem.* **95**, 2127–2129 (1991).

Ro84 E. A. Rohlfing, D. M. Cox and A. Kaldor, "Production and Characterization of Supersonic Carbon Cluster Beams," *J. Chem. Phys.* **81**, 3322–3330 (1984).

Sa93 C. H. Sah, "Combinatorial Construction of Fullerene Structures," *Croat. Chem. Acta* (in press).

Sc88 T. G. Schmalz, W. A. Seitz, D. J. Klein and G. E. Hite, "Elemental Carbon Cages," *J. Am. Chem. Soc..* **110**, 1113–1127 (1988).

Sm92 R. E. Smalley, "Self-Assembly of the Fullerenes," *Acc. Chem. Res.* **25**, 98–105 (1992).

Sm93 A. L. Smith, D. R. Kirklin, Y. W. Hui and D. Li, Submitted to *J. Phys. Chem.*

St84 W. V. Steele, R. D. Chirico, N. K. Smith, W. E. Billups, P. R. Elmore and A. E. Wheeler, "Standard Enthalpy of Formation of Buckminsterfullerene," *J. Phys. Chem.* **96**, 4731–4733 (1992).

Ta90* R. Taylor, J. P. Hare, A. K. Abdul-Sada and H. W. Kroto, "Isolation, Separation and Characterisation of the Fullerenes C_{60} and C_{70}: The Third Form of Carbon," *J. Chem. Soc., Chem. Commun.* 1423–1425 (1990).

Vo91 G. von Helden, M. T. Hsu, P. R. Kemper and M. T. Bowers, "Structures of Carbon Cluster Ions from 3 to 60 Atoms: Linears to Rings to Fullerenes," *J. Chem. Phys.* **95**, 3835–3837 (1991).

Vo93* G. von Helden, N. G. Gotts and M. T. Bowers, "Experimental Evidence for the Formation of Fullerenes by Collisional Heating of Carbon Rings in the Gas Phase," *Nature* **363**, 60–63 (1993).

We88* F. D. Weiss, J. L. Elkind, S. C. O'Brien, R. F. Curl and R. E. Smalley, "Photophysics of Metal Complexes of Spheroidal Carbon Shells," *J. Am. Chem. Soc* **110**, 4464–4465 (1988).

We89 D. E. Weeks and W. G. Harter, "Rotation-Vibration Spectra of Icosahedral Molecules. II. Icosahedral Symmetry, Vibrational Eigenfrequencies, and Normal Modes of Buckminsterfullerene," *J. Chem. Phys.* **90**, 4744–4771 (1989).

Xi92* Q. Xie, E. Pérez-Cordero and L. Echegoyen, "Electrochemical Detection of C_{60}^{6-} and

C_{70}^{6-}: Enhanced Stability of Fullerides in Solution," *J. Am. Chem. Soc.* **114**, 3978–3980 (1992).

Ya91 C. S. Yannoni, P. P. Bernier, D. S. Bethune, G. Meijer and J. R. Salem, "NMR Determination of Bond Lengths in C_{60}," *J. Am. Chem. Soc.* **113**, 3190–3192 (1991).

Reprinted with permission from Nature
Vol. 318, No. 6042, pp. 162–163, 14 November 1985
© 1985 Macmillan Magazines Limited

C₆₀: Buckminsterfullerene

H. W. Kroto[*], **J. R. Heath, S. C. O'Brien, R. F. Curl & R. E. Smalley**

Rice Quantum Institute and Departments of Chemistry and Electrical Engineering, Rice University, Houston, Texas 77251, USA

Fig. 1 A football (in the United States, a soccerball) on Texas grass. The C_{60} molecule featured in this letter is suggested to have the truncated icosahedral structure formed by replacing each vertex on the seams of such a ball by a carbon atom.

During experiments aimed at understanding the mechanisms by which long-chain carbon molecules are formed in interstellar space and circumstellar shells[1], graphite has been vaporized by laser irradiation, producing a remarkably stable cluster consisting of 60 carbon atoms. Concerning the question of what kind of 60-carbon atom structure might give rise to a superstable species, we suggest a truncated icosahedron, a polygon with 60 vertices and 32 faces, 12 of which are pentagonal and 20 hexagonal. This object is commonly encountered as the football shown in Fig. 1. The C_{60} molecule which results when a carbon atom is placed at each vertex of this structure has all valences satisfied by two single bonds and one double bond, has many resonance structures, and appears to be aromatic.

The technique used to produce and detect this unusual molecule involves the vaporization of carbon species from the surface of a solid disk of graphite into a high-density helium flow, using a focused pulsed laser. The vaporization laser was the second harmonic of Q-switched Nd:YAG producing pulse energies of ~30 mJ. The resulting carbon clusters were expanded in a supersonic molecular beam, photoionized using an excimer laser, and detected by time-of-flight mass spectrometry. The vaporization chamber is shown in Fig. 2. In the experiment the pulsed valve was opened first and then the vaporization laser was fired after a precisely controlled delay. Carbon species were vaporized into the helium stream, cooled and partially equilibrated in the expansion, and travelled in the resulting molecular beam to the ionization region. The clusters were ionized by direct one-photon excitation with a carefully synchronized excimer laser pulse. The apparatus has been fully described previously[2-5].

The vaporization of carbon has been studied previously in a very similar apparatus[6]. In that work clusters of up to 190 carbon atoms were observed and it was noted that for clusters of more than 40 atoms, only those containing an even number of atoms were observed. In the mass spectra displayed in ref. 6, the C_{60} peak is the largest for cluster sizes of >40 atoms, but it is not completely dominant. We have recently re-examined this system and found that under certain clustering conditions the C_{60} peak can be made about 40 times larger than neighbouring clusters.

Figure 3 shows a series of cluster distributions resulting from variations in the vaporization conditions evolving from a cluster distribution similar to that observed in ref. 3, to one in which C_{60} is totally dominant. In Fig. 3c, where the firing of the vaporization laser was delayed until most of the He pulse had passed, a roughly gaussian distribution of large, even-numbered clusters with 38–120 atoms resulted. The C_{60} peak was largest but not dominant. In Fig. 3b, the vaporization laser was fired at the time of maximum helium density; the C_{60} peak grew into a feature perhaps five times stronger than its neighbours, with the exception of C_{70}. In Fig. 3a, the conditions were similar to those in Fig. 3b but in addition the integrating cup depicted in Fig. 2 was added to increase the time between vaporization and expansion. The resulting cluster distribution is completely dominated by C_{60}, in fact more than 50% of the total large cluster abundance is accounted for by C_{60}; the C_{70} peak has diminished in relative intensity compared with C_{60}, but remains rather prominent, accounting for ~5% of the large cluster population.

Our rationalization of these results is that in the laser vaporization, fragments are torn from the surface as pieces of the planar graphite fused six-membered ring structure. We believe that the distribution in Fig. 3c is fairly representative of the nascent distribution of larger ring fragments. When these hot ring clusters are left in contact with high-density helium, the clusters equilibrate by two- and three-body collisions towards the most stable species, which appears to be a unique cluster containing 60 atoms.

When one thinks in terms of the many fused-ring isomers with unsatisfied valences at the edges that would naturally arise from a graphite fragmentation, this result seems impossible: there is not much to choose between such isomers in terms of stability. If one tries to shift to a tetrahedral diamond structure, the entire surface of the cluster will be covered with unsatisfied valences. Thus a search was made for some other plausible structure which would satisfy all sp^2 valences. Only a spheroidal structure appears likely to satisfy this criterion, and thus Buckminster Fuller's studies were consulted (see, for example, ref. 7). An unusually beautiful (and probably unique) choice is the truncated icosahedron depicted in Fig. 1. As mentioned above, all valences are satisfied with this structure, and the molecule appears to be aromatic. The structure has the symmetry of the icosahedral group. The inner and outer surfaces are covered with a sea of π electrons. The diameter of this C_{60} molecule is ~7 Å, providing an inner cavity which appears to be capable of holding a variety of atoms[8].

Assuming that our somewhat speculative structure is correct, there are a number of important ramifications arising from the existence of such a species. Because of its stability when formed under the most violent conditions, it may be widely distributed in the Universe. For example, it may be a major constituent of circumstellar shells with high carbon content. It is a feasible constituent of interstellar dust and a possible major site for

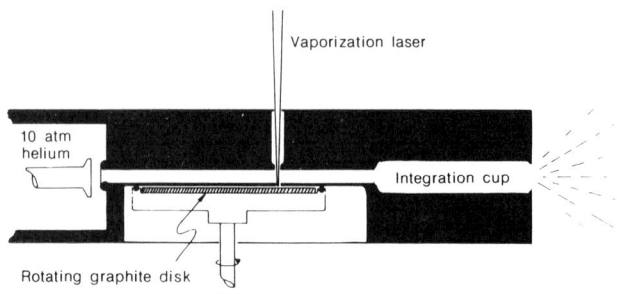

Fig. 2 Schematic diagram of the pulsed supersonic nozzle used to generate carbon cluster beams. The integrating cup can be removed at the indicated line. The vaporization laser beam (30–40 mJ at 532 nm in a 5-ns pulse) is focused through the nozzle, striking a graphite disk which is rotated slowly to produce a smooth vaporization surface. The pulsed nozzle passes high-density helium over this vaporization zone. This helium carrier gas provides the thermalizing collisions necessary to cool, react and cluster the species in the vaporized graphite plasma, and the wind necessary to carry the cluster products through the remainder of the nozzle. Free expansion of this cluster-laden gas at the end of the nozzle forms a supersonic beam which is probed 1.3 m downstream with a time-of-flight mass spectrometer.

[*] **Permanent address:** School of Chemistry and Molecular Sciences, University of Sussex, Brighton BN1 9QJ, UK.

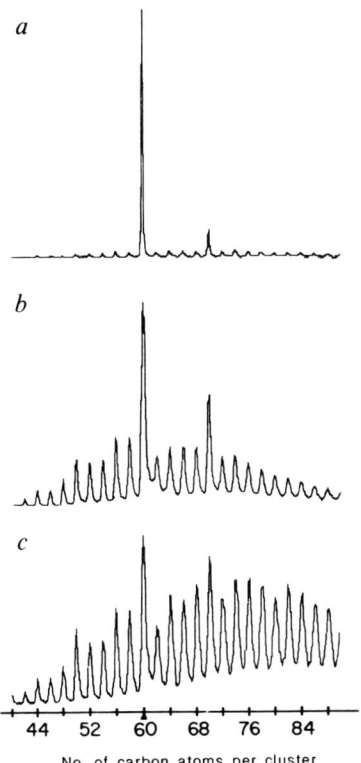

Fig. 3 Time-of-flight mass spectra of carbon clusters prepared by laser vaporization of graphite and cooled in a supersonic beam. Ionization was effected by direct one-photon excitation with an ArF excimer laser (6.4 eV, 1 mJ cm^{-2}). The three spectra shown differ in the extent of helium collisions occurring in the supersonic nozzle. In c, the effective helium density over the graphite target was less than 10 torr—the observed cluster distribution here is believed to be due simply to pieces of the graphite sheet ejected in the primary vaporization process. The spectrum in b was obtained when roughly 760 torr helium was present over the graphite target at the time of laser vaporization. The enhancement of C_{60} and C_{70} is believed to be due to gas-phase reactions at these higher clustering conditions. The spectrum in a was obtained by maximizing these cluster thermalization and cluster–cluster reactions in the 'integration cup' shown in Fig. 2. The concentration of cluster species in the especially stable C_{60} form is the prime experimental observation of this study.

surface-catalysed chemical processes which lead to the formation of interstellar molecules. Even more speculatively, C_{60} or a derivative might be the carrier of the diffuse interstellar lines[9].

If a large-scale synthetic route to this C_{60} species can be found, the chemical and practical value of the substance may prove extremely high. One can readily conceive of C_{60} derivatives of many kinds—such as C_{60} transition metal compounds, for example, C_{60}Fe or halogenated species like $C_{60}F_{60}$ which might be a super-lubricant. We also have evidence that an atom (such as lanthanum[8] and oxygen[1]) can be placed in the interior, producing molecules which may exhibit unusual properties. For example, the chemical shift in the NMR of the central atom should be remarkable because of the ring currents. If stable in macroscopic, condensed phases, this C_{60} species would provide a topologically novel aromatic nucleus for new branches of organic and inorganic chemistry. Finally, this especially stable and symmetrical carbon structure provides a possible catalyst and/or intermediate to be considered in modelling prebiotic chemistry.

We are disturbed at the number of letters and syllables in the rather fanciful but highly appropriate name we have chosen in the title to refer to this C_{60} species. For such a unique and centrally important molecular structure, a more concise name would be useful. A number of alternatives come to mind (for example, ballene, spherene, soccerene, carbosoccer), but we prefer to let this issue of nomenclature be settled by consensus.

We thank Frank Tittel, Y. Liu and Q. Zhang for helpful discussions, encouragement and technical support. This research was supported by the Army Research Office and the Robert A. Welch Foundation, and used a laser and molecular beam apparatus supported by the NSF and the US Department of Energy. H.W.K. acknowledges travel support provided by SERC, UK. J.R.H. and S.C.O'B. are Robert A. Welch Predoctoral Fellows.

Received 13 September; accepted 18 October 1985.

1. Heath, J. R. *et al. Astrophys. J.* (submitted).
2. Dietz, T. G., Duncan, M. A., Powers, D. E. & Smalley, R. E. *J. chem. Phys.* **74**, 6511–6512 (1981).
3. Powers, D. E. *et al. J. phys. Chem.* **86**, 2556–2560 (1982).
4. Hopkins, J. B., Langridge-Smith, P. R. R., Morse, M. D. & Smalley, R. E. *J. chem. Phys.* **78**, 1627–1637 (1983).
5. O'Brien, S. C. *et al. J. chem. Phys.* (submitted).
6. Rohlfing, E. A., Cox, D. M. & Kaldor, A. *J. chem. Phys.* **81**, 3322–3330 (1984).
7. Marks, R. W. *The Dymaxion World of Buckminster Fuller* (Reinhold, New York, 1960).
8. Heath, J. R. *et al. J. Am. chem. Soc.* (in the press).
9. Herbig, E. *Astrophys. J.* **196**, 129–160 (1975).

Reprinted with permission from Chemical Physics Letters
Vol. 125, No. 5,6, pp. 459–464, 18 April 1986

ELECTRONIC STRUCTURE AND BONDING IN ICOSAHEDRAL C_{60}

R.C. HADDON, L.E. BRUS and Krishnan RAGHAVACHARI

AT&T Bell Laboratories, Murray Hill, NJ 07974, USA

Received 18 December 1985; in final form 3 February 1986

The Hückel molecular orbital theory for non-planar conjugated organic molecules has been applied to study the electronic structure and properties of the proposed icosahedral geometry of C_{60}. The results support the suggestion that C_{60} may be the first example of a spherical aromatic molecule. The molecule is calculated to have a stable closed shell singlet ground electronic state.

1. Introduction

Carbon clusters have been the focus of many experimental studies using a variety of techniques [1–4]. Recently, Kroto, Heath, O'Brien, Curl and Smalley [5] have obtained a remarkably stable cluster containing 60 carbon atoms and have proposed a highly symmetric truncated icosahedral structure ("football or soccerball" like) for this molecule [5,6]. These authors have proposed the name "Buckminsterfullerene" for this unusual structure of C_{60}.

The proposed icosahedral structure of C_{60} raises a number of particularly interesting questions regarding the electronic structure of the molecule. Perhaps most intriguing is the spherical shape of the non-planar conjugated system which is (formally) solely composed of sp^2 hybridized carbon atoms. The molecule is comprised of 20 six-membered rings (6-MR) and 12 five-membered rings (5-MR), such that all the atoms are identical. Each atom is at the vertex of one 5-MR and two 6-MRs. There are two independent bond types: 30 bonds which lie solely in 6-MRs, while 60 bonds form the edges of both a 5- and a 6-MR. Particularly noteworthy is the observation that the molecule is non-alternant [7], but that as a result of the symmetry of the molecule the charge densities are all equal. Thus the proposed structure for C_{60} appears to be unique among non-alternant hydrocarbons in this respect.

Some computed properties of C_{60} are compared with those of related molecules in table 1. The choice of benzene, the [∞]annulene [11,12] and graphite [13] as reference molecules is clear. The hydrocarbon $C_{56}H_{22}$ or quateranthrene, is the graphite-like fragment closest in molecular size to C_{60}. Two sets of results are reported for C_{60}. The first set (first column of numbers in the table under C_{60}) refers to calculations with standard Hückel molecular orbital (HMO) theory*, without correction for non-planarity. The second set (final column of numbers in the table) was calculated with the three-dimensional HMO (3D HMO) method [15] in which the HMO resonance integrals (β) are adjusted for the effects of non-planarity by use of the π-orbital axis vector (POAV) analysis [16, 17] (discussed below). Although unsophisticated, the HMO theory [7,18–21] provides a useful starting point for molecular comparisons and the correlation of properties.

2. Planar HMO treatment

The π-bond energy per carbon (or electron) in C_{60} is slightly less than that of graphite, but greater than that of the graphite fragment ($C_{56}H_{22}$). The two-dimensional structures (graphite, $C_{56}H_{22}$ and C_{60}) on the right of table 1 are favored in this comparison

---

* A referee has drawn our attention to previous calculations of the eigenvalue spectrum of icosahedral C_{60} [14].

Table 1
HMO results [a]

Property	C_6H_6	$[\infty]A$ [b]	$C_{56}H_{22}$ [c]	Graphite [d]	C_{60} planar [e]	C_{60} 3D [f]
E_π [g]	8.000	–	81.820	–	93.162	81.673
E_π/C [h]	1.333	1.273	1.461	1.576	1.553	1.361
E_π/B [i]	1.333	1.273	1.121	1.051	1.035	0.907
P [j]	0.667	0.637	0.560	0.525	0.518	0.454
λ [k]	1.0	∞	1.016	u)	0.893	1.019
F_r [l]	0.399	0.459	0.566	0.156	0.179	0.371
L_r^+ [m]	2.536	–	1.620	–	2.767	2.426
$L_r^\cdot$ [n]	2.536	–	1.620	–	2.629	2.305
L_r^- [o]	2.536	–	1.620	–	2.490	2.182
E_{HOMO} [p]	1.0	0.0	0.051	0.0	0.618	0.542
E_{LUMO} [q]	–1.0	0.0	–0.051	0.0	–0.139	–0.121
ΔE [r]	2.0	0.0	0.102	0.0	0.757	0.664
IP (eV) [s]	8.81	–	6.07	–	7.70	7.45
IP (eV) [t]	9.55	–	7.20	–	8.60	8.41

a) Averaged where necessary. In units of β, except where noted.   b) Infinite annulene, $(CH)_\infty$.   c) Quateranthrene.
d) Infinite sheet, C_∞.   e) Hypothetical planar case (resonance integral of β).
f) POAV1/3D HMO analysis with correction of resonance integrals ($\rho\beta$) for non-planarity ($\rho = 0.877$).
g) Total π-electron energy.   h) π-electron energy per carbon (electron).   i) π-electron energy per bond.   j) Bond order (unitless).
k) Highest eigenvalue of bond–bond polarizability matrix (β^{-1}).   l) Free valence (unitless).
m) Localization energy for electrophilic attack.   n) Localization energy for radical attack.
o) Localization energy for nucleophilic attack.   p) Energy of highest occupied molecular orbital.
q) Energy of lowest unoccupied molecular orbital.   r) Energy gap.   s) Ionization potential, from relationship given in ref. [8].
t) Ionization potential, from relationship given in ref. [9].   u) Stable to distortion [10].

over the linearly conjugated systems (benzene and annulene). This is principally due to the fact that the former structures have more bonds (1.5 bonds per carbon) than the latter forms (1 bond per carbon). As noted above, C_{60} has two classes of bonds, and the calculated bond orders are found to be: 0.476 (5/6-MR) and 0.601 (6/6-MR). Applying the conventional relationship [7,18–21] between bond orders and bond lengths we calculate C_{60} to possess 30 bonds of length 1.405 Å (6/6-MR) and 60 bonds of length 1.426 Å (5/6-MR), and the most appropriate Kekulé representation is shown in fig. 1. Particularly important is the prediction that C_{60} will not undergo a distortion to a point group of lower symmetry as a result of the second-order Jahn–Teller effect [19]. It has been shown [22] that by diagonalizing the bond–bond polarizability matrix, it is possible to ascertain the likely symmetry of distortion (eigenvectors) and the energy gain (eigenvalues, λ) due to second-order Jahn–Teller distortion. In fact, the

largest eigenvalue of the bond–bond polarizability matrix (table 1), indicates that C_{60} has less tendency to break symmetry than benzene. This contrasts with the large aromatic annulenes [11,23,24] where bond alternation is thought to occur in the vicinity of [18]annulene ($\lambda_{critical} \approx 1.8$).

The free valence (F_r) provides an index of the unsatisfied valence or residual bonding power at a particular atom (r) in a molecule [7,18–21]. The localization indices provide a measure of the energetic destabilization of the π system when an atom (r) is removed from conjugation by sp^3 bond formation with an incoming electrophilic (L_r^+), radical ($L_r^\cdot$), or nucleophilic (L_r^-) reagent [7,18–21]. It is apparent that all of these indices indicate a high chemical stability for C_{60}. This primarily results from the absence of perimeter atoms (two nearest neighbors); as all of the atoms in C_{60} are "interior", chemical reaction should be inhibited. The geometry (discussed below), however, does predispose C_{60} toward chemical attack when

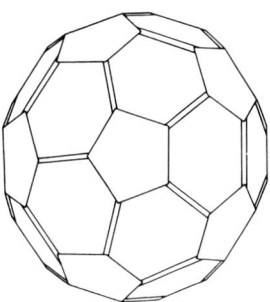

Fig. 1. A hemisphere of C_{60} showing the principal Kekulé structure.

comparison is made with planar conjugated systems as the necessary rehybridization for a tetrahedral transition state is already well advanced in the ground state of C_{60} (see below).

The non-alternant character of C_{60} (and the presence of 5-MRs) is reflected in the asymmetric distribution of the highest occupied molecular orbitals (HOMO) and lowest unoccupied molecular orbitals (LUMO) about the zero of energy (fig. 2). The MO energy levels indicate an ionization and oxidation potential comparable to naphthalene (E_{HOMO} = 0.618) and very different from that of graphite. The calculations suggest an exceptionally high electron affinity (facile reduction). For comparison, decacyclene reversibility undergoes addition of six electrons [25] in solution (the lowest three vacant molecular orbitals lie at −0.333, −0.333 and −0.425 β). In C_{60} there are three MOs at −0.139 β and three at −0.382 β, and thus it is likely that the molecule will *add up to twelve*

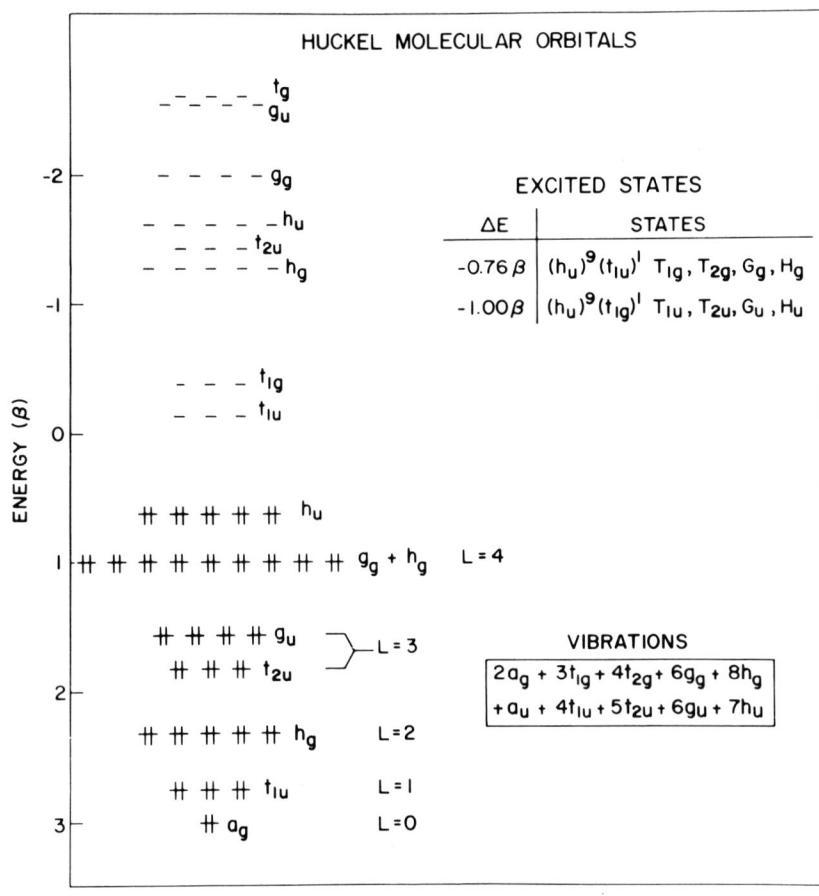

Fig. 2. HMO energy level diagram for C_{60} (unscaled β, see text).

electrons under suitable conditions. The calculated HOMO–LUMO energy gap (ΔE) lies between the values found for anthracene ($\Delta E = 0.828\beta$) and tetracene ($\Delta E = 0.590\,\beta$), although the non-alternant azulene [26,27] ($\Delta E = 0.877\,\beta$) may be a more appropriate reference point for C_{60}.

3. Three-dimensional HMO treatment

It is now appropriate to consider the way in which the idealized results discussed above are modified when full account is taken of the spherical nature of the π-electron system. The molecule lends itself to the recently developed POAV/3D HMO analysis [15–17]. In the POAV approach [16,17] the σ bonds are assumed to lie along the internuclear axes, and the orbital orthogonality relationships are used to solve for the π-orbital hybridization and direction. It has been shown that this analysis provides the logical and natural bridge between the $\sigma–\pi$ separability assumed in planar conjugated systems and the realistics of π bonding in non-planar situations. The 3D HMO method extends this analysis by utilizing the local POAV hybrids to calculate Slater overlap integrals between nearest neighbors in the π-electron network. By comparison with the overlap integral for the same bond with "perfect" π-orbital overlap a reduced resonance integral ($\rho\beta$) may be evaluated which is subsequently used in the (3D) HMO calculation on the molecule. The POAV analysis is non-parametric, and merely requires the molecular coordinates. In the present case we assumed a common carbon–carbon bond length of 1.4 Å. In fact, the molecule has two degrees of freedom, and although the 5-MRs and 6-MRs separately maintain equal bond angles ($108°$ and $120°$, respectively), the molecule is expected to exhibit two diferent bond lengths, so that the 6-MRs become somewhat bond-alternate and the 5-MRs remain bond-equalized. For the present qualitative discussion, the preceding approximation is expected to be adequate.

The POAV1 analysis [16] of the bond-equalized structure leads to $s^{0.093}p$ hybridization for the π orbital and $sp^{2.278}$ hybridization for the three σ orbitals. The angle between the σ and π orbitals ($\theta_{\sigma\pi}$) is found to be $101.6°$, and hence the σ bonds may be considered to be inclined at an angle of $11.6°$ below a tangential plane to the surface (this angle is $0°$ in

graphite and $19.5°$ in diamond). Thus the rehybridization in C_{60} is extremely high – much larger than the values found in the most strained bridged annulenes [16,17]. Nevertheless, C_{60} is found to have perfect π-orbital alignment (unlike most non-planar conjugated organic systems), and the reduction in the overlap integrals in C_{60} results from the inclination of the orbitals with respect to the bond axis rather than torsional misalignment. Calculating the Slater overlap integral between the local π hybrids gives $\rho = 0.877$ for the π bonds in C_{60} (reduced resonance integral = $0.877\,\beta$).

As we are treating all bonds on the same basis, it is merely necessary to scale the appropriate quantities for the hypothetical planar structure by ρ to obtain the 3D HMO results (last column of table 1). It is clear from table 1 that the π system is destabilized by this perturbation. Nevertheless, the reactivity indices (F_r, L_r) suggest that the molecule should retain sufficient resistance to chemical attack to be isolable. The only caveat to this predictoni concerns the rehybridization, which makes the ground state atoms of C_{60} approach the transition state for reaction (formation of a fourth σ bond) much more closely than in normal planar conjugated organic molecules. The molecules is still expected to be stable to a reduction in symmetry, but the bond lengths are now predicted to be 1.418 (6/6-MR) and 1.436 Å (5/6-MR).

It is apparent that our calculations support the proposal by Kroto et al. [5] that C_{60} may qualify as the first example of a spherical aromatic molecule. The π-electron energy per carbon (or per electron) is 86% of that of graphite and 93% of that of a size-consistent graphite fragment. Of pivotal importance in this regard is the fact that all of the σ bonds are fully satisfied with little strain (ring bond angles of $108°$ and $120°$), after rehybridization. In comparison, the (dehydrogenated) graphite fragment (derived from $C_{56}H_{22}$) would have 22 unsatisfied valences. Clearly C_{60} by its topology is a superbly adapted cluster geometry for the bonding characteristics of carbon.

4. Electronic and vibrational spectroscopy

The HMO scheme predicts the HOMO to be completely filled (fig. 2) and therefore the ground state of C_{60} should be diamagnetic ^1A_g. The icosahedral group is close to spherical symmetry and one can clearly recognize the parentage of the low-lying MOs in spherical harmonics (Y_{LM} where L is the orbital angular momentum quantum number). In particular the S, P, and D ($L = 0$, 1 and 2) spherical harmonics become a_g, t_{1u}, and h_g representations in I_h. The atomic orbital coefficients in the three lowest MOs are simply the corresponding spherical harmonics evaluated at the angular positions of the carbon nuclei. The $L = 3$ functions split into t_{2u} and g_u as shown. Interestingly, the two components of the $L = 4$ level remain accidentally degenerate at the elementary level of HMO theory. Above $L = 4$, the various components mix strongly. Low-L states have the lowest energies since they have the fewest number of nodes in the wavefunction. The molecular orbitals in C_{60} provide a three-dimensional analogy to those found in the annulenes in two dimensions with respect to nodal structure.

The two lowest excited configurations are each fifteen-fold degenerate, $(h_u)^9(t_{1u})^1$ and $(h_u)^9(t_{1g})^1$. These configurations yield eight excited states, each of which is at least three-fold degenerate as shown in fig. 2. The symmetry is so high that among these only T_{1u} is allowed in one-photon absorption from the ground state and only H_g is allowed in two-photon absorption. T_{1u} lies in the second configuration and it is likely that optically forbidden states resulting from the first configuration will lie below it even when more accurate calculations are performed.

The low-lying electronically forbidden degenerate states could undergo Jahn–Teller distortion and acquire some oscillator strength. In general, we expect that vibronic intensity borrowing interactions will be weak, because the electronic states are delocalized over many atoms, and the carbon framework is relatively rigid. There are only two a_g vibrational normal modes in C_{60}, and we would expect only modest Franck–Condon progressions in these modes in absorption and fluorescence for allowed transitions. Fluorescence will be depolarized, and photoselection effects absent, from states of either electronic or vibronic T_{1u} symmetry. We conjecture that excited-state radiationless transitions should be relatively slow because excited-state geometries will be very similar to the ground-state geometry, and there are no

high-frequency C–H accepting modes.

Planar aromatic hydrocarbons with high symmetry and many rings often have intensely allowed $\pi-\pi^*$ transitions with oscillator strengths of ≥ 1, and absorption coefficients $\epsilon \geq 10^5$ ℓ/mol cm [28]. Elementary Hückel theories indicate that maximum oscillator strengths scale with the size of the π-electron system. Fully allowed $A_g \rightarrow T_{1u}$ transitions in C_{60} could be exceedingly strong. The $A_g \rightarrow T_{1u}$ transition in the second configuration lies at an excitation energy of $\approx -1.0\beta$. This is about the same Hückel excitation energy as predicted for anthracene, whose S_1 state lies in the near ultraviolet. This crude estimate ignores substantial shifts effected from interelectronic repulsion and configuration interaction. The high symmetry ensures that the electronic spectrum will be "molecule-like" with discrete, well separated transitions.

The high symmetry of C_{60} also causes the expected infrared and Raman spectroscopy to be unusually simple. While there are 174 normal modes ($3N-6$), the infrared spectrum should show only four fundamentals corresponding to the four three-fold degenerate t_{1u} vibrations. The non-resonant spontaneous Raman spectrum should also have an interesting structure. The molecule has only two degrees of freedom that conserve I_h symmetry. These motions are sphere breathing (symmetric bond stretching), and antisymmetric stretching of bonds between 6-MRs out of phase with bonds between 5- and 6-MRs. These a_g motions, especially the antisymmetric mode, should be prominent in the spontaneous Raman spectrum, based upon experience with planar aromatic hydrocarbons. There are also h_g modes that are allowed in the Raman spectrum. The relative displacements of carbon atoms in a_g, t_{1u}, and h_g normal modes is simply given by the spherical harmonics Y_{LM} for $L = 0$, 1, and 2.

5. Conclusion

Icosahedral C_{60} appears to be the first example of a spherical aromatic molecule. The molecule is predicted to be stable, isolable, and capable of accepting up to twelve electrons in solution. The carbon hybridization is calculated to be about $s^{0.09}p$ for the π orbitals and $sp^{2.28}$ for the σ system with approximately 90% of the resonance stabilization of an equivalent

464

planar graphite system. The electronic and vibrational spectra should be comparatively simple due to the high symmetry of the molecule.

References

[1] N. Furstenau and F. Hillenkamp, Intern. J. Mass Spectrom. Ion Phys. 37 (1981) 135.

[2] E.A. Rohling, D.M. Cox and A. Kaldor, J. Chem. Phys. 81 (1984) 3322.

[3] L.A. Bloomfield, M.E. Geusic, R.R. Freeman and W.L. Brown, Chem. Phys. Letters 121 (1985) 33.

[4] M.E. Geusic, T.J. McIlrath, M.F. Jarrold, L.A. Bloomfield, R.R. Freeman and W.L. Brown, to be published.

[5] H.W. Kroto, J.R. Heath, S.C. O'Brien, R.F. Curl and R.E. Smalley, Nature 318 (1985) 162.

[6] J.R. Heath, S.C. O'Brien, Q. Zhang, Y. Liu, R.F. Curl, H.W. Kroto, F.K. Tittel and R.E. Smalley, J. Am. Chem. Soc., to be published.

[7] C.A. Coulson, B. O'Leary and R.B. Mallion, Hückel theory for organic chemists (Academic Press, New York, 1978) ch. 6.

[8] R. Boschi, E. Clar and W. Schmidt, J. Chem. Phys. 60 (1974) 4406.

[9] A. Streitwieser Jr., J. Am. Chem. Soc. 82 (1960) 4123.

[10] T. Anno and L. Coulson, Proc. Roy. Soc. A264 (1969) 165.

[11] R.C. Haddon, J. Am. Chem. Soc. 101 (1979) 1722.

[12] R.C. Haddon, K. Raghavachari and M.-H. Whangbo, J. Am. Chem. Soc. 106 (1984) 5364.

[13] C.A. Coulson and R. Taylor, Proc. Phys. Soc. (London) A65 (1952) 815.

[14] D.A. Bochvar and E.G. Gal'pern, Dokl. Akad. Nauk SSSR Chem. Engl. 209 (1972) 239;
A.D.J. Haymet, Chem. Phys. Letters 122 (1985) 421;
D.J. Klein, to be published;
M.D. Newton, to be published.

[15] R.C. Haddon, to be published.

[16] R.C. Haddon and L.T. Scott, Pure Appl. Chem. 58 (1986) 137.

[17] R.C. Haddon, Chem. Phys. Letters 125 (1986) 231;
J. Am. Chem. Soc., to be published.

[18] A. Streitwieser Jr., Molecular orbital theory for organic chemists (Wiley, New York, 1961).

[19] L. Salem, The molecular orbital theory of conjugated systems (Benjamin, New York, 1966).

[20] M.J.S. Dewar, The molecular orbital theory of organic chemistry (McGraw-Hill, New York, 1969).

[21] E. Heilbronner and H. Bock, The HMO model and its applications, Vols. 1–3 (Wiley, New York, 1976).

[22] G. Binsch, E. Heilbronner and J.N. Murrell, Mol. Phys. 11 (1966) 305.

[23] H. Baumann and J.F.M. Oth, Helv. Chim. Acta 65 (1982) 1885.

[24] R.C. Haddon and K. Raghavachari, J. Am. Chem. Soc. 107 (1985) 280.

[25] J. Heinze, Angew. Chem. Intern. Ed. Engl. 23 (1984) 831.

[26] J. Michl and E. Thulstrup, Tetrahedron 32 (1976) 205.

[27] L.T. Scott, M. Oda and I. Erden, J. Am. Chem. Soc. 107 (1985) 7213.

[28] J.B. Birks, ed., Organic molecular photophysics, Vol. 1 (Wiley, New York, 1973).

Reprinted with permission from Chemical Physics Letters
Vol. 181, No. 2,3, pp. 105–111, 21 June 1991
© 1991 Elsevier Science Publishers

Theoretical studies of the fullerenes: C_{34} to C_{70}

David E. Manolopoulos, Jonathan C. May and Sarah E. Down

Department of Chemistry, University of Nottingham, University Park, Nottingham NG7 2RD, UK

Received 7 March 1991; in final form 6 April 1991

An interesting correlation is found between the kinetic stabilities of the fullerenes C_{34} to C_{70}, as determined by simple Hückel theory HOMO–LUMO energy separations, and the intensities of photoionisation signals from carbon clusters produced by laser vaporisation of graphite. This correlation provides further circumstantial evidence that the observed C_{34} to C_{70} clusters are indeed fullerenes, closed carbon cages containing only five- and six-membered rings.

1. Introduction

Carbon clusters containing up to about 100 atoms are now routinely studied by irradiating graphite with a high-power vaporising laser, cooling the resulting carbon vapour in a helium jet, and photoionising the clusters so obtained for detection by time-of-flight mass spectrometry [1–3]. Intriguing mass spectra are often seen in these experiments, the fine details of which depend on the vaporisation and photoionisation conditions used. One such spectrum, taken in the C_{34} to C_{70} mass range from the early results of Kaldor and co-workers [1], is reproduced in fig. 1. This particular spectrum is not unusual, being similar to those seen both for mild clustering conditions with low laser flux photoionisation (ref. [2], fig. 2c), and extensive clustering conditions with high laser flux photoionisation (ref. [4], fig. 2d). Other experimental conditions can give markedly different spectra, so fig. 1 should be viewed with some care. In particular, while clusters with an odd number of carbon atoms are hardly ever seen in this mass range, the ion signals from C_{50}, C_{60} and C_{70} can be significantly enhanced [2,3].

It is now often speculated that the clusters which give rise to the ion signals in fig. 1 might be fullerenes, closed carbon cages containing only five- and six-membered rings [4,5]. Such structures are expected to be stable on chemical grounds, because they perfectly satisfy the valence requirements of sp² hybridised carbon atoms without the σ bonding strain

of three- and four-membered rings, and on geodesic grounds, because several highly symmetric fullerene isomers have already been found which can evenly disperse whatever bond angle deformation strain energy remains [5]. In fact the particularly stable C_{60} and C_{70} clusters, buckminsterfullerene [2] and fullerene-70 [6], have recently been isolated from evaporated graphite in macroscopic amounts [6–8]. Their ¹³C NMR spectra have been measured, and are fully consistent with I_h and D_{5h} symmetry fullerene structures, respectively, which have been anticipated for some time [2,4,5].

Closed carbon cages with n atoms have a total of $3n/2$ σ bonds, and since this number must be an integer we know that all closed carbon cages have an even number of atoms. The absence of any ion signal in fig. 1 corresponding to a cluster with an odd number of atoms is therefore already well explained by the fullerene idea. However, in order to provide even stronger circumstantial evidence that the observed carbon clusters are indeed fullerenes, one has also to explain why certain "magic number" clusters [4,5], including in particular C_{60}, are produced with such high yield. This *can* be done qualitatively in terms of adjacencies between five-membered rings (C_{50} is the smallest fullerene for which it is possible to make a structure with no more than two adjacent five-membered rings; C_{60} and C_{70} are the smallest two fullerenes for which it is possible to make structures without any adjacent five-membered rings at all [5,9]),

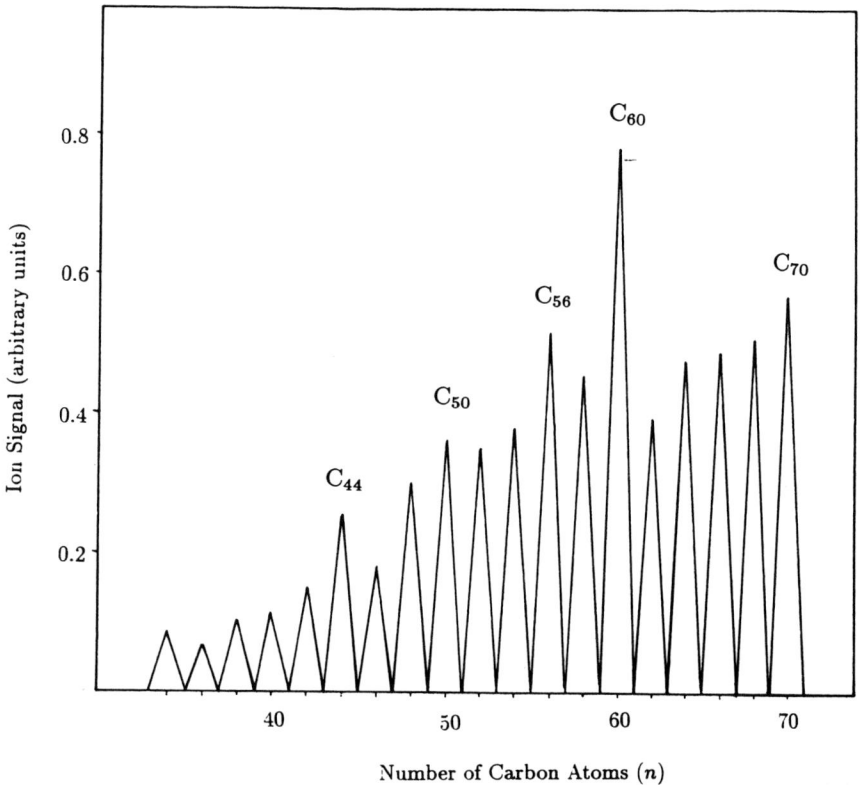

Fig. 1. Photoionisation signals from carbon clusters produced by laser vaporisation of graphite. Selected experimental results from ref. [1].

but no more quantitative explanation has yet been proposed.

Clearly, if any such quantitative explanation is to be found, it must take account of the fact that there can be a large number of distinct fullerene isomers for each given C_n. There are, for example, at least 1812 distinct fullerene isomers for C_{60} alone (vide infra). Icosahedral C_{60}, which was first discovered by intuition [2], does indeed happen to be the most stable of these. However, since the majority of carbon clusters do *not* admit such remarkably symmetric closed-shell structures, the most stable C_n fullerene isomer is generally more difficult to find. Section 2 describes a practical solution to this problem, which amounts to no more than a direct computer search of all possible fullerene structures for each given C_n. Section 3 then offers a semi-quantitative explanation for fig. 1, based on the relative kinetic stabilities of the fullerenes C_{34} to C_{70}.

2. Hückel calculations

While large scale ab initio geometry optimisation calculations on the fullerenes are currently impractical, except perhaps in a few high-symmetry cases [10–13], one can at least find the most stable fullerene structure for given C_n within Hückel π-electron theory [14,15]. In fact this turns out to be quite straightforward, at least for small n, because every possible fullerene structure can be "peeled" like an orange.

Fig. 2 shows a 2D representation of icosahedral C_{60}, in which the 32 faces are numbered according to their order in this "orange peel" scheme. Note in particular that each face after the first borders its immediate predecessor, so that the "peel comes off the orange" in a single continuous spiral. In fact the entire Hückel theory adjacency matrix for the structure can be reconstructed from the order of the penta-

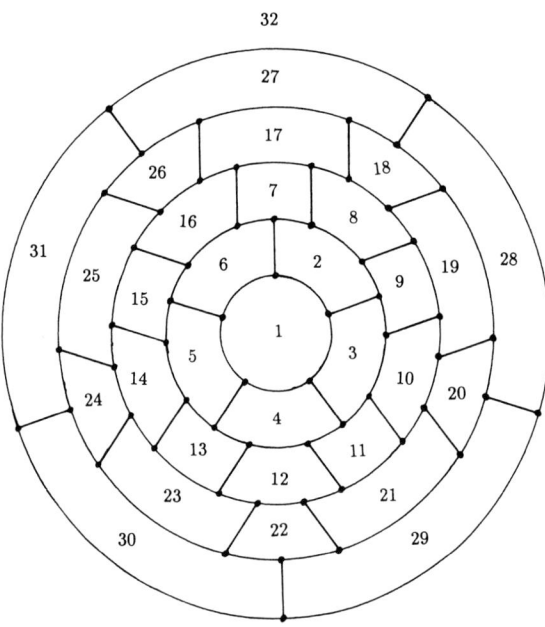

Fig. 2. 2D representation (molecular graph) of icosahedral C_{60}, with faces labelled according to an 'orange peel' scheme. The pentagonal face labelled 32 lies on the opposite side of the molecule to the pentagonal face labelled 1.

gonal and hexagonal faces in this spiral, and so one can represent icosahedral C_{60} still more compactly using a "face spiral" of the form

$$(I_h)C_{60} = 56666656565656566565656565666665 .$$

Moreover similar face spirals can be found for all other fullerenes mentioned in the literature to date. Face spirals for the specific C_{28}, C_{32}, C_{50} and C_{70} structures described by Kroto [4,5], for example, are given by

$$(T_d)C_{28} = 5556565655555556 ,$$

$$(D_3)C_{32} = 555656565565656555 ,$$

$$(D_{5h})C_{50} = 5656665566556566656665655656 ,$$

and

$$(D_{5h})C_{70}$$

$$= 5666666566565656566566565666565666566 .$$

These equations can easily be verified by building a molecular model of each fullerene, progressively adding each new face in the spiral at a time.

Of course this correspondence between face spirals and distinct fullerene structures is not one-to-one. We could equally well, for instance, have started the icosahedral C_{60} face spiral of fig. 2 on a hexagon, and then proceeded directly to either a hexagonal or a pentagonal face. However, while the existence of several different face spirals for each distinct fullerene structure may well be a nuisance, in that it leads to more work, it is not a serious problem. All that really matters is that every possible fullerene structure can, in light of the above samples, be represented by at least one face spiral of the form shown in fig. 2.

Since the fullerenes C_n have $(n/2+2)$ faces, of which 12 are pentagonal and the remaining $(n/2-10)$ hexagonal [9], we need only consider a total of

$$\frac{(n/2+2)!}{12!(n/2-10)!}$$

distinct face spirals for each given C_n. For n less than about 70 it is practical to conduct a direct computer search of these spirals, and to calculate the Hückel molecular orbital energy levels of those which do indeed close to form fullerenes. (In fact the vast majority of face spirals do *not* close to form fullerenes, and so can immediately be dropped from the search.) The thermodynamically (kinetically) most stable C_n fullerene structure, according to Hückel theory, is then that with the largest resonance energy (HOMO–LUMO energy separation) so obtained.

Proceeding in this way we have found that the most stable of all 1812 distinct C_{60} fullerene structures does indeed have icosahedral (I_h) symmetry [#1], with no adjacent pentagonal faces, and corresponds to the face spiral shown in fig. 2. (In other words Hückel theory really does *predict* that the fullerene C_{60} is icosahedral.) The most stable C_{50} fullerene has D_3 symmetry, with no more than two adjacent pentagonal

---

[#1] Here two C_n fullerene structures are regarded as distinct if they have different Hückel theory resonance energies and/or HOMO–LUMO energy separations. This definition discounts the fact that identical resonance energies and HOMO–LUMO energy separations might conceivably be obtained for two different C_n fullerene molecular graphs.

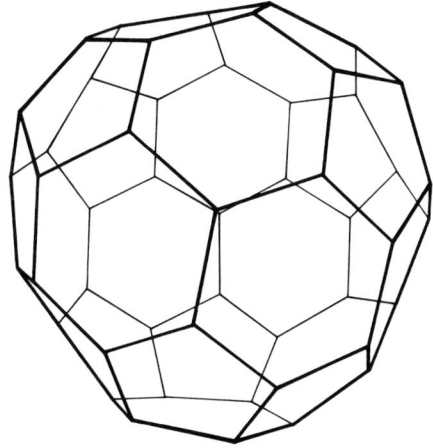

Fig. 3. Line drawing of the most stable (D₃ symmetry) C_{50} structure, viewed approximately along its threefold symmetry axis.

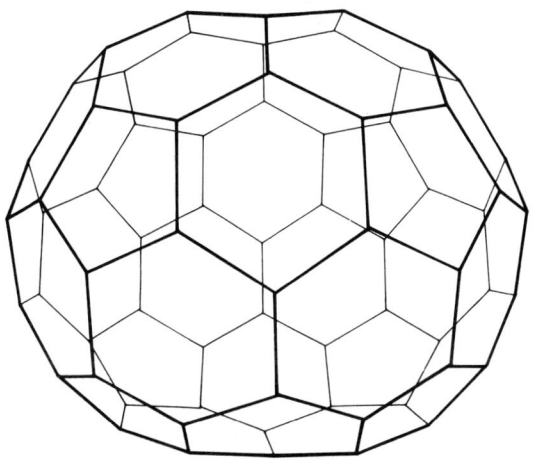

Fig. 4. Line drawing of the most stable (D₅ₕ symmetry) C_{70} structure, viewed approximately from a point within its horizontal plane [4–6].

faces, and is shown in fig. 3. This structure has both a higher resonance energy per atom (-0.541β versus -0.539β), and a larger HOMO–LUMO energy separation (-0.468β versus -0.103β), than the D₅ₕ symmetry C_{50} structure described by Kroto [4,5]. The most stable C_{70} fullerene has D₅ₕ symmetry, with no adjacent pentagonal faces, and is shown in fig. 4. (See also fig. 4 of the recent paper by Taylor and co-workers [6], which contains the first experimental confirmation of D₅ₕ symmetry for C_{70} by ^{13}C NMR.) In all three of these examples the fullerene with max-

imum resonance energy per atom also has maximum HOMO–LUMO energy separation, and so is both thermodynamically and kinetically the most stable.

3. Kinetic stabilities

Returning now to the graphite laser vaporisation results in fig. 1, we should first explain why we think it is appropriate to concentrate on kinetic, rather than thermodynamic, stability. In fact there are two main reasons for this, one of which is physical and the other theoretical.

The physical reason is complicated by the fact that mass spectra like that in fig. 1 are obtained under a variety of different experimental conditions [1,2,4]. When the clustering conditions and the photoionising laser flux are both mild, as they are for example in ref. [2] (fig. 2c), it is likely that each mass peak corresponds directly to a parent cluster ion. In this instance, since the clustering conditions are only mild, one would not expect thermodynamic equilibrium between the various parent clusters to have been reached. When the clustering conditions and the photoionising laser flux are both extreme, as they are in ref. [4] (fig. 2d), the vast majority of mass peaks (C_{60} and C_{70} being the only two exceptions) come from larger parent clusters which have fragmented on photoionisation, and one would be even less likely to expect these fragment clusters to be in thermodynamic equilibrium. However, since the spectra from these two very different physical processes are broadly similar, and both are similar to the independent results (obtained under yet another different set of experiment conditions [1]) in fig. 1, one can only conclude that *some* special property of the C_{34} to C_{70} clusters must be determining all three. It is there that *kinetic* stability, by which we mean stability with respect to the activated complex of any further clustering or fragmentation reaction, would appear to be the natural choice.

The theoretical reason is that the maximum Hückel theory resonance energies per atom of the fullerenes increase more or less monotonically from C_{20} to the graphite limit, reflecting the increased aromaticity associated with a larger proportion of six-membered rings. Clearly, then, thermodynamic equilibrium between the fullerenes will never account for any ion signal intensity pattern remotely like that in fig. 1, or indeed like any other experimental spectrum we have seen [1–4]. By contrast, the maximum Hückel theory HOMO–LUMO energy separations of the fullerenes go to zero both at C_{20} *and* at the graphite limit, and so must have a maximum somewhere between. Given our present results, in conjunction with earlier theoretical work on larger icosahedral symmetry fullerenes [16,17], we are convinced that this maximum lies at C_{60}. Kinetic stability, as determined in Hückel theory by HOMO–LUMO energy separation, is therefore far more likely to account for the spectrum in fig. 1. (In fact there *are* small local fluctuations in the maximum resonance energy, due essentially to the occurrence of special high symmetry structures like icosahedral C_{60}, and these fluctuations do tend to parallel local fluctuations in the

maximum HOMO–LUMO energy separation. However, as noted previously by Fowler [16], both simple Hückel [18] and MNDO [19] methods predict D_{5h} symmetry C_{70} to be thermodynamically *more* stable than icosahedral C_{60}.)

Fig. 5, in light of these arguments, shows the maximum HOMO–LUMO energy separations of the fullerenes C_{34} to C_{70}. It is probably worth noting here that a large HOMO–LUMO energy separation is conventionally associated with high kinetic stability because it is energetically unfavourable to add electrons to a high-lying LUMO, and to extract electrons from a low-lying HOMO, and so to form the activated complex of any potential reaction. This basic argument becomes all the more valid in the carbon and helium atmosphere of a graphite laser-vaporisation experiment, if only because the π molecular

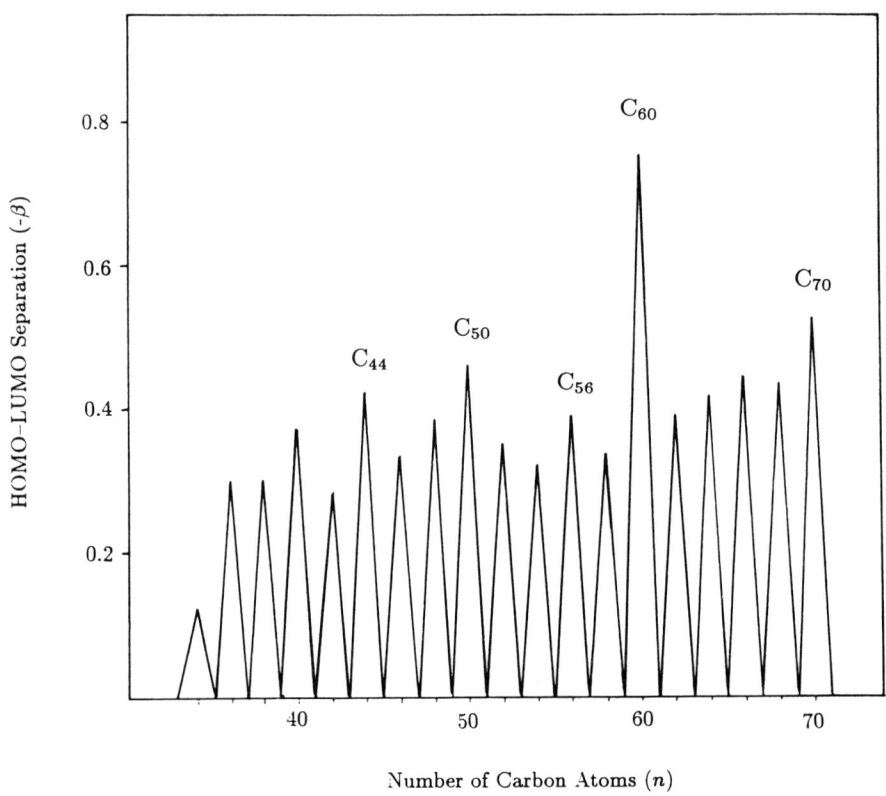

Fig. 5. Maximum HOMO–LUMO energy separations of the fullerenes. Presents Hückel molecular orbital results.

110

orbital barycentres of all reactive species in the environment are the same [#2].

A glance at figs. 1 and 5 shows that there is indeed a significant correlation between the maximum HOMO–LUMO energy separations of the fullerenes and the intensities of photoionisation signals from carbon clusters produced by laser vaporisation of graphite. The correlation is not perfect, since for example the photoionisation signal from C_{56} is larger relative to that from C_{50} than the Hückel calculations, taken in conjunction with the above arguments, would suggest. However, given that Hückel theory simply provides a crude first-order description of π-electron structure, that no attempt has been made to correct for departures from perfect sp^2 hybridisation caused by non-planarity [9,21], and that one would not in any case expect direct proportionality between kinetic stability as determined by HOMO–LUMO energy separation and concentration as measured by photoionisation signal intensity, the correlation is very good.

Finally, and in addition to the obvious caveats regarding Hückel theory just mentioned, we should again stress that the graphite laser-vaporisation results in fig. 1 can change markedly under different experimental conditions. Of course this is not inconsistent with our reasoning, being fully compatible with kinetic control. Indeed what is particularly interesting, in view of fig. 5, is that the C_{44}, C_{50}, C_{56}, C_{60} and C_{70} mass peaks are pronounced relative to their neighbours in every experimental spectrum we have seen [1–4,22].

4. Concluding remarks

Despite its obvious limitations, simple Hückel molecular orbital theory is still widely used as a

---

[#2] In fact the Hückel theory LUMOs of all fullerenes in the C_{34} to C_{70} range, with the sole exception of icosahedral C_{60}, are either slightly bonding or non-bonding. We shall ignore this added complication here, though, and continue to assume that a large HOMO–LUMO energy separation can reasonably be taken as an indication of high kinetic stability in the carbon and helium environment. The non-bonding Hückel LUMO of $(D_{5h})C_{70}$ does indeed become properly anti-bonding at a more sophisticated level [20], and one would expect a similar shift for slightly bonding Hückel LUMOs of other fullerenes.

starting point for the correlation of molecular properties [14]. We might therefore conclude that the correlation between the HOMO–LUMO energy separations in fig. 5 and the graphite laser vaporisation results in fig. 1 is no accident. Rather, by optimising the HOMO–LUMO energy separations over all possible fullerene structures for each C_n, we have at least been looking in the right place. In short, then, our results provide further circumstantial evidence that the observed C_{34} to C_{70} clusters are indeed fullerenes, closed carbon cages containing only five- and six-membered rings.

It is interesting to note that the very special role of C_{60} among the fullerenes has, according to our arguments, nothing to do with thermodynamic stability. Buckminsterfullerene is special because it has a higher HOMO–LUMO energy separation, and hence greater kinetic stability, than any other fullerene we have seen.

Acknowledgement

It is a pleasure to acknowledge several very helpful discussions concerning this work with Dr. Peter Sarre, Professor John Simons, Dr. Frank Palmer, and Professor Odile Eisenstein. The calculations described in this Letter were performed on a VAX computer at the University of Nottingham Cripps Computing Centre.

References

[1] E.A. Rohlfing, D.M. Cox and A. Kaldor, J. Chem. Phys. 81 (1984) 3322.

[2] H.W. Kroto, J.R. Heath, S.C. O'Brien, R.F. Curl and R.E. Smalley, Nature 318 (1985) 162.

[3] D.M. Cox, D.J. Trevor, K.C. Reichmann and A. Kaldor, J. Am. Chem. Soc. 108 (1986) 2457.

[4] H.W. Kroto, Science 242 (1988) 1139.

[5] H.W. Kroto, Nature 329 (1987) 529.

[6] R. Taylor, J.P. Hare, A.K. Abdul-Sada and H.W. Kroto, J. Chem. Soc. Chem. Commun. (1990) 1423.

[7] W. Krätschmer, K. Fostiropoulos and D.R. Huffmann, Chem. Phys. Letters 170 (1990) 167.

[8] W. Krätschmer, L.D. Lamb, K. Fostiropoulos and D.R. Huffmann, Nature 347 (1990) 354.

[9] T.G. Schmalz, W.A. Seitz, D.J. Klein and G.E. Hite, J. Am. Chem. Soc. 110 (1990) 1113.

[10] R.L. Disch and J.M. Schulman, Chem. Phys. Letters 125 (1986) 465.

[11] H.P. Lüthi and J. Almlöf, Chem. Phys. Letters 135 (1987) 357.

[12] P.W. Fowler, P. Lazzeretti and R. Zanazi, Chem. Phys. Letters 165 (1990) 79.

[13] G.E. Scuseria, Chem. Phys. Letters 176 (1991) 423.

[14] J.N. Murrell, S.F.A. Kettle and J.M. Tedder, The chemical bond (Wiley, New York, 1978) ch. 9.

[15] F.A. Cotton, Chemical applications of group theory (Wiley, New York, 1971) ch. 7.

[16] P.W. Fowler, Chem. Phys. Letters 131 (1986) 444.

[17] D.J. Klein, W.A. Seitz and T.G. Schmalz, Nature 323 (1986) 703.

[18] P.W. Fowler and J. Woolrich, Chem. Phys. Letters 127 (1986) 78.

[19] M.D. Newton and R.E. Stanton, J. Am. Chem. Soc. 108 (1986) 2469.

[20] P.W. Fowler, private communication (1991).

[21] R.C. Haddon, L.E. Brus and K. Raghavachari, Chem. Phys. Letters 125 (1986) 459.

[22] Y.A. Yang, P. Xia, A.L. Junkin and L.A. Bloomfield, Phys. Rev. Letters 66 (1991) 1205.

Reprinted with permission from Chemical Physics Letters
Vol. 170, No. 2,3, pp. 167–170, 6 July 1990

The infrared and ultraviolet absorption spectra of laboratory-produced carbon dust: evidence for the presence of the C_{60} molecule

W. Krätschmer [a], K. Fostiropoulos [a] and Donald R. Huffman [b]

[a] *Max-Planck-Institut für Kernphysik, P.O. Box 103980, D-6900 Heidelberg, Federal Republic of Germany*
[a] *Department of Physics, University of Arizona, Tucson, AZ 85721, USA*

Received 1 May 1990

In carbon smoke samples prepared from vaporized graphite at elevated quenching gas pressures (e.g. > 100 Torr He) new absorption features have been observed in the infrared (the strongest at 1429, 1183, 577, and 528 cm^{-1}). Broader features also have been observed in the ultraviolet (the strongest at 340, 270, and 220 nm). By studying ^{13}C-enriched samples we have shown that the infrared absorptions are produced by large, pure carbon molecules. The evidence supports the idea that the features are produced by the icosahedral C_{60} molecule.

1. Introduction

The C_{60} molecule was discovered as a peak in the mass spectra of quenched carbon vapor [1,2]. In order to explain the stability of this molecule, Kroto et al. [2] proposed the highly symmetric soccer-ball-like molecular shape, coined Buckminsterfullerene (a truncated icosahedron with point group I_h). Because of its unusually high symmetry C_{60} is expected to have only four infrared active vibrational modes (of species T_{1u}) which calculations have shown are expected to occur at about 1600 ± 200, 1300 ± 200, 630 ± 100, and 500 ± 50 cm^{-1} [3–6], the spread indicating the author-to-author variation in the calculated line positions. The presence of strong electronic transitions at 260 nm and shorter wavelengths has also been predicted [7]. Several authors have commented on the importance of producing enough C_{60} for carrying out absorption spectroscopy, but so far the only known absorption feature is a weak electronic transition detected by depletion spectroscopy at 386 nm [8]. In order to further confirm the existence of the soccer-ball-shaped molecule and to investigate its role in both interstellar and terrestrial chemistry, knowledge of its absorption spectra is vitally needed.

We have recently reported the discovery of four strong lines in the infrared and three broader features in the near ultraviolet, observed in absorption spectra of laboratory-produced carbon smoke [9]. The infrared lines have also been noticed by other investigators [10]. The aim of this communication is to report new spectroscopic results on isotopically modified samples which support the idea that the reported features in fact originate from a pure and massive carbon molecule which very likely is the soccer-ball-shaped C_{60} molecule.

2. Experimental procedures and results

Carbon smoke particles were produced by evaporating graphite rods by resistive heating in a conventional glass bell jar evaporator filled with an inert quenching gas. The carbon vapor nucleates in the presence of an inert quenching gas to form smoke particles, which can be collected on substrates. In order to produce ^{13}C dust particles, we made rods from commercially available, isotopically enriched carbon powder. In making the rods the powder was compressed (at about 1 kbar) and heated (to about 1300°C) for a few minutes within a quartz-glass tube.

168

The smoke produced by vaporizing the rods was collected either on transparent substrates for transmission measurements or on gold-coated glass surfaces for reflection measurements. The latter method avoids the complicating interference fringes that arise from transparent substrates when spectra are taken at high resolution. It also avoids unwanted substrate absorptions. Infrared measurements were made with an FTIR spectrometer (Bruker 113V) at 2 or 0.3 cm^{-1} resolution, and visible–ultraviolet measurements (in the range shortward of 600 nm) were taken with grating instrument (PE-330) at 2 nm resolution.

The spectra from carbon dust obtained from natural carbon (about 99% ^{12}C) are shown in fig. 1. At pressures of about 10 Torr, the infrared spectra consist of a featureless background which steadily rises towards shorter wavelengths. Broad humps at 1600–1200 cm^{-1} and 900–500 cm^{-1} are superimposed on this rise. These humps originate from stretching and bending modes of carbon within the highly distorted graphite structure of the dust particles. Infrared features which have been observed at 1588 and 868 cm^{-1} only in highly oriented single crystals of graphite [11] are absent in our spectra. In overall shape, our spectra are very similar to those published in the literature [12]. However, for samples produced at higher quenching gas pressures (100 Torr for example), we noticed four prominent peaks which emerge out of the essentially unchanged continuum, along with a number of weaker lines. Also, in the ultraviolet, new features appear which seem to be correlated with the new infrared features. The infrared spectrum obtained with 99% ^{13}C at the same quenching gas pressure shows that the four strong bands, as well as a number of the weaker lines, are displaced by a constant fraction (0.9625). This value is almost precisely the square root of the ratio of the ^{12}C to the ^{13}C masses, as expected for pure carbon.

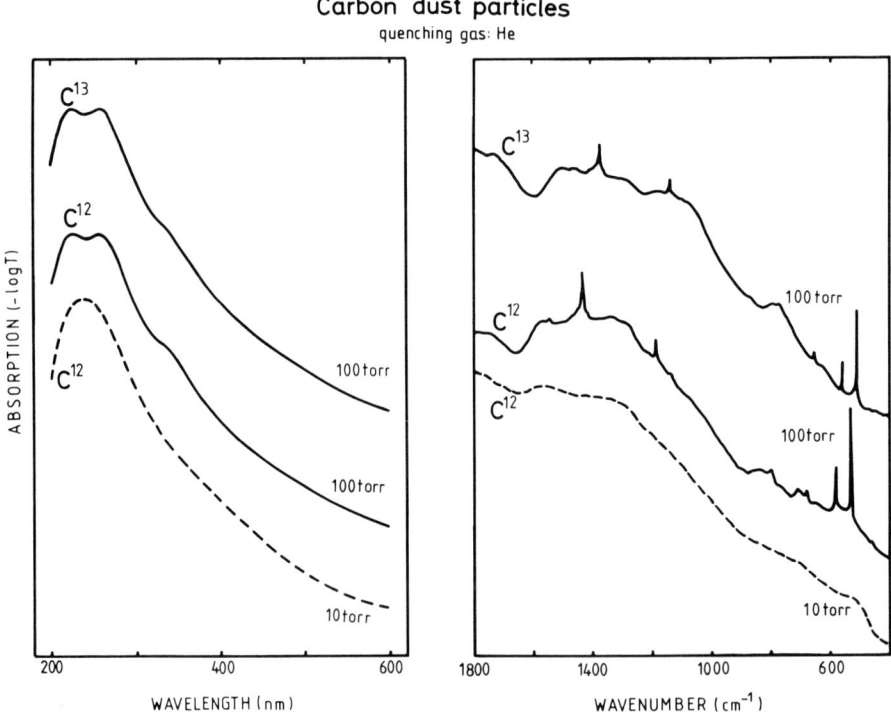

Carbon dust particles
quenching gas: He

Fig. 1. The ultraviolet, visible, and infrared spectra of laboratory-produced carbon dust particles produced using different quenching gas pressures and isotopic compositions (99% ^{12}C or ^{13}C respectively). As an evaluation of the displacement of the infrared lines in ^{13}C samples shows, the carrier of the infrared features which appear at higher quenching gas pressures must be a pure carbon molecule. For clarity, the spectra are vertically displaced.

Table 1
The infrared line positions (rounded to full wavenumbers) and integrated strength of lines which show similar frequency ratios in ^{12}C and ^{13}C dust samples. For the weaker features, this list is not complete

Obs. freq. 99% ^{12}C (cm^{-1})	Obs. freq. 99% ^{13}C (cm^{-1})	Relative intensity	Width 99% ^{12}C (cm^{-1})	Width 90% ^{13}C (cm^{-1})
1429	1375	100	9.2	15.8
1183	1138	20	4.6	6.2
796	765	4		
675	649	6		
643	619	3		
577	556	20	3.3	4.0
565	545	3		
536	516	4		
528	508	70	3.0	3.5

This is strong evidence that the line carriers in our spectra are pure carbon molecules. Table 1 lists the observed IR line positions for the 99% ^{12}C and ^{13}C cases.

The UV spectra for 100 Torr quenching gas pressure and different isotopic compositions remain essentially unchanged. The extra features maintain their positions at 340, 270, and 220 nm (fig. 1). It was observed that the three ultraviolet features and the four infrared lines always appear together and form a characteristic pattern which is not intensity correlated with the background continuum. Thus the infrared and ultraviolet features both appear to originate from a new carbon species which is formed at higher gas pressures along with the dust which produces the continuum. The appearance of new spectral features in our experiments as the quenching gas pressure is increased is reminiscent of the appearance of the strong C_{60} peak in the mass spectra of the previous experiments [2] when the vaporizing laser pulse was timed to occur at the maximum density of the He quenching gas pulse.

Infrared spectroscopy performed at 0.3 cm^{-1} resolution on 90% ^{13}C samples displayed an increased line width for all the four main features but no essential change in the line shapes. Table 1 lists the observed widths. For a light carbon molecule one would expect line broadening caused by the various possible isotopic modifications (isotopomers) of the molecule. Instead we noted that the increase in linewidth (e.g. for the 528 cm^{-1} feature) is rather lim-

ited. It thus appears from these data that the carrier of the infrared features is a quite large molecule or group of large molecules.

In addition to the four strong infrared bands, table 1 lists five weak bands which also show the isotope shifts expected of pure carbon. There may also be more weaker bands. It is possible that these may also arise from the same C_{60} molecule, but with its symmetry disturbed by the carbon impurity mass present in approximately 1% abundance. The disturbed symmetry may be causing some disallowed modes to become observable. Other symmetry distortions may occur by interaction of C_{60} with the carbon grains. One may also speculate on the possibility that these additional bands are from the C_{70} molecule, which appears to a lesser extent in the mass spectra under conditions that produce a strong C_{60} peak [2].

In a search of the spectral range from about 100 to 5000 cm^{-1}, as well as in the near infrared and visible, we detected no other features. Based on the ratio of absorption strengths in the infrared lines to the infrared continuum one can estimate that the abundance of the C_{60} molecule is of the order of 1% of the total sample.

3. Conclusions

Correlated absorption features in the ultraviolet and in the infrared, which were recently reported for the first time in our previous work, appear to be caused by pure carbon in view of the isotope shift observed in the infrared spectra. The number of strong bands in the vibrational region of the infrared is four, which agrees in number with that predicted by the symmetry of the soccer-ball-shaped C_{60} molecule. Vibrational frequencies noted in this paper are also in acceptable agreement with published theoretical values. The ultraviolet features are in rough agreement with the expected absorption frequencies of the same C_{60} molecule. We also note a qualitative correspondence between the high quenching gas pressure required to produce our newly observed bands and the high entraining gas pressure and relatively long clustering times involved in the previous mass spectrometric works. For these reasons we believe we have produced the first reported sample of Buckminsterfullerene in sufficient quantities for doing infrared and ultraviolet absorption spectroscopy. For future studies of the properties of C_{60} it may be possible to extract bulk quantities of this molecule from smoke samples.

Acknowledgement

We gratefully acknowledge the assistance of Mr. Bernd Wagner. One of us (DRH) expresses appreciation to the Alexander von Humboldt Stiftung for support in the form of a Senior US Scientist Award during the early part of this work.

References

[1] E.A. Rohlfing, D.M. Cox and A. Kaldor, J. Chem. Phys. 81 (1984) 3322.

[2] H.W. Kroto, J.R. Heath, S.C. O'Brien, R.F. Curl and R.E. Smalley, Nature 318 (1985) 162.

[3] D.E. Weeks and W.G. Harter, J. Chem. Phys. 90 (1989) 4744.

[4] R.E. Stanton and M.D. Newton, J. Phys. Chem. 92 (1988) 2141.

[5] S.J. Cyvin, E. Brendsdal, B.N. Cyvin and J. Brunvoll, Chem. Phys. letters 143 (1988) 377.

[6] Z.C. Wu, D.A. Jelski and T.F. George, Chem. Phys. Letters 137 (1987) 291.

[7] S. Larsson, A. Volosov and A. Rosén, Chem. Phys. Letters 137 (1987) 501.

[8] J.R. Heath, R.F. Curl and R.E. Smalley, J. Chem. Phys. 87 (1987) 4236.

[9] W. Krätschmer, K. Fostiropoulos and D.R. Huffman, in: Dusty objects in the universe, eds. E. Bussoletti and A.A. Vittone (Kluwer, Dordrecht, 1990), in press.

[10] J. Hare, private communication (1990).

[11] R.J. Nemanich, G. Lucovsky and S.A. Solin, Solid State Commun. 23 (1977) 117.

[12] E. Bussoletti, L. Colangeli and V. Orofino, in: Experiments on cosmic dust analogues, eds. E. Bussoletti, C. Fusco and G. Longo (Kluwer, Dordrecht, 1988) p. 63.

Reprinted with permission from Nature
Vol. 347, No. 6271, pp. 354–358, 27 September 1990
© 1990 Macmillan Magazines Limited

Solid C₆₀: a new form of carbon

W. Krätschmer*, Lowell D. Lamb†, K. Fostiropoulos* & Donald R. Huffman†

* Max-Planck-Institut für Kernphysik, 6900 Heidelberg, PO Box 103980, Germany
† Department of Physics, University of Arizona, Tucson, Arizona 85721, USA

A new form of pure, solid carbon has been synthesized consisting of a somewhat disordered hexagonal close packing of soccer-ball-shaped C_{60} molecules. Infrared spectra and X-ray diffraction studies of the molecular packing confirm that the molecules have the anticipated 'fullerene' structure. Mass spectroscopy shows that the C_{70} molecule is present at levels of a few per cent. The solid-state and molecular properties of C_{60} and its possible role in interstellar space can now be studied in detail.

FOLLOWING the observation that even-numbered clusters of carbon atoms in the range C_{30}–C_{100} are present in carbon vapour[1], conditions were found[2-4] for which the C_{60} molecule could be made dominant in the large-mass fraction of vapourized graphite. To explain the stability of the molecule, a model was proposed of an elegant structure in which the carbon atoms are arranged at the 60 vertices of a truncated icosahedron, typified by a soccer ball. The structure, dubbed buckminsterfullerene[2] because of its geodesic nature, has been the subject of several theoretical stability tests[5,6] and has been discussed widely in the literature. Calculations of many physical properties have been made, including electron energies[7-9], the optical spectrum[9], vibrational modes[10-15], and the electric and magnetic properties[16,17]. There has been speculation on the possible chemical and industrial uses of C_{60} (ref. 2), and on its importance in astrophysical environments[18-20]. Until now, it has not been possible to produce sufficient quantities of the material to permit measurement of the physical properties, to test the theoretical calculations, or to evaluate the possible applications.

Some of us have recently reported evidence[21,22] for the presence of the C_{60} molecule in soot condensed from evaporated graphite. The identification was based primarily on the observed isotope shifts of the infrared absorptions when ^{12}C was replaced by ^{13}C, and on comparison of the observed features with theoretical predictions. The measured infrared and ultraviolet absorption bands were superimposed on a rather large continuum background absorption from the graphitic carbon which comprised ≥95% of the sample. Here we report how to extract the carrier of the features from the soot, how to purify it, and evidence that the material obtained is in fact primarily C_{60}.

Method of production

The starting material for our process is pure graphitic carbon soot (referred to below as simply soot) with a few per cent by weight of C_{60} molecules, as described in refs 21, 22. It is produced by evaporating graphite electrodes in an atmosphere of ~100 torr of helium. The resulting black soot is gently scraped from the collecting surfaces inside the evaporation chamber and dispersed in benzene. The material giving rise to the spectral features attributed to C_{60} dissolves to produce a wine-red to brown liquid, depending on the concentration. The liquid is then separated from the soot and dried using gentle heat, leaving a residue of dark brown to black crystalline material. Other non-polar solvents, such as carbon disulphide and carbon tetrachloride, can also dissolve the material. An alternative con-

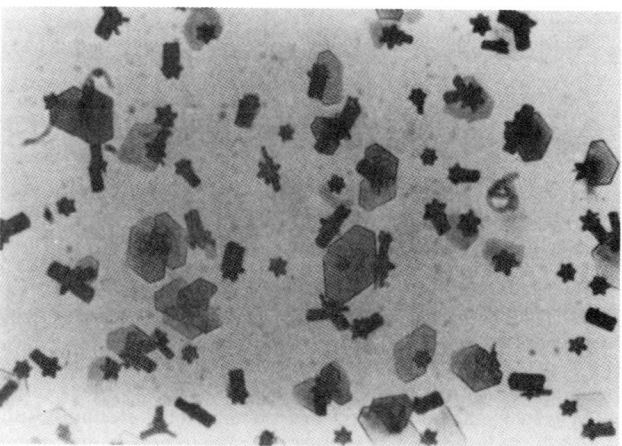

FIG. 1 Transmission micrograph of typical crystals of the C_{60} showing thin platelets, rods and stars of hexagonal symmetry.

centration procedure is to heat the soot to 400 °C in a vacuum or in an inert atmosphere, thus subliming the C_{60} out of the soot (W. Schmidt, personal communication). The sublimed coatings are brown to grey, depending on the thickness. The refractive index in the near-infrared and visible is about two. To purify the material, we recommend removing the ubiquitous hydrocarbons before the concentration procedure is applied (for example, by washing the initial soot with ether). Thin films and powder samples of the new material can be handled without special precautions and seem to be stable in air for at least several weeks, although there does seem to be some deterioration with time for reasons that are as yet unclear. The material can be sublimed repeatedly without decomposition. Using the apparatus described, one person can produce of the order of 100 mg of the purified material in a day.

Studies by optical microscopy of the material left after evaporating the benzene show a variety of what appear to be crystals—mainly rods, platelets and star-like flakes. Figure 1 shows a micrograph of such an assemblage. All crystals tend to exhibit six-fold symmetry. In transmitted light they appear red to brown in colour; in reflected light the larger crystals have a metallic appearance whereas the platelets show interference colours. The platelets can be rather thin and are thus ideally suited for electron-diffraction studies in an electron microscope (see the inset in Fig. 3).

Mass spectroscopy

The material has been analysed by mass spectrometry at several facilities. All mass spectra have a strong peak at mass 720 a.m.u., the mass of C_{60}. Significant differences in the spectra occur only at masses lower than 300 a.m.u. Most of these differences seem to originate from the different ionization techniques and in the different methods of desorbing molecules from the sample. Mass spectra recorded at low and high resolution are shown in Fig. 2. The spectra were obtained using a time-of-flight secondary-ion mass spectrometer[23] and a C_{60}-coated stainless-steel plate. In the mass range above 300 a.m.u., the spectrum is dominated by C_{60} ions and its fragments (even-numbered clusters of atomic carbon), and C_{70} ions. In this sample, the ratio of C_{70} to C_{60} is

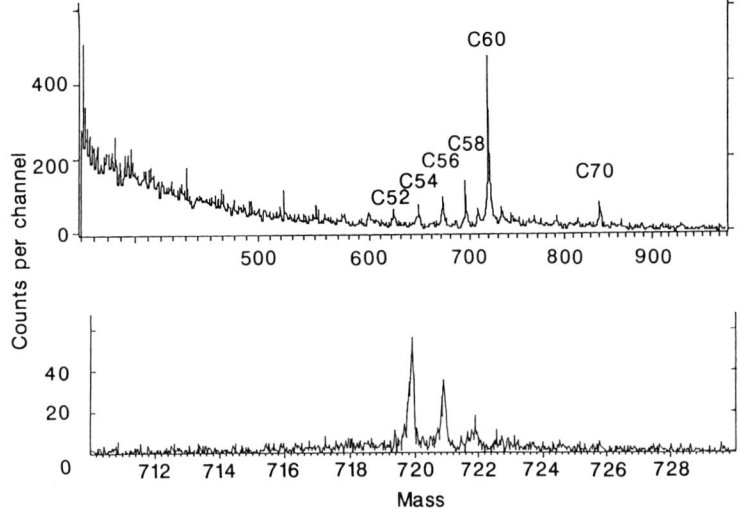

FIG. 2 Low-resolution (top) and high-resolution time-of-flight mass spectra of positive ions obtained from coatings of solid C_{60}. A 5-keV Ar$^+$ ion beam was used to sputter and ionize the sample. The isotope pattern (bottom) is approximately that expected for C_{60} molecules composed of ^{12}C and ^{13}C isotopes of natural abundance.

~0.1. The high-resolution mass spectrum shows approximately the expected isotope pattern for C_{60}. The increasing background in the low-resolution mass spectrum is not produced by the sample—such backgrounds also occur in blank measurements on uncoated stainless-steel substrates.

So far, the cleanest mass spectra have been obtained when the material was evaporated and ionized in the vapour phase by electrons. In such spectra the low-mass background is substantially reduced and the entire mass spectrum is dominated by C_{60} ions and its fragments. The ratio of C_{70} to C_{60} in these mass spectra is ~0.02 and seems to be smaller than that shown in Fig. 2. Both ratios are of the order of those reported from laser-evaporation experiments[2,3]. We assume, as previously suggested[24], that the C_{70} molecule also has a closed-cage structure, either elongated[24] or nearly spherical[25]. Further details of the mass spectroscopy of the new material will be published elsewhere.

Structure

To determine if the C_{60} molecules form a regular lattice, we performed electron and X-ray diffraction studies on the individual crystals and on the powder. A typical X-ray diffraction pattern of the C_{60} powder is shown in Fig. 3. To aid in comparing the electron diffraction results with the X-ray results we have inset the electron diffraction pattern in Fig. 3. From the hexagonal array of diffraction spots indexed as shown in the figure, a d spacing of 8.7 Å was deduced corresponding to the (100) reciprocal lattice vector of a hexagonal lattice. The

most obvious correspondence between the two types of diffraction is between the peak at 5.01 Å of the X-ray pattern and the (110) spot of the electron diffraction pattern, which gives a spacing of ~5.0 Å. Assuming that the C_{60} molecules are behaving approximately as spheres stacked in a hexagonal close-packed lattice with a c/a ratio of 1.633, d spacings can be calculated. The results are shown in Table 1. The values derived from this interpretation are $a = 10.02$ Å and $c = 16.36$ Å. The nearest-neighbour distance is thus 10.02 Å. For such a crystal structure the density is calculated to be 1.678 g cm^{-3}, which is consistent with the value of 1.65 ± 0.05 g cm^{-3} determined by suspending crystal samples in aqueous $GaCl_3$ solutions of known densities. Although the agreement shown in Table 1 is good, the absence of the characteristically strong (101) diffraction of the hexagonal close-packed structure, and the broad continuum in certain regions suggest that the order is less than perfect. Further, X-ray diffraction patterns from carefully grown crystals up to 500 μm in size with well developed faces yielded no clear spot pattern (in contrast to the electron diffraction pattern on micrometre-sized crystals). It therefore appears that these larger crystals do not exhibit long-range periodicity in all directions.

A likely explanation for these facts lies in the disordered stacking of the molecules in planes normal to the c axis. It is well known that the positions taken by spheres in the third layer of stacking determines which of the close-packed structures occurs, the stacking arrangement in a face-centred cubic structure being ABCABC... whereas that in a hexagonal close-

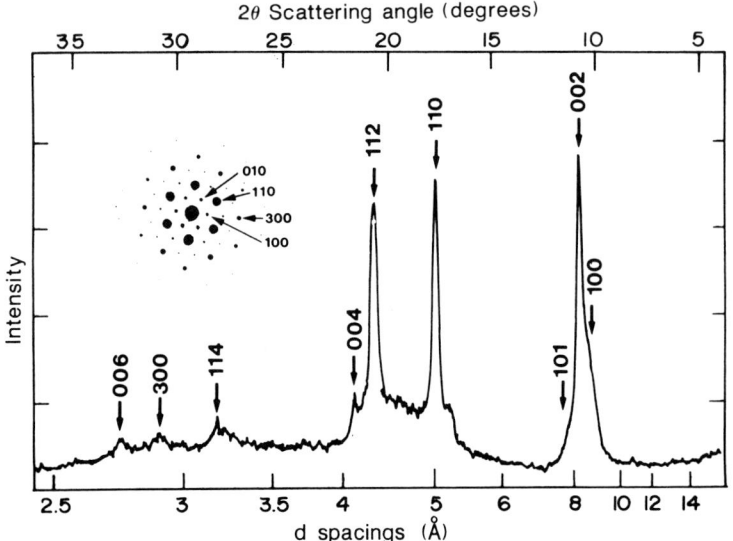

FIG. 3 X-ray diffraction pattern of a microcrystalline powder of C_{60}. Inset (upper left) is a single-crystal electron diffraction pattern indexed with Miller indices compatible with the X-ray pattern. The pattern is from a thin platelet such as those in Fig. 1 with the electron beam perpendicular to the flat face.

TABLE 1 X-ray diffraction results

Measured 2θ (deg)	Measured d spacing (Å)	Calculated d spacing (Å)	Assignment (hkl)
10.2 shoulder	8.7	8.68	(100)
10.81	8.18	8.18	(002)
		7.68	(101)
17.69	5.01	5.01	(110)
20.73	4.28	4.28	(112)
21.63	4.11	4.09	(004)
28.1	3.18	3.17	(114)
30.8	2.90	2.90	(300)
32.7	2.74	2.73	(006)

Assignments for a hexagonal lattice using $a = 10.02$ Å, $c = 16.36$ Å. $(1/d^2) = \frac{4}{3}[(h^2 + hk + k^2)/a^2] + l^2/c^2$.

packed structure is ABABAB... If the stacking sequence varies, the X-ray lines owing to certain planes will be broadened by the disorder whereas other lines will remain sharp. Such disordered crystalline behaviour was observed long ago in the hexagonal close-packed structure of cobalt[26-28] where X-ray diffraction lines such as (101), (102) and (202) were found to be substantially broadened by the stacking disorder. Reflections from planes such as (002) remain sharp because these planes have identical spacings in the face-centred cubic and hexagonal close-packed structures. For the planes producing broadened diffraction peaks because of this kind of disorder, the following condition for the Miller indices (hkl) has been shown to apply[27,29]: $h - k = 3t \pm 1$ (where t is an integer) and $l \neq 0$. None of these broadened reflections are apparent in the X-ray pattern of Fig. 3. This may explain the weakness of the characteristically strong (101) peak. Whether or not this stacking disorder is related to the presence of the possibly elongated C_{70} molecule has yet to be determined.

In small crystals at least, the C_{60} molecules seem to assemble themselves into a somewhat ordered array as if they are effectively spherical, which is entirely consistent with the hypothesis that they are shaped like soccer balls. The excess between the nearest-neighbour distance (10.02 Å) and the diameter calculated for the carbon cage itself (7.1 Å) must represent the effective van der Waals diameter set by the repulsion of the π electron clouds extending outward from each carbon atom. Because the van der Waals diameter of carbon is usually considered to be 3.3–3.4 Å the packing seems a little tighter than one might expect for soccer-ball-shaped C_{60} molecules. The reason for this has not yet been determined.

In summary, our diffraction data imply that the substance isolated is at least partially crystalline. The inferred lattice constants, when interpreted in terms of close-packed icosahedral C_{60}, yield a density consistent with the measured value. Further evidence that the molecules are indeed buckminsterfullerene and that the solid primarily consists of these molecules comes from the spectroscopic results.

Spectroscopy

The absorption spectra of the graphitic soot[21,22] showed evidence for the presence of C_{60} in macroscopic quantities. Following the purification steps described above the material can be studied spectroscopically with the assurance that the spectra are dominated by C_{60}, with some possible effects from C_{70}. Samples were prepared for spectroscopy by subliming pure material onto transparent substrates for transmission measurements. Depending on the pressure of helium in the sublimation chamber, the nature of the coatings can range from uniform films (at high vacuum) to coatings of C_{60} smoke (sub-micrometre microcrystalline particles of solid C_{60}) with the particle size depending to some extent on the pressure.

Figure 4 shows the transmission spectrum of an ~2-μm-thick C_{60} coating on a silicon substrate. The infrared bands are at the same positions as previously reported[21,22], with the four most

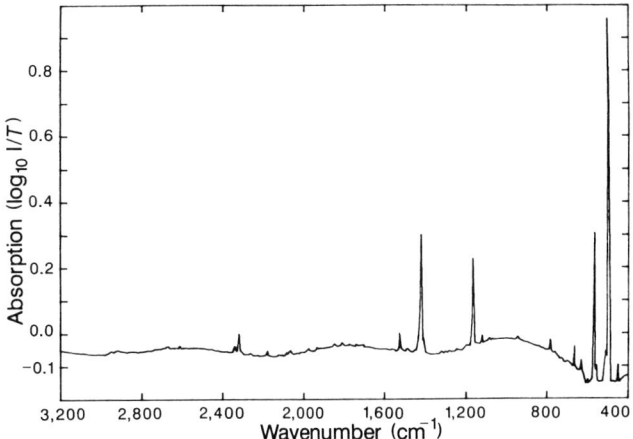

FIG. 4 Infrared absorption spectrum of a coating, ~2 μm thick, of solid C_{60} on a silicon substrate, referenced to a clean silicon substrate. Apparent negative absorptions are due to the coating acting in part as a non-reflecting layer.

intense lines at 1,429, 1,183, 577 and 528 cm^{-1}; here, however, there is no underlying continuum remaining from the soot. In many of our early attempts to obtain pure C_{60}, there was a strong band in the vicinity of 3.0 μm, which is characteristic of a CH-stretching mode. After much effort this contaminant was successfully removed by washing the soot with ether and using distilled benzene in the extraction. The spectrum in Fig. 4 was obtained when the material cleaned in such a manner was sublimed under vacuum onto the substrate. The spectrum shows very little indication of CH impurities. Vibrational modes to compare with the measured positions of the four strong bands have been calculated by several workers[10-15]. As noted previously, the presence of only four strong bands is expected for the free, truncated icosahedral molecule with its unusually high symmetry. Also present are a number of other weak infrared lines which may be due to other causes, among which may be absorption by the C_{70} molecule or symmetry-breaking produced (for example) by isotopes other than ^{12}C in the C_{60} molecule or by mutual interaction of the C_{60} molecules in the solid. Weaker features at ~2,330 and 2,190 cm^{-1}, located in the vicinity of the free CO_2 and CO stretching modes, may imply some attachment of the CO_2 or CO to a small fraction of the total number of C_{60} molecules. Another notable feature is the peak at 675 cm^{-1}, which is weak in the thin-film substrates but almost as strong as the four main features in the crystals. We suspect that this vibrational mode may be of solid state rather than molecular origin.

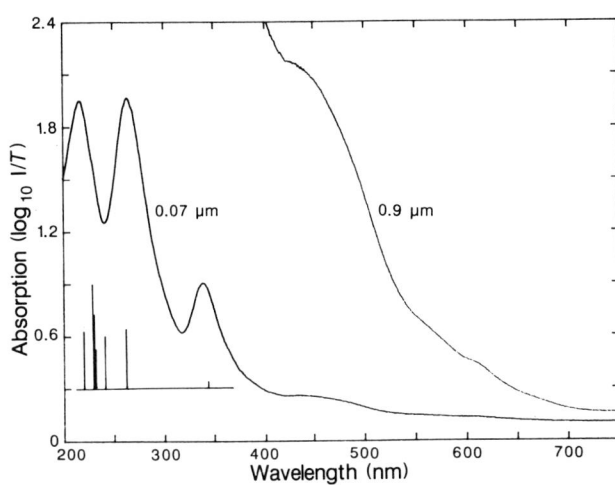

FIG. 5 Visible–ultraviolet absorption spectra of two thicknesses of solid C_{60} on quartz. The calculated[9] positions and relative oscillator strengths for allowed transitions of C_{60} are shown on the bottom.

29

Figure 5 shows an absorption spectrum taken on a uniform film coated on a quartz glass substrate. The ultraviolet features are no longer obscured by the graphitic carbon background as in our previous spectra[22]. Broad peaks at 216, 264 and 339 nm dominate the spectra. Weaker structures show up in the visible, including a plateau between ~460 and 500 nm and a small peak near 625 nm. At the bottom of Fig. 5 we have shown positions and relative oscillator strengths taken from Larsson, Volosov and Rosén[9] calculated for the C_{60} molecule. They also reported a variety of forbidden bands with the lowest energy ones in the vicinity of 500 nm. There seems to be a rough correspondence between our measurements on solid films and the allowed transitions predicted for the molecule. The possibility exists, however, that one or more of the absorption features shown in Fig. 5 are due to C_{70}. We still do not observe a band at 386 nm in our films, as observed[30] using a laser depletion spectroscopy method and attributed to the C_{60} molecule. Quite similar spectra to that in Fig. 5 have been recorded for microcrystalline coatings deposited at helium pressures of 100 torr, for example. The peaks occur at the slightly shifted positions of 219, 268 and 345 nm.

Possible interstellar dust

The original stimulus for the work[2] that led to the hypothesis of the soccer-ball-shaped C_{60} molecule, buckminsterfullerene, was an interest in certain unexplained features in the absorption and emission spectra of interstellar matter. These include an intense absorption band at 217 nm which has long been attributed to small particles of graphite[31], a group of unidentified interstellar absorption bands in the visible that have defied explanation for more than 70 years[31,32], and several strong emission bands attributed to polycyclic aromatic hydrocarbons[33,34]. Based on the visible and infrared absorption spectra of Figs 4 and 5, we do not see any obvious matches with the interstellar features. The ultraviolet band at 216–219 nm has a similar peak wavelength to an interstellar feature, although the other strong bands of the spectrum have no interstellar counterparts. As the influence of C_{70} absorptions on the spectrum is not yet known, a conclusive comparison with the 217-nm interstellar band is difficult. We note that the visible–ultraviolet spectrum presented here is characteristic of a solid, rather than of free molecules. In addition, these new results do not relate directly to absorption in the free C_{60}^+ molecular ion, which has been envisaged[19] to explain the diffuse interstellar bands. Nevertheless, these data should now provide guidance for possible infrared detection of the C_{60} molecule, if it is indeed as ubiquitous in the cosmos as some have supposed.

Summary

To our method for producing macroscopic quantities of C_{60}, we have added a method for concentrating it in pure solid form. Analyses including mass spectroscopy, infrared spectroscopy, electron diffraction and X-ray diffraction leave little doubt that we have produced a solid material that apparently has not been reported previously. We call the solid fullerite as a simple extension of the shortened term fullerene, which has been applied to the large cage-shaped molecules typified by buckminsterfullerene (C_{60}). The various physical and chemical properties of C_{60} can now be measured and speculations concerning its potential uses can be tested.   □

Received 7 August; accepted 7 September 1990.

1. Rohlfing, E. A., Cox, D. M. & Kaldor, A. J. chem. Phys. **81**, 3322–3330 (1984).
2. Kroto, H. W., Heath, J. R., O'Brien, S. C., Curl, R. F. & Smalley, R. E. Nature **318**, 162–163 (1985).
3. Zhang, Q. L. et al. J. phys. Chem. **90**, 525–528 (1986).
4. Liu, Y. et al. Chem. Phys. Lett. **126**, 215–217 (1986).
5. Newton, M. D. & Stanton, R. E. J. Am. chem. Soc. **108**, 2469–2470 (1986).
6. Lüthi, H. P. & Almlöf, J. Chem. Phys. Lett. **135**, 357–360 (1987).
7. Satpathy, S. Chem. Phys. Lett. **130**, 545–550 (1986).
8. Haddon, R. C., Brus, L. E. & Raghavachari, K. Chem. Phys. Lett. **125**, 459–464 (1986).
9. Larsson, S., Volosov, A. & Rosén, A. Chem. Phys. Lett. **137**, 501–504 (1987).
10. Wu, Z. C., Jelski, D. A. & George, T. F. Chem. Phys. Lett. **137**, 291–294 (1987).
11. Stanton, R. E. & Newton, M. D. J. phys. Chem. **92**, 2141–2145 (1988).
12. Weeks, D. E. & Harter, W. G. Chem. Phys. Lett. **144**, 366–372 (1988).
13. Weeks, D. E. & Harter, W. G. J. chem. Phys. **90**, 4744–4771 (1989).
14. Elser, V. & Haddon, R. C. Nature **325**, 792–794 (1987).
15. Slanina, Z. et al. J. molec. Struct. **202**, 169–176 (1989).
16. Fowler, P. W., Lazzeretti, P. & Zanasi, R. Chem. Phys. Lett. **165**, 79–86 (1990).
17. Haddon, R. C. & Elser, V. Chem. Phys. Lett. **169**, 362–364 (1990).
18. Kroto, H. Science **242**, 1139–1145 (1988).
19. Kroto, H. W. in Polycyclic Aromatic Hydrocarbons and Astrophysics (eds Léger, A. et al.) 197–206 (Reidel, Dordrecht, 1987).
20. Léger, A., d'Hendecourt, L., Verstraete, L. & Schmidt, W. Astr. Astrophys. **203**, 145–148 (1988).
21. Krätschmer, W., Fostiropoulos, K. & Huffman, D. R. in Dusty Objects in the Universe (eds Bussoletti, E. & Cittone, A. A.) (Kluwer, Dordrecht, in the press).
22. Krätschmer, W., Fostiropoulos, K. & Huffman, D. R. Chem. Phys. Lett. **170**, 167–170 (1990).
23. Steffens, P., Niehuis, E., Friese, T. & Benninghoven, A. Ion Formation from Organic Solids (ed. Benninghoven, A.) Ser. chem. Phys. Vol. 25, 111–117 (Springer-Verlag, New York, 1983).
24. Kroto, H. W. Nature **329**, 529–531 (1987).
25. Schmalz, T. G., Seitz, W. A., Klein, D. J. & Hite, G. E. J. Am. chem. Soc. **110**, 1113–1127 (1988).
26. Hendricks, S. B., Jefferson, M. E. & Schultz, J. F. Z. Kristallogr. **73**, 376–380 (1930).
27. Edwards, O. S., Lipson, H. & Wilson, A. J. C. Nature **148**, 165 (1941).
28. Edwards, O. L. & Lipson, H. Proc. R. Soc. A**180**, 268–277 (1942).
29. Houska, C. R., Averbach, B. L. & Cohen, M. Acta Metal. **8**, 81–87 (1960).
30. Heath, J. R., Curl, R. F. & Smalley, R. E. J. chem. Phys. **87**, 4236–4238 (1987).
31. Huffman, D. R. Adv. Phys. **26**, 129–230 (1977).
32. Herbig, E. Astrophys. J. **196**, 129–160 (1975).
33. Léger, A. & Puget, J. L. Astr. Astrophys. Lett. **137**, L5–L8 (1984).
34. Allamandola, L. J., Tielens, A. G. & Barker, J. R. Astrophys. J. **290**, L25–L28 (1985).

ACKNOWLEDGEMENTS. W.K. and K.F. thank our colleagues F. Arnold, J. Kissel, O. Möhler, G. Natour, P. Sölter, H. Zscheeg, H. H. Eysel, B. Nuber, W. Kühlbrandt, M. Rentzea and J. Sawatzki. L.D.L. and D.R.H. thank our colleagues J. T. Emmert, D. L. Bentley, W. Bilodeau, K. H. Schramm and D. R. Luffer. D.R.H. thanks the Alexander von Humboldt Stiftung for a senior US Scientist award. We also thank H. W. Kroto and R. F. Curl for discussions.

Reprinted with permission from Chemical Physics Letters
Vol. 174, No. 3,4, pp. 219–222, 9 November 1990
© 1990 Elsevier Science Publishers

The vibrational Raman spectra
of purified solid films of C_{60} and C_{70}

Donald S. Bethune, Gerard Meijer, Wade C. Tang and Hal J. Rosen

IBM Research Division, Almaden Research Center, 650 Harry Road, San Jose, CA 95120-6099, USA

Received 11 September 1990

A technique to produce samples consisting primarily of C_{60} and C_{70} by fractional sublimation of carbon soot was found and used to produce solid films of these molecules. Film compositions were determined using a surface analytical mass spectrometric technique. Vibrational Raman spectra of the purified films were measured and vibrational lines of both C_{60} and C_{70} are identified. A C_{60} line at 273 cm^{-1} is observed, in agreement with theoretical predictions for the lowest frequency H_g "squashing" mode of Buckminsterfullerene. The two strongest C_{60} lines, found at 1469 and 497 cm^{-1}, can consistently be assigned to the two totally symmetric A_g modes on the basis of their frequencies and measured depolarization ratios.

1. Introduction

The notion of a 60-atom, pure carbon molecule with the structure of a soccerball – a spherical shell made up of 20 hexagons and 12 pentagons arranged as a truncated icosahedron – was conceived of by several authors as a purely theoretical possibility as early as 1970 (for this history, see ref. [1]). It was independently hit upon by Kroto et al. [2] in their effort to understand the extraordinary abundance and inertness of C_{60}, observed in pioneering carbon cluster beam experiments carried out at Exxon [3] and Rice [2]. Buckminsterfullerene, as the molecule was dubbed, captured the imaginations of theorists and experimentalists alike, leading to intense efforts to understand the properties of such a molecule. While theorists have produced numerous studies of the electronic, vibrational, and rotational properties of Buckminsterfullerene [4–9] [1], experimentalists have had a more difficult time trying to measure these properties during the brief span of time between the creation of these clusters in a laser-produced carbon plasma and the end of their flight through a molecular beam machine. Until recently the only spectroscopic feature attributed to C_{60} was a single line at

---

[1] Ref. [5] contains an extensive set of references to theoretical work up to about 1989.

386 nm measured by Heath et al. [10].

This situation is now changing rapidly. In a recent Letter, Krätschmer, Fostiropoulos and Huffman reported UV–VIS and IR spectra taken of carbon dust deposits produced by resistively heating graphite under 100 Torr of He [11]. At that pressure (but not at 10 Torr) four sharp lines appeared on top of broad absorption features. The number, frequencies, intensities and isotope shifts of these lines led the authors to believe that their sample contained significant quantities of Buckminsterfullerene ($\approx 1\%$), and that they had observed its infrared spectrum.

At about the same time, we developed an alternative method for depositing carbon clusters on surfaces using laser ablation of graphite under a static inert gas atmosphere [12]. We used a surface analytical mass spectrometric technique to analyze the deposited material. In our apparatus, a laser desorption jet-cooling mass spectrometer, species present on a surface are brought into the gas phase by laser desorption and are entrained in a supersonic Ar expansion. This cools and transports the sample material into a high vacuum region where it is laser ionized between the acceleration plates of a time-of-flight mass spectrometer. The apparatus has been fully described in a recent publication [13]. Under the conditions we use, this technique samples carbon clusters *already present on the surface* and does not produce them in the desorption process. This has been demonstrated by carrying out an isotope scrambling experiment [12].

Using this surface analytical technique, we were also able to verify that soot deposits made as described by Krätschmer, Fostiropoulos and Huffman [11] indeed contain a significant fraction of C_{60} [14]. This result strongly supports their conclusion that the sharp lines seen in their carbon dust IR spectrum are indeed due to this species.

In this paper we report the production of purified solid films of C_{60} and C_{70} by fractional sublimation of carbon soot. Film compositions were determined using the mass spectrometric technique described above. Vibrational Raman spectra of these films were obtained and vibrational lines of both C_{60} and C_{70} are identified. A C_{60} line at 273 cm^{-1} is observed, in agreement with theoretical predictions for the lowest frequency H_g "squashing" mode of Buckminsterfullerene. The two strongest C_{60} lines, found at 1469 and 497 cm^{-1}, can consistently be assigned to the two totally symmetric A_g modes on the basis of their frequencies and measured depolarization ratios.

2. Sample preparation

Carbon dust produced by resistively heating 99.99% pure graphite bars under 100 Torr of He quenching gas, as described by Krätschmer et al. [11], was collected on a glass disc. The deposited material was scraped from the disc (in air), yielding a black powder. Approximately 20 mg of this crude material was loaded into a small stainless steel cell equipped with a 2 mm inner diameter nozzle, and this was placed in a vacuum chamber pumped to $\approx 10^{-5}$ Torr. In a preliminary run, the cell was heated to 460°C to bake out higher vapor pressure contaminants. Two more runs which we label by a and b were made, with the same soot sample in the cell. In these runs the cell was held at 500° and 600°C, respectively, for several minutes, with cooled suprasil slides placed ≈ 3 mm from the nozzle. In each case a 3 mm wide strip of tungsten foil was fixed to the slide, with its edge ≈ 1 mm off center. The films produced varied in color from light brown at the edges to dark brown in the center. Visible interference fringes near the periphery of the deposit indicated that the layer

of material was at least several microns thick.

The composition of the samples on the tungsten strips was determined using the surface mass spectrometer. A KrF laser (60 μJ in a 0.25 mm spot) was used for desorption, and ionization was accomplished with an ArF laser (200 μJ in a 1.5 mm spot). The mass spectra of the films from runs a and b are displayed in fig. 1. Spectrum a shows that sample a consists almost entirely of C_{60}. The observed C_{60}^+/C_{70}^+ peak ratio is 12. Small peaks of C_{58}^+, C_{56}^+ and C_{54}^+ also appear. Spectrum b is similar, except that the C_{60}^+/C_{70}^+ peak ratio is reduced to 2.1. Therefore sample b, produced at the higher cell temperature, has six times more C_{70} relative to C_{60} than sample a. The mysterious long tails to the high mass sides of both the C_{60} and C_{70} peaks are not yet understood, but they may be reaction products with air. The variation of the "tail signals" with ArF laser fluence suggests that these species are one-photon ionized and

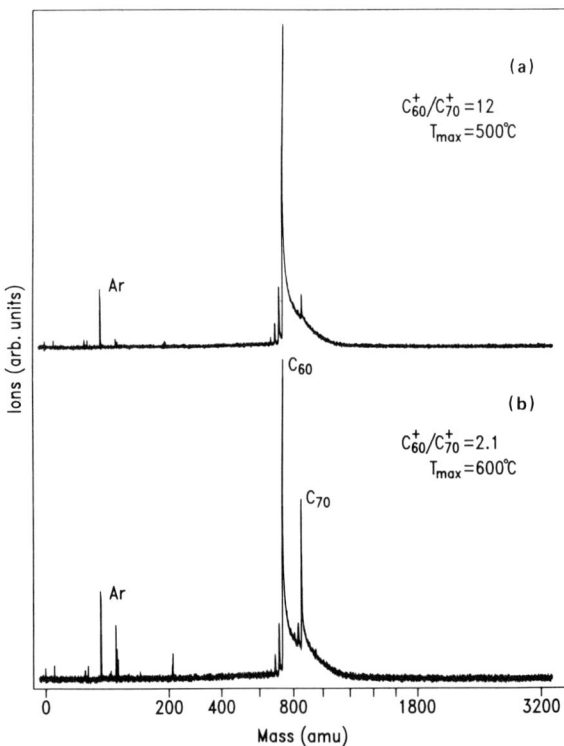

Fig. 1. Time-of-flight mass-spectra of purified films of C_{60} and C_{70} on tungsten foil, taken using the laser-desorption jet-cooling mass spectrometer. The horizontal axis is linear in time-of-flight and the corresponding (nonlinear) mass scale is indicated. The maximum sublimation cell temperatures, T_{max} are indicated.

are therefore detected with high sensitivity. It is possible that these species are responsible for the observed color of the samples, since C_{60} and C_{70} are not expected to have absorptions in the visible spectral region [15,16].

3. Raman spectra

Raman spectra of the samples on the suprasil slides were taken in air at room temperature with a micro-Raman spectrometer. The Ar ion laser beam (140 μW at 514 nm) was focused to a 13 μm diameter spot on the sample film. A back-scattering geometry was used, and the instrumental resolution was ≈ 9 cm^{-1}. Unpolarized spectra obtained from samples a and b are shown in figs. 2a and 2b, respectively. In

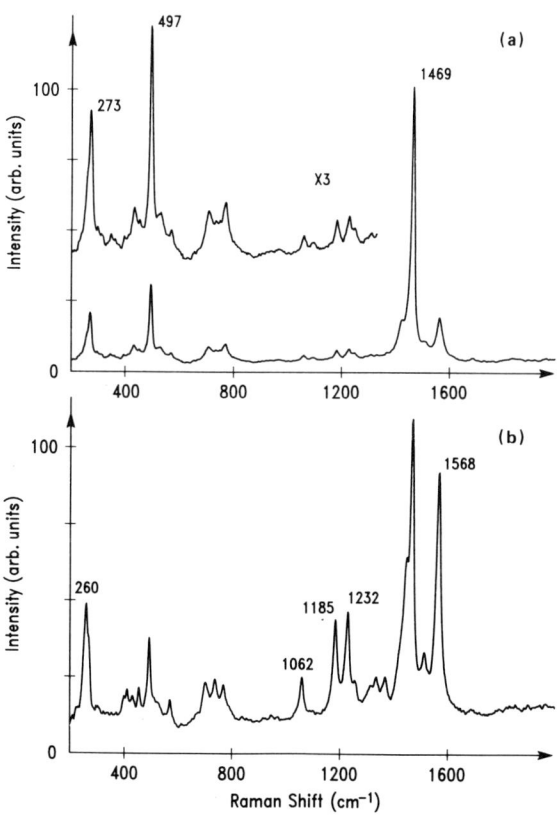

Fig. 2. Unpolarized Raman spectra of purified films of C_{60} and C_{70} on suprasil substrates with a and b corresponding to mass spectra a and b shown in fig. 1. The labelled lines in spectra a and b are assigned to C_{60} and C_{70}, respectively. In both spectra the intensity of the line at 1469 cm^{-1} is normalized to 100.

Table 1

Observed Raman line positions (± 2 cm^{-1}), intensities relative to the 1469 cm^{-1} C_{60} lines ($\equiv 100$), depolarization ratios ($\rho \equiv I_{\perp}/I_{\parallel}$), and assignments. I_a and I_b refer to intensities in spectra a and b in fig. 2

ν (cm^{-1})	I_a	I_b	$\rho (\pm 0.02)$	Assignments
260	7	34		C_{70}
273	17	17		C_{60}, H_g
413		9		
435	5	6		
457		9		
497	27	27	0.16	C_{60}, A_g
571	2	9		
705		13		
711	4			
739		13		
773	6	13		
1062	2	14	0.23	C_{70}
1185	4	34	0.19	C_{70}
1232	4	36	0.19	C_{70}
1336		11		
1370		11		
1430	13			
1448		32		
1469	100	100	0.11	C_{60}, A_g
1513	3	15		
1568	15	88	0.24	C_{70}

spectrum a four strong lines and numerous weaker ones are apparent. In spectrum b, obtained from the sample with the higher C_{70}/C_{60} ratio, additional bands appear and some of the features already recognizable in spectrum a increase markedly in intensity. The positions and intensities of the observed lines in these spectra are summarized in table 1. In both spectra the intensity of the line at 1469 cm^{-1} is normalized to 100. In going from spectrum a to b, some lines keep the same relative intensity, whereas others grow by roughly a factor 6. The three strongest lines in spectrum a, at 1469, 497 and 273 cm^{-1}, belong to the first group, and are therefore assigned to C_{60}. The lines indicated in spectrum b at 1568, 1232, 1185, 1062 and 260 cm^{-1} belong to the second group and are assigned to C_{70}. Note that the lowest C_{60} and C_{70} lines overlap. Additional weaker lines at 1513 and 571 cm^{-1} are also likely to be due to C_{70} vibrations. Between 700 and 800 cm^{-1} there is a feature which seems to include overlapping lines of both C_{60} and C_{70}. These lines cannot be deconvolved from these spectra. No lines were observed in the ranges 90–200 and 2000–2500 cm^{-1}.

222

For the strongest non-overlapped lines in spectrum b we measured the depolarization ratio $\rho \equiv I_\perp / I_\parallel$, where $I_\perp$ and $I_\parallel$ are the intensities of the Raman scattered light polarized perpendicular and parallel to the incident laser polarization. As seen from table 1 the two strongest C_{60} lines are highly polarized, strongly suggesting that they are associated with totally symmetric vibrational modes. We are assuming the C_{60} molecules are randomly oriented. In fact they may even be undergoing rotational motion in the solid films at 300 K.

At this point it is instructive to compare the observed Raman frequencies of C_{60} to the predicted frequencies of the 10 Raman-active modes obtained assuming the truncated icosahedral structure [5–8]. The lowest frequency vibration of this structure is the Raman active H_g "squashing" mode, an almost entirely radial distortion of the spherical structure into an ellipsoid. For this mode the various calculations are in striking agreement with each other, predicting a value of 273 ± 10 cm^{-1} [5–8]. Furthermore, this frequency is well below the next Raman active mode, which all calculations place above 400 cm^{-1}. In this picture, therefore, the observed C_{60} line at 273 cm^{-1} is due to the lowest frequency H_g mode. The polarization data suggest that the C_{60} lines observed at 1469 and 497 cm^{-1} may well be due to totally symmetric modes. In this case they would necessarily be the two A_g modes, corresponding to the double-bond stretching "pentagonal pinch" mode and the "breathing" mode [6]. The frequencies calculated for these modes are in the range 1627–1830 and 510–660 cm^{-1}, respectively. Considering that the calculated frequencies are likely to be too high [7], as also suggested by a comparison of calculated IR frequencies of Buckminsterfullerene to the experimental IR data [11], the observed Raman lines at 1469 and 497 cm^{-1} can consistently be assigned to the A_g modes.

For C_{70}, only one set of vibrational frequencies has been calculated, assuming a structure with D_{5h} symmetry [5]. Comparison of our data with this more complicated predicted spectrum has not yet been attempted.

4. Conclusions

We have shown that C_{60} and C_{70} can be purified by fractional sublimation of carbon soot. In this way purified films of a known composition were produced and Raman spectra of these films were taken. Vibrational frequencies of both C_{60} and C_{70} are found from these spectra. A strong low frequency mode of C_{60} is found at 273 cm^{-1}. The two strongest C_{60} Raman lines at 1469 and 497 cm^{-1} have very low depolarization ratios, and are likely to be due to totally symmetric modes. These data, together with the recently published IR spectrum of C_{60} [11], can be consistently interpreted in terms of the vibrational spectrum calculated for the truncated icosahedral structure of Buckminsterfullerene. With techniques to produce macroscopic amounts of purified C_{60} now available, many measurements to confirm this structure will undoubtedly soon be made. But it looks already as if the notion of a 60-atom, pure carbon molecule with the structure of a soccerball is realized in nature.

References

[1] H. Kroto, Science 242 (1988) 1139.
[2] H.W. Kroto, J.R. Heath, S.C. O'Brien, R.F. Curl and R.E. Smalley, Nature 318 (1985) 162.
[3] E.A. Rohlfing, D.M. Cox and A. Kaldor, J. Chem. Phys. 81 (1984) 3322.
[4] W.G. Harter and D.E. Weeks, J. Chem. Phys. 90 (1989) 4727.
[5] Z. Slanina, J.M. Rudzinski, M. Togasi and E. Osawa, J. Mol. Struct. THEOCHEM 202 (1989) 169.
[6] D.E. Weeks and W.G. Harter, J. Chem. Phys. 90 (1989) 4744.
[7] R.E. Stanton and M.D. Newton, J. Phys. Chem. 92 (1988) 2141.
[8] Z.C. Wu, D.A. Jelski and T.F. George, Chem. Phys. Letters 137 (1987) 291.
[9] S.J. Cyvin, E. Brendsdal, B.N. Cyvin and J. Brunvoll, Chem. Phys. Letters 143 (1988) 377.
[10] J.R. Heath, R.F. Curl and R.E. Smalley, J. Chem. Phys. 87 (1987) 4236.
[11] W. Krätschmer, K. Fostiropoulos and D.R. Huffman, Chem. Phys. Letters 170 (1990) 167.
[12] G. Meijer and D.S. Bethune, J. Chem. Phys., in press.
[13] G. Meijer, M.S. de Vries, H.E. Hunziker and H.R. Wendt, Appl. Phys. B, in press.
[14] G. Meijer and D.S. Bethune, Chem. Phys. Letters 175 (1990), in press.
[15] I. Laszlo and L. Udvardi, J. Mol. Struct. THEOCHEM 183 (1989) 271.
[16] S. Larsson, E. Volosov and A. Rosen, Chem. Phys. Letters 137 (1987) 501.

Reprinted with permission from J. Chem. Soc., Chemical Communications,
Issue 20, pp. 1423–1425, 1990
© 1990 Royal Society of Chemistry

Isolation, Separation and Characterisation of the Fullerenes C_{60} and C_{70}: The Third Form of Carbon

Roger Taylor, Jonathan P. Hare, Ala'a K. Abdul-Sada and Harold W. Kroto

School of Chemistry and Molecular Sciences, University of Sussex, Brighton BN1 9QJ, UK

Pure samples of the species C_{60} (Buckminsterfullerene) and C_{70} (fullerene-70) have been prepared, and their structures characterised by their mass and ^{13}C NMR spectra; the results indicate the existence of a family of stable fullerenes, thus confirming that carbon possesses a third form in addition to diamond and graphite.

Following the proposal in 1985 that the ultra-stable molecule, C_{60}, Buckminsterfullerene, forms spontaneously when carbon is laser vaporised,[1] much effort has been expended on finding support for the original suggestion.[2,3] In early 1990 Kraetschmer et al,[4,5] provided spectroscopic evidence for the presence of trace amounts of C_{60} in smoke obtained from a carbon arc; their observation of four weak but distinct IR bands being consistent with theory.[6–10] We confirmed this result; however, mass spectrometry showed that the material from similarly processed carbon gave rise to strong peaks at m/z 720 and 840, commensurate with C_{60} and C_{70}. Kraetschmer and co-workers[11] have now obtained unequivocal evidence for the structure of C_{60}, thus confirming our original proposal that it is Buckminsterfullerene.[1] We describe

here the results of a parallel study in which pure C_{60} and C_{70} are isolated, separated and characterised. These results not only confirm the results of Kraetschmer et al. on C_{60} but show that C_{70} is also a closed cage and thus provide compelling evidence for the stability of the fullerene family in general.[12–14] The separation of new forms of an element by chromatography is believed to be unique.

Graphite rods were resistively heated in a vessel to deposit carbon smoke under Ar between 50—100 mbar. The films that result from this rudimentary production technique tend to be of somewhat variable quality. However, the four IR bands, first seen by Kraetschmer et al.,[4,5] are usually observed weakly; at best they sometimes appear at about half the

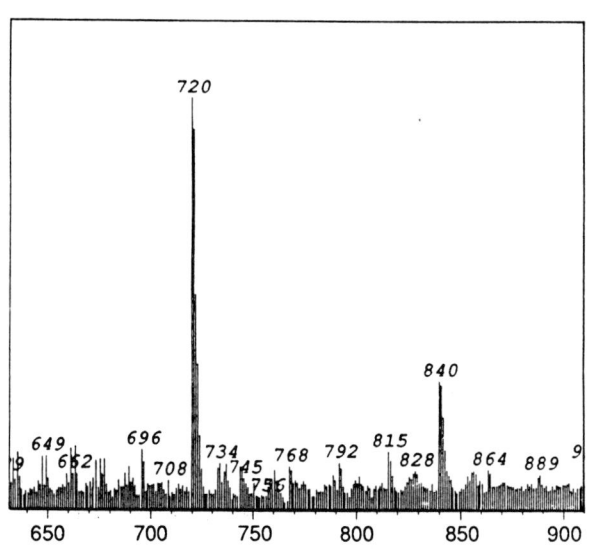

Fig. 1 Mass spectrum of extract from discharge processed carbon

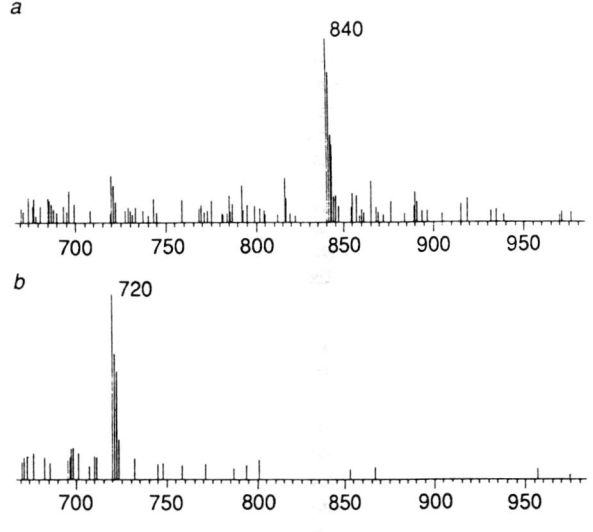

Fig. 2 *a* Mass spectrum of the separated red solid. *b* Mass spectrum of the mustard coloured solid

intensity/signal-to-noise in their spectrum.[4,5] Solid black soot-like material is collected by scraping it from all parts of the apparatus.[15] This material when placed in benzene readily gives rise to a red solution which can be decanted from the black insoluble soot-like material.

The mass spectrum of the smoke produced directly without further processing gives strong peaks at $m/z = 720$ and 840. However, the material which is solvent extracted gives the spectrum shown in Fig. 1. This should be compared with the laser vaporisation results[1-3] and shows that the fullerenes C_n ($n = 62, 64, 66, 68, etc.$) are also present and apparently stable in air.

Solvent extraction of batches of these carbon deposits gave plum-coloured solutions which yielded, after solvent removal, a black-brown solid in up to 8% yield. Chromatographic separation (alumina, hexane) then yielded pure C_{60} and C_{70} in a ratio of approximately 5:1, and together these comprised about 15% of the extract; the total yield was 8 mg. In one of the two chromatographically separated samples the dominant peak appears at $m/z = 720$ and in the other it appears at $m/z = 840$ (Figs. 2 b and a, respectively).

C_{60} is a mustard-coloured solid that appears brown or black with increasing film thickness. It is soluble in the common organic solvents, especially aromatic hydrocarbons, and gives beautiful magenta-coloured solutions. C_{70} is a reddish brown solid, and thicker films are greyish black; its solutions are port-wine red. Solutions of mixtures of C_{60} and C_{70} are red due to C_{70} being more intensely coloured. Both compounds are crystalline with high m.p.'s (>280 °C). Although both compounds are very soluble in, e.g., benzene (C_{60} is the more soluble), they are nevertheless slow to dissolve reflecting the excellent close packing achieved by the spherical and near-spherical molecular structures. Crystals of C_{60} are both needles and plates, the latter being a mixture of squares, triangles and trapezia; the needles consist of a series of overlapping plates.

The ^{13}C NMR spectrum for C_{60} (18 h integration) consists of a single line (Fig. 3a), as required, at 142.68 ppm, and unaltered by proton decoupling. This is significantly downfield from the peaks for the corresponding positions in naphthalene (133.7 ppm), acenaphthylene (128.65 ppm), and benzo[g,h,i]fluoranthene (126.85, 128.05 and 137.75 ppm).[16] This is not unexpected since strain produces downfield shifts which may be attributed to strain-induced hybridisation changes, as shown for example by the ^{13}C peaks for the bridgehead carbons in tetralin (136.8 ppm), indane (143.9 ppm) and benzcyclobutene (146.3 ppm).[16]

Fig. 3c shows the ^{13}C NMR spectrum for C_{70} [run in the presence of Cr(acac)$_3$. Hacac = pentane-2,4-dione, which produces a ca. 0.12 ppm upfield shift of the peaks] consisting of five lines as required, at 150.07, 147.52, 146.82, 144.77 and 130.28 ppm, also unaffected by proton decoupling. The number of lines (five) is compelling evidence for the fullerene-70 structure[12,13] depicted in Fig. 4. Fig. 3b shows the spectrum of a mixed sample in which C_{60} is much reduced. It is clear from this diagram that fullerene-70 possesses five sets of inequivalent carbon atoms with an $n_a : n_b : n_c : n_d : n_e$ ratio of 10:10:20:20:10 (sighting along vertical planes as indicated in the diagram), and this is precisely the ratio of the line intensities observed in the spectrum. The peak at 130.28 ppm can reasonably be assigned to the equatorial ring of ten carbon atoms (e in Fig. 4), since these correspond to the tertiary carbons in pyrene which appear at 124.6 ppm;[16] strain again produces a downfield shift of the peaks. The peaks at 144.77 and 147.52 ppm arise from carbons d and c (Fig. 4). It is probable that the peak at 144.77 ppm is due to the carbon atoms d since models indicate these to be less strained than carbon atoms c. Similarly we suggest that the lines at 150.07 and 146.82 ppm be assigned to type a and b carbon nuclei respectively as indicated in Fig. 3c.

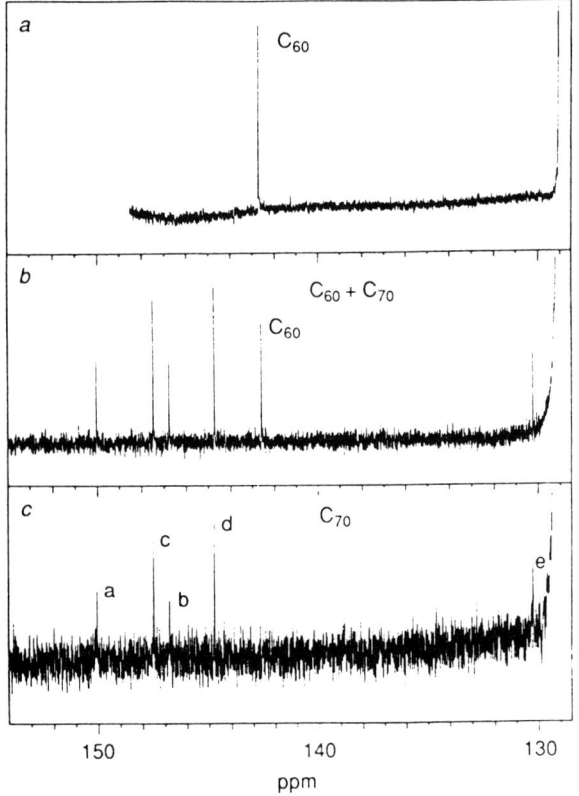

Fig. 3 a ^{13}C NMR spectrum of C_{60}, Buckminsterfullerene. b ^{13}C NMR spectrum of a mixed sample in which C_{60} is much reduced. c ^{13}C NMR spectrum of C_{70}, fullerene-70. The line assignments given are based on the observed intensities and semi-quantitative strain arguments, and are subject to confirmation. The wing of the intense benzene solvent signal lies at the far right hand side

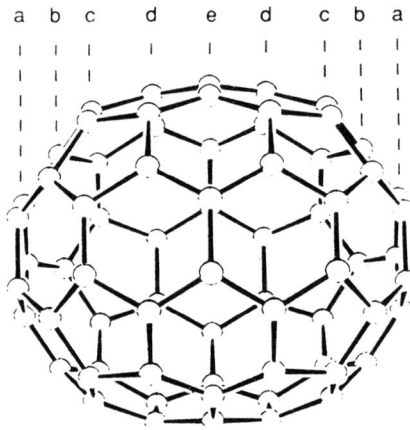

Fig. 4 Schematic diagram of fullerene-70 (based on the diagram in ref. 10). The five sets of identical carbon nuclei a–e lie in the vertical planes as indicated

The mass spectrum of C_{60} (obtained by the FAB method using m-nitrobenzyl alcohol as the matrix) exhibits the parent ion at $m/z = 720$ with complete absence of a peak at $m/z = 840$, showing the sample to be free from C_{70}, confirming the NMR result. The mass spectrum of C_{70}, similarly obtained, shows the main peak at $m/z = 840$, but also a small peak at $m/z = 720$ which we attribute to fragmentation of C_{70} to C_{60}. Analogous fragmentation of fullerenes has been indicated by earlier work.[17]

The electronic spectra for the compounds (UV in hexane and VIS in benzene) are quite different. C_{60} shows peaks at 202.5, 208.3, 211.9, 215.9, 230(sh), 257.6(main), 329.7, 405.7, 408.6, 500(sh), 540.0, 571.0, 596.8, 604.0 and 625(sh) nm. For C_{70} the peaks are at 205.6, 207.3, 210.2(main), 232(sh), 254.8, 322.2, 334.4, 360.0, 378.7, 473.6, 550.0, 600.0, 617.0 and 644.0 nm.

The IR spectra of the separated materials deposited on a KBr plate have been obtained. In the spectrum of the solid obtained from the magenta solution, which gave a very strong m/z 720 mass peak, four strong peaks consistent with the observation of Kraetschmer *et al.* are observed. Normal mode calculations[6-10] predict only four active bands, adding compelling support for the assignment of Kraetschmer *et al.* and supported by the measurements here. The spectrum of fullerene-70 shows many more lines. The calculations of Osawa and co-workers predict 26 allowed vibrations.[14]

We thank Wolfgang Kraetschmer and Michael Jura for their advice and encouragement in the earlier phase of these experiments. We also thank Tony Avent, Steven Firth, Tony Greenway, Jim Hanson, Gerry Lawless, Ahmit Sakar and David Walton of this Laboratory and Paul Scullion of V.G. Analytical Ltd., for valuable assistance. We should also like to thank SERC and British Gas for a CASE studentship to J. P. H.

Received, 10th September 1990; Com. 0/04111D

References

1 H. W. Kroto, J. Heath, S. C. O'Brien, R. F. Curl and R. E. Smalley, *Nature (London)*, 1985, **318**, 162.

2 H. W. Kroto, *Science*, 1988, **242**, 1139.

3 R. F. Curl and R. F. Smalley, *Science*, 1988, **242**, 1017.

4 W. Kraetschmer, K. Fostiropoulos and D. R. Huffman, *Dusty Objects in the Universe*, eds. E. Bussoletti and A. A. Vittone, Kluwer, Dordrecht, 1990.

5 W. Kraetschmer, K. Fostiropoulos and D. R. Huffmann, *Chem. Phys. Lett.*, 1990, **170**, 167.

6 M. D. Stanton and R. E. Newton, *J. Phys. Chem.*, 1988, **92**, 2141.

7 Z. C. Wu, D. A. Jelski and T. F. George, *Chem. Phys. Lett.*, 1988, **137**, 291.

8 D. E. Weeks and W. G. Harter, *Chem. Phys. Lett.*, 1988, **144**, 366.

9 D. E. Weeks and W. G. Harter, *J. Chem. Phys.*, 1989, **90**, 4744.

10 Z. Slanina, J. M. Rudzinski, M. Togaso and E. Osawa, *J. Mol. Struct.*, 1989, **202**, 169.

11 W. Kraetschmer, L. D. Lamb, K. Fostiropoulos and D. R. Huffmann, *Nature (London)*, in the press.

12 H. W. Kroto, *Nature (London)*, 1987, **329**, 529.

13 T. G. Schmalz, W. A. Seitz, D. J. Klein and G. E. Hite, *J. Am. Chem. Soc.*, 1988, **110**, 1113.

14 J. M. Rudzinski, Z. Slanina, M. Togaso and E. Osawa, *Thermochim. Acta*, 1988, **125**, 155.

15 W. Kraetschmer, personal communication.

16 E. Breitmaier, G. Haas and W. Voelter, *Atlas of ^{13}C NMR Data*, Heyden, London, 1979.

17 S. C. O'Brien, J. R. Heath, R. F. Curl and R. E. Smalley, *J. Chem. Phys.*, 1985, **88**, 220.

Reprinted with permission from The Journal of Physical Chemistry
Vol. 94, No. 24, pp. 8630–8633, 1990
© 1990 American Chemical Society

Characterization of the Soluble All-Carbon Molecules C_{60} and C_{70}

Henry Ajie,[†] Marcos M. Alvarez,[†] Samir J. Anz,[†] Rainer D. Beck,[†] François Diederich,[†]
K. Fostiropoulos,[‡] Donald R. Huffman,[§] Wolfgang Krätschmer,[‡] Yves Rubin,[†] Kenneth E. Schriver,[†]
Dilip Sensharma,[†] and Robert L. Whetten*,[†]

*Department of Chemistry and Biochemistry, University of California, Los Angeles, California 90024-1569;
Max-Planck-Institut für Kernphysik, 6900 Heidelberg, Box 103980, F.R.G.; and Department of Physics,
University of Arizona, Tucson, Arizona 85721 (Received: October 3, 1990)*

We report on the further physical and chemical characterization of the new forms of molecular carbon, C_{60} and C_{70}. Our results demonstrate a high yield of production (14%) under optimized conditions and reveal only C_{60} and C_{70} in measurable quantity, in an 85:15 ratio. These two new molecular forms of carbon can be completely separated in analytical amounts by column chromatography on alumina. Comparison among mass spectra obtained by the electron impact, laser desorption, and fast atom bombardment (FAB) methods allows a clear assessment of the composition of the mixed and pure samples, and of the fragmentation and double ionization patterns of the molecules. In addition, spectroscopic analyses are reported for the crude mixture by ^{13}C NMR and by IR spectroscopy in KBr pellet, and for pure C_{60} and C_{70} in solution by UV–vis spectroscopy.

Introduction

In a surprising recent development, Krätschmer et al.[1] have shown that certain all-carbon molecules are produced in large quantities in the evaporation of graphite and can be isolated as soluble, well-defined solids. The major species was identified as molecular C_{60} through mass spectrometry and by comparison of the infrared spectrum with theoretical predictions for the celebrated truncated-icosahedron structure, which had earlier been proposed to account for cluster beam observations.[2] The solid material, described as a new form of elemental carbon in a nearly pure state, has a disordered *hcp* lattice of packed quasi-spherical molecules, but determination of the precise molecular structure awaits diffraction from well-ordered crystals.

Kroto et al.[3] have followed this announcement with a partial chemical separation of the soluble all-carbon molecules generated by the same procedure. They used mass spectrometric evidence to conclude that other air-stable C_n molecules are present (n = 62, 64, 66, 68, 70). They reported that chromatographic separation yields C_{60} and C_{70} in a 3:1 ratio, in contrast to the 2–10% of C_{70} estimated in ref 1. The reported ^{13}C NMR spectrum of the C_{60} fraction, in particular, evidently confirms the existence of a species with all 60 carbon atoms chemically equivalent (proposed structures as shown in Chart I).

This paper describes the further physical and chemical characterization of these two new forms of molecular carbon.[4] Our results include the high-yield production (14%) of soluble material under optimized conditions, consisting of only C_{60} and C_{70} in measurable quantity. These have been separated in analytical amounts by column chromatography and have been characterized in pure or mixed forms by a combination of electron impact, fast atom bombardment (FAB), and laser desorption mass spectrometry. Spectroscopic characterization is reported including the ^{13}C NMR spectrum and the infrared absorption spectrum for the crude

CHART I

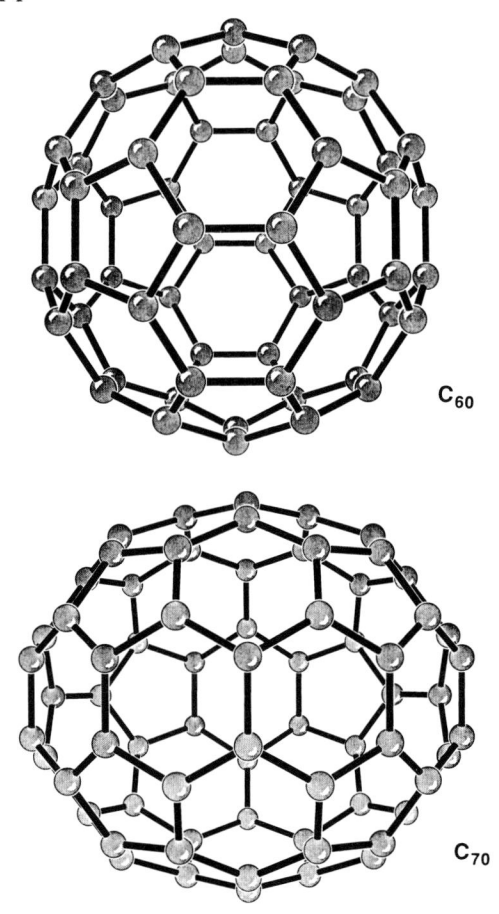

C_{60}

C_{70}

C_{60}/C_{70} mixture, and the UV–vis spectrum of pure C_{60} and C_{70} in solution. All five peaks of C_{70} in the ^{13}C NMR spectrum are

---

[†] University of California, Los Angeles.
[‡] Max-Planck-Institut für Kernphysik.
[§] University of Arizona.

unambiguously found.

Preparation Method and Chromatographic Separation

Two sets of samples have been used in the experiments described below, i.e., one prepared in Heidelberg,[1] used in the earlier stage of this work, and samples newly collected in Los Angeles. We now describe in more detail the method used and yields obtained in the latter preparations. The samples have been prepared following closely the method described by Krätschmer et al.[1] A carbon rod is evaporated by resistive heating under partial helium atmosphere (0.3 bar). The best results were obtained from high-uniformity graphite rods (Poco Graphite, Inc., Type DFP-2, <4 μm grain size, 0.8 μm average pore size). The rods are 1/8 in. diameter and emit a faint gray-white plume when heated by a current of 140–180 A. The sootlike material, collected from the partial evaporation of rods onto a glass shield surrounding the electrodes, is extracted with boiling benzene or toluene to give, after filtration, a brown-red solution. Evaporation yields a brown-black crystalline material in 14% yield (30 mg, identified below to be C_{60} and C_{70}). This yield is higher than reported earlier (ca. 1%,[1b] or "up to 8 percent"[3]) and is believed to be related to the graphite quality and the higher He pressure used. Further evaluation of these methods and yields will follow in a subsequent paper. Chromatographic filtration of the concentrated solution of "crude" material on silica gel with benzene can be performed, but the material obtained remains identical, in all aspects (same C_{60}/C_{70} ratio, no other constituents), to the crude material obtained from the benzene extractions.

Separation of the mixture of C_{60}/C_{70} proved to be a challenging task, particularly because of the poor solubility of the material in most organic solvents. While the solubility in benzene is about 5 mg/mL at 25 °C, the compound is soluble with difficulty at the same temperature in chloroform, dichloromethane, tetrachloromethane, diethyl ether, tetrahydrofuran, n-hexane, n-pentane, and n-octane. The mixture of C_{60}/C_{70} dissolves appreciably better in boiling cyclohexane, from which small black cubes crystallize out on cooling. The material did not melt below 360 °C in a sealed tube; the resulting sample redissolves in benzene, showing no sign of decomposition.

Analytical thin-layer chromatography on silica gel indicated some separation with n-hexane or with n-pentane as eluents, but not with cyclohexane. Analytical HPLC performed with hexanes (5-μm Econosphere silica, Alltech/Applied Science) gave a satisfactory separation (retention times 6.64 and 6.93 min for C_{60} and C_{70}, respectively, at a flow rate of 0.5 mL/min; detector wavelength, 256 nm), indicating the content of C_{70} to be approximately 15% for the Los Angeles samples. Two other minor peaks, possibly other unidentified C_n species, were observed (retention times 5.86 and 8.31 min), but they constituted less than 1.5% of the total mass.

Column chromatography with hexanes on flash silica gel gave a few fractions of C_{60} with ≥95% purity, as determined by HPLC, along with later fractions containing mixtures of C_{60}/C_{70} in various ratios. Because of the poor solubility of C_{60} and C_{70} in these alkanes, only limited amounts of pure C_{60} were made available this way, in insufficient quantity for a reliable ^{13}C NMR spectrum to be measured (see below). However, column chromatography on neutral alumina with hexanes gave an *excellent* separation in analytical quantities. Thus, pure fractions containing C_{60} (99.85%) and C_{70} (>99%) were obtained, as indicated by mass spectrometric measurements described below.

(1) (a) Krätschmer, W.; Lamb, L. D.; Fostiropoulos, K.; Huffman, D. R. *Nature* **1990**, *347*, 354. (b) Krätschmer, W.; Fostiropoulos, K.; Huffman, D. R. *Chem. Phys. Lett.* **1990**, *170*, 167.

(2) (a) Kroto, H. W.; Heath, J. R.; O'Brien, S. C.; Curl, R. F.; Smalley, R. E. *Nature* **1985**, *318*, 162. (b) Curl, R. F.; Smalley, R. E. *Science* **1988**, *242*, 1017.

(3) Private communication from H. W. Kroto.

(4) D. Bethune et al. have also reported a partial separation by sublimation and Raman spectra of C_{60} and C_{70}; Bethune, D. S.; Meijer, G.; Tang, W. C.; Rosen, H. J. *Chem. Phys. Lett.*, in press.

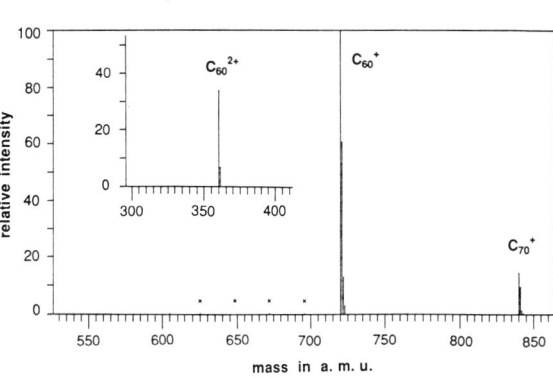

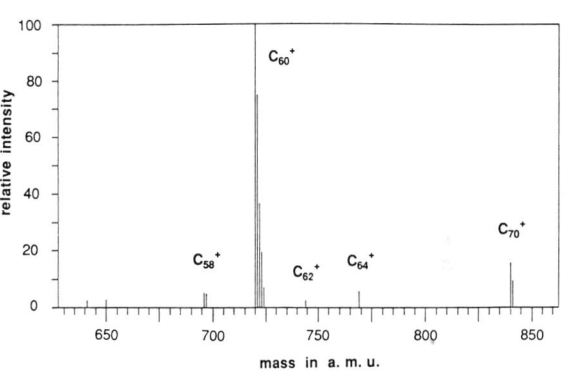

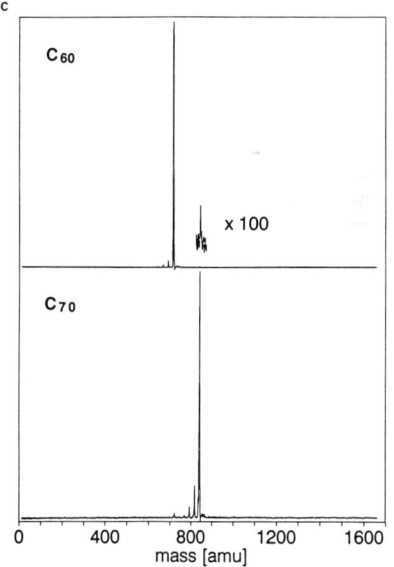

Figure 1. (A) EI MS spectrum of the C_{60} and C_{70} mixture at 70 eV, with a source temperature of 340 °C. Peaks marked with an × are for the ions of C_{58}, C_{56}, C_{54}, and C_{52} at m/z = 696, 672, 648, and 624, respectively. The insert shows the C_{60}^{2+} ion of the same sample. The C_{70}^{2+} ion which also appears in the spectrum is not shown. (B) FAB MS spectrum of the C_{60} and C_{70} mixture with NOBA as the matrix. (C) Laser desorption mass spectra of *pure* C_{60} (above) and C_{70} (below).

Mass Spectrometric Characterization

The 70-eV EI mass spectrometric measurements of the C_{60}/C_{70} mixture's vapor at a source/probe temperature of 340 °C gave exceptionally clean spectra for the samples (Figure 1A). The C_{60}^+ ion at m/z = 720 and the C_{70}^+ ion (m/z = 840) are dominant, each with precisely the expected isotopic patterns. The next larger peaks are the doubly charged species C_{60}^{2+} and C_{70}^{2+}, at about one-third of the base-peak intensities, indicating that they must be very stable species as well. At lower ionization energy (16 eV), the doubly charged peaks disappear. From several EI spectra, the average ratio of C_{60} to C_{70} was determined to be 87:13, in striking accordance with the HPLC and the ^{13}C NMR estimations (see below). It seems possible that previous estimates[1a] of the C_{70} content ($\sim$2%) of the crude C_{60} samples might have resulted from selective sublimation during the preparation of samples for IR and mass spectrometry. To examine this possibility, we have recorded the apparent abundances of the two molecules as the source/probe temperature is increased from 240 to 340 °C and find that the composition of the vapor changes from >99% C_{60} to the above ratio between 260 and 320 °C. A full analysis of the sublimation properties of these species will be reported separately.

The 70-eV EI spectrum of Figure 1A also contains peaks of C_{58}, C_{56}, C_{54}, and C_{52} at m/z = 696, 672, 648, and 624, respectively, corresponding to the expected fragmentation pattern.[5] Because they fail to appear at 16 eV, they must presumably result from fragmentation of the C_{60}^+ ion, yet their low abundance (0.8% summed together) suggests a very high stability to the carbon cage of C_{60}.

The fast atom bombardment (FAB) mass spectra of the *same* material were again astonishingly simple (Figure 1B), revealing only C_{58}, C_{62}, and C_{64} species besides the C_{60}^+ and the C_{70}^+ ions. No peaks were found beyond m/z = 840/841/842 (up to m/z = 1200), thus indicating the absence of higher mass C_n species. In agreement with the EI spectra of the high-temperature vapor, the FAB results indicated the C_{60}/C_{70} ratio to be 87:13. However, the minority species are more abundant than in the EI spectra and are probably the result of intermolecular reactions/fragmentations in the *m*-nitrobenzyl alcohol (NOBA) matrix used in FAB MS.

The laser desorption method is less destructive than the FAB method, but shares the feature of not requiring a continuous, long-time heating of the sample.[6b] Accordingly, the very clean LD mass spectra of C_{60} and C_{70} (Figure 1C) were obtained by time-of-flight analysis of the ions desorbed when 266-nm laser pulses were directed into a pulsed He jet flowing over the sample. On the crude mixture of C_{60}/C_{70}, it was found that all features of (0–2000 amu) except C_{60} and C_{70} vanished at the lowest laser fluences, and these exhibited the same ratio (88:12), within experimental uncertainty, as found in the high-temperature EI mass spectra. When applied to the samples of C_{60} and C_{70} separated by column chromatography on alumina, the C_{60} fraction was found to have a purity of 99.85%, with C_{70} as the residual. On the other hand, the C_{70} fraction had a purity of >99%; the minor peaks, corresponding to C_{68}, C_{66}, C_{64}, and C_{60}, are found by careful analysis of the laser-fluence dependence to all be fragmentation products of C_{70}. In particular, the relative intensity of the C_{60} peak increased directly with increasing laser fluence, thus demonstrating that it results from the fragmentation of C_{70}.

Carbon-13 NMR Spectra

The crude samples of C_{60}/C_{70} obtained from our two sources were independently investigated by ^{13}C NMR and gave identical results. Since it was expected that the spin–lattice relaxation time of the ^{13}C nuclei would be quite long, an inversion recovery experiment was performed to obtain a rough estimate of T_1. Thus, it was determined that T_1 for C_{60} was ≥20 s. The samples of C_{60}/C_{70} were dissolved in an excess of benzene-d_6 and evaporated at 25 °C until saturation was achieved. Using a 30-deg pulse and a 20-s pulse delay, a total of 5780 accumulations obtained over 32 h on a Bruker AM 360 instrument (90.56 MHz) gave a spectrum with an acceptable signal/noise ratio, showing clearly the presence of C_{60} and C_{70} only (Figure 2B). Thus, the peak corresponding to C_{60} is observed at 143.2 ppm (cf. ref 3), and the peaks at 130.9, 145.4, 147.4, 148.1, and 150.7 ppm are attributed to C_{70}, the number of carbons and the 10/20/10/20/10 peak ratio being as expected for the proposed molecular structure.

(5) Radi, P. P.; Bunn, T. L.; Kemper, P. R.; Molchan, M. E.; Bowers, M. T. *J. Chem. Phys.* **1988**, *88*, 2809.

(6) (a) For the gas-phase synthesis of another carbon allotrope, C_{18}, see: Diederich, F.; Rubin, Y.; Knobler, C. B.; Whetten, R. L.; Schriver, K. E.; Houk, K. N.; Li, Y. *Science* **1989**, *245*, 1088. (b) For a preparation of C_{60} and C_{70} starting from molecular precursors, see: Rubin, Y.; Kahr, M.; Knobler, C. B.; Diederich, F.; Wilkins, C. L. *J. Am. Chem. Soc.*, in press.

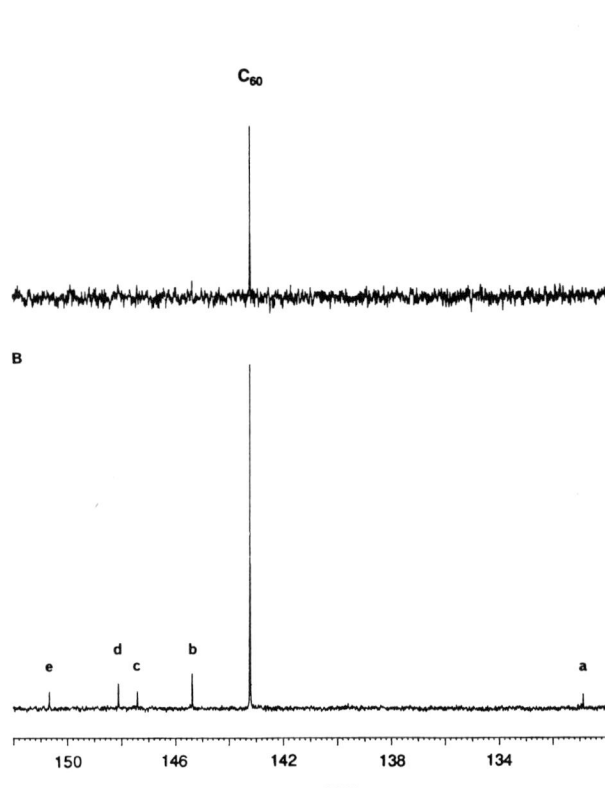

Figure 2. (A) ^{13}C NMR spectrum of a C_{60} and C_{70} mixture in benzene-d_6 at 296 K after 484 accumulations with a 2-s delay between 30-deg pulses. (B) Same sample after 5780 accumulations with a 20-s delay between pulses. Peaks labeled a, b, c, d, and e are assigned to C_{70}. Both spectra are plotted from 130 to 152 ppm.

The ^{13}C NMR spectrum shown in Figure 2A was performed on the same sample with only 484 accumulations with a 2-s delay between pulses. Thus it is possible to see the single C_{60} peak in a very short time during the experiment. This demonstrates that erroneous interpretations about the purity of the sample can be made if too few accumulations or a less sensitive instrument is used.

With relaxation times for the ^{13}C nuclei of C_{60} and C_{70} being probably very similar, an estimate of the ratio of the two compounds (Figure 2B) by comparison of the peak heights was expected to give a good estimate of the composition of the compound. Thus, the ratio was determined to be 82:18, in reasonable accordance with the HPLC and mass spectrometric determinations described above.

As expected, the proton NMR spectra of the samples dissolved in benzene-d_6 were devoid of any absorptions besides the C_6D_5H peak at 7.15 ppm and a few impurities at 0.3–1.4 ppm, also present in neat solvent.

Optical Absorption Spectra

In ref 1, the optical absorption spectrum in the ultraviolet and visible region was reported for the sublimed C_{60}/C_{70} mixture as a neat solid film. In our work, the spectra of pure C_{60} and C_{70} were recorded in *n*-hexane. Figure 3 shows the absorption spectra in the 200–800-nm region for C_{60} (99.85% purity) and C_{70} (>99%) at 25 °C. Compared with the spectrum of ref 1, one observes small hypsochromic shifts of the peak maxima of C_{60} and alterations in relative intensities as a result of removing the C_{70} contaminant. In addition, the spectrum of pure C_{70} appears to be distinct from any previously reported, including the brief list of maxima given for a sample of unstated purity in ref 3.

Absorption by C_{60} begins with an abrupt onset of 635 nm, followed by several bands of varying width (centered at 621, 598, 591, 568, 540, and 492 nm), and a highly transparent region at 420–440 nm. The structure of the visible absorption spectrum

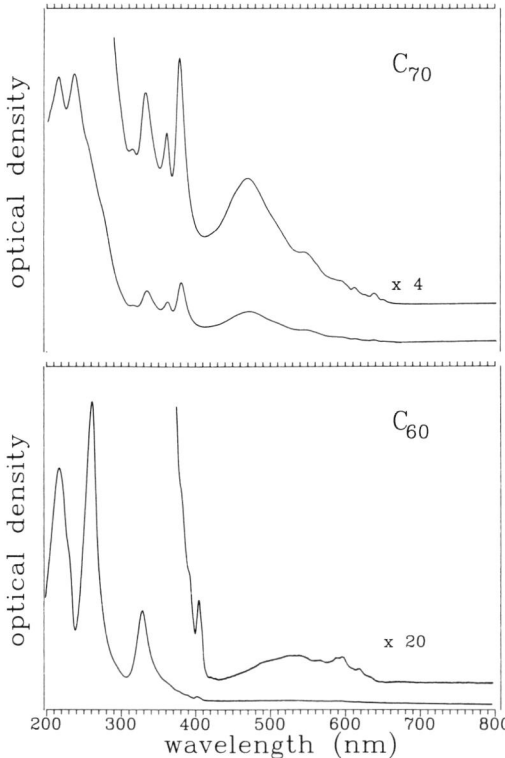

Figure 3. (A, top) Electronic absorption spectrum of dilute C_{70} in hexanes at 25 °C. The insert is the spectrum of the same sample at 4× concentration. (B, bottom) Electronic absorption spectrum of dilute C_{60} in hexanes at 25 °C. The insert is the spectrum of the same sample at 20× concentration.

is suggestive of vibrational structure from one or two forbidden electronic transitions. The combination of transparency in this blue region and in the red (>635 nm) gives dilute solutions a distinct purple color to the eye. A second onset leading to stronger absorption occurs in the form of a band at 404 nm, with a shoulder at 408 nm. These are followed by distinct shoulders at 396, 391, 377, and 365 nm, also suggestive of vibrational structure, appearing on a strong rise toward the first major maximum at 328 nm. The ultraviolet region is dominated by this feature and two other strong broad bands peaking at 211 and 256 nm, the former with a shoulder at 227 nm.

Based on these results, it seems unlikely that neutral C_{60} is the carrier of the interstellar 220-nm absorption band,[7] as the 255-

and 330-nm peaks are not concurrently observed.

Absorption by C_{70} begins with a weak onset band at 650 nm, followed by a series of peaks (637, 624, 610, 600, and 594 nm) superimposed on a gradually rising continuum leading to stronger maxima at 544 and 469 nm. A broad minimum covers the blue-violet region, and maxima of intermediate strength appear in the near-ultraviolet region at 378, 359, and 331 nm. Following a weaker maximum at 313 nm, very strong absorptions appear with three shoulders leading to the dominant bands at 236 and 215 nm. Dilute C_{70} solutions are orange-red in color. Detailed comparison of the spectra indicates no coincidences between C_{70} and C_{60} bands, as would be expected from the purities stated above.

The FT-IR spectrum of the C_{60} and C_{70} mixture performed with a conventional KBr pellet showed the four strongest bands for C_{60} at 1430, 1182, 577, and 527 cm^{-1}, as observed previously by Krätschmer et al.[1] In addition, the strong peak at 673 cm^{-1} mentioned by these authors was present, together with a smaller peak at 795 cm^{-1}.

Conclusion

The new molecular forms of carbon, C_{60} and C_{70}, were prepared following the method of Krätschmer et al.[1] in a consistently high yield (14%).[1] The benzene-soluble material extracted from the graphite evaporation product is predominantly constituted of C_{60} and C_{70}. Three independent methods, namely HPLC, mass spectroscopy, and ^{13}C NMR, demonstrate that these two compounds are present in a ratio near 85:15. The two compounds can be separated by column chromatography on alumina, allowing as for now the purification of minute quantities of pure C_{60} and C_{70}. Support for the proposed symmetrical cage structures (fullerenes) of C_{60} and C_{70} is inferred from the simplicity of the ^{13}C NMR spectra and the strong presence of the C_{60}^{2+} and C_{70}^{2+} ions. Attempts at the X-ray determination of the C_{60} molecular structure are now actively pursued.[6]

Note Added in Proof. The soluble byproduct (HPLC retention time 8.31 min) has since been isolated and determined to be C_{84} by laser desorption MS. The ^{13}C NMR spectrum of *pure* C_{70} (in 1,1,2,2-tetrachloroethane-d_2) has *only* the five lines at 131.0, 145.4, 147.5, 148.2, and 150.8 ppm.

Acknowledgment. We thank Prof. B. Dunn and his students for their generous help with the carbon evaporator, C. B. Knobler for several attempts at the diffractometer and for discussions about the crystallization of the compound, J. M. Strouse for her assistance in the ^{13}C NMR experiments, and H. W. Kroto and D. Bethune for communication of unpublished results. The work in Los Angeles was supported by the National Science Foundation and by Office of Naval Research contracts to RLW and FND.

(7) Huffman, D. R. *Adv. Phys.* **1977**, *26*, 129.

Reprinted with permission from The Journal of Physical Chemistry
Vol. 95, pp. 11–12, 1991
© 1991 American Chemical Society

Photophysical Properties of C_{60}

James W. Arbogast, Aleksander P. Darmanyan, Christopher S. Foote,* Yves Rubin,
François N. Diederich, Marcos M. Alvarez, Samir J. Anz, and R. L. Whetten

Department of Chemistry and Biochemistry, University of California, Los Angeles,
Los Angeles, California 90024 (Received: November 20, 1990)

A number of important photophysical properties of C_{60} have been determined, including its lowest triplet-state energy (near 33 kcal/mol), lifetime, and triplet–triplet absorption spectrum. The triplet state is formed in near quantitative yield and produces a very high yield of singlet oxygen by energy transfer. C_{60} does not react with singlet molecular oxygen and quenches it only slowly by an unknown mechanism. These results are discussed in terms of the unusual geometry of this molecule.

The recent isolation and purification of the intriguing spherical all-carbon molecule, C_{60}, in high yields is an astonishing accomplishment.[1-3] This compound is produced in relatively high yield along with lesser amounts of its congeners C_{70} and C_{84} in the carbon arc and is readily purified by chromatography. We report that C_{60} possesses interesting photophysical properties, summarized in Table I, which are most likely due to the unique geometry of this compound.

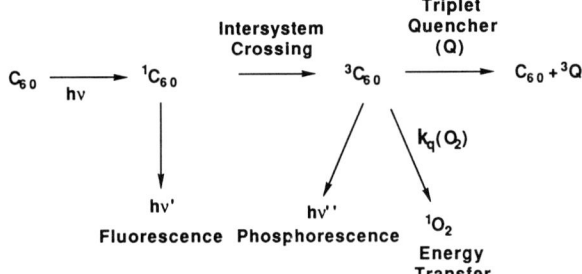

The longest wavelength absorption maximum[3] of C_{60} lies at 620 nm, corresponding to a singlet energy (E_S) of 46.1 kcal/mol. No fluorescence from C_{60} was detected at room temperature in either hexane (UV excitation) or benzene (visible excitation). Phosphorescence has not yet been observed; however, the triplet is formed in high yield (see below).

The triplet–triplet absorption spectrum for C_{60} is shown in Figure 1. The extinction coefficient at 480 nm ($\epsilon_T - \epsilon_{S_0}$), 2.4×10^3 M^{-1} cm^{-1}, was estimated by the method of Bensasson and Land[6] by comparison with the T–T absorption of acridine.[7] The triplet lifetime under our experimental conditions is 40 ± 4 μs. The triplet state of C_{60} is efficiently quenched by $^3O_2$; in air-saturated C_6H_6, the lifetime is 330 ± 25 ns. This yields a quenching rate constant by oxygen of $k_q(O_2) = 2 \times 10^9$ M^{-1} s^{-1}, which is typical for aromatic hydrocarbons.

TABLE I: Photophysical Properties of C_{60}[a]

E_S	46.1 kcal/mol[b]
E_T	37.5 ± 4.5 kcal/mol[c]
ϵ_T (480 nm)	$(2.8 \pm 0.2) \times 10^3$ M^{-1} cm^{-1}
τ_T	40 ± 4 μs[d]
$k_q(O_2)$	$(1.9 \pm 0.2) \times 10^9$ M^{-1} s^{-1}
$\Phi_{^1O_2}$(355 nm)	0.76 ± 0.05[e]
$\Phi_{^1O_2}$(532 nm)	0.96 ± 0.04[f]
$k_q(^1O_2)$	$(5 \pm 2) \times 10^5$ M^{-1} s^{-1}[g]

[a] All measurements in C_6H_6 or C_6D_6 at room temperature, with concentrations of C_{60} $\sim 3 \times 10^{-5}$ and $\sim 3 \times 10^{-4}$ M (OD $\sim$ 0.5) at λ_{exc} = 355 and 532 nm, respectively. [b] Calculated from the lowest energy absorbance maximum at 620 nm (see ref 3). [c] Average of the triplet energy of TPP (33.0 kcal/mol) and anthracene (42 kcal/mol); however, the low quenching rate with TPP suggests the actual energy is closer to the level of TPP. [d] In argon-saturated solution. [e] Average of five determinations with TPP ($\Phi_\Delta = 0.62$)[4] and acridine ($\Phi_\Delta = 0.84$)[5] standards. [f] Average of three determinations with TPP ($\Phi_\Delta = 0.62$)[4] as standard. [g] Total of physical and chemical quenching of $^1O_2$ by C_{60} (TPP, 532 nm). The maximum concentration of C_{60} was limited by saturation.

The energy level of the triplet (E_T) was estimated by triplet–triplet energy transfer. C_{60} quenches the triplet states of both acridine ($E_T = 45.3$ kcal/mol)[8] and anthracene ($E_T = 42$ kcal/mol)[8] efficiently, with rate constants of $(9.3 \pm 1.1) \times 10^9$ and $(6.1 \pm 1.0) \times 10^9$ M^{-1} s^{-1}, respectively. The rate constant of quenching of the tetraphenylporphine (TPP) triplet state ($E_T = 33.0$ kcal/mol)[9] by C_{60} is significantly lower, $k_q = (3.5 \pm 0.3) \times 10^7$ M^{-1} s^{-1}. The reason for the slower quenching in this case is probably endothermic energy transfer. Energy transfer from the C_{60} triplet to produce the rubrene triplet ($E_T = 26.6$ kcal/mol)[10] is diffusion-controlled. Thus, we conclude that the energy level of the triplet state of C_{60} lies near that of TPP and well below anthracene, i.e., 33 kcal/mol $\leq E_T <$ 42 kcal/mol (Table I).

The low (≤ 750 M^{-1} cm^{-1}) extinction coefficients for the absorption spectrum[3] of C_{60} between 492 and 620 nm indicate that this band, probably $S_0 \rightarrow S_1^*$, is connected with a strongly symmetry-forbidden transition. Similarly, low extinction coefficients are associated with the longest wavelength $T_1 \rightarrow T_2$ absorption.

The small S–T splitting in C_{60} ($\Delta E_{S_1T} \sim 9$ kcal/mol) is probably a result of the large diameter of the molecule and the resulting small electron–electron repulsion energy.[11] This small splitting, the very low value of the fluorescence rate constant, and the expected large spin–orbital interaction in the spherical C_{60} explain why intersystem crossing (ISC) is a dominant process.

C_{60} produces singlet oxygen in large quantities as measured by $^1O_2$ luminescence at 1268 nm.[12] The quantum yield ($\Phi_{^1O_2}$) is 0.76

(1) Haufler, R. E.; Conceicao, J.; Chibante, L. P. F.; Chai, Y.; Byrne, N. E.; Flanagan, S.; Haley, M. M.; O'Brien, S. C.; Pan, C.; Xiao, Z.; Billups, W. E.; Ciufolini, M. A.; Hauge, R. H.; Margrave, J. L.; Wilson, L. J.; Curl, R. F.; Smalley, R. E. *J. Phys. Chem.* **1990**, *94*, 8634.

(2) Krätschmer, W.; Fostiropoulos, K.; Huffman, D. R. *Chem. Phys. Lett.* **1990**, *170*, 167. Krätschmer, W.; Lamb, L. D.; Fostiropoulos, K.; Huffman, D. R. *Nature* **1990**, *347*, 354. Taylor, R.; Hare, J. P.; Abdul-Sada, A. K.; Kroto, H. W. *J. Chem. Soc., Chem. Commun.* **1990**, 1423.

(3) Ajie, H.; Alvarez, M. M.; Anz, S. A.; Beck, R. D.; Diederich, F.; Fostiropoulos, K.; Huffman, D. R.; Krätschmer, W.; Rubin, Y.; Schriver, K. E.; Sensharma, D.; Whetten, R. L. *J. Phys. Chem.* **1990**, *94*, 8630.

(4) Schmidt, R.; Afshari, E. *J. Phys. Chem.* **1990**, *94*, 4377.

(5) Redmond, R. W.; Braslavsky, S. E. *Chem. Phys. Lett.* **1988**, *148*, 523.

(6) Bensasson, R. L.; Land, E. J. *Trans. Faraday Soc.* **1971**, *67*, 1904.

(7) Determined by excitation with a Nd:YAG laser pulse at 355 nm in argon-saturated C_6H_6. The extinction coefficient for C_{60} was obtained by comparing the triplet–triplet absorption of acridine and C_{60} under the same experimental conditions. The extinction coefficient for acridine at λ_{max} is 2.43 $\times 10^4$ M^{-1} cm^{-1}.[6]

(8) Birks, J. B. *Photophysics of Aromatic Molecules*; Wiley-Interscience: New York, 1970.

(9) McLean, A. J.; McGarvey, D. J.; Truscott, T. G.; Lambert, C. R.; Land, E. J. *J. Chem. Soc., Faraday Trans.* **1990**, *86*, 3075.

(10) Darmanyan, A. P. *Chem. Phys. Lett.* **1982**, *86*, 405.

(11) McGlynn, S. P.; Azumi, T.; Kinoshita, M. *Molecular Spectroscopy of the Triplet State*; Prentice-Hall: Englewood Cliffs, NJ, 1969.

12

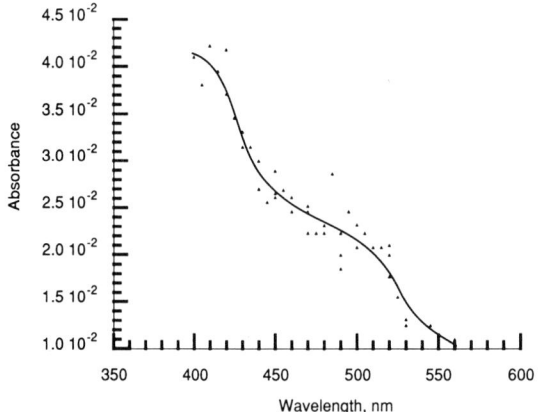

Figure 1. Triplet–triplet absorption spectrum of C_{60} in benzene.

$\pm$ 0.05 (355 nm) and 0.96 $\pm$ 0.04 (532 nm). Formation of $^1O_2$ occurs by energy transfer from the highly populated C_{60} triplet state to molecular oxygen. These values represent a lower limit for the quantum yield of triplet production (Φ_T).

(12) Ogilby, P. R.; Foote, C. S. *J. Am. Chem. Soc.* **1982**, *104*, 2069.

The lower singlet oxygen quantum yield at 355 nm appears to be outside the experimental error and is currently being investigated. One possible explanation is that there is a lower intersystem crossing yield from upper singlet excited states than from S_1. The higher singlet states may be connected with additional deactivation channels which compete with ISC.

C_{60} also quenches singlet oxygen with an approximate rate constant $k_q(^1O_2) = (5 \pm 2) \times 10^5$ M^{-1} s^{-1}, as shown by a small shortening of the lifetime of the singlet oxygen luminescence with a nearly saturated solution of C_{60} in C_6D_6. The mechanism of quenching by C_{60} is presently unknown; chemical reaction is unlikely because no loss of starting material or formation of new product (by UV–visible absorption spectra) occurs following hundreds of laser pulses under O_2 at either excitation wavelength.

C_{60} is a potent generator of singlet oxygen. Its very high singlet oxygen yield and inertness to photooxidative destruction suggests a strong potential for photodynamic damage to biological systems. Thus, the degree to which C_{60} is present in the environment becomes a very important question.

Acknowledgment. This work was supported by NSF Grants CHE89-11916 and CHE 89-21133 (F.N.D. and R.L.W.) and NIH Grant GM-20080.

Reprinted with permission from The Journal of Physical Chemistry
Vol. 95, No. 21, pp. 8402–8409, 1991
© 1991 American Chemical Society

Resilience of All-Carbon Molecules C_{60}, C_{70}, and C_{84}: A Surface-Scattering Time-of-Flight Investigation

Rainer D. Beck, Pamela St. John, Marcos M. Alvarez, François Diederich, and Robert L. Whetten*

Department of Chemistry and Biochemistry, University of California, Los Angeles, California 90024-1569
(Received: April 2, 1991)

Ion beam scattering experiments on the larger carbon molecules ($C_{60}^{\pm}$, C_{70}^{+}, C_{84}^{+}) demonstrate their exceptionally high stability with respect to impact-induced fragmentation processes. The charged molecules are formed by ultraviolet laser desorption of high-purity molecular samples into a pulsed helium jet. Extracted ions impact Si(100) or graphite(0001) in a high-resolution ion beam/surface collider with mass time-of-flight and angular analysis. Collisions are highly inelastic processes: A large fraction of the entire perpendicular momentum component is lost, and 60 ± 20% of the parallel component is either lost or exchanged. No more than 10% of the incident ions are returned, which is attributed to neutralization during the collision event. In contrast to all molecular ions (benzene and naphthalene cations) and clusters (alkali-metal halides), these molecules exhibit no evidence for impact-induced fragmentation, even at impact energies exceeding 200 eV. In the case of C_{60}^{-}, both the intact parent ion and ejected electrons are detected, with the latter becoming dominant above 120 eV impact energy. C_{60}^{+} is found to have an exceptionally low energy threshold for inducing sputtering processes of adsorbed overlayers on graphite. Some of these results may be interpretable in terms of the unique structural-energetic characteristics of the fullerene family. The results are compared to recent computer simulations of the impact event, which predict high resilience for these molecules.

I. Introduction

The recent discovery of methods of isolating[1] and purifying[2-4] large quantities of molecular allotropic forms of carbon has made it possible to determine the physical and chemical properties of these fascinating new molecules.[5] Prior to this development, we carried out preliminary surface-collision experiments on a number of charged carbon molecules, C_N^{+}, as obtained from laser ablation of graphite in a helium-jet source. More recently, starting with pure molecular solids[6] C_N (N = 60, 70, 76, 84, ...) in hand, we[7] and others[8,9] have worked to obtain pulsed, mass-selected beams

with orders of magnitude greater intensity and stability using ultraviolet laser *desorption* of the solid molecular films. Because of the unique shapes and bonding of these molecules—cage structures, like those shown in Figure 1, that can be loosely thought of as closed graphitic sheets[10]—and their reputed ultrahigh stability under extreme energization,[11] it seemed likely that their response to impact with solid surfaces would be very unusual and interesting. In the new experiments reported here, in which $C_{60}^{\pm}$, C_{70}^{+}, and C_{84}^{+} are collided against graphite and silicon over a range of energies, this supposition is confirmed. In concurrent research, Callahan et al.[12] have collided C_{60}^{2+} against polished stainless steel, and Mowrey et al.[13] have conducted extensive simulations of the collisions of (neutral) C_{60} with graphite and with H-terminated diamond. In the latter work, it has been predicted that collision

(1) Krätschmer, W.; Fostiropoulos, K.; Huffman, D. R. *Chem. Phys. Lett.* **1990**, *170*, 167. Krätschmer, W.; Lamb, L. D.; Fostiropoulos, K.; Huffman, D. R. *Nature* **1990**, *347*, 354.

(2) Taylor, R.; Hare, J. P.; Abdul-Sada, A. K.; Kroto, H. W. *J. Chem. Soc., Chem. Commun.* **1990**, 1423–5.

(3) Bethune, D. S.; Meijer, G.; Tang, W. C.; Rosen, H. J. *Chem. Phys. Lett.* **1990**, *174*, 219.

(4) Aije, H.; Alvarez, M. M.; Anz, S. J.; Beck, R. D.; Diederich, F.; Fostiropoulos, K.; Huffman, D. R.; Krätschmer, W.; Rubin, Y.; Schriver, K. E.; Sensharma, K.; Whetten, R. L. *J. Phys. Chem.* **1990**, *94*, 8630.

(5) For a collection of early reports, see: *Mater. Res. Soc. Proc.* **1991**, *206*.

(6) Diederich, F.; Ettl, R.; Rubin, Y.; Whetten, R. L.; Beck, R. D.; Alvarez, M. M.; Anz, S. J.; Sensharma, D.; Wudl, F.; Khemani, K. C.; Koch, A. *Science* **1991**, *252*, 548.

(7) Whetten, R. L.; Alvarez, M. M.; Anz, S. J.; Schriver, K. E.; Beck, R. D.; Diederich, F. N.; Rubin, Y.; Ettl, R.; Foote, C. S.; Darmanyan, A. P.; Arbogast, J. W. *Mater. Res. Soc. Proc.* **1991**, *206*, 639.

(8) Haufler, R. E.; Wang, L.-S.; Chibante, L. P. F.; Jin, C.; Conceicao, J. J.; Chai, Y.; Smalley, R. E. *Chem. Phys. Lett.*, in press.

(9) Bethune, D. S.; deVries, M.; et al., private communication.

(10) Kroto, H. W.; Heath, J. R.; O'Brien, S. C.; Curl, R. F.; Smalley, R. E. *Nature* **1985**, *318*, 162. For reviews, see: Smalley, R. E.; Curl, R. F. *Science* **1988**, *242*, 1017. Kroto, H. W. *Science* **1988**, *242*, 1139.

(11) O'Brien, S. C.; Heath, J. R.; Curl, R. F.; Smalley, R. E. *J. Chem. Phys.* **1988**, *88*, 220.

(12) McElvany, S. W.; Ross, M. M.; Callahan, J. H. *Mater. Res. Soc. Proc.* **1991**, *206*, 697.

(13) Mowrey, R. C.; Brenner, D. W.; Dunlap, B. I.; Mintmire, J. W.; White, C. T. *Mater. Res. Soc. Proc.* **1991**, *206*, 357; *J. Phys. Chem.* **1991**, *95*, 7138. Mowrey, R. C. Private communication.

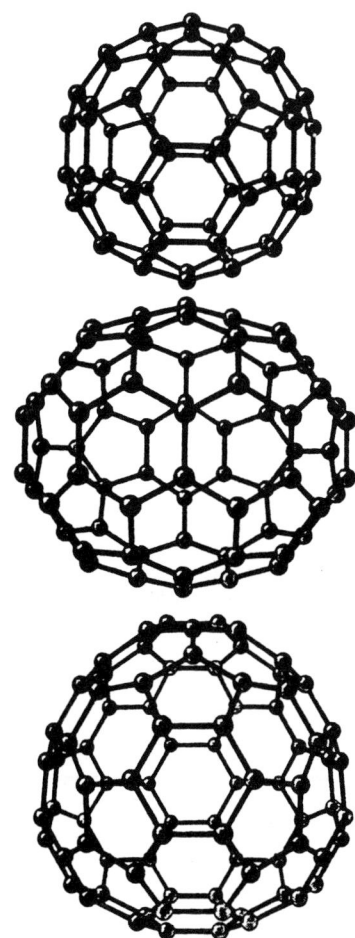

Figure 1. Structures of the all-carbon molecules C_{60} and C_{70} (known) and C_{84} (proposed).

events are highly inelastic, but largely nonreactive and nonfragmenting, although the long-time consequences of such heating could not be studied. These simulations predict a remarkable resilience of the structure of C_{60}: despite collision energies of 150 eV that during impact deform the structure until it is nearly planar, the hollow-cage structure is retained upon recoil. It will now prove possible to establish some experimental correspondences to these results.

II. Experimental Method

Molecular Carbon Samples. These have prepared as described previously.[6,7,14] Briefly, a high-quality graphite rod is evaporated by resistive heating under an helium atmosphere. The collected material is dissolved in boiling toluene, and the insoluble carbon fraction is separated by filtration. The resulting yield of several hundred milligrams of soluble material consists of C_{60}, C_{70}, and higher molecules,[6] which can each be obtained in pure form by repeated liquid-gravity chromatography on alumina. The resulting pure molecular solids are soluble (several mg/cm³) to a limited extent in a variety of solvents, so that a thick film of material, suitable as a laser-desorption target, can be formed by dipping the substrate in the solution and rapidly evaporating the solvent.[15]

Laser-Desorption Mass Spectrometry. Thick films of pure or mixed molecular carbon are deposited on a 3 mm diameter steel or tantalum rod. The rod is mounted in a high-vacuum chamber (Figure 2) within a short nozzle assembly,[15] which in turn mounts on the faceplate of a pulsed gas valve. The rod is continuously rotated and translated by a motor mounted exterior to the vacuum chamber, in order to continually expose fresh surface to the laser. To obtain a laser-desorption mass spectrum, the radiation (<1 mJ) from the Nd:YAG laser fourth harmonic (266 nm) or from an excimer laser (193 or 248 nm) is lightly focused through a perpendicular channel in the nozzle assembly onto the rod surface. The laser-radiation pulse is synchronous with the gas pulse, typically 10 bar of helium through a 0.5-mm valve orifice. Desorbed molecules, charged and neutral, are swept along by the helium as the jet expands freely into vacuum, is skimmed, and enter the ion-extraction region of a reflectron-type time-of-flight mass spectrometer. The ($\pm$)-charged molecules are extracted perpendicular to the helium beam by a high-voltage pulse and enter a second high-vacuum chamber, separating according to mass. The entire mass spectrum is accumulated by repeated pulses (typically 100–2000) using waveform digitizing and averaging electronics. A complete range of pulse fluences is explored to ensure the softest possible desorption–ionization event, which can only occur with ultraviolet radiation, where the molecules are strongly absorbing and easily ionizing.[4,15] Two instruments have been used, the first (not shown) a reflectron instrument designed for this work, and the other an ion beam surface-scattering instrument (Figure 2) designed for charged-cluster beam experiments and operated in the reflectron (collision-free) mode.

Molecule–Surface Collisions. In the cluster–surface collider (Figure 2),[16] the molecules desorbed in the source are entrained in and cooled by a helium nozzle-jet expansion, to yield an intense, pulsed beam. Charged molecules are mass-gated[17] so that only a selected cluster size n enters the collision chamber. This packet is retarded to the desired momentum and angle of impact; the scattered, charged particles are accelerated in the same field toward the detector. The graphite surface (HOPG) is cleaned by several hours heating at 450 °C, which removes the hydrocarbon overlayer, as verified by observing the intensity of ions sputtered from the adsorbate. The Si surface is heated similarly, but no attempt is made to remove the oxide passivation layer. In operation, the beam energy and surface angle are set to collect a certain fraction of scattered fragments. Because a substantial fraction of the parallel momentum is lost or exchanged during the collision, the surface angle must be set away from the specular angle in order to collect in an unbiased manner both the intact scattered ion and its fragments.[16] Charged fragments are mass-analyzed through their flight-time in the secondary flight region, because the instrument operating voltage greatly exceeds the collision energy. The uncertainty in time of collision is typically $10^{-8}n^{1/2}$ ns (the peak width in the reflectron mode). The rms collision energy spread is typically no greater than $\sigma_E = 5$ eV.

III. Results and Analysis

A. Laser Desorption Mass Spectra. Figure 3 shows a set of laser-desorption mass spectra of purified molecular carbon samples. As reported in an earlier Letter,[4] C_{60}^+ and C_{70}^+ are detected in a virtually fragmentation-free manner by ultraviolet-laser-desorption mass spectrometry (LD-MS), and we have used this as one of a complementary set of methods for the quantitative analysis of molecular carbon mixtures.[7] At higher fluences, the desorption–ionization yield increases, at the expense of introducing fragmentation and bimolecular reactions (Figure 3c). Under these conditions, the transient heating of the solid, as readily calculated by using published optical densities[7] and measured laser fluences, can reach temperatures in excess of 1000 K, so that solid-state reaction processes are expexted to compete with the sublimation–desorption processes. The remaining frames of Figure 3 show laser-desorption mass spectra of other fractions from the chromatography that are rich in the larger carbon molecules. Repeated

(14) Allemand, P.-M.; Koch, A.; Wudl, F.; Rubin, Y.; Diederich, F.; Alvarez, M. M.; Anz, S. J.; Whetten, R. L. *J. Am. Chem. Soc.* **1991**, *113*, 1050.

(15) Schriver, K. E., Ph.D. Thesis, University of California, 1990. Diederich, F.; Rubin, Y.; Knobler, C. B.; Whetten, R. L.; Schriver, K. E.; Houk, K. N.; Li, Y. *Science* **1989**, *245*, 1088.

(16) Beck, R. D.; St. John, P.; Whetten, R. L. *J. Chem. Phys.*, submitted for publication. Beck, R. D.; St. John, P.; Homer, M. L.; Whetten, R. L. *Mater. Res. Soc. Proc.* **1991**, *206*, 341. For cold cluster ion extraction methods, see: Zheng, L.-S.; Brucat, P. J.; Pettiette, C. L.; Yang, S.; Smalley, R. E. *J. Chem. Phys.* **1985**, *83*, 4273.

(17) Beck, R. D.; Whetten, R. L. *Rev. Sci. Instrum.*, submitted for publication.

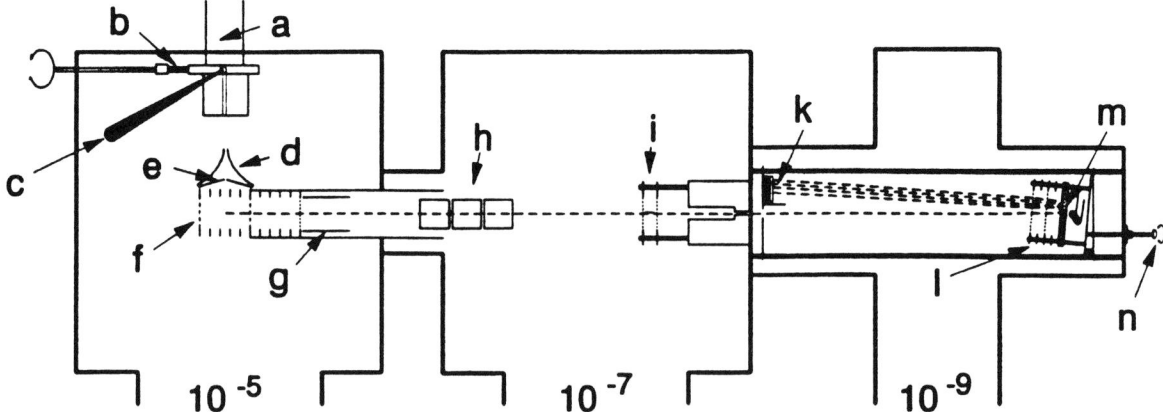

Figure 2. Scale diagram of the cluster-beam collider apparatus, consisting of three chambers. The source chamber (left) contains the pulsed gas valve a containing helium at 8 bar pressure, the rod (coated with molecular carbon), and drive shaft b inserted into the nozzle block, which contains a perpendicular channel for the ablation laser beam c. The jet is skimmed by the mechanical skimmer d, with the resulting beam shape determined by slits e, prior to entering the pulsed extraction field region f. The extracted ion's trajectory (horizontal dashed line) passes into the second chamber (middle, base pressure $<10^{-7}$ Torr), where it is steered by the horizontal deflection plates g, focused in the far field by the einzel lens h, and mass-filtered by the temporal mass gate i, consisting of a thin charged wire on the beam axis whose voltage is zeroed only at the time window when the desired cluster-mass packet passes. The scattering chamber (at right, base pressure $<10^{-9}$ Torr) contains the retarding field screens l, the scattering surface m (heated radiatively from behind), and a mechanical drive n to set the angle with respect to the fixed detector at k.

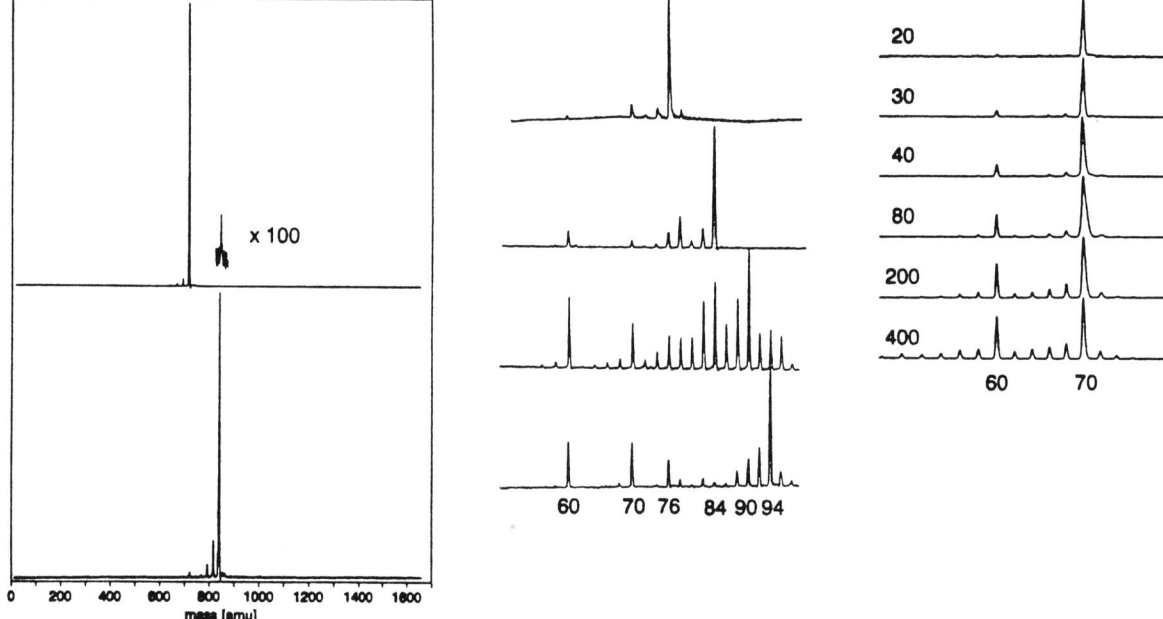

Figure 3. Laser-desorption time-of-flight mass spectra of molecular carbon samples. (a, left) Pure C_{60} and C_{70} samples. The ions are produced by 50 μJ pulses (over a 1-mm² area) of 266 nm radiation. Small peaks at lower masses are due to fragmentation occurring during desorption. (b, center) Enriched samples of the larger fullerenes.[6] (c, right) Effect of desorbing laser pulse fluence on the mass spectrum of pure C_{70} samples (pulse energies indicated in μJ per pulse).

chromatography, as described elsewhere,[6] is capable of purifying these. Furthermore, more effective solvents have recently been shown to be capable of extracting much larger carbon molecules from the evaporated graphite material.

Because the laser-desorption method avoids the need for high fluences above the ablation threshold, the beams generated are much more intense at the desired mass than are those generated by laser ablation of graphite and are much more stable as well. We have made no attempt to quantify the enhancement in intensity. However, Haufler et al.[8] have quoted intensities from a similar source "several orders of magnitude higher", and our qualitative observations are consistent with this.

B. Collisions of Positively Charged Carbon Molecules. In our experiments, we have collided C_{60}^+ with graphite and silicon over a range of conditions, and C_{70}^+ and C_{84}^+ with graphite only over a narrower range of instrument parameters. In the following, we focus attention on C_{60}^+. None of these molecular ions shows a surface-induced dissociation (SID)-type fragmentation pattern, even under conditions where all other molecules (e.g., benzene or naphthalene)[16,18] or clusters (large sodium fluoride clusters)[19]

yield very distinctive and extensive fragmentation.

Figure 4 shows the integrated ion intensity of the C_{60}^+ molecule as a function of energy, obtained from a series of time-of-flight profiles like those shown in Figure 5. The ion intensity is plotted for two angles. At the specular or reflectron angle, the intensity is constant until the surface voltage is sufficiently low for impact to occur and then declines suddenly, then more gradually. (Analysis of the time-of-flight peak profiles distinguishes between colliding and noncolliding events.) The scattered-ion signal is strongest at an angle near 3° from specular, corresponding to a loss of 60% of the parallel momentum in a 10° full-angle instrument. At this angle, the intensity of scattered ions increases abruptly and reaches a maximum at an energy near 30 eV, or significantly different from zero. It seems possible that lower

(18) For review of the surface-induced dissociation method applied to polyatomic ions, see: Cooks, R. G.; Ast, T.; Mabud, Md. A. *Int. J. Mass Spectrom. Ion Processes*, in press. For a recent comprehensive review of molecule-surface hyperthermal collision processes and scattering, see: Amirav, A. *Comments At. Mol. Phys.*, invited review, in press.

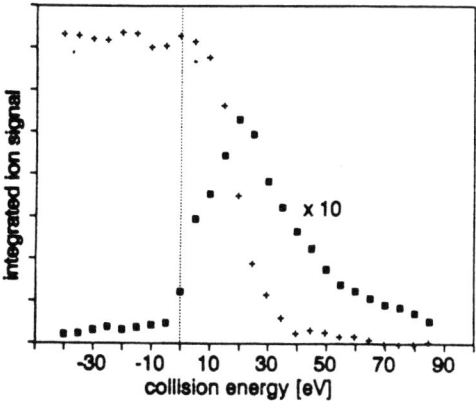

Figure 4. Integrated yield of C_{60}^+ versus collision energy (on graphite) for two surface angles. At the specular angle (5°, crosses) the signal due to unscattered C_{60} falls off quickly when collisions start. The "tail" toward higher collision energy is due to C_{60} scattered from the surface. Tilting the surface toward the detector (8°; solid squares) suppresses the signal of unscattered C_{60}^+ and enhances the scattering signal, indicating a loss of parallel momentum.

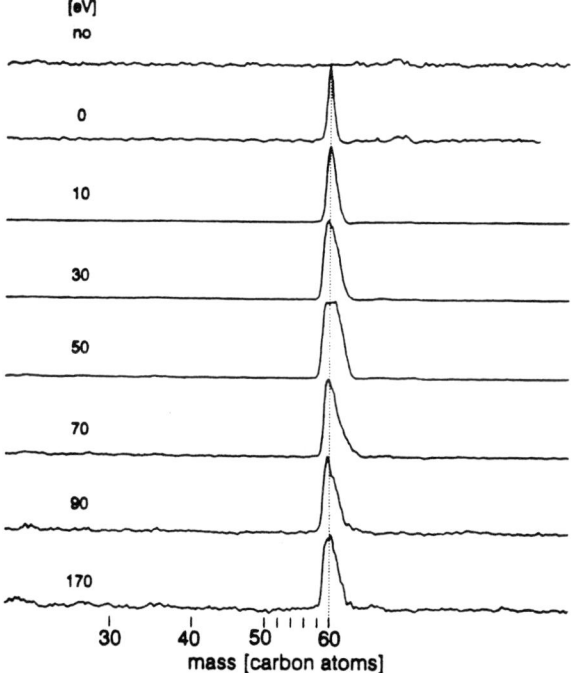

Figure 5. Time-of-flight profiles of C_{60}^+ scattering from graphite(0001). No charged C_n fragments ($n < 60$) are detected at any of the probed collision energies (0–170 eV). The C_{60}^+ peak (always scaled to equal height) broadens and shifts (see Figure 6) due to the inelasticity of the collision.

energy collisions might lead to sticking or to an enhanced neutralization probability. At the maximum, the total yield of scattered ions is approximately 7% of the absolute intensity of the incoming beam. (This value must be regarded as a lower bound, because of the limited collection efficiency at a particular angle.) This yield is typical of organic molecular ions in SID experiments[18] and is attributed to the effectiveness of electron-transfer neutralization upon impact with low work function surfaces. (When the incident molecule has a lower ionization potential, as in the case of charged alkali-metal halide clusters, the yield is much higher.[19])

The profiles shown in Figure 5 also reveal a shift and a broadening of the time-of-flight peak, indicating the inelasticity of the collision. More specifically, the change in the time of flight

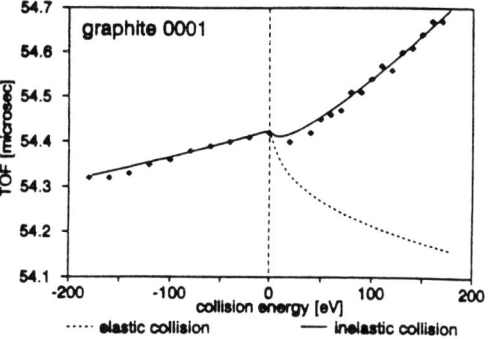

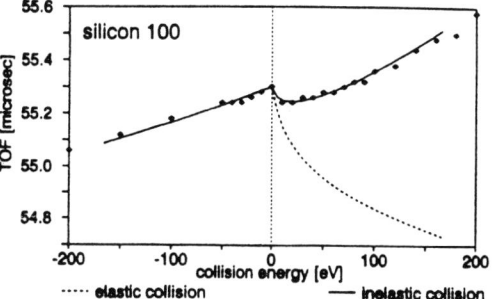

Figure 6. Comparison between the simulated time-of-flight and measurements for C_{60}^+ colliding with graphite (top) and silicon (bottom). In both cases the experimental data points agree well with the calculation assuming totally inelastic scattering from the surface.

is related to the loss of the perpendicular component of the momentum (Figure 2) and can be modeled precisely using the apparatus parameters. In Figure 6 we have taken the maximum in the time-of-flight profile and plotted it against the collision energy, for impact with both graphite and silicon surfaces, along with instrument curves assuming either totally inelastic (upper branch) or totally elastic (lower branch) scattering. It is clear that the observations are not distinct from that predicted for totally inelastic collisions. With the limited resolution presently available, it would appear that most of the perpendicular momentum is lost during the collision. For C_{60}^+ against Si, this has been repeated with both clean and hydrocarbon-contaminated surfaces and shows no substantial differences. The energy associated with the perpendicular momentum, ranging from 0 to 170 eV in Figure 5, is therefore available for heating the target material and the molecule. However, it is not clear that the parallel momentum is all similarly available or contains a substantial elastic component.

In the case of C_{60}^+, we have also noticed an anomalously high yield of sputtering from adsorbed overlayer material on either graphite or silicon, as compared to other incident ions. Figure 7 shows the pattern and intensity of sputter ions at increasing impact energies, and Figure 8 illustrates the decrease in sputtering intensity subsequent to heating of the surface. The low threshold and high intensity of this channel might be associated with the unusual combination of its high mass (720 amu) and high ionization potential (7.6 eV).[20]

C. Collisions of C_{60}^-. By reversing the potentials of all the electrostatic elements in the collider apparatus, and floating the detector, we have extracted C_{60}^-, a high electron affinity ion (2.7 eV) that has been the subject of a number of earlier and concurrent experiments.[21] As in the case of the positively charged clusters, the most striking result (Figure 9) is the absence of any significant SID fragmentation pattern. At the same energies, collision of various strongly bound negatively charged alkali-metal halide clusters (like $Na_{n-1}F_n^-$) with Si(100) results in specific fragmentation at lowest energies, and increasingly extensive nonspecific fragmentation at higher energies.[19]

(19) Beck, R. D.; St. John, P.; Homer, M. L.; Whetten, R. L. *Science* **1991**, *253*, 879.

(20) Zimmerman, J. A.; Eyler, J. R.; Bach, S. B. H.; McElvany, S. W. *J. Chem. Phys.* **1990**.

(21) Yang, S. H.; Pettiette, C. L.; Conceicao, J.; Cheshnovsky, O.; Smalley, R. E. *Chem. Phys. Lett.* **1987**, *139*, 233. See also ref 8.

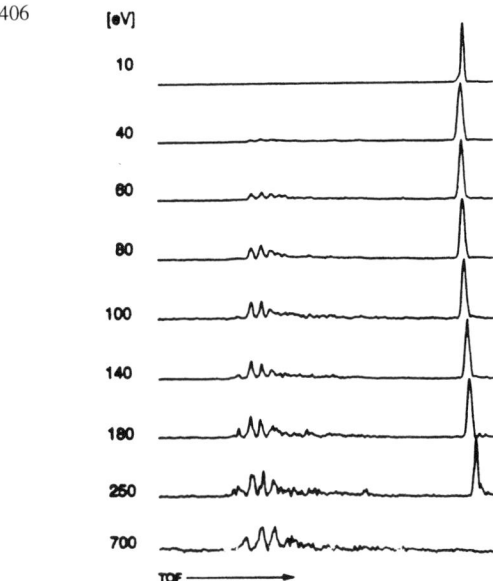

Figure 7. Time-of-flight profiles for C_{60}^+ scattering from Si(100). As in Figure 5, no large C_n fragments are observed at any collision energy. At low mass, small hydrocarbon fragments ($C_nH_x^+$) are observed due to sputtering from a oil film present on the uncleaned surface (see Figure 8).

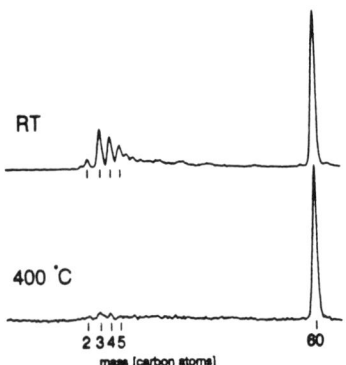

Figure 8. Effect of heating the surface (1 h at 400 °C) on the sputtering signal from C_{60}^+ at 100 eV. The scattered C_{60}^+ signal is essentially unchanged by the elevated surface temperature. Heating for 12 additional hours produces no additional changes.

Also apparent in Figure 9 are two other major features. The first has an onset near 110–120 eV and arrives at the detector essentially simultaneously with the collision time. This signal corresponds to electrons ejected promptly upon collision. At 150 eV this signal becomes comparable in intensity to that of C_{60}^-.

The second major feature corresponds to a mass of 1.0 amu and is therefore assigned to the H$^-$ ion, ejected from the surface adsorbate. It is particularly strong for very high energy impact (near 1 keV). Other, weaker features at longer times remain unassigned.

D. Comparison with Molecules (Benzene, Naphthalene) and Clusters (AHCs). Because the most important result described above is a negative one, i.e., an absence of detectable fragmentation induced by impact, it is necessary to confirm that identical experiments performed on somewhat comparable systems will exhibit the expected fragmentation. The most relevant experiments carried out on the same instrument are those on aromatic hydrocarbons benzene and naphthalene and on larger clusters. Impact has been extensively described as a method for imparting a large, relatively well-defined energy into an ion, a hypothesis that is thoroughly supported by fragmentation patterns in which specific fragments appear at energies correlated with known fragmentation energies.[18]

As an example, Figure 10 shows the time-of-flight profiles for the impact-induced fragmentation profiles of naphthalene, $C_{10}H_8^+$,

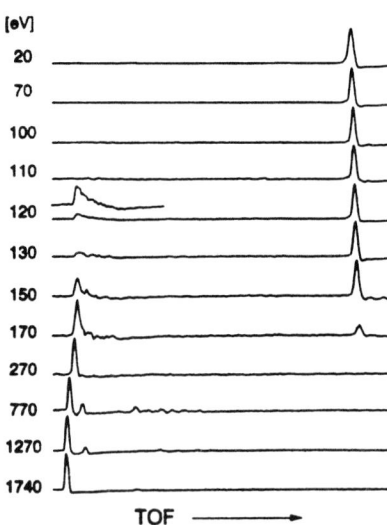

Figure 9. Time-of-flight profiles for C_{60}^- scattering from Si(100). At 120 eV, a signal due to electrons starts to be observable; this signal sharpens and increases in relative intensity with increasing collision energy. Other weaker peaks are due to light ions (like H$^-$) sputtered from adsorbates on the surfaces.

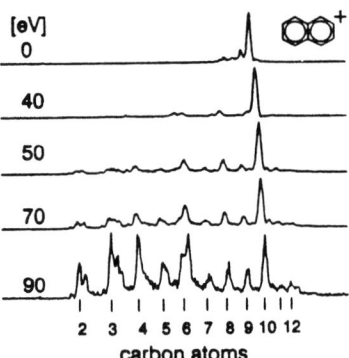

Figure 10. Surface-induced dissociation (SID) profiles for naphthalene ($C_{10}H_8^+$) including signal due to pick-up of one or more methyl groups from the surface.

over the range 0–90 eV. Clear peaks corresponding to hydrocarbon fragments as small as C_2^+ grow in with increasing impact energy, until the parent ion (or $C_{10}H_x^+$) is less than 5% of the scattered ion intensity at 90 eV. At higher energies, no parent ion can be observed. Naphthalene also undergoes pickup reactions with the overlayer (one or two CH$_x$ groups). That C_{60} is undergoing similar reactions cannot be entirely excluded at the present resolution.

IV. Discussion and Conclusions

The primary result of this work is that impact of stable all-carbon molecular ions $C_{60}^\pm$, C_{70}^+, and C_{84}^+ against crystalline surfaces leads to intact scattering over the entire energy range 0–200 eV. A negligible yield of charged fragments is obtained, under conditions where all other molecular ions and clusters that have been investigated give rise to intense fragmentation patterns characteristic of the molecular structure. For example, one might have expected to observe the same sequence of $-C_2$ losses found when trapped ions are electronically excited by ultraviolet lasers.[11] The indication that the collisions are highly inelastic makes it seem quite unlikely that the carbon molecules are not energized by the collision. Earlier work on smaller molecules indicates that 10–20% of the impact energy can be expected to be imparted to the internal modes of the scattered ion,[18] and our more recent work on larger systems (clusters) shows that 20–35% is not unusual.[19] The MD simulations of Mowrey et al. give values in the 25–30% range for the C_{60}/diamond impact event (see below), and Landman and co-workers have found very similar numbers for different systems (alkali-metal halides impacting graphite) but sharing the same size scale.[22]

It is therefore implausible that the scattered ions are insufficiently heated for fragmentation to occur. The question then becomes one of the *time scale* of the fragmentation process. In this instrument the acceleration time of $C_{60}^{\pm}$ is typically 2 μs, 1.5 kV potential drop over 1–2 cm. Fragmentation of metastable ions occurring after this time is invisible with the present arrangement. By increasing the impact energy indefinitely, the metastable decay rate should increase so that it is possible to see fragment peaks even with this limitation. However, at energies above 250 eV the total yield of charged ions is too weak for mass analysis. Unfortunately, it is not simple to establish connections with past experiments on fragmentation kinetics of $C_{60}^{\pm}$. For example, Radi et al.[23] have studied the metastable decay of very hot C_{60}^{+} sputtered from graphite using tandem mass spectrometry; however, the clusters that undergo decay might not necessarily have the closed-cage structures of the stable molecules used here. Among the work on multiphoton fragmentation, O'Brien et al.[11] have made some estimates of energy deposition required to observe fragmentation starting with internally cold ions, but often the ultraviolet multiphoton absorption is an inherently uncontrolled method of energy deposition. The situation with regard to modeling the fragmentation kinetics is made worse by the absence of information, experimental (e.g., pyrolysis) or theoretical, on the barrier to unimolecular fragmentation of these molecules. As a crude estimate, taking $-C_2$ loss to be the lowest channel, the net energy required to produce this highly energetic species from C_{60} could be estimated by noting that four C–C bonds (of 3/2 order) must be broken, possibly requiring 10–16 eV. However, a sizable barrier to the reverse process (C_2 attack upon C_{58}) could result in a higher number.

In the case of C_{60}^{-}, an additional low-energy channel is opened, as ejection of the electron requires only 2.7 eV. We have assigned the "zero-mass" peak to ejected electrons, although it was not possible to prove whether it comes from the incident ion or from the silicon surface. The onset at 110 eV is interesting, however, because the time-of-flight peak profile is not sharp, but rather has a tail toward longer times. This is characteristic of metastable decay during the acceleration period in a time-of-flight mass spectrometer. We propose that the hot, scattered C_{60}^{-} decays by emission of the electron over a range of later times, as in *thermionic emission* from hot solids.[24] Maruyama et al.[25] have recently discussed the evidence that much larger carbon clusters, heated under vacuum by repeated laser pulses, undergo electron emission by just such a mechanism. Under higher energy impact, the rate of thermionic emission increases, as that the peak becomes progressively sharper, closer to the apparatus limit. If such a mechanism is valid, and the microscopic theory of thermionic emission could prove valid, then the observed rates might serve as a thermometer for the impact-heated cluster. As a crude example, using the macroscopic expression[24] for the rate, with an assumed surface area of 2 nm^2, yields the result for the mean time to emit an electron

$$\tau_{\mathrm{ti}} = (5 \times 10^{-8} \text{ s})T^{-2} \exp(\phi/k_{\mathrm{B}}T)$$

where T is the temperature (in kelvin) and ϕ is the work function, taken here as the electron affinity. In the experiment, a 120-eV impact yields a metastable e$^-$ signal on the $\tau_{\mathrm{exp}} = 300$ ns time scale. Using an estimated 0.25 fraction energy converted to heating the cluster (see above) implies an internal temperature jump of 2000 K, which when combined with $\phi = 2.75$ eV yields a $\tau_{\mathrm{ti}} = 150$ ns, in agreement with observations. We state this result here primarily to indicate once again that impact almost surely deposits a very large quantity of energy into the scattered molecule.

That impact of these carbon molecules deposits large quantities of energy but exhibits negligible amount of fragments is unique in our experience and also seems unprecedented in the molecular-ion surface-induced dissociaton literature. Some larger molecules, such as PAHs, do show a small amount of parent ion scattering at energies exceeding 100 eV, but always show an extensive series of fragments.[18]

The weakest part of the present investigation is its inability to reveal the processes occurring at the surface. Neutralization, sticking (physisorption), and chemical reaction may all be responsible for the undetected fraction. We have carried out unsuccessful experiments attempting to reionize any neutralized, scattered molecules (intact or fragmented) using excimer-laser radiation above the surface. This approach deserves further investigation. There also remains the possibility, rather unlikely in our view, that unknown chemical reaction processes will make it impossible to observe the molecules that are sufficiently energized to result in fragments.

Complementary insight into these questions comes from two other sources. The first is an experiment by McElvany, Ross, and Callahan,[12] who collided C_{60}^{2+} (formed by electron impact of sublimed C_{60} solid) against a stainless steel surface, and observed only singly charged C_{60}^{+} (and $C_{60}H^{+}$) as charged scattering products. This demonstrates the importance of neutralization processes in the scattering of carbon molecular ions. [Collisions of C_{60}^{+} with stainless steel at up to 60 eV energy are also mentioned, and no fragmentation was detected.] Second, Mowrey and co-workers have carried out extensive molecular dynamics simulations of the high-energy impact processes of C_{60} with diamond (111) and, more recently, with graphite (0001).[13] We now describe this latter work in greater detail in order to establish some correspondences to our own work.

Figure 11 shows a typical nonreactive (200 eV) trajectory from their work of C_{60} with hydrogen-terminated diamond. Empirical many-body potentials are used for all interactions—intra-C_{60}, intra-solid, and C_{60}-solid. The solid and cluster temperatures are initially 300 K, and the collisions range from 150 to 250 eV. Clearly, the impact with this very rigid surface—much closer in its elastic properties to Si than graphite—leads to a tremendous deformation of the cluster, yet it rebounds into its original shape, vibrationally excited. This kind of response, which is familiar in the macroscopic world, is best characterized as that of a *resilient* object (resilient: bouncing or springing back into shape after being stretched, bent, or (especially) compressed).[26] It seems possible that the unique ability of these molecules to withstand and rebound from high-energy impact is associated with the resilience inherent in this class of structures. Quantitatively, Mowrey et al. find that at lower energies, 150 eV, no fragmentation or surface reaction is observed. The collisions are highly inelastic, with ca. 20% of the energy in recoil, 25–30% in heating the cluster, and the residual heating the surface. At higher energies, 200 eV, nonreactive scattering is still the major event (86%), with very small amounts of sticking and –H or –CH pickup observed. At 250 eV, one-third of the events are nonreactive scattering, one-third involve reactive adhesion to the surface (nonscattering), and other events include cluster breakup to C_{57} and C_{59} as well as formation of $C_{60}H$ and $C_{61}H$. Some of the events, in which a transient bond is formed with the surface, have almost zero recoil energy, and so could account for the low recoil energies observed in experiments. The experimental decrease observed in scattering at higher energies might be explained in terms of the reactive adhesion process. However, neutralization and other electronic processes cannot be treated within this molecular dynamics formalism. The recoiling, nonreactive clusters are highly energized, as 30–60 eV internal energy corresponds to an internal temperature of 2000–4000 K, so it seems remarkable that (in the experiment) they can survive 10^{-6} s without fragmentation. Yet this is also precisely the con-

(22) Landman, U.; et al., private communication of unpublished results.

(23) Radi, P.; Hsu, M. T.; Rincon, M. E.; Kemper, P. R.; Bowers, M. T. *Chem. Phys. Lett.* **1990**, *174*, 223.

(24) For a textbook description of the rate theory of thermionic emission: Ashcroft, N. W.; Mermin, N. D. *Solid State Physics*; Saunders: Philadelphia, 1976; pp 362–4.

(25) Maruyama, S.; Lee, M. Y.; Haufler, R. E.; Chai, Y.; Smalley, R. E. *Z. Phys. D*, in press. Even, U.; deLange, P. J.; Jonkman, H. T.; Kommandeur, J. *Phys. Rev. Lett.* **1986**, *56*, 965.

(26) *Webster's New World Dictionary*, 3rd ed. Resilience: the ability to bounce or spring back into shape .... The ability to recover strength ... quickly. Resilient: bouncing or springing back into shape after being stretched, bent, or (especially) compressed.

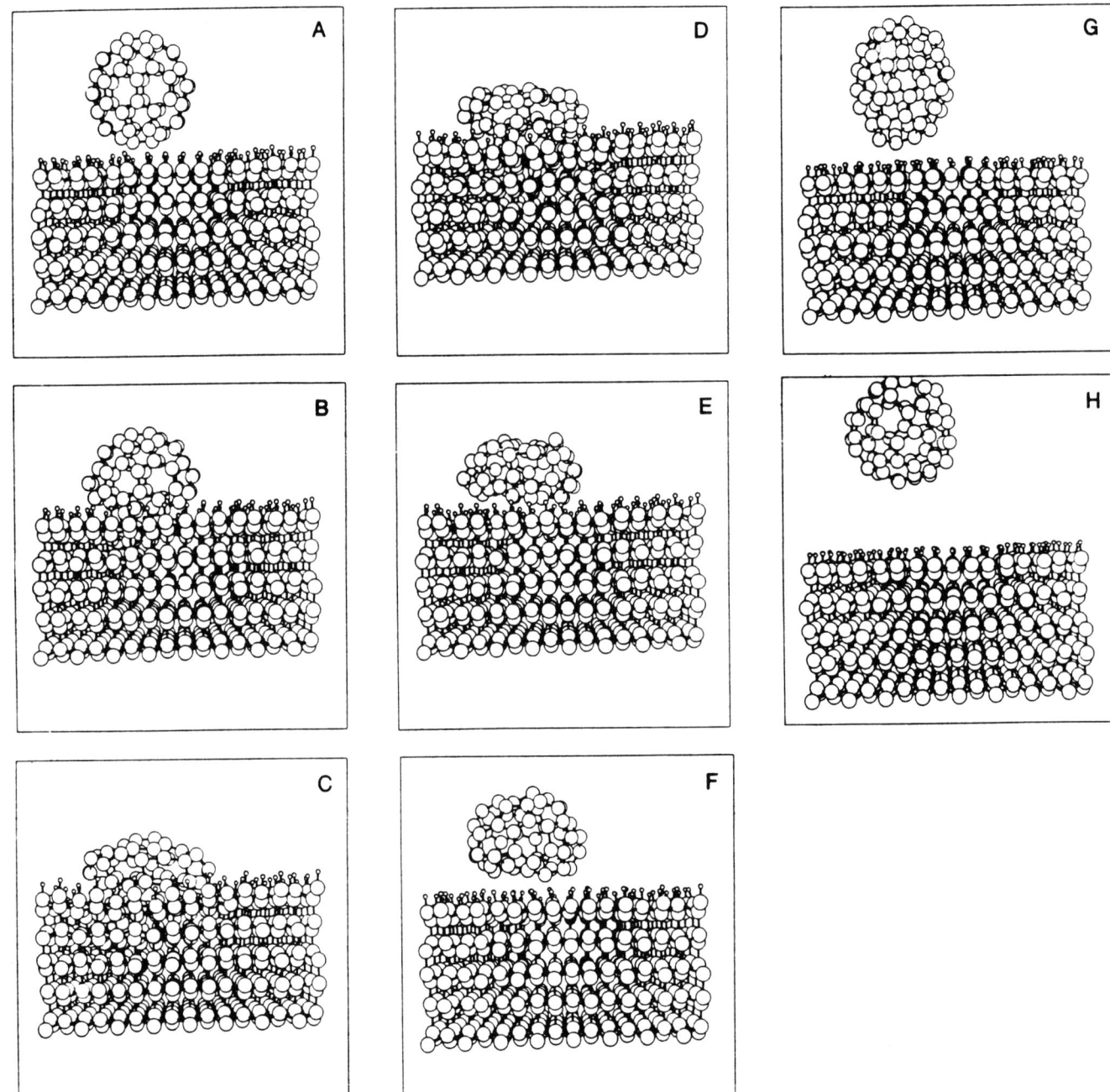

Figure 11. Molecular dynamics simulation (Mowrey et al.[13]) of a 200 eV impact encounter between neutral C_{60} and a diamond surface (111) terminated by hydrogen atoms (small spheres). From upper left in 0.2-ps intervals: (A) At $t = 0$, just prior to impact, the snapshot of incoming C_{60} and diamond reflects thermal disorder at 300 K. (B) $t = 0.2$ ps, just at impact, the lower half of the molecule deforms toward planarity. (C) and (D) at $t = 0.4$–0.6 ps, a bilayerlike C_{60} is maximally deformed, and the diamond-lattice is crumpled three layers deep. (E) $t = 0.8$ ps, recoil begins. (F) $t = 1.0$ ps, after a contact time of 0.8–1.0 ps, the intact C_{60} lifts off, and the surface relaxes. (G) $t = 1.2$ ps, now in full flight with a speed near 3×10^5 cm/s (ca. 30 eV), the recoiled C_{60} shape has passed from oblate at 1.0 ps. (F), through near-spherical at 1.1 ps (not shown) to the strongly prolate-ellipsoid structure shown. (H) Finally, at 1.4 ps, a snapshot shows C_{60} at the spherical point of its quadrupolar vibration, with the entire bonding network still clearly intact. Despite absorbing typically 50 eV energy, corresponding to a kinetic temperature of 3500 K, the ball remains intact while exhibiting large-amplitude quadrupolar deformations (harmonic frequency 270 cm^{-1}, corresponding to $t = 0.13$ ps, but looks more like 0.5 ps at this highly anharmonic energy).

clusion reached by Smalley and co-workers[11] on the basis of an entirely different excitation process (ultraviolet multiphoton absorption).

In conclusion, an ion beam surface-scattering investigation of the impact-induced processes of cold, stable carbon molecules $C_{60}^{\pm}$, C_{70}^{+}, and C_{84}^{+} with graphite and silicon has revealed high inelasticity but no evidence for impact-induced fragmentation at energies up to 200 eV. This is in accord with simulations showing that the structural integrity of C_{60} neutral is maintained in such collisions, despite highly efficient transfer (approaching 30%) of energy into the recoiling molecule. This result may thus be taken as evidence of the remarkable resilience of the closed-cage structures of these molecules. The only impact-induced process

observed in these experiments is the ejection of electrons; the time-resolved signature of this peak at its appearance threshold is consistent with a thermionic ejection of the electron from the scattered C_{60}^{-}.

Since completing this work, we have been informed of very interesting preliminary results of J. Callahan, V. Wysocki, and co-workers[27] in which the singly charged products of collisions of C_{60}^{2+} with stainless steel are observed. At an impact energy of 270 eV, corresponding to the upper end of our experimental range, C_{60}^{+} is joined by C_{58}^{+} and C_{56}^{+} in ca. 10:3.5:2.5 ratio, in

(27) Callahan, J. H., private communication to Whetten, R. L., received on 4 April 1991.

apparent contradiction to our results. However, three points should be considered: (i) the doubly charged ion is probably a little less stable to begin with (Coulomb repulsion of 1–2 eV across the molecular framework); (ii) neutralization itself imparts an energy up to the second ionization potential (9.7 eV)[12] into the molecule; and (iii) the C_{60}^{2+} is formed by high-energy electron-impact ionization of vapor sublimed from a hot source (up to 800 °C), likely imparting a large amount of electronic energy into some fraction of the nascent ions. Therefore, we feel that there is no necessary contradiction with our results obtained using jet-cooled singly charged molecules. [Extensive fragmentation of C_{60}^{2+} has also been observed in Xe CID experiments under multiple-collision conditions.] Second, our subsequent work includes the following: (i) we have succeeded in laser reionization experiments to detect the scattered neutrals and found under lowest fluence conditions that only $C_{60}^{(0)}$ is returned (no definitive evidence for neutral fragments); (ii) photofragmentation of scattered ions (under single-photon conditions) show definitively that a large amount of energy is deposited into impact-heated C_{60}^+ by the higher energy collisions; and (iii) velocity distributions of the scattered neutrals have been measured by time-delayed laser reionization time-of-flight measurements, and confirm that a substantial fraction of the incident translational energy is lost in most collisions. This work will be reported separately.[28]

Acknowledgment. This research was funded by the Office of Naval Research. We acknowledge crucial discussions of surface scattering with (the late) R. B. Bernstein. The surface materials used in this work were provided by R. S. Williams. Assistance in preparation of the molecular carbon samples was obtained from S. Anz, R. Ettl, and F. Ettl. For providing information prior to publication, special thanks goes to Mark Ross, Steven McElvany, and John Callahan, and to B. Dunlap and R. Mowrey of the Naval Research Laboratory for providing results on impact simulations, including permission to quote their results and to reprint Figure 11.

Registry No. C_{60}, 99685-96-8; C_{70}, 115383-22-7; C_{84}, 134847-08-8.

(28) Beck, R. D.; Yeretzian, C.; St. John, P.; Whetten, R. L., to be published.

Reprinted with permission from Nature
Vol. 363, pp. 60–63, 6 May 1993
© 1993 Macmillan Magazines Limited

Experimental evidence for the formation of fullerenes by collisional heating of carbon rings in the gas phase

Gert von Helden, Nigel G. Gotts & Michael T. Bowers*

Department of Chemistry, University of California, Santa Barbara,
California 93106, USA

THE discovery[1-6] of the spherical carbon cage compound buckminsterfullerene (C_{60}) and the recent development of methods to produce it in bulk[7] have led to an explosion in research in the physical and chemical properties of this unique species[8,9]. Nevertheless, the question of the formation mechanism of C_{60} (or of the other fullerenes) is still far from settled. We have shown elsewhere that carbon clusters in the gas phase develop from linear chains to planar ring systems to fullerenes as their size increases. One can easily envisage the transformation from chains to rings, but how the three-dimensional near-spherical fullerenes evolve from large planar rings is not obvious. Here we show that 'heating' these large ring systems above their 'melting' point leads to 100% fullerene formation accompanied by the evaporation of a small carbon fragment (C_1 or C_3 for odd systems and C_2 for even systems). We propose a mechanism, based on these data, for efficient C_{60} production in carbon arcs.

Early suggestions for the formation mechanism of fullerenes[1-4] focused on curling up of graphitic sheets that grow by the addition of C_2 units. More recently, Heath[14] has proposed the 'fullerene road' scenario, in which fullerenes appear (according to an unspecified mechanism) around C_{40} and grow by sequential addition of C_2 units, eventually stopping at C_{60} where further addition becomes difficult. Smalley[15], in contrast has favoured the 'pentagon road', which is really a variant of the earlier curling-graphitic-sheet model, except that pentagons appear, because of annealing, in just the right positions so that closure to form C_{60} occurs naturally and defects, overshooting and so forth are minimized. 'Ring-stacking' is another recent model[16], where an appropriate number of rings of different sizes are stacked, eventually forming C_{60}. Rubin et al.[17] and

McElvany et al.[18] have shown by mass spectrometry that (presumably fullerene) C_{60} and C_{70} can be formed in the gas phase from C_{18}, C_{24} and C_{30} cyclic all-carbon precursors by ion-molecule reactions. These studies suggest a cationic pathway in the formation process from condensation of large ring systems. Curl[19] has recently summarized and analysed some of these models.

We have recently developed a gas-phase ion chromatography technique[20] and applied it to carbon cluster cations[10,11] and anions[11-13]. A pulse of mass-selected cluster ions is injected into a high-pressure drift cell filled with 2–5 torr of helium. The ionic mobilities of different isomeric structures depend on their different collision cross-sections with He, and the isomers are therefore separated while drifting through the cell, under the influence of a weak electric field. The absolute value of the ionic mobility for a given cluster together with computer simulations often allows unambiguous determination of the cluster

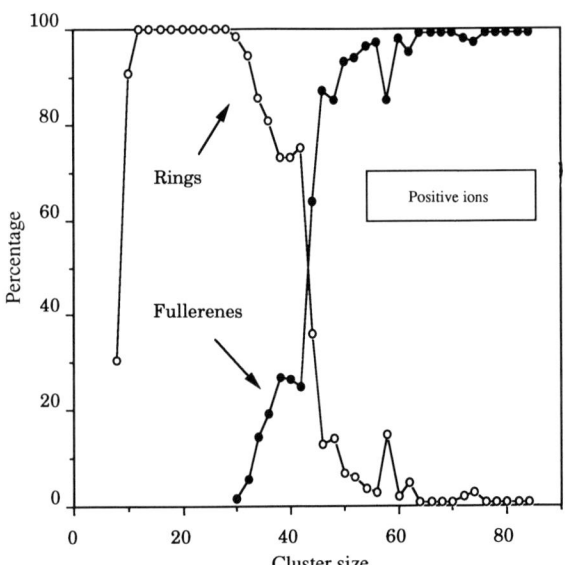

FIG. 1 Plots of percentage of planar rings and fullerenes against carbon cluster size. Experimental details are given in the text and in refs 10–13. Only the even cluster ions are shown for clarity. The odd cluster positive ions yield curves with a similar shape, but the planar rings decrease less quickly as size increases, and make up about 40% of the total near C_{60}^+.

---

* To whom correspondence should be addressed.

structure[12,13]. The method also allows us to determine the carbon cluster isomer abundances as a function of cluster size. We have found that small clusters are linear; near C_{10}, planar monocyclic rings appear, which gradually give way to planar bicyclic rings between C_{20} and C_{30} (refs 12, 13); finally planar tricyclic rings evolve in the C_{30}s and dominate the ring systems thereafter[13]. Fullerenes first appear at C_{30} for the cations and dominate by C_{50}. For negative ions the planar multicyclic rings dominate to well above C_{60}. In Fig. 1 these trends are quantified by plotting the fraction of planar rings and fullerenes against cluster size for clusters made in our laser vaporization source.

Here we focus on the transition region between C_{30}^+ and C_{40}^+, as fullerenes first appear in this range. The progression linear → monocyclic rings → bicyclic rings → tricyclic rings can readily be understood: at a certain minimum size, energized linear molecules can close and form a ring and then these rings can react with each other to form bicyclic and tricyclic rings. But where do the fullerenes come from? We find no evidence[11,12] for fullerenes or 'graphitic' or 'cup'-like fullerene precursors in our chromatography data for ions below C_{30}^+, yet we see large abundances of fullerenes for larger clusters[13]. There seems to be no simple way the families of planar cyclic rings can suddenly switch to fullerenes simply by adding new members.

It occurred to us that our source might not be providing long enough periods of high temperatures for structural 'annealing'

to occur. We therefore tried a variant of an experiment successfully applied on silicon clusters[21]. To generate the data shown in Fig. 1, a narrow pulse of mass-selected ions was injected into the chromatography cell at very low energy (2 to 20 eV), and the ions were rapidly thermalized by the 5 torr of He present. In our 'annealing' studies, in contrast, the ions were intentionally injected at fairly high voltages, creating a transient internal energy increase that can allow isomerization, or even collision-induced dissociation. In both cases, the arrival time distribution (ATD) at the detector gives the isomer distribution when the quadrupole (following the chromatography cell) is tuned to the appropriate mass. The results for C_{39}^+ and C_{40}^+ are shown in Fig. 2a and b respectively. At low injection energies (Fig. 2a, panel A), the C_{39}^+ ATD is dominated by planar bicyclic and tricyclic ring systems, but at high injection energies those C_{39}^+ ions that survive (those not dissociated by the internal energy surge) are converted essentially 100% to the monocyclic ring isomer (Fig. 2a, panel B). The three-dimensional ring and fullerene are essentially unaffected. Clearly the monocyclic ring is the most stable planar ring structure. This result does not give any information on fullerene formation, but by scanning the quadrupole mass filter following the chromatography cell, we noticed that a substantial fraction of the C_{39}^+ ions injected at 150 eV had been collisionally dissociated.

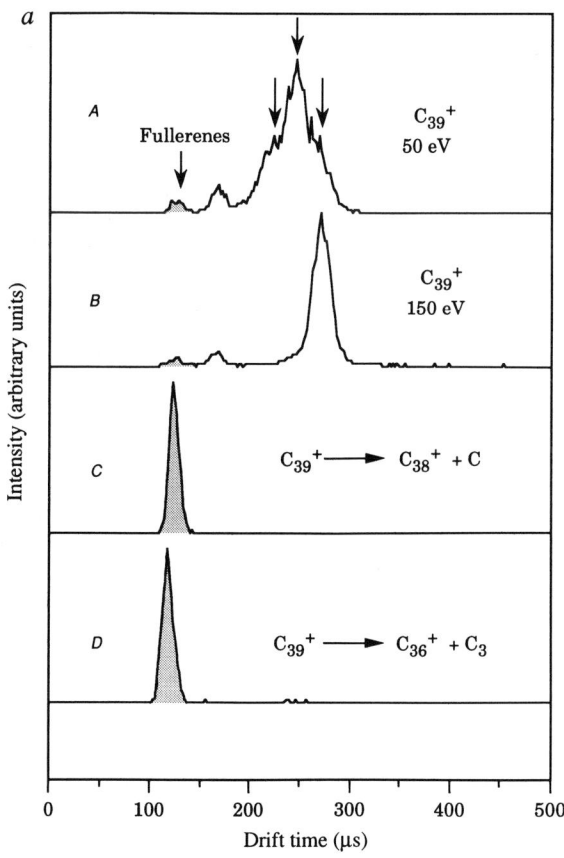

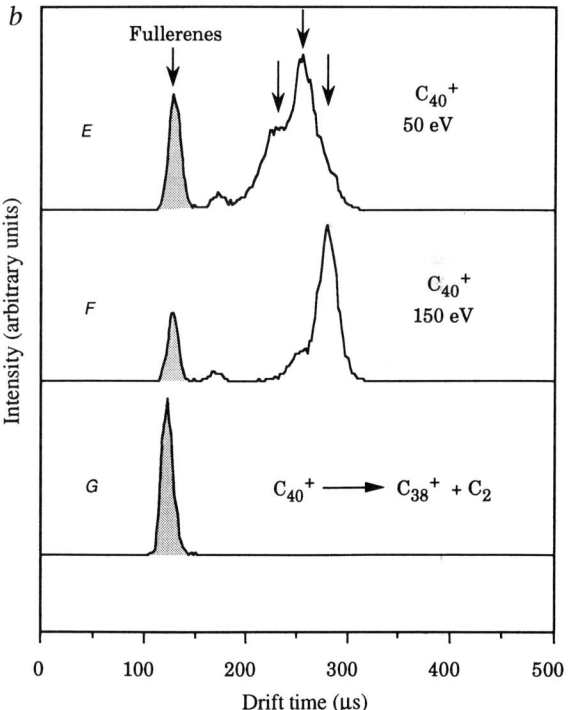

ring systems have been 'annealed' to the monocyclic ring. C, ATD for C_{38}^+. These C_{38}^+ ions are formed from 150 eV injection of C_{39}^+ ions into the chromatography cell, which undergo transient collisional heating followed by dissociation. Note that 100% of the C_{38}^+ ions formed are fullerenes even though the C_{39}^+ parent ions were almost exclusively planar rings. D, ATD of C_{36}^+ ions formed from collision-induced dissociation of C_{39}^+. These C_{36}^+ ions are essentially 100% fullerene. b, Results for C_{40}^+. E, ATD of C_{40}^+ injected into the chromatography cell at 50 eV is given. This ATD is very similar to one obtained by injection at much lower voltages. The arrows over the composite peak at longest times indicate the presence of monocyclic, bicyclic and tricyclic ring systems. The shaded peak at shortest times (near 120 μs) is the C_{40}^+ fullerene peak. F, the ATD for surviving C_{40}^+ ions injected at 150 eV. Annealing of the planar ring systems leads predominantly to the monocyclic ring isomer. G, ATD for C_{38}^+ ions. These are obtained by injection of C_{40}^+ ions into the chromatography cell, followed by loss of C_2. Clearly, the resultant C_{38}^+ ions are 100% fullerene.

FIG. 2 Arrival time distributions (ATD; ion chromatograms) obtained under different conditions (see text and refs 10–13). a, Results for C_{39}^+. A, C_{39}^+ injected at 50 eV into the cell. This distribution is essentially identical to those injected at 2–3 eV. The three arrows over the large composite peak indicate the presence of monocyclic planar rings (longest time) bicyclic planar rings (middle) and tricyclic planar rings (shortest time of this peak). The small shaded peak near 120 μs is the C_{39}^+ fullerene. The remaining peak near 160 μs is a three-dimensional ring structure. b, ATD for surviving C_{39}^+ injected into the chromatography cell at 150 eV. The bicyclic and tricyclic

62

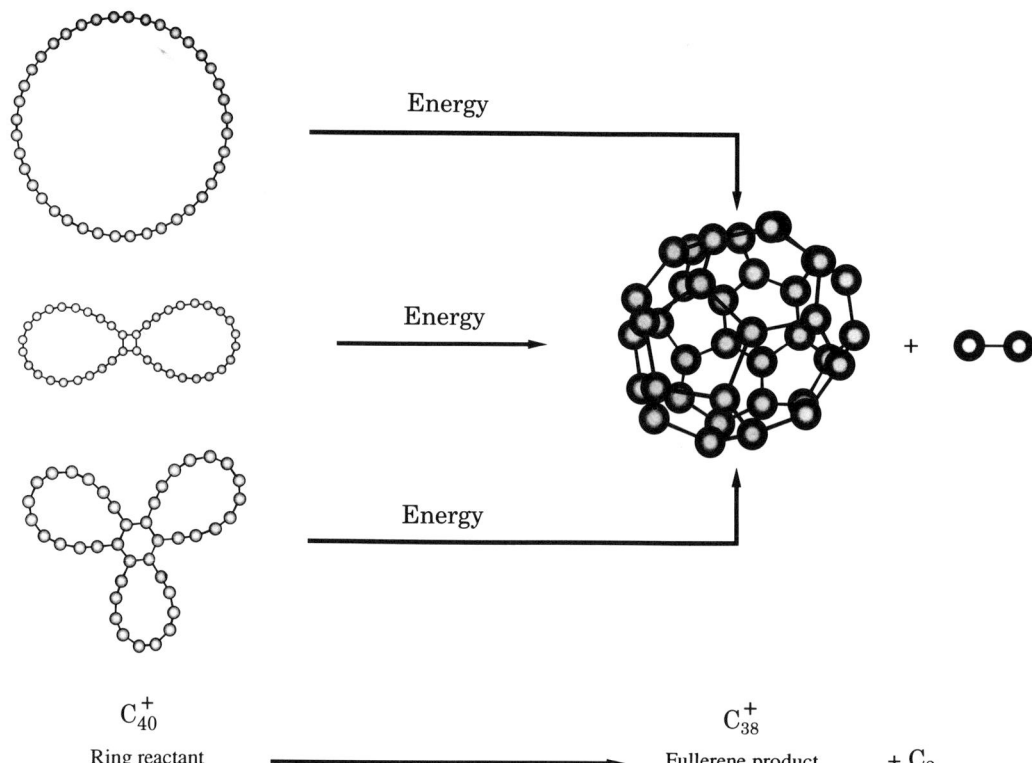

FIG. 3 Planar monocyclic, bicyclic and tricyclic C_{40}^+ ions form C_{38}^+ fullerene plus C_2 when collisionally heated above the isomerization barrier for fullerene formation (see Fig. 2b). All structures shown are minima on the potential energy surface, according to semi-empirical PM3 electronic structure calculations. The bicyclic and tricyclic rings have lower isomerization barriers to fullerene formation than the monocyclic ring.

C_{40}^+

Ring reactant

C_{38}^+

Fullerene product   + C_2

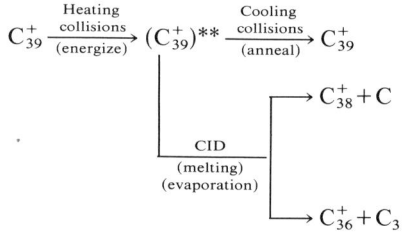

The isomer chromatograms of the C_{36}^+ and C_{38}^+ fragments are given in Fig. 2a (panels C and D). The results are unambiguous: in both instances, the products are essentially 100% fullerene. This process is represented pictorially in Fig. 3, where C_{38}^+ fullerene is created by heating C_{40}^+ ring systems.

It seems that when the C_{39}^+ or C_{40}^+ planar multicyclic ring ions gain insufficient energy to rearrange to a fullerene, they rearrange by breaking one or two bonds and form the more stable monocyclic ring structure. This is not common in the ion source because when these multi-ring species are first formed (presumably through ring–ring collisions), rapid cooling quenches the cluster products and results in a metastable isomer distribution. By injecting this distribution into the chromatography cell at high voltage, however, we induce a strong transient heating pulse. Once the C_{39}^+ or C_{40}^+ planar ring ions are collisionally 'heated' above the threshold for isomerization to a fullerene, rapid bond breaking and rearrangement occurs and the cluster begins to relax to its most stable form. At this point the internal energy in the cluster can be enormous. For example, the calculations of ref. 22 indicate that conversion of bicyclic planar C_{36} to a C_{36} fullerene liberates ~9 eV, and conversion of a tricyclic planar C_{60} ring system to C_{60} liberates ~45 eV. If one adds to this the internal energy obtained in the injection pulse[21] (5–20 eV), vibrational temperatures near 4,000 K are indicated; this can be compared with the carbon melting temperature of 3,740 K. The superheated C_{39}^+ 'fullerene' ion cools itself by 'evaporating' either a C atom or a C_3 molecule, leaving behind a C_{38}^+ or C_{36}^+ fullerene, whereas C_{40}^+ loses C_2 to form C_{38}^+ fullerene.

This fullerene formation process starts gradually with C_{33}^+, where a small amount of C_{30} fullerene is made by loss of C_3. Theory[22,23] predicts that fullerenes become more stable than planar rings near C_{30}, in support of our observations. The process rapidly accelerates with increasing size, as is evident from Figs 1 and 2. This result is consistent with the fact that for carbon cations larger than C_{40}^+, the fullerene isomer dominates (Fig. 1). We conclude from these data that the barrier for transforming planar ring systems to fullerenes decreases with size, a result consistent with the rapid increase in stability of fullerenes relative to rings in this size range[22].

A comment about the C_{60} formation in carbon arc reactors is in order. Experiments[24-27] using C_{12}/C_{13} graphite unambiguously showed that C_{60} is synthesized starting from carbon atoms. Efficient synthesis of C_{60} occurs only for a narrow range of experimental parameters (pressure, rod feed rate and so on) so the kinetics must be fairly specific. Our chromatography experiments seem to have ruled out a graphitic-like growth mechanism, such as the 'pentagon road', as we have shown[10-13] that carbon initially forms linear chains, followed by planar ring systems as it grows. We have shown here that collisional heating of large planar ring systems induces isomerization to fullerenes. It seems possible (or even probable) that in arc reactors the atomic carbon 'soup' diffuses out of the arc, and initially forms chains and then rings. The carbon density immediately surrounding the arc may be high enough that these rings rapidly coalesce to sizes larger than C_{60} and spontaneously rearrange to fullerenes because of the low rearrangement barrier for large systems and the hot environment. Evaporation of small particles from the newly formed fullerene would preferentially yield C_{60}, the most stable species in the immediate size range. It is reasonable that the growth of planar rings to the size that will eventually form C_{60} would be maximized under specific experimental conditions in the arc reactor, otherwise smaller (or larger) fullerenes would be formed, as shown in Fig. 1. The abundance of C_{60} could be further enhanced by growth of smaller fullerenes, formed from isomerization of smaller ring systems, via the 'fullerene road', although adequate supplies of small carbon fragments and several successive steps are required for this to occur.

The data presented here unlock the mystery of how fullerenes appear near C_{30} in laser desorption experiments when no graphitic building blocks are present at smaller sizes. The ring → fullerene isomerization mechanism might also help explain the rather surprising propensity for C_{60} formation in carbon arc rectors. □

Received 12 November 1992; accepted 6 April 1993.

1. Kroto, H., Heath, J. R., O'Brien, S. C., Curl, R. F. & Smalley, R. E. *Nature* **318,** 162–163 (1985).
2. Heath, J. R. et al. *J. Am. chem. Soc.* **107,** 7779–7780 (1985).
3. Zhang, Q. et al. *J. phys. Chem.* **90,** 525–528 (1986).
4. Liu, Y. et al. *Chem. Phys. Lett.* **126,** 215–217 (1986).
5. Krätschmer, W., Fostiropolos, K. & Huffman, D. R. *Chem. Phys. Lett.* **170,** 167–170 (1990).
6. Krätschmer, W., Lamb, L. D., Fostiropolos, K. & Huffman, D. R. *Nature* **347,** 354–358 (1990).
7. Haufler, R. E. et al. *J. phys. Chem.* **94,** 8634–8636 (1990).
8. *Acct. chem. Res. spec. Issue Buckminster Fullerenes* **25**(3), 97–175 (1992).
9. *Fullerenes: Synthesis Properties and Chemistry of Large Carbon Clusters* (eds Hammond, G. S. & Kuck, V. J.) (Am. chem. Soc. Symp. Series No. 481, Washington DC, 1991).
10. von Helden, G., Hsu, M.-T., Kemper, P. R. & Bowers, M. T. *J. chem. Phys.* **95,** 3835–3837 (1991).
11. von Helden, G., Kemper, P. R., Gotts, N. & Bowers, M. T. *Science* **259,** 1300–1302 (1993).
12. von Helden, G., Hsu, M.-T., Gotts, N., Kemper, P. R. & Bowers, M. T. *Chem. Phys. Lett.* **204,** 15–22 (1993).
13. von Helden, G., Hsu, M.-T., Gotts, N., Kemper, P. R. & Bowers, M. T. *J. phys. Chem.* (submitted).
14. Heath, J. R. in *Fullerenes, Synthesis, Properties and Chemistry of Large Carbon Clusters* (eds Hammond, G. S. & Kuck, V. J.) 1–27 (Am. chem. Soc. Symp Series No. 481, Washington DC, 1991).
15. Smalley, R. E. *Acct. chem. Res.* **25,** 97–105 (1992).
16. Wakabayashi, T. & Achiba, Y. *Chem. Phys. Lett.* **190,** 465–468 (1992).
17. Rubin, Y., Kahr, M., Knobler, C. B., Diederich, F. & Wilkins, C. *J. Am. chem. Soc.* **113,** 495–500 (1991).
18. McElvany, S. W., Ross, M. M., Goroff, N. S. & Diederich, F. *Science* **259,** 1594–1596 (1993).
19. Curl, R. F. *Phil. Trans. R. Soc. Series A* (in the press).
20. Kemper, P. R. & Bowers, M. T. *J. phys. Chem.* **95,** 5134–5146 (1991); *J. Am. chem. Soc.* **112,** 3231–3232 (1990).
21. Jarrold, M. F. & Honea, E. C. *J. Am. chem. Soc.* **114,** 459–464 (1992).
22. Feyereisen, M., Gutowski, M., Simons, J. & Almlöf, J. *J. chem. Phys.* **96,** 2926–2932 (1992).
23. Jing, X. & Chelikowsky, J. R. *Phys. Rev.* B**46,** 15,503 (1992).
24. Meijer, G. & Bethune, D. S. *J. chem. Phys.* **93,** 7800–7802 (1990).
25. Johnson, R. D., Yannoni, C. S., Salem, J. & Bethune, D. S. *Mater. Res. Soc. Proc.* **206,** 715 (1991).
26. Hawkins, J. M., Meyer, A., Loren, S. & Nunlist, R. *J. Am. chem. Soc.* **113,** 9394–9395 (1991).
27. Ebbesen, T. W., Tabuchi, J. & Tanigaki, K. *Chem. Phys. Lett.* **191,** 336–338 (1992).

ACKNOWLEDGEMENTS. We thank the Air Force Office of Scientific Research and the U.S. NSF for support. G.V.H. acknowledges IBM for a graduate fellowship.

Reprinted with permission from the Journal of the American Chemical Society
Vol. 114, No. 10, pp. 3978–3980, 1992

Electrochemical Detection of C_{60}^{6-} and C_{70}^{6-}: Enhanced Stability of Fullerides in Solution

Qingshan Xie, Eduardo Pérez-Cordero, and Luis Echegoyen*

Department of Chemistry
University of Miami
Coral Gables, Florida 33124
Received February 6, 1992

Although theory predicts that the LUMO of C_{60} should be able to accept at least six electrons to form diamagnetic C_{60}^{6-},[1-3] the latter species has so far eluded direct characterization in solution.[4-9] The existence of the hexaanionic molecule in the solid state is inferred from the formation of species such as K_6C_{60}, which have been well characterized by a variety of techniques, including solid-state ^{13}C NMR spectroscopy.[10] Part of the impetus behind the present work stemmed from the desire to generate and detect C_{60}^{6-} in a relatively stable environment in order to confirm the theoretical predictions.[1-3] The development of a general method capable of generating stable C_{60}^{n-} species, where $n = 1$–6, was another important driving force behind the present work.

There have been several reports concerning the electrochemical properties of C_{60} (and C_{70}).[4-9] One recent report by Wudl et al. described the reversible, three-electron electrochemical reduction leading to C_{60}^{3-}.[4] After this communication, Dubois and Kadish reported the observation of an additional reduction wave for C_{60} and the electrochemical formation and detection of C_{70}^{4-}.[4,5] The most recent and, to our knowledge, the only report of five reversible reduction processes for C_{60}, leading to C_{60}^{5-}, was also reported recently by Kadish et al.[6] These authors also studied the electrochemical properties of C_{70}, but no voltammetric data were presented for this compound.[6] They pointed out that several new peaks appeared after the fourth reduction of C_{70}, but that none could be unambiguously assigned to C_{70}^{5-}.

A wider expansion of the available potential window down to -3.3 V vs Fc/Fc^+ is reported in this communication. This was accomplished by the use of a mixed solvent system and low temperature. On the basis of supporting electrolyte and fullerene solubility considerations, the optimal solvent composition was between 15 and 20% by volume of acetonitrile in toluene. These new conditions have allowed the first observation of the sixth

(1) Haddon, R. C.; Brus, L. E.; Raghavachari, K. *Chem. Phys. Lett.* **1986**, *125*, 459.

(2) (a) Haymet, A. D. *Chem. Phys. Lett.* **1985**, *122*, 421. (b) Disch, R. L.; Schulman, J. N. *Chem. Phys. Lett.* **1986**, *125*, 465.

(3) Scuseria, G. E. *Chem. Phys. Lett.* **1991**, *176*, 423.

(4) Allemand, P.-M.; Koch, A.; Wudl, F.; Rubin, Y.; Diederich, F.; Alvarez, M. M.; Anz, S. J.; Whetten, R. L. *J. Am. Chem. Soc.* **1991**, *113*, 1050.

(5) Dubois, D.; Kadish, K. M.; Flanagan, S.; Haufler, R. E.; Chibante, L. P. F.; Wilson, L. J. *J. Am. Chem. Soc.* **1991**, *113*, 4364.

(6) Dubois, D.; Kadish, K. M.; Flanagan, S.; Wilson, L. J. *J. Am. Chem. Soc.* **1991**, *113*, 7773.

(7) Haufler, R. E.; Conceicao, J.; Chibante, L. P. F.; Chai, Y.; Byrne, N. E.; Flanagan, S.; Haley, M. M.; O'Brien, S. C.; Pan, C.; Xiao, Z.; Billups, W. E.; Ciufolini, M. A.; Hauge, R. H.; Margrave, J. L.; Wilson, L. J.; Curl, R. F.; Smalley, R. E. *J. Phys. Chem.* **1990**, *94*, 8634.

(8) Allemand, P.-M.; Srdanov, G.; Koch, A.; Khemani, K.; Wudl, F.; Rubin, Y.; Diederich, F.; Alvarez, M. M.; Anz, S. J.; Whetten, R. L. *J. Am. Chem. Soc.* **1991**, *113*, 2780.

(9) Cox, D. M.; Behal, S.; Disko, M.; Gorun, S. M.; Greaney, M.; Hsu, C. S.; Kollin, E. B.; Millar, J.; Robbins, J.; Robbins, W.; Sherwood, R. D.; Tindall, P. *J. Am. Chem. Soc.* **1991**, *113*, 2940.

(10) Tycko, R.; Dabbagh, G.; Rosseinsky, M. J.; Murphy, D. W.; Fleming, R. M.; Ramirez, A. P.; Tully, J. C. *Science* **1991**, *253*, 884.

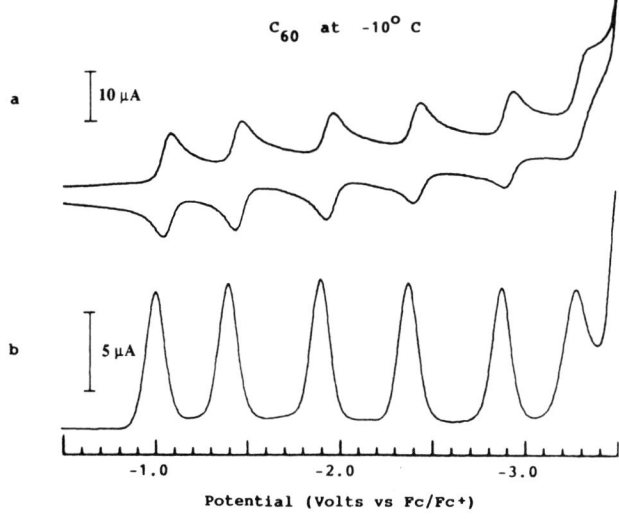

C$_{60}$ at -10° C

a

|10 μA|

b

|5 μA|

-1.0   -2.0   -3.0

Potential (Volts vs Fc/Fc+)

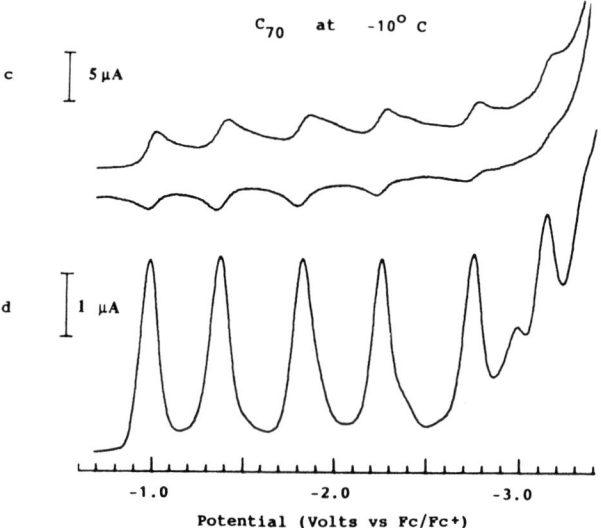

C$_{70}$ at -10° C

c

|5 μA|

d

|1 μA|

-1.0   -2.0   -3.0

Potential (Volts vs Fc/Fc+)

Figure 1. Reduction of C$_{60}$ and C$_{70}$ in CH$_3$CN/toluene at -10 °C using (a and c) cyclic voltammetry at a 100 mV/s scan rate and (b and d) differential pulse voltammetry (50-mV pulse, 50-ms pulse width, 300-ms period, 25 mV/s scan rate).

electron reduction for C$_{60}$ and of the fifth and sixth electron reductions for C$_{70}$, leading to the formation of C$_{60}^{6-}$ and C$_{70}^{6-}$. None of these had been previously observed. The highest temperature allowing resolution of the sixth reduction of C$_{60}$ is approximately 5 °C. For C$_{70}$ it is possible to observe the fifth and sixth reductions even at 25 °C. An added advantage of the present protocol is that these multiple reductions are all reversible and observed at very slow potential scan rates (100 mV/s) as opposed to those required in ref 6, ~20 V/s.[11]

(11) Samples of C$_{60}$ and C$_{70}$ were obtained by following the general procedures reported by Ajie et al.[13] The C$_{70}$ sample was further purified by double recrystallization from benzene. The electrochemical cell used was designed and constructed in-house and was very similar to that described in ref 12. Typically, a 7–8 × 10^{-4} M solution of C$_{60}$, or 4–5 × 10^{-4} M C$_{70}$, was placed in only the working compartment of the high-vacuum cell. TBAPF$_6$, 0.1 M, was used as supporting electrolyte. Highly purified acetonitrile and toluene from Aldrich were dried over P$_2$O$_5$, deaerated by repeated freeze–pump–thaw cycles, pumped to 10^{-5}–10^{-6} mmHg, and vapor transferred directly into the electrochemical cell.[12] The ratio of acetonitrile to toluene was 1:5.4 by volume. Voltammograms (cyclic and differential pulse) were recorded using a BAS-100 electrochemical analyzer interfaced with a Houston Instruments HIPLOT DMP-40 apparatus. Ohmic resistance was compensated 100% in all cases. A conventional three-electrode configuration was used, with a 3-mm-diameter glassy carbon electrode as the working electrode, a platinum counter electrode, and a silver wire as a pseudoreference. All measurements were recorded under high vacuum. Low-temperature experiments were performed by immersion of the cell in an external bath whose temperature was monitored.

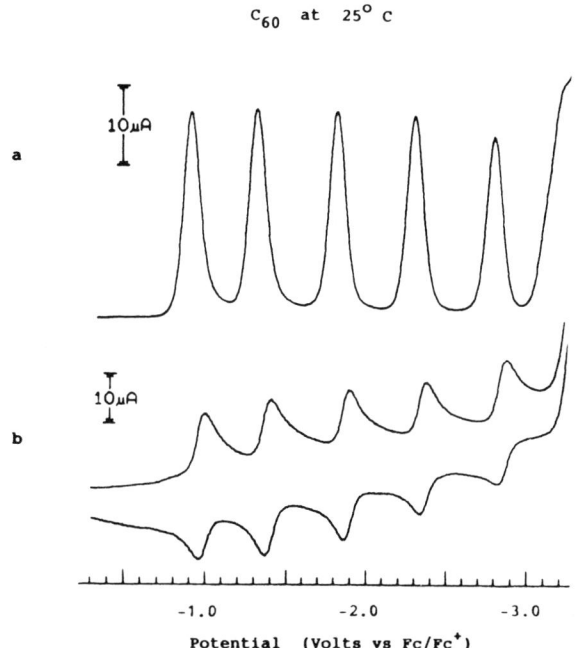

C$_{60}$ at 25° C

a

|10 μA|

b

|10 μA|

-1.0   -2.0   -3.0

Potential (Volts vs Fc/Fc+)

Figure 2. (a) Differential pulse voltammetry (80-mV pulse, 50-ms width, 200-ms pulse period, 10 mV/s scan rate) and (b) cyclic voltammetry at 100 mV/s of C$_{60}$ in CH$_3$CN/toluene at 25 °C. Note that the sixth reduction is almost evident in the DPV.

Parts a and b of Figure 1 show the cyclic and differential pulse voltammograms of C$_{60}$ at -10 °C, respectively. Chemical and electrochemical reversibilities are evident from this figure. The potentials measured, $E_{1/2}$, relative to Fc/Fc$^+$ were -0.98, -1.37, -1.87, -2.35, -2.85, and -3.26 V. In all cases successive potential values are closer than the corresponding ones measured by Kadish et al.[6] The significance of this observation must be related to the solvent composition.[4]

Parts c and d of Figure 1 show the -10 °C voltammograms, cyclic and differential pulse, for C$_{70}$. This C$_{70}$ sample contained some C$_{60}$ as an impurity, as judged by its mass spectrum. This is also evident from the voltammograms shown in Figure 1. The waves corresponding to C$_{60}$ are observed as small shoulders to the right of the C$_{70}$ waves. These diverge from the C$_{70}$ ones as the potential becomes more negative. What is important to note from the voltammograms in Figure 1c,d is the fact that six reversible reduction waves are observed for C$_{70}$ for the first time. The $E_{1/2}$ values for these six reduction processes, relative to Fc/Fc$^+$, are -0.97, -1.34, -1.78, -2.21, -2.70, and -3.07 V. There is only one wave which does not seem to correspond well to any of either C$_{60}$ or C$_{70}$, at -2.94 V. It is 90 mV more negative than the closest C$_{60}$ wave, which appears at -2.85 V. All other observed waves, including the -3.26-V wave, can be accounted for rather unequivocally by C$_{60}$ and C$_{70}$.

In conclusion, this work presents the first reversible generation of C$_{60}^{6-}$. More importantly, this was accomplished under the condition of very slow scan rates. It also presents the first observation of C$_{70}^{5-}$ and C$_{70}^{6-}$. All of these fulleride species appear to be stable in solution, especially at -10 °C, under high vacuum, and in the time scale of the voltammetric experiments. Even at room temperature, C$_{60}^{5-}$ appears to be stable, but spectroscopic confirmation is still necessary. Figure 2 shows the voltammograms of C$_{60}$ under conditions identical with those used for Figure 1, except that the temperature was 25 °C. It seems evident from Figure 2 that it will be possible to generate C$_{60}^{5-}$ at room temperature in order to study its properties in solution. The unusual stabilities and clean voltammetric data indicate that it will probably

(12) Echegoyen, L.; DeCian, A.; Fischer, J.; Lehn, J.-M. *Angew. Chem., Int. Ed. Engl.* **1991**, *30*, 838.

(13) Ajie, H.; Alvarez, M. M.; Anz, S. J.; Beck, R. D.; Diederich, F.; Fostiropoulos, K.; Huffman, D. R.; Krätschmer, W.; Rubin, Y.; Schriver, K. E.; Sensharma, D.; Whetten, R. L. *J. Phys. Chem.* **1990**, *94*, 8630.

be possible to use these electrosynthetic methods to generate fullerides for ion-pairing studies by ESR spectroscopy. Such work is currently underway.

Acknowledgment. We thank the Chemistry Division of the National Science Foundation (Grant No. CHE-9011901) for support of this work. We also thank Professor Fulin Zuo and Dr. Tingsheng Li of the Physics Department of the University of Miami for the carbon soot sample from which the C_{60} and C_{70} were extracted.

II

HIGHER FULLERENES

The first paper which coined the term Buckminsterfullerene discussed only C_{60}.[Kr85*] However, similar carbon cage structures exist for a large range of sizes, to which the generic name fullerenes is commonly applied. In this chapter, we discuss stable fullerenes larger than C_{60}.

The most abundant higher fullerene is C_{70}. As we have discussed in Chapter I, any fullerene has twelve pentagonal rings, and those structures in which the pentagons are isolated from one another are particularly stable. It may come as a surprise that geometry does not allow isolated pentagon structures between C_{60} and C_{70}, but this is the case.[Kr87, Ma91a*, Sa93] In molecular beam experiments, C_{70} was always seen to accompany C_{60}, even under conditions where other sizes were significantly less abundant.[Kr85*, He85] In soot produced by the Krätschmer-Huffman technique, extraction of fullerenes soluble in benzene or toluene yields 15–30% C_{70}.[Aj90*, Di92]

The hypothesized ([He85]) isolated-pentagon structure of C_{70} is constructed (conceptually) from that of C_{60} by slicing it across a five-fold axis, rotating one half by 36°, and adding a "waist" of five hexagons. This structure was confirmed by ^{13}C nuclear magnetic resonance (NMR) very soon after the soot-extraction technique was invented. Five lines were observed, with intensities in the ratio 1:2:1:2:1, corresponding to the five different carbon atom sites, containing 10, 20, 10, 20, and 10 atoms, respectively.[Ta90*, Aj90*] A stronger confirmation of this molecular structure, with direct evidence for the assignment of chemical shifts for the specific sites, was made by an advanced NMR technique which is sensitive to the coherent excitation of two adjoining atoms.[Jo91a]

With the topology and the symmetry of the C_{70} molecule established, we can seek a greater level of detail. There are eight bond lengths and eleven angles; only twelve of these parameters are independent within the D_{5h} molecular symmetry. (Molecular symmetry groups such as D_{5h} are defined in any standard chemistry textbook.) Two Hartree-Fock calculations of the molecular structure are in substantial agreement.[Ba91a, Sc91]

The first experimental determination of the C_{70} structure was based on electron diffuse scattering from a thin film of C_{70}.[Mc92] This experiment showed significant disagreement with the predictions cited above, claiming considerable pinching in at the waist.

Crystallographic determinations of the atomic structure of C_{70} have generally been thwarted by the large degree of orientational disorder and polyphase character observed. However, there are two notable exceptions. There is an organometallic derivative, in which an iridium complex is chemically attached to the C_{70} molecule, stabilizing it within the crystal.[Ba91b] A published crystal structure of $C_{70}S_{48}$ is even more remarkable for the absence of orientational disorder, because there is no chemical bonding between the fullerene molecule and S_8 rings which comprise the rest of the structure.[Ro92] The C_{70} geometry in these two crystals is much closer to the Hartree-Fock calculations than to the electron scattering determination.

The first Hückel molecular orbital theory for C_{70} showed lower degeneracies of both the HOMO and the LUMO than for C_{60}.[Fo86] According to this calculation, the C_{70} HOMO contains two electrons, *vs.* ten for C_{60}, and the six-fold degenerate LUMO of C_{60} splits into a two-fold LUMO and a four-fold 2nd LUMO. This is easily understood: the lower symmetry of the molecule breaks orbital degeneracy. More sophisticated calculations actually interchange the two lowest unoccupied molecular orbitals.[Ba91a, Sc91] Experimentally, it is observed in photoemission that the valence band (derived from the HOMO) edge is split and broadened, relative to C_{60}.[Jo91b, We92]

There is some disagreement among the electrochemical experiments probing the LUMO. The paper reprinted in this volume indicates six reduction steps, and an unexplained seventh step.[Xi92*]

The splitting of the LUMO from the 2nd LUMO would have to be very small to be reconciled with six reduction steps, nearly equally spaced. There is disagreement with some of the earlier papers (reporting fewer reduction steps), cited in that reference.[Xi92*]

Solid C_{70} at room temperature has dynamical orientational disorder, revealed by NMR spectroscopy.[Ty91*] Vaughan et al. found that, at high temperatures, solid C_{70} was a mixture of two close packed lattices, face-centered-cubic (fcc) and hexagonal-close-packed (hcp).[Va91] The nearest neighbor distances in the two observed lattices were equal, 10.61 ± 0.01 Å, but larger than the 10.02 Å of C_{60}, implying that the elongated C_{70} molecule was highly (although not necessarily isotropically) disordered at 440 K. Subsequent x-ray diffraction experiments on vapor grown single crystals have confirmed the existence of separate fcc and hcp high temperature crystals.[Ve92, Va93]

In view of the elongated shape of the C_{70} molecule, there is a variety of possible phases, differing in their molecular packing and the degree of orientational order. Indeed, the first experiments observed thermodynamic signatures of phase transitions at 276 K and 337 K.[Va91] Single crystal x-ray experiments have shown several intermediate phases, [Ve92, Gr92, Va93] but their structural identification is not yet complete.

Proceeding to even higher masses, from 5% to 30% of the soluble soot extract consists of fullerenes heavier than C_{70}, depending on the solvent used. There is an excellent early review of the isolation and characterization of fullerenes beyond C_{70} by Diederich and Whetten.[Di92] Considerable development was required to discover combinations of chromatography column and eluent capable of making a clean separation of the higher fullerenes.[Di91a] Their structures have been determined by comparing their NMR spectra with the results of theoretical searches for possible fullerene structures. As we shall see, this enterprise becomes more complicated with increasing fullerene size, because the number of possible isomers grows very rapidly.

The first fullerene to be so characterized was C_{76}.[Et91*] The mass identification of that fraction was made by laser desorption mass spectrometry. Its NMR spectrum showed 19 lines of nearly equal intensity, implying that there were 19 distinct atomic sites, and therefore a point group

with exactly four (76/19) elements. In the meantime, Manolopoulos had continued the enumeration of fullerene structures beyond C_{70} [Ma91a*] with a detailed study of the possible isomers of C_{76}.[Ma91b] The total number of C_{76} isomers is daunting, but we have seen that there is good justification for limiting attention to only those isomers with isolated pentagons. In that case, there are only two candidates, having D_2 and T_d symmetries. The latter case is incompatible with 19 distinct sites, and so one can be certain of the structural assignment, based only on counting the NMR lines! A remarkable feature of this structure is that it is chiral; distinct left- and right-handed stereo-isomers exist. Recently, the chiral enantiomers have been separated, to a purity of 97%, by an asymmetric oxmilation reaction which apparently attacks a specific bond.[Ha93]

The stability of the D_2 over the T_d isomer is easily understood, because the latter has an open electronic shell in elementary Hückel theory. This degeneracy of the HOMO and LUMO is expected to greatly destabilize the candidate open-shell isomer.

C_{76} is also the first fullerene capable of donating electrons as well as accepting them.[Li92*] As discussed in Chapter I, electrochemical experiments on C_{60} and C_{70} showed only reversible reduction (C_n^-) steps, not oxidation (C_n^+).[Xi92*] Both oxidation and reduction waves were observed from C_{76} in solution. This was partly anticipated by the relatively large energy of the highest occupied molecular orbital (HOMO) predicted from semi-empirical molecular orbital calculations.[Et91*] Consequently, the chemistry of C_{76} is expected to differ significantly from C_{60} and C_{70}, including the possibility of novel reactions through cationic fullerene intermediates.

Continuing to higher isomers, the story becomes more complex. High performance chromatographic separation yielded two distinct C_{78} fractions, in a 5:1 ratio.[Di91b*] Theoretically, there are five isolated-pentagon isomers of C_{78}.[Fo91] As with C_{76}, it was possible to assign structures to the two fractions based only on counting the NMR lines. The majority fraction was found to have a C_{2v} symmetry with 21 distinct sites, and the minority component was D_3, with 13 distinct sites. This latter structure is chiral.

In another laboratory, samples of C_{78} were found to contain a third isomer.[Ki92*] Based on the assignments of NMR lines from the first C_{78} study, it could be determined that the new isomer had a different C'_{2v} structure, with 22 sites. Furthermore, this new isomer was the most abundant in their sample. That group reported abundances of $C_{2v}:D_3:C'_{2v}$ in the ratio 2:2:5, compared with the earlier result 5:1:0. Such widely differing abundances from different preparations seem surprising, but it may be that the use of different chromatographic eluents (hexane-toluene by UCLA and carbon disulfide by Tokyo Metropolitan University) led to the different observed samples. Another group has also observed NMR lines which it tentatively assigned to the same C'_{2v} isomer.[Ta92]

Another surprise offered by C_{78} was that theory had predicted that a fourth isomer, with D_{3h} symmetry, should be the most stable, based on having the largest HOMO-LUMO gap.[Fo91] Indeed, of the five isolated pentagon isomers of C_{78}, only the unobserved D_{3h} has a truly closed-shell electronic structure (in molecular orbital theory) with an antibonding LUMO. The observation of fullerenes with pseudo-closed shell structures (bonding LUMO, as distinct from closed shell structures, which have a nonbonding or antibonding LUMO) indicates that the difference between closed and pseudo-closed shell electronic structure and the magnitude of the Hückel theory gap are not the only determinants of stability. It may be that steric strain is comparably important. A kinetic explanation has also been advanced,[Di91b*] but the full picture is by no means clear.

The next observed fullerene is C_{82}. It is again interesting to note that of two different laboratories, using essentially similar techniques, one has observed this molecule,[Ki92*] while the other has not.[Di91a, Di92] Again, the most obvious difference is in the chromatographic eluent. There are nine possible isolated-pentagon isomers predicted for C_{82},[Ma91c, Ma92b] and between three and six of them are observed.[Ki92*] The dominant fraction is one of three possible isomers of C_2 symmetry, but the assignment of the remaining lines is more ambiguous. From the standpoint of molecular orbital theory, none of the isomers of C_{82} has a properly closed shell (nonbonding or antibonding LUMO), although eight have pseudo-closed shells.[Ma91c]

According to several accounts, C_{84} is the most abundant fullerene after C_{60} and C_{70}. Its NMR spectrum indicates only two different isomers, in a 2:1 ratio. This seems all the more surprising in view of the (theoretical) existence of 24 isolated-pentagon isomers.[Ma92a] The less abundant is uniquely determined as D_{2d}; the more abundant could be any one of four with symmetry D_2.[Ki92*] Semi-empirical electronic structure calculations show that there are two isomers with significantly lower energies than all others: one particular D_2 and the D_{2d}.[Zh92] More sophisticated semi-empirical [Ra92] and local density functional [Wa92] calculations confirm the stability of those two isomers. Indeed, it has been pointed out that if the two isomers have identical energy and if they are allowed to equilibrate, they would be present in the ratio of 2:1, due to the two enantiomers of the D_2 form.[Ma92c]

The electrochemistry of C_{84} has been investigated. Five reversible reductions but no oxidations were observed.[Me93] One must note that this experiment was performed on a mixture of the two isomers, which may have slightly distorted the observed electrochemical potentials.

It is worthwhile to consider briefly fullerenes that have not been observed. Based on the information available, the factors that are relevant to fullerene stability are: (1) isolated pentagon structure; (2) the closed or pseudo-closed lowest unoccupied molecular orbital; and (3) the steric strain of the molecule. There are isolated pentagon isomers for all even fullerenes above C_{70}, but there are no published reports of the isolation of C_{72}, C_{74}, or C_{80}. The single isolated pentagon isomer of C_{72} is known to have a closed electronic shell, based on the "leapfrog" transformation.[Ma92a] However, it is highly strained.[Fo93a, Fo93b] The one isolated-pentagon C_{74} and six of the seven isolated-pentagon isomers of C_{80} all have pseudo-closed shells.[Ma91c] This is sufficient to stabilize the observed C_{78}, so we are left with no simple Hückel-theory explanation for the absence of these two fullerenes.

Pressing on to still larger fullerenes, there have been mass spectra showing fullerenes well beyond C_{200} in laser desorption cluster sources,[Ma90] and extracted from Krätchmer-Huffman soot.[Pa91, Sm92, Di92, Pa92] It is interesting to speculate on the shape of these giant fullerenes. Roughly spherical fullerenes would be expected, extrapolating from the observed fullerenes C_{60} to C_{84}. On the other hand, the observation of carbon nanotubes and their relation to fullerenes (discussed further in Chapter VII) suggests that capped tubes should be considered as a candidate for the most stable state of large, closed carbon clusters. One must also consider the role of growth kinetics, inasmuch as it is clear that the observed fullerene distributions are not in thermodynamic equilibrium.

While it has not been possible to determine the structure of any giant fullerene molecules to the atomic level, it is known from scanning tunneling microscopy that there are fullerenes as large as 20 Å in diameter (approximately C_{330}). Interestingly, all of the solvent-extracted fullerenes reported in this STM study are essentially spherical.[La92]

There have been several theoretical comparisons of the binding energy of spherical vs. capsule-shaped fullerenes. In general, such calculations start with an assumed configuration of atoms, and then relax their positions to minimize the energy. We mention two here: a quantum monte-carlo pseudopotential calculation of balls, tubes and capsules starting from assumed configurations of up to 240 atoms,[Ad92] and a realistic atomic potential starting with as many as 10^4 atoms, distributed on a spherical shell.[Ma93] In both calculations, the spherical fullerenes are found to be significantly more stable than other configurations.

Another possible structure for the giant fullerenes is a multi-shelled fullerene "onion." Ugarte has created such structures by irradiating graphitic or amorphous carbon in an electron microscope.[Ug92] The concentric shells appear to be nearly spherical, and the distance between successive shells is close to that of graphite, 3.34 Å. Calculations show that a multi-shelled onion is more stable than a single, roughly spherical fullerene above a critical size of about 6000 atoms.[Ma93] This is significantly larger than any fullerenes observed in the gas phase, but well below the size of the onions imaged by Ugarte.

Similar structures are observed by Dravid et al. in the remains of a carbon arc discharge.[Dr93] This indicates that the special conditions of electron-beam heating are not necessary to produce multi-shelled onions. Dravid et al. also point out that some of these onions contain unterminated graphene sheets.

In conclusion, it is clear that there are many interesting phenomena and problems to be understood in the fullerenes beyond C_{60}. In particular, as the size of the fullerene grows larger, one's viewpoint changes from that of a molecule to that of a small bulk system. We can certainly expect further developments on the formation, stability, and other properties of large and giant fullerenes.

Bibliography

References marked with * are reprinted in the present volume.

Aj90* H. Ajie et al., "Characterization of the Soluble All-Carbon Molecules C_{60} and C_{70}," J. Phys. Chem. **94**, 8630–8633 (1990).

Ad92 G. B. Adams, O. F. Sankey, J. B. Page, M. O'Keefe and D. A. Drabold, "Energetics of Large Fullerenes: Balls, Tubes and Capsules," Science **256**, 1792–1795 (1992).

Ba91a J. Baker, P. W. Fowler, P. Lazzeretti, M. Malagoli and R. Zanasi, "Structure and Properties of C_{70}," Chem. Phys. Lett. **184**, 182–186 (1991).

Ba91b A. L. Balch, V. J. Catalano, J. W. Lee, M. M. Olmstead, and S. R. Parkin, "(η^2-C_{70})Ir(CO)Cl(PPh$_3$)$_2$: The Synthesis and Structure of an Organometallic Derivative of a Higher Fullerene," J. Am. Chem. Soc. **113**, 8953–8955 (1991).

Di91a F. Diederich et al., "The Higher Fullerenes: Isolation and Characterization of C_{76}, C_{84}, C_{90}, C_{94}, and $C_{70}O$, an Oxide of D_{5h}-D_{70}," Science **252**, 548–551 (1991).

Di91b* F. Diederich et al., "Fullerene Isomerism: Isolation of C_{2v}-C_{78} and D_3-C_{78}," Science **254**, 1768–1770 (1991).

Di92 F. Diederich and R. L. Whetten, "Beyond C_{60}: The Higher Fullerenes," Acc. Chem. Res. **25**, 119–126 (1992).

Dr93 V. P. Dravid et al., "Buckytubes and Derivatives: Their Growth and Implications for Buckyball Formation," Science **259**, 1601–1604 (1993).

Et91* R. Ettl, I. Chao, F. Diederich and R. L. Whetten, "Isolation of C_{76}, a Chiral (D_2) Allotrope of Carbon," Nature **353**, 149–153 (1991).

Fo86 P. W. Fowler and J. Woolrich, "π-Systems in Three Dimensions," Chem. Phys. Lett. **127**, 78–83 (1986).

Fo91 P. W. Fowler, R. C. Batten and D. E. Manolopoulos, "The Higher Fullerenes: A Candidate for the Structure of C_{78}," J. Chem. Soc. Faraday Trans. **87**, 3103–3104 (1991).

Fo93a P. W. Fowler, personal communication to P. W. Stephens (1993).

Fo93b P. W. Fowler and D. E. Manolopoulos, An Atlas of Fullerenes (Oxford University Press, to be published, 1993).

Gr92 M. A. Green, M. Kurmoo, P. Day and K. Kikuchi, "Structural Phase Transformations in C_{70}," J. Chem. Soc., Chem. Commun. 1676–1677 (1992).

Ha93 J. M. Hawkins and A. Meyer, "Optically Active Carbon: Kinetic Resolution of C_{76} by Asymmetric Osmylation," *Science* **260**, 1918–1920 (1993).

He85 J. R. Heath, *et al.*, "Lanthanum Complexes of Spheroidal Carbon Shells," *J. Am. Chem. Soc.* **107**, 7779–7780 (1985).

Jo91a R. D. Johnson, G. Meijer, J. R. Salem and D. S. Bethune, "2D Nuclear Magnetic Resonance Study of the Structure of the Fullerene C_{70}," *J. Am. Chem. Soc.* **113**, 3619–3621 (1991).

Jo91b M. B. Jost *et al.*, "Occupied and Unoccupied Electronic States of solid C_{70} with Comparison to C_{60}," *Chem. Phys. Lett.* **184**, 423–427 (1991).

Ki92* K. Kikuchi *et al.*, "NMR Characterization of Isomers of C_{78}, C_{82}, and C_{84} Fullerenes," *Nature* **357**, 142–145 (1992).

Kr85* H. W. Kroto, J. R. Heath, S. C. O'Brien, R. F. Curl and R. E. Smalley, "C_{60}: Buckminsterfullerene," *Nature* **318**, 162–163 (1985).

Kr87 H. W. Kroto, "The Stability of the Fullerenes C_n, with $n = 24, 28, 32, 36, 50, 60$, and 70," *Nature* **329**, 529–531 (1987).

La92 L. D. Lamb *et al.*, "Extraction and STM Imaging of Spherical Giant Fullerenes," *Science* **255**, 1413–1416 (1992).

Li92* Q. Li, F. Wudl, C. Thilgen, R. L. Whetten and F. Diederich, "Unusual Electrochemical Properties of the Higher Fullerene, Chiral C_{76}," *J. Am. Chem. Soc.* **114**, 3994–3996 (1992).

Ma90 S. Maruyama, L. R. Anderson and R. E. Smalley, "Direct Injection Supersonic Cluster Beam Source for FT-ICR Studies of Clusters," *Rev. Sci. Instrum.* **61**, 3686–3693 (1990).

Ma91a* D. E. Manolopoulos, J. C. May and S. E. Down, "Theoretical Studies of the Fullerenes: C_{34} to C_{70}," *Chem. Phys. Lett.* **181**, 105–111 (1991).

Ma91b D. E. Manolopoulos, "Proposal of a Chiral Structure for the Fullerene C_{76}," *J. Chem. Soc. Faraday Trans.* **87**, 2861–2862 (1991).

Ma91c D. E. Manolopoulos and P. W. Fowler, "Structural Proposals for Endohedral Metal-Fullerene Complexes," *Chem. Phys. Lett.* **187**, 1–7 (1991).

Ma92a D. E. Manolopoulos and P. W. Fowler, "Molecular Graphs, Point Groups, and Fullerenes," *J. Chem. Phys.* **96**, 7603–7614 (1992).

Ma92b D. E. Manolopoulos, P. W. Fowler and R. P. Ryan, "Hypothetical Isomerisations of LaC_{82}," *J. Chem. Soc. Faraday Trans.* **88**, 1225–1226 (1992).

Ma92c D. E. Manolopoulos, P. W. Fowler, R. Taylor, H. W. Kroto and D. R. M. Walton, "An End to the Search for the Ground State of C_{84}?" *J. Chem. Soc. Faraday Trans.* **88**, 3117–3118 (1992).

Ma93 A. Maiti, C. J. Barbec and J. Bernholc, "Structure and Energetics of Single and Multilayer Fullerene Cages," *Phys. Rev. Lett.* **70**, 3023–3026 (1993).

Mc92 D. R. McKenzie, C. A. Davis, D. J. H. Cockayne, D. A. Muller and A. M. Vassallo, "The Structure of the C_{70} Molecule," *Nature* **355**, 622–624 (1992).

Me93 M. S. Meier, T. F. Guarr, J. P. Selegue and V. K. Vance, "Elevated Temperature Gel Permeation Chromatography and Electrochemical Behaviour of the C_{84} Fullerene," *J. Chem. Soc., Chem. Commun.*, 63–65 (1993).

Pa91 D. H. Parker *et al.*, "High-Yield Synthesis, Separation, and Mass-Spectrometric Characterization of Fullerenes C_{60} to C_{266}," *J. Am. Chem. Soc.* **113**, 7499–7503 (1991).

Pa92 D. H. Parker *et al.*, "Fullerenes and Giant Fullerenes: Synthesis, Separation, and Mass Spectrometric Characterization," *Carbon* **30**, 1167–1182 (1992).

Ra92 K. Raghavachari, "Ground State of C_{84}: Two Almost Isoenergetic Isomers," *Chem. Phys. Lett.* **190**, 397–400 (1992).

Ro92 G. Roth and P. Adelmann, "The Crystal Structure of $C_{70}S_{48}$: The First *a-priori* Structure Determination of a C_{70}-Containing Compound," *J. de Physique I (Paris)*, **2**, 1541–1548 (1992).

Sa93 C. H. Sah, "Combinatorial Construction of Fullerene Structures," *Croat. Chem. Acta* (in press).

Sc91 G. E. Scuseria, "The Equilibrium Structure of C_{70}: An *Ab-Initio* Hartree-Fock Study," *Chem. Phys. Lett.* **180**, 451–456 (1991).

Sm92 C. Smart *et al.*, "Extraction of Giant Fullerene Molecules, and their Subsequent Solvation in Low Boiling Point Solvents," *Chem. Phys. Lett.* **188**, 171–176 (1992).

Ta90* R. Taylor, J. P. Hare, A. K. Abdul-Sada and H. W. Kroto, "Isolation, Separation and Characterisation of the Fullerenes C_{60} and C_{70}: The Third Form of Carbon," *J. Chem. Soc., Chem. Commun.* (1990) 1423–1425.

Ta92 R. Taylor, G. J. Langley, T. J. S. Dennis,

H. W. Kroto and D. R. M. Walton, "A Mass Spectrometric-NMR Study of Fullerene-78 Isomers," *J. Chem. Soc., Chem. Commun.* 1043–1046 (1992).

Ty91*   R. Tycko, *et al.*, "Solid State Magnetic Resonance Spectroscopy of Fullerenes," *J. Phys. Chem.* **95**, 518–520 (1991).

Ug92   D. Ugarte, "Curling and Closure of Graphitic Networks Under Electron-Beam Irradiation," *Nature* **359**, 707–709 (1992).

Va91   G. B. M. Vaughan *et al.*, "Orientational Disorder in Solvent-Free Solid C_{70}," *Science* **254**, 1350–1353 (1991).

Va93   G. van Tendeloo *et al.*, "Structural Phase Transitions in C_{70}," *Europhys. Lett.* **21**, 329–334 (1993).

Ve92   M. A. Verheijen *et al.*, "The Structure of Different Phases of Pure C_{70} Crystals," *Chemical Physics* **166**, 287–297 (1992).

Wa92   X. Q. Wang, C. Z. Wang, B. L. Zhang and K. M. Ho, "Structural and Electronic Properties of C_{84}: A First-Principles Study," *Phys. Rev. Lett.* **69**, 69–72 (1992).

We92   J. H. Weaver, "Electronic Structures of C_{60}, C_{70}, and the Fullerides: Photoemission and Inverse Photoemission Studies," *J. Phys. Chem. Solids* **53**, 1433–1447 (1992).

Xi92*   Q. Xie, E. Pérez-Cordero and L. Echegoyen, "Electrochemical Detection of C_{60}^{6-} and C_{70}^{6-}: Enhanced Stability of Fullerides in Solution," *J. Am. Chem. Soc.* **114**, 3978–3980 (1992).

Zh92   B. L. Zhang, C. Z. Wang and K. M. Ho, "Search for the Ground-State Structure of C_{84}," *J. Chem. Phys.* **96**, 7183–7185 (1992).

Reprinted with permission from Nature
Vol. 353, pp. 149–153, 12 September 1991
© 1991 Macmillan Magazines Limited

Isolation of C_{76}, a chiral (D_2) allotrope of carbon

Roland Ettl, Ito Chao, François Diederich*
& Robert L. Whetten*

Department of Chemistry and Biochemistry, University of California,
Los Angeles, California 90024-1569, USA

A LONG search for molecular allotropic forms of carbon[1-4] culminated in the discovery[5-7] of a method for preparing large quantities of the C_{60} molecule and the subsequent confirmation[8,9] of its cage-like truncated-icosahedral structure[1]. The C_{70} molecule prepared by the same method was later also isolated and found to have the predicted cylindrical (D_{5h}) structure. Incomplete chromatographic separation of the large molecules C_{76}, C_{78} and C_{84} was achieved at the same time[10,11], and small quantities of highly enriched samples were later isolated[12]. Preliminary NMR and spectroscopic studies of these carbon clusters failed, however, to provide evidence for the symmetrical structures predicted earlier[13]. Here we report the isolation and characterization of the C_{76} molecule, a third molecular form of carbon following C_{60} and C_{70}. We isolated and purified substantial quantities of this species using the extraction technique of ref. 6 together with chromatography. The ^{13}C NMR spectrum consists of 19 lines of essentially equal intensity, confirming a compact, cage-like fullerene structure. Among the several thousand possible structures for C_{76}, theoretical calculations by Manolopolous[14] predict a chiral structure with D_2 symmetry, consisting of a spiralling, double-helical arrangement of edge-sharing pentagons and hexagons. Our NMR spectrum is uniquely consistent with this remarkable structure.

We isolated C_{76} following methods previously described, in a three-step procedure. (1) The crude toluene-soluble material is separated chromatographically into the fractions C_{60}, C_{70} (with residual C_{60}), and higher fullerenes (with residual C_{70}). (2) After 100 mg of higher fullerenes have been collected, they are separated on an alumina column with hexane/toluene elution, to yield three fractions: (I), C_{70}, (II), C_{76} and C_{78} (with a little C_{70} and C_{84}); and (III), the major component C_{84} (with C_{76}, C_{78} and $C_{70}O$), thus isolating 30 mg C_{76} containing C_{78} as

* To whom correspondence should be addressed.

FIG. 1 Characteristics of molecular C_{76}. *a*, HPLC traces of molecular carbon samples (1) after column-chromatography, with C_{76}, C_{78} and C_{84} peaks (in order of increasing retention time) prominent, and (2) after preparative HPLC separation, with only the C_{76} peak remaining (see text for details). *b*, A mass spectrum of C_{76} after the final stage of purification. The plot is a positive-ion laser-desorption (4.0 eV; 0.1 mJ cm^{-2}) time-of-flight mass spectrum which shows, in addition to the dominant C_{76} peak and its normal mass-spectrometric fragment C_{74}, minor impurity peaks C_{78} and C_{84} at the 1–2% level, and a small C_{60} peak remaining as contaminant from an earlier calibration run. Deep-ultraviolet photoionization of laser-desorbed neutrals yields a similar spectrum. *c*, Optical spectrum of C_{76} (in liquid dichloromethane solvent at ambient temperature) across the ultraviolet (UV), visible (VIS) and near-infrared (NIR) spectral range. The inset shows an amplified view of the spectrum in the green to near-infrared region, emphasizing the absorption onset near 1.4 eV. Absolute extinction coefficients are as follows (sh stands for shoulder):

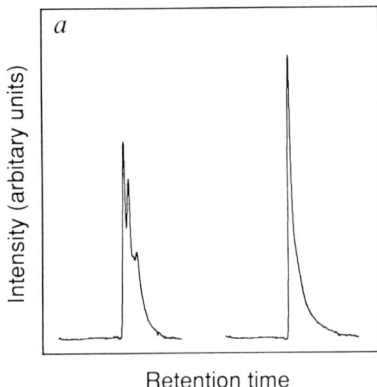

Retention time

A	760 nm	[1.63 eV]	2,620 l mol^{-1} cm^{-1}
B	715	[1.73]	3,080
C	572	[2.17]	2,600
D	524 (sh)	[2.37]	3,800
E	452 (sh)	[2.74]	10,840
F	408	[3.04]	16,280
G	356 (sh)	[3.48]	21,680
H	331	[3.75]	28,150
I	290	[4.28]	54,880
J	242	[5.12]	107,850

d, Infrared absorption spectrum of C_{76} in CS_2 solution recorded in a NaCl cell. Bands below 500 cm^{-1} and above 1,450 cm^{-1} are obscured by absorption by the window and solvent, respectively. The strongest band (near 700 cm^{-1}) is broadened and truncated by saturation.

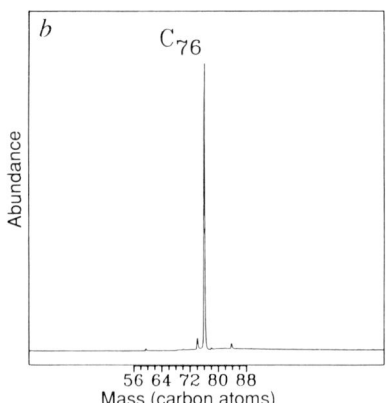

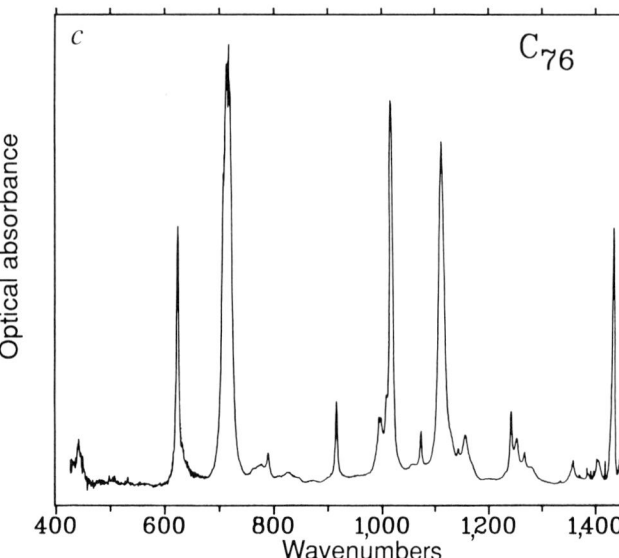

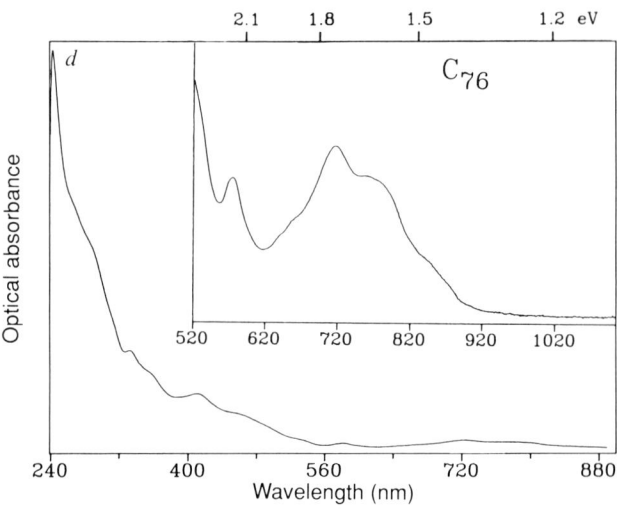

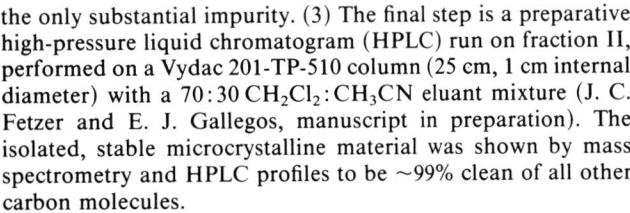

the only substantial impurity. (3) The final step is a preparative high-pressure liquid chromatogram (HPLC) run on fraction II, performed on a Vydac 201-TP-510 column (25 cm, 1 cm internal diameter) with a 70:30 CH_2Cl_2:CH_3CN eluant mixture (J. C. Fetzer and E. J. Gallegos, manuscript in preparation). The isolated, stable microcrystalline material was shown by mass spectrometry and HPLC profiles to be ~99% clean of all other carbon molecules.

Routine characterization is demonstrated in Fig. 1, which shows HPLC traces of the mixture and the pure material, together with a mass spectrum of the latter and the optical absorption spectra of C_{76} in the infrared region (400–1,500 cm^{-1})

and the near-infrared/visible/ultraviolet region. An unusual feature is the near-infrared absorption, which has an onset near 890 nm (1.39 eV) and very large peak intensities (ε_{max} near 3,000 l mol^{-1} cm^{-1}) considering the cubic dependence of absorption on frequency. Both the solution phase and the crystalline form of the material are yellow-green when optically thin.

We have obtained structural information from ^{13}C NMR spectra (Fig. 2), taken in CS_2 solvent, in which C_{76} has a solubility greater than 1 mg ml^{-1} at room temperature. In contrast to our previous report, we obtain an interpretable spectrum after only 1,000 sweeps (4 s sweep time with Cr(acac)$_3$ as relaxant) because of the high solubility, and the spectrum shown was obtained after 38,000 scans. The spectrum consists of a series of 19 distinct lines of near-equal intensity, as confirmed by repeated scans, spanning the chemical shift region 130–150 p.p.m. Table 1 lists the line positions and their integrated intensities. An expanded view of the spectrum shows no other resonances in the ^{13}C NMR range, confirming high sample purity. The lines neatly span the same region as the five-line spectrum of C_{70} (refs 8, 10), and the centres of gravity of the three molecules are close: 143.2, 145.0 and 142.7 for C_{60}, C_{70} and C_{76} respectively. These facts point strongly to a fullerene structure.

The 19 lines of equal intensity evidently correspond to 19 distinct environments occupied each by four symmetry-equivalent carbon atoms. This strongly restricts the choice of structure. It implies a point-group consisting of four, and only four, elements (C_4, S_4, D_2, C_{2v}, and C_{2h}), with none of the atoms lying on a symmetry element. Fullerene structures, consisting only of pentagonal and hexagonal rings, are inconsistent with C_4, and also with C_{2v} when the atoms are excluded from the symmetry axis and mirror planes. Structures of C_{2h} symmetry require a belt of $4m$ hexagons around the mirror plane, and no plausible structure of this type exists. Here we use the strict definition of 'fullerene': a closed, hollow network of 12 pentagonal and m hexagonal rings for a C_{20+2m} molecule. Extensive experimentation with models has led to one plausible structure, illustrated in Fig. 3, which we believe to be unique. In particular, this structure has four groups of 'pyrene' sites—intersection of three six-membered rings—accounting for the four lines at chemical shifts well below 140 p.p.m. in perfect analogy to the description[15] of C_{70}. We now describe the structure in greater detail.

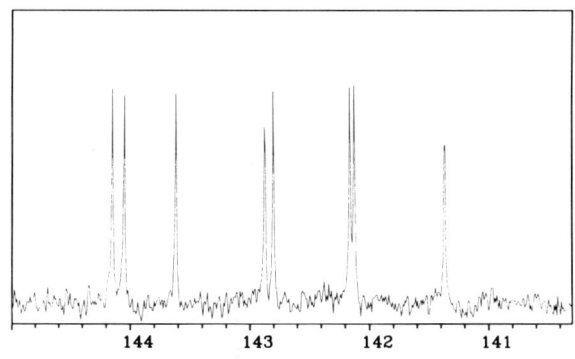

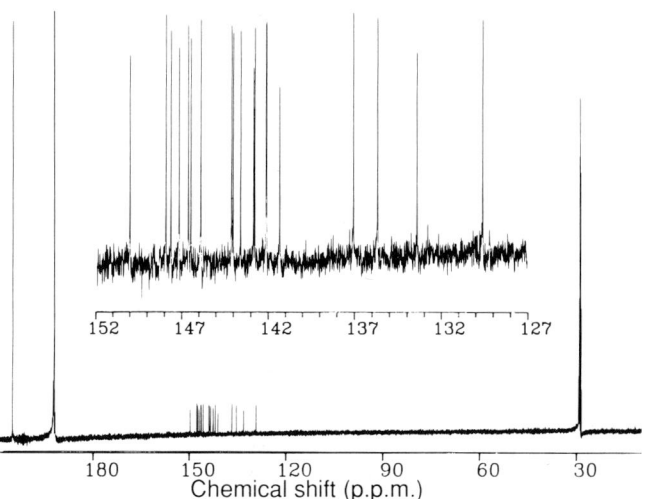

FIG. 2 ^{13}C-NMR spectra of 1.5 mg C_{76} in CS_2, taken at 125.6 MHz, with acetone as an internal lock. (The acetone peaks are near 206 and 40 p.p.m., and the solvent peak is near 190 p.p.m.) The inset is an expanded view of the fullerene region. Closely spaced lines in the region 145.0–140.3 p.p.m. are expanded in the box above.

The molecule D_2-C_{76} has a fullerene structure; that is, it consists of 28 six-membered rings (6MR) and 12 five-membered rings (5MR), and it also obeys the pentagon isolation rule[1] by maximizing the separation of pentagons. Its relation to the C_{60} structure (in which each 5MR is surrounded by six 6MRs, and each 6MR is surrounded by alternating 5MRs and 6MRs) is most easily visualized through the following construction, illustrated in Fig. 3b. Take as each of two polar caps the pyracylene unit (the base unit of C_{60}), and surround them with alternating 5MRs and 6MRs to obtain the bold region of structure I. These last 5MRs in turn each require two additional hexagons to satisfy the pentagon-isolation rule, completing I. The resulting structures act as caps in C_{76}, and are each rather large fragments of C_{60} (ten 6MRs and four 5MRs). The crucial step (structure II) is the assembly of these caps at the edges of the last-mentioned 6MRs (9–12 and their equivalents); this requires that they be staggered by 20° about the long symmetry axis (a). The residual space around the perimeter is composed of two 6MRs and two 5MRs on each side.

A second, holistic view is gained by examining the connectivity of the rings in the assembled structure. As shown in Fig. 3d, it can be decomposed into a double-helix structure of edge-sharing pentagons and hexagons. The two strands are tied at the ends of the long axis by edge-sharing hexagons on the pyracylene bridges mentioned above. This view emphasizes the intrinsic helicity of the structure, which has no chiral centre.

A local, chemical view of the structure is obtained by considering the atomic environments (see Fig. 3b and c) of which there are three kinds. The most distinctive ('pyrene'-like) are the carbon atoms lying at the intersection of three hexagons, like those in the centre of a pyrene molecule. On each of the flattest faces (looking down the c-axis) there is a chain of six such sites snaking about a helix and crossing the axis, as shown by the shaded sites in Fig. 3c, structure C. On each of the most curved faces (looking down the b axis; Fig. 3c, structure B), there is a bonded pair of such atoms crossing the axis. These combine to give four groups of four atoms, and are assigned to the upfield (most shielded) peaks in the ^{13}C NMR spectrum. The second group encompasses the 'pyracylene' sites, labelled a to i in Fig. 3b, which are neatly grouped on the caps of the structure, each of which has six 5MRs and giving rise to nine pyracylene bridges with two sites each. All sites in C_{60} are like this, and the 40 sites in C_{70} having the least shielding are also like this. By analogy,

TABLE 1 Line positions in the ^{13}C NMR spectrum of C_{76} compared with C_{60} and C_{70}.

C_{76} δ (p.p.m.)	intensity	Type	C_{60}	C_{70}
				150.8 (10) [pc]
150.03	(3.6)	[pc]*		
				148.3 (20) ["]
147.96	(4.3)	["]		
				147.8 (10) ["]
147.66	(4.1)	["]		
147.19	(3.8)	["]		
146.65	(4.1)	["]		
146.50	(3.8)	["]		
145.92	(4.0)	["]		
				144.4 (20) [cor]
144.14	(4.0)	["]		
144.05	(4.1)	["]		
143.61	(4.0)	[cor]†		
			143.2 (60) [pc]	
142.86	(3.6)	["]		
142.79	(4.1)	["]		
142.15	(4.1)	["]		
142.11	(3.5)	["]		
141.35	(4.0)	["]		
137.06	(4.4)	[pyrene]		
135.67	(4.4)	["]		
133.40	(3.8)	["]		
				130.8 (10) [pyrene]
129.56	(4.2)	["]		

* Pyracylene.

† Corrannulene.

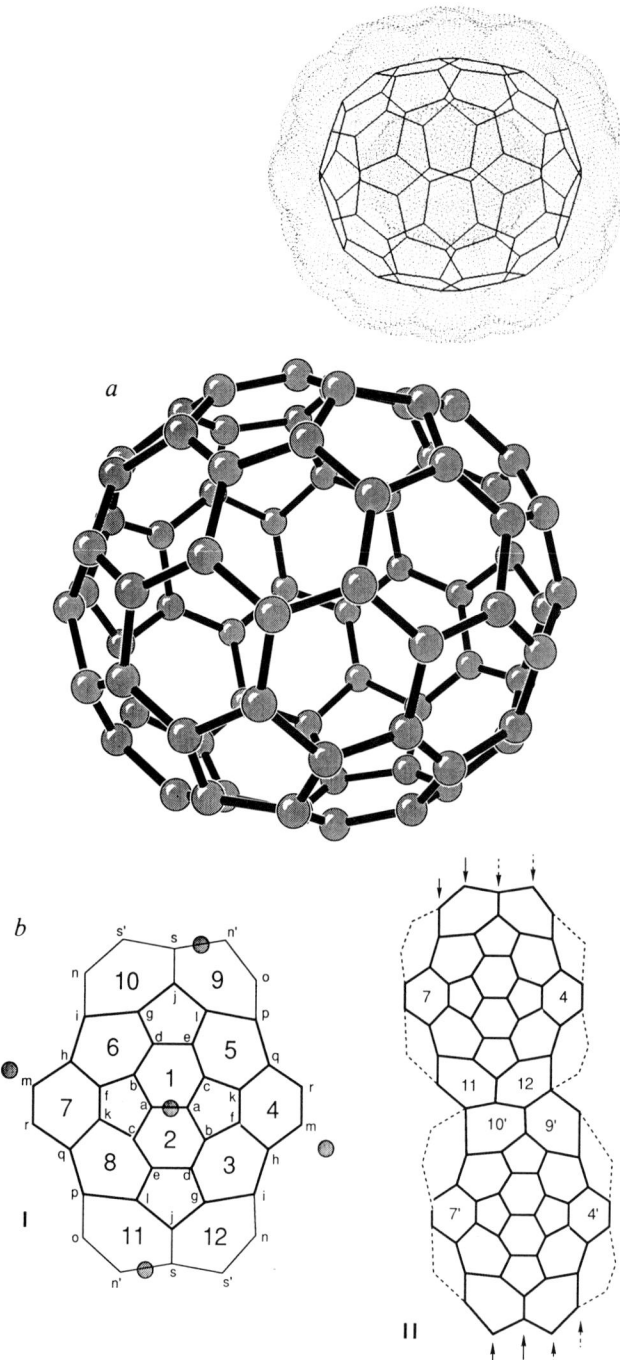

a

b

I

II

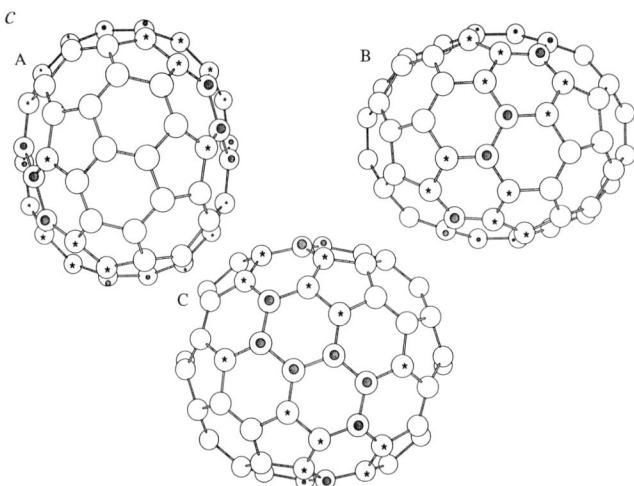

c

A

B

C

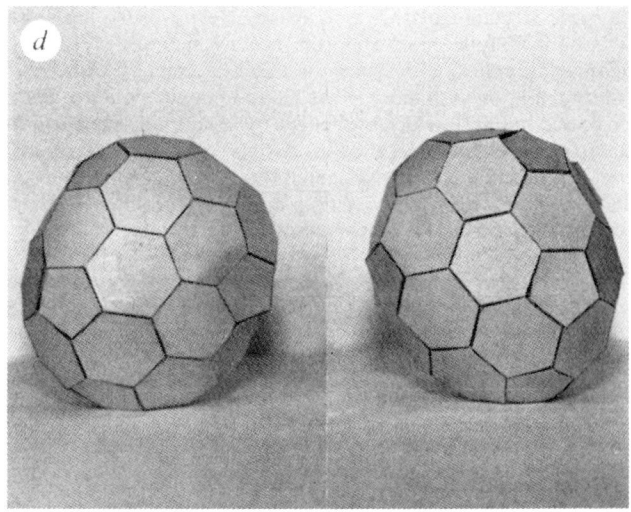

d

FIG. 3 Illustrations of the structure of C_{76}. *a*, Optimized structure of D_2-C_{76}, with the long (*a*) axis horizontal, and looking down the short (*c*) axis. Inset: photograph of the simulated structure, with dot-cloud representing the interior and exterior van der Waals surfaces. *b*, Construction of D_2-C_{76} from the pyracylene caps (bold area of structure I). The labelling of atoms (*a*)–(*s*) corresponds to the discussion in the text. *c*, Views along each of the three symmetry axes of D_2-C_{76}. The shaded atoms are pyrene-type sites (*p, q, r, s* in *b*), to which the downfield peaks of the ^{13}C-NMR are assigned. The starred atoms are corrannulene sites (*j–o*), and the remainder are pyracylene sites (*a–i*). *d*, Photograph of paper models, showing the left- and right-handed forms.

we propose that the nine resonances above 144 p.p.m. correspond to these sites. The third kind of site, 'corrannulene', also lies at the intersection of one 5MR and two 6MR, but is not a pyracylene bridge (that is, it is not connected to another 5MR). There are 24 of these (six types labelled *j* to *o*) forming a disconnected boundary around the caps, and we propose that these sites correspond to the six closely spaced resonances in the 141.3–143.6 p.p.m. region.

Thus the NMR spectrum can be assigned in detail to the proposed structure. Two-dimensional NMR could be used to establish this connectivity, as in the case of C_{70} (ref. 15).

We have examined the structure theoretically in several ways. First, we have performed Hückel molecular orbital calculations at the naïve level (a common β for all bonds), which reveal a small but significant HOMO–LUMO gap (0.344 β, or less than half of the 0.757 β found for C_{60} (ref. 17)) and a large delocalization energy (0.5551 β per atom, as compared with 0.5527 β for

C_{60} and the limiting value of 0.5743 for graphite). These values are in perfect agreement with those reported separately by Manolopoulos[14]. The HOMO–LUMO transition is allowed $b_1 \rightarrow a$. A reasonably optimized structure has been found through molecular mechanics calculations, and used as the initial structure in semi-empirical molecular orbital calculations (AM1-Gaussian 90)[18-20], within which the structure was further optimized. The important results are as follows.

(1) The D_2 symmetry is preserved with unconstrained optimization (Fig. 3). The bond lengths range from 1.37 to 1.47 Å, and show an alternation similar to that described for C_{60}. The dimensions along the three symmetry axes (*a–c*) are $D_a =$ 8.79 Å, $D_b = 7.64$ Å, $D_c = 6.68$ Å, as compared with 7.0 Å for C_{60}. The density should correspondingly be 25–30% lower than that of C_{60}. (2) The estimated standard heat of formation (relative to graphite) is $\Delta H_f = +14.75$ kcal mol^{-1} atom^{-1}, somewhat lower (more stable) than the estimate for C_{60} (+16.21 kcal

$mol^{-1}\,atom^{-1}$) at the same level of theory. (3) The estimate for the HOMO–LUMO gap is large, about 80% of that calculated for C_{60} (5.4 compared with 6.7 eV, although these greatly over-estimate the optical transition energies), and the locations of the HOMO (-8.9 eV) and LUMO (-3.5 eV) indicate that it should be both a better donor and a better acceptor than C_{60} (corresponding locations -9.6 eV and -2.9 eV).

The electronic spectrum in the near-infrared is consistent with this theoretical description. The onset at 1.39 eV and first maximum or shoulder near 1.5 eV can be compared to the values for C_{60} (2.0 and 2.2 eV), and the transitions are strong enough to be fully allowed. Magnetic properties would also be of interest for comparison with those of C_{60} and C_{70} and when considering the aromaticity of fullerenes[21-23]. Consistent with the relatively low symmetry of the structure, the infrared spectrum provides a very complex fingerprint, with irregular structure dominated by strong bands at 660, 747, 1,035 1,127 and 1,434 cm^{-1}. This spectrum should provide a sensitive test for theoretical calculations based on model potentials. We anticipate that these and higher-level calculations could be used to predict and explain interesting electronic and optical properties of the molecular and pure-solid materials.

We have shown that the third molecular form of carbon belongs to the fullerene family, as anticipated, but that it has highly unusual symmetry properties, as expressed in its point group (D_2) and more specifically in its chirality, which is based on helicity in the absence of a single chiral centre. These properties are interesting, unprecedented and largely unexpected. Recently, Fowler[16] has put forward a structural proposal for C_{84}, another abundant higher fullerene, having the same symmetry. Close analysis of this D_2-C_{84} structure (ref. 14, and P. Labastie, H.-P. Cheng and R.L.W., manuscript in preparation) reveals that it has the same kind of helicity as D_2-C_{76}. It remains to be seen whether this is the natural form of C_{84}. There is also the interesting question of why these helical forms might be preferred to the cylindrical or cubic forms discussed earlier[13].

The chiral nature of D_2-C_{76} implies that the two enantiomeric forms could be separated to yield elemental optically active materials. The optical properties, such as coefficients for optical rotation, may have interesting applications.

Much remains to be done to characterize this material, its chemical, electronic and magnetic properties, its detailed structure, its solid-state properties and so on. At first inspection it appears to be one of the most fascinating molecular architectures ever seen in nature. We expect that with the worldwide attention given to scaling up production of C_{60}, gram-size quantities of C_{76} will be available in the next few years. □

Received 19 August; accepted 22 August 1991.

1. Kroto, H. W., Heath, J. R., O'Brien, S. C., Curl, R. F. & Smalley, R. E. Nature 318, 162–164 (1985).
2. Curl, R. F. & Smalley, R. E. Science 242, 1017–1022 (1988).
3. Kroto, H. W. Science 242, 1139–1145 (1988).
4. Diederich, F. et al. Science 245, 1088–1090 (1989).
5. Kratschmer, W., Fostiropolous, K. & Huffman, D. R. Chem. Phys. Lett. 170, 167–170 (1990).
6. Kratschmer, W., Lamb, L. D., Fostiropolous, K. & Huffman, D. R. Nature 347, 354–358 (1990).
7. Bethune, D. S., Meijer, G., Tang, W. C. & Rosen, H. J. Chem. Phys. Lett. 174, 219–222 (1990).
8. Taylor, R., Hare, J. P., Abdul-Sada, A. K. & Kroto, H. W. J. Chem. Soc. chem. Commun. 1423–1425 (1990).
9. Johnson, R. J., Meijer, G. & Bethune, D. S. J. Am. chem. Soc. 112, 8983–8985 (1990).
10. Aije, H. et al. J. phys. Chem. 94, 8630–8633 (1990).
11. Whetten, R. L. et al. Mater. Res. Soc. Symp. Proc. 206 (in the press).
12. Diederich, F. et al. Science 252, 548–551 (1991).
13. Fowler, P. W., Cremona, J. E. & Steer, J. I. Theor. chim. Acta 73, 1–26 (1988).
14. Manolopoulos, D. E. J. chem. Soc. Faraday Trans. (in the press).
15. Johnson, R. D., Meijer, G., Salem, J. R. & Bethune, D. S. J. Am. chem. Soc. 113, 3619–3621 (1991).
16. Fowler, P. W. J. chem. Soc. Faraday Trans. 87, 1945–1946 (1991).
17. Haddon, R. C., Brus, L. E. & Raghavachari, K. Chem. Phys. Lett. 125, 459 (1986).
18. Dewar, M. J. S., Zoebisch, E. G., Healy, E. F. & Stewart, J. J. P. J. Am. chem. Soc. 107, 3902 (1985).
19. Frisch M. J. et al. GAUSSIAN-90, rev. 1, (Gaussian Inc., Pittsburgh, 1990).
20. Rudzinski, J. M., Slanina, Z., Togasi, M., Iizuka, T. Thermochim. Acya 125, 155–162 (1988).
21. Ruoff, R. S. J. Phys. Chem. 95, 3457–3459 (1991).
22. Haddon, R. C. et al. Nature 350, 46–47 (1991).
23. Fowler, P. Nature 350, 20–21 (1991).

ACKNOWLEDGEMENTS. We thank F. Ettl and S. J. Anz for help in carbon production and purification, H.-P. Cheng for assistance in modelling this structure, P. Labastie for carrying out the Hückel calculations, D. Sensharma, M. M. Alvarez and B. A. DiCamillo for mass spectrometric analyses, N. Goroff for assistance with the infrared spectroscopy, P. W. Fowler for communications and helpful correspondence and R. E. Smalley for providing a copy of ref. 14. We thank the NSF and ONR for funding.

Reprinted with permission from the Journal of the American Chemical Society
Vol. 114, No. 10, pp. 3994–3996, 1992
© 1992 American Chemical Society

Unusual Electrochemical Properties of the Higher Fullerene, Chiral C_{76}

Q. Li and Fred Wudl*

*Institute for Polymers and Organic Solids and
Departments of Chemistry and Physics
University of California at Santa Barbara
Santa Barbara, California 93106*

Carlo Thilgen, Robert L. Whetten, and François Diederich*

*Department of Chemistry and Biochemistry
University of California at Los Angeles
Los Angeles, California 90024-1569*

*Received December 16, 1991
Revised Manuscript Received March 13, 1992*

Approximately a year ago, it was discovered that the fullerenes C_{60} and C_{70} have an exceptionally high electron affinity. Two,[1a] three,[1b,c] four,[1d] and five reversible reduction steps[1e] have been recorded for C_{60}, and the two fullerenes were found to have essentially the same redox properties.[1b] No *reversible* oxidation waves were observed in the cyclic voltammograms of both carbon spheres. An explanation for the high electron affinity in terms of pyracyclene units was proposed and shown to be a good model for predicting the chemical reactivity of C_{60}.[1b] Recently, some of us reported the isolation in pure form and structural characterization of three higher fullerenes, chiral C_{76}[2] (Figure 1) and the two isomers C_{2v}-C_{78} and D_3-C_{78}.[3] The onset of the electronic absorption by these compounds occurs at much lower energy than the optical absorption onset measured for C_{60} and C_{70}, and the result of calculations on the electronic structure of the higher fullerenes was used to predict that C_{76} would be both a better donor and a better acceptor than C_{60}.[2,4] Since milligram quantities of the title fullerene have become available in the recent past, it was imperative to determine its electrochemical properties by cyclic voltammetry (CV).

For the isolation of the higher fullerenes, C_{60} and C_{70} were first removed by two chromatographic runs on neutral alumina with hexanes/toluene (95:5) as the eluent. The higher fullerene fraction was subsequently separated in three sequential runs on a 25 × 2.5 cm Vydac 201-TP C18 reversed-phase column with acetonitrile/toluene (50:50) as the eluent, giving pure C_{76}, C_{2v}-C_{78}, and D_3-C_{78} as well as a C_{84} fraction containing two isomers.[3,5-7]

(1) (a) Haufler, R. E.; Conceicao, J.; Chibante, L. P. F.; Chai, Y.; Byrne, N. E.; Flanagan, S.; Haley, M. M.; O'Brien, S. C.; Pan, C.; Xiao, Z.; Billups, W. E.; Ciufolini, M. A.; Hauge, R. H.; Margrave, J. L.; Wilson, L. J.; Curl, R. F.; Smalley, R. E. *J. Phys. Chem.* **1990**, *94*, 8634–8636. (b) Allemand, P. M.; Koch, A.; Wudl, F.; Rubin, Y.; Diederich, F.; Alvarez, M. M.; Anz, S. J.; Whetten, R. L. *J. Am. Chem. Soc.* **1991**, *113*, 1050–1051. (c) Cox, D. M.; Bahal, S.; Disko, M.; Gorun, S. M.; Greaney, M.; Hsu, C. S.; Kollin, E. B.; Millar, J.; Robbins, J.; Sherwood, R. D.; Tindall, P. *J. Am. Chem. Soc.* **1991**, *113*, 2940–2944. (d) Dubois, D.; Kadish, K. M.; Flanagan, S.; Haufler, R. E.; Chibante, L. P. F.; Wilson, L. J. *J. Am. Chem. Soc.* **1991**, *113*, 4364–4366. (e) Dubois, D.; Kadish, K. M.; Flanagan, S.; Wilson, L. J. *J. Am. Chem. Soc.* **1991**, *113*, 7773–7774.

(2) Ettl, R.; Chao, I.; Diederich, F.; Whetten, R. L. *Nature (London)* **1991**, *353*, 149–153.

(3) Diederich, F.; Whetten, R. L.; Thilgen, C.; Ettl, R.; Chao, I.; Alvarez, M. M. *Science (Washington, D.C.)* **1991**, *254*, 1768–1770.

(4) Cheng, H.-P.; Whetten, R. L. *Chem. Phys. Lett.*, in press. See also: Manolopoulos, D. E. *J. Chem. Soc., Faraday Trans.* **1991**, *87*, 2861–2862.

(5) Diederich, F.; Whetten, R. L. *Acc. Chem. Res.* **1992**, *25*, 119–126.

(6) Vydac C18 reversed-phase column: The Separations Group, 17434 Mojave St., Hesperia, CA 92345.

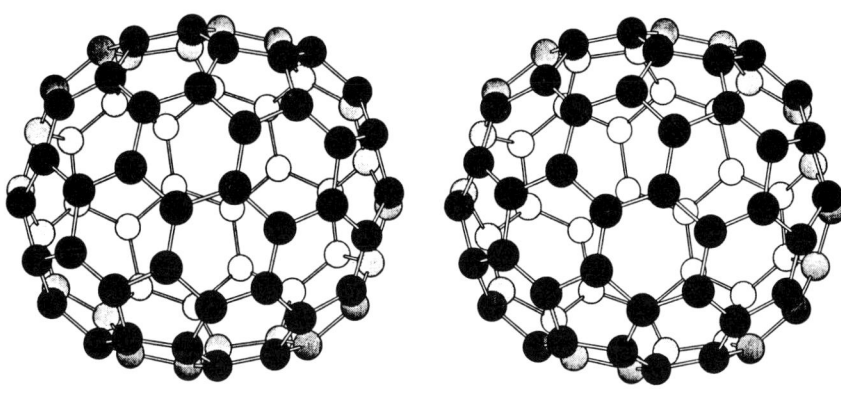

C_{76}

Figure 1. Stereodrawing of C_{76}.

Table I. Half-Cell Potentials[a]

compd	scan rate (mV/s)	half-cell potential (mV)		
		THF	PhCN	CH_2Cl_2
C_{76}	100	+354,[b,f] −64,[g] −560,[h] −1161,[i] −1605,[j] −2112[k]	−246,[h] −674,[i] −1224[j]	−287,[h] −620,[i] −1065[j]
C_{76}	1000	+349,[b,f] −46,[g] −564,[h] −1155,[i] −1606,[j] −2110[k]	−240,[h] −670,[i] −1200,[j] −1535[k]	+1360,[c,e] +905,[c,f] +500,[c,g] −283,[h] −628,[i] −1705,[j] −1495[k]
C_{60}	100	−228,[h] −826,[i] −1418,[j] −1916[k]	−392,[h] −812,[i] −1298[j]	−486,[h] −854,[i] −1339[j]
Fc[d]	100	+620	+524	+503
rest potentials of C_{76} (mV)		−299	+125	+39

[a] Half-cell potentials are defined as $E_1 = {}^1/_2(E_{p,red} + E_{p,ox})$; 0.1 M Bu_4NBF_4, electrolyte; Ag/AgCl, ref. [b] Due to adsorption (peak separation <59 mV). [c] Irreversible oxidation peak. [d] Fc = ferrocene. [e–k] Oxidation couples corresponding to the given potentials: e, (+3/+2); f, (+2/+1); g, (+1/0); h, (0/−1); i, (−1/−2); j, (−2/−3); k, (−3/−4).

Figure 2 shows typical cyclic voltammograms of C_{76} and, for comparison, of tetrathiafulvalene (TTF) and C_{60}, recorded in THF with a commercial BAS 100A apparatus. Table I shows the solvent and scan rate effect on the position of the voltammetric waves. All values are relative to Ag/AgCl with internal ferrocene for calibration. The following are salient features:

(1) In sharp contrast to C_{60} and C_{70}, in THF, the higher fullerene exhibits both reversible reduction *and* oxidation waves. In addition to four reduction steps, the CV of C_{76} shows one reversible oxidation step.[8] We propose that the extended linear benzo- and cyclopentadieno-annellated acene substructures present in the higher fullerene but absent in C_{60} and C_{70} are responsible for the observed oxidation at remarkably low potential.

(2) The reduction waves of C_{76} do not correspond to those of C_{60} (and C_{70}). The only feature that the higher fullerene has in common with C_{60} and C_{70} is that there is no apparent correlation of the CV waves with solvent polarity.

(3) Whereas high-level calculations using density-functional theory[4] predicted that the HOMO–LUMO gap in C_{76} should be on the order of 1.07 eV, the "electrochemical gap" found by us is only approximately 0.4 eV.

(4) In THF, the higher fullerene shows a CV wave at a potential which is less positive than the first wave of TTF and should

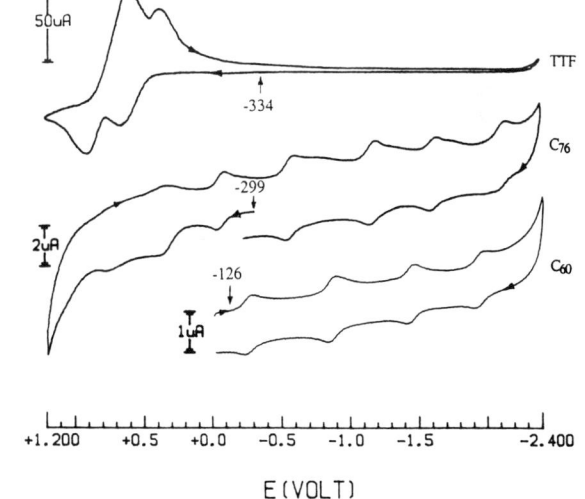

Figure 2. Cyclic voltammograms of TTF, C_{76}, and C_{60} in THF with 0.1 M Bu_4N^+ BF_4^- at ambient temperature. The arrows point to rest potentials. Working and counter electrodes were Pt, the geometric area of the working electrode was 0.0314 cm², the reference electrode was Ag/AgCl, and the scan rate was 1000 mV/s (Table I). The concentration of C_{76} is 0.7 mM; that of C_{60} is <0.5 mM (solubility problems prevented accurate determination).

therefore be more easily oxidized.

(5) In benzonitrile (scan rate 100 mV/s), the first reduction wave of C_{76} is more positive than that of C_{60} by 146 mV. In CH_2Cl_2, compared to C_{60}, the corresponding wave is more positive by 199 mV. In these two solvents, the oxidation wave is shifted to more positive potentials than in THF and, furthermore, becomes irreversible.

Thus, the higher fullerene C_{76} is expected to exhibit organic electron donor properties, and acceptor-doped materials should be stable. This exciting possibility is currently being actively pursued.

(7) For other work on higher fullerenes, see: (a) Diederich, F.; Ettl, R.; Rubin, Y.; Whetten, R. L.; Beck, R.; Alvarez, M.; Anz, S.; Sensharma, D.; Wudl, F.; Khemani, K. C.; Koch, A. *Science (Washington, D.C.)* **1991**, *252*, 548–551. (b) Fetzer, J. C.; Gallegos, E. J. *J. Polycyclic Aromat. Compd.*, in press. (c) Parker, D. H.; Wurz, P.; Chatterjee, K.; Lykke, K. R.; Hunt, J. E.; Pellin, M. J.; Hemminger, J. C.; Gruen, D. M.; Stock, L. M. *J. Am. Chem. Soc.* **1991**, *113*, 7499–7503. (d) Kikuchi, K.; Nakahara, N.; Honda, M.; Suzuki, S.; Saito, K.; Shiromaru, H.; Yamauchi, K.; Ikemoto, I.; Kuramochi, T.; Hino, S.; Achiba, Y. *Chem. Lett.* **1991**, 1607–1610. (e) Kikuchi, K.; Nakahara, N.; Wakabayashi, T.; Honda, M.; Matsumiya, H.; Moriwaki, T.; Suzuki, S.; Shiromaru, H.; Saito, K.; Yamauchi, K.; Ikemoto, I.; Achiba, Y. *Chem. Phys. Lett.*, in press. (f) Ben-Amotz, D.; Cooks, R. G.; Dejarme, L.; Gunderson, J. C.; Hoke, S. H., II.; Kahr, B.; Payne, G. L.; Wood, J. M. *Chem. Phys. Lett.* **1991**, *183*, 149–152.

(8) It is possible that another, more difficult oxidation at higher positive potential will be observable under other conditions, which will have to be developed in the future.

Acknowledgment. This work was supported by the Office of Naval Research (F.D., R.L.W.), the National Science Foundation (F.W.; Grants DMR 8820933, DMR 9111097, CHE 8908323), NATO (C.T.), and the Grand-Duchy of Luxembourg (C.T.).

Reprinted with permission from Science
Vol. 254, pp. 1768–1770, 20 December 1991
© 1991 The American Association for the Advancement of Science

Fullerene Isomerism: Isolation of C_{2v}-C_{78} and D_3-C_{78}

FRANÇOIS DIEDERICH,* ROBERT L. WHETTEN, CARLO THILGEN, ROLAND ETTL, ITO CHAO, MARCOS M. ALVAREZ

Early reports on the formation of the higher fullerenes C_{76}, C_{78}, C_{84}, C_{90}, and C_{94} by resistive heating of graphite stimulated theoretical calculations of possible cage structures for these all-carbon molecules. Among the five fullerene structures with isolated pentagons found for C_{78}, a closed-shell D_{3h}-isomer was predicted to form preferentially. Two distinct C_{78}-isomers were formed in a ratio of ~5:1 and could be separated by high-performance liquid chromatography. The carbon-13 nuclear magnetic resonance (NMR) spectrum of the major isomer is uniquely consistent with a C_{2v}-structure. The NMR data also support a chiral D_3-structure for the minor isomer. The isolation of specifically these two isomers of C_{78} provides insight into the stability of higher fullerene structures and into the mechanism for fullerene formation in general.

WHEN MACROSCOPIC QUANTITIES of C_{60} (*1*) and the oblong D_{5h}-C_{70} were isolated (*2*) and characterized in pure form (*3–9*), the isolation of enriched samples of C_{76}/C_{78} and C_{84} from the soot obtained by resistive heating of graphite under inert atmosphere was also reported (*10*). Improved chromatographic separation methods later led to the isolation of highly enriched samples of these all-carbon molecules (*11–14*), but it was only recently (*15*) that a third molecular form of carbon, C_{76}, could be isolated and characterized in pure form. The ^{13}C NMR spectrum of this molecule was uniquely consistent with a chiral D_2-symmetrical fullerene structure that obeys the isolated pentagon rule (IPR). This structure had also been predicted in theoretical work by Manolopoulos (*16*) who applied a spiral algorithm (*17*) combined with the IPR to the search for reasonable fullerene structures. Fowler and Manolopoulos subsequently found the five IPR-satisfying structures for C_{78} (Table 1) (*18*) and predicted on the basis of qualitative molecular orbital (MO) theory that the closed-shell D_{3h}-isomer would be the most stable isomer.

We isolated the C_{78}-fraction from the toluene extract of the carbon produced through resistive heating of graphite under helium by methods previously described (*11, 15*): (i) the crude extract was separated chromatographically on neutral alumina into the fractions C_{60}, C_{70}, and higher fullerenes; (ii) the higher fullerenes (0.1 g) were separated on a second alumina column with hexane-toluene elution to yield three fractions, (I) C_{70}, (II) C_{76} and C_{78}, and (III) the major component C_{84}; and (iii) the final step was threefold purification by high-performance liquid chromatography (HPLC) of fraction II, performed on a Vydac 201-TP-510 column (C_{18} reversed phase, 250 mm, 10-mm internal diameter) (*13, 19*). After most of the C_{76} was removed with a 70:30 CH_2Cl_2-CH_3CN eluant mixture, chromatography with a 1:1 toluene-CH_3CN mixture provided two highly enriched fractions of two C_{78}-isomers. A final run with toluene-CH_3CN led to the fractions shown by the HPLC profiles in Fig. 1, A to C. In HPLC assays with various solvent mixtures, the purity was assessed to be >98%. The first fraction yielded ~2 mg of the major isomer, and the second fraction yielded ~0.2 mg of the minor C_{78}-isomer. Mass spectrometric analysis (laser desorption time-of-flight) of each fraction (Fig. 1, D and E) verified that both contain material of C_{78} chemical formula with >98% purity.

The structures of the two isomers were elucidated by ^{13}C NMR spectroscopy. The spectrum of the purified major isomer, recorded in CS_2 with $Cr(acac)_3$ (acac, acetylacetonate) as a relaxant after 71,000 scans, shows 18 large peaks of nearly equal intensity and 3 small peaks, each with one-half the intensity of the major ones (Fig. 2 and Table 2). This spectrum is thus uniquely consistent with one of the two C_{2v}-symmetrical, IPR-satisfying fullerene structures calculated by Manolopoulos and Fowler for C_{78} (Table 1) (*18*). As in C_{70} (*20*) and C_{76} (*15*), each C atom in C_{78} lies in one of three distinct environments. Five resonances, four of higher and one of lower intensity, originate from C atoms lying at the intersection of three hexagons ("pyrene"-like) and show the largest

Department of Chemistry and Biochemistry, University of California, Los Angeles, CA 90024–1569.

*To whom correspondence should be addressed.

Table 1. The five IPR-satisfying isomeric structures of C_{78}, the number of independent resonances in their ^{13}C NMR spectra, the gas-phase heats of formation ΔH_f° (298 K), and the differences in ΔH_f° calculated with the molecular mechanics program MM3. (MM3 calculates a heat of formation of 573.8 kcal mol^{-1} for C_{60}.)

Structure	^{13}C NMR lines (intensity)	ΔH_f° (kcal mol^{-1})	$\Delta(\Delta H_f^\circ)$ (kcal mol^{-1})
C_{2v}	18 (4 C) + 3 (2 C)	695.7	1.4
D_3	13 (6 C)	697.8	3.5
C_{2v}'	17 (4 C) + 5 (2 C)	694.3	0.0
D_{3h}	5 (12 C) + 3 (6 C)	702.3	8.0
D_{3h}'	5 (12 C) + 3 (6 C)	697.8	3.5

Fig. 1. (**A** to **C**) HPLC profiles of (A) a mixture of C_{76} (retention time t_R = 9.08 min), C_{2v}-C_{78} (t_R = 9.67 min), and D_3-C_{78} (t_R = 10.30 min); (B) pure C_{2v}-C_{78}, and (C) pure D_3-C_{78}. Experimental conditions: Vydac 201 TP 510 C-18 reversed-phase column of dimensions 25 cm by 1 cm, particle size, 5 μm; eluant, toluene-acetonitrile 1:1; pressure, 1000 psi; flow rate, 6 ml/min; and ultraviolet detection at 310 nm. (**D** and **E**) Laser-desorption time-of-flight mass spectra recorded for (D) C_{2v}-C_{78} and (E) D_3-C_{78}. Spectrum D corresponds to the HPLC profile in (B) and spectrum E corresponds to (C). In addition to C_{76}, C_{74}, and C_{72} peaks resulting from C_2-fragment losses, spectrum D contains minor C_{60} and C_{70} peaks from previous calibration runs and spectrum E shows C_{84} as a minor impurity in the D_3-C_{78} isomer.

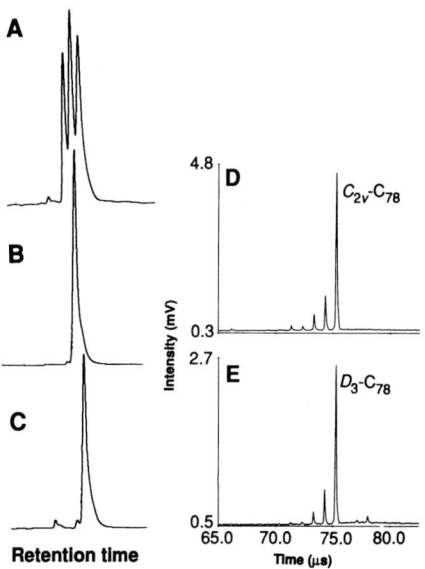

shielding near 130 ppm. A total of nine resonances, eight of higher and one of lower intensity, originate from C atoms in the bonds between two five-membered rings (5MR) in pyracylene substructures. In analogy to C_{70} and C_{76}, these signals should be the most deshielded near 150 ppm. The remaining seven peaks, six of higher and one of lower intensity, are due to C atoms at corannulene sites, as defined by the intersection of one 5MR and two 6MRs but not connected to another 5MR. The symmetry elements and the location of the three different types of C atoms in C_{2v}-C_{78} are shown in Fig. 3, A to C. The molecule is depicted with a remarkably flat face shaped by a coronene subunit facing the viewer (Fig. 3A), in a view perpendicular to the symmetry plane (Fig. 3B), and in a view onto one of the two identical caps (Fig. 3C). If a paper model of C_{2v}-C_{78} is put on its flat coronene base, it resembles a helmet or a turtle's shell.

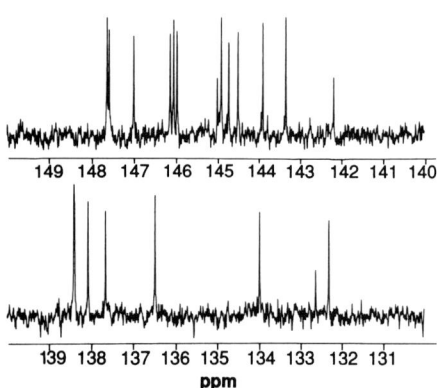

C_{2v}-C_{78}

The small amount of minor C_{78}-isomer that we isolated was insufficient to record a ^{13}C NMR spectrum of the pure compound. However, after the spectra of pure C_{76} and C_{2v}-C_{78} were known, structural evidence for the minor C_{78} isomer was obtained by subtraction from the spectrum of a concentrated CS_2 solution containing a mixture of C_{76}, C_{2v}-C_{78}, and the minor

isomer of C_{78} in a ratio of ~5:5:1. This spectrum showed 13 resonances of approximately equal intensity in addition to the 19 equally intensive resonances of D_2-C_{76} (15) and the 21 resonances of C_{2v}-C_{78} (Table 2). Of the five calculated IPR-satisfying fullerene structures for C_{78}, only the chiral D_3-structure (Table 1) is in agreement with the observed spectrum. A more detailed analysis of the NMR data provides additional support for the assignment of the D_3-structure to the minor C_{78}-isomer. A view onto one of the three symmetry-related, smoothly curved faces in D_3-C_{78} that are capped by two triphenylene caps is shown in Fig. 3D. It indicates both the location of the pyrene- and corannulene-type C sites that are assembled on each face. There are three different pyrene-type sites in the molecule, and three distinctively upfield-shifted resonances between 132 and 135 ppm are observed in the spectrum. We tentatively assign the four signals between 136 and 141 ppm to the four corannulene-type sites and the six residual, least-shielded resonances to the six pyracylene-type sites in the molecule.

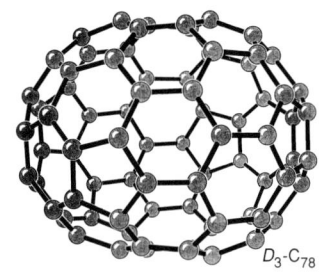

D_3-C_{78}

The two C_{78}-isomers differ considerably in their optical properties. A solution of the C_{2v}-isomer (in CH_2Cl_2) is chestnut brown, whereas a solution of the D_3-compound has a more golden-yellow color. The electronic absorption spectra of the two molecules are distinctively different (Fig. 4). Whereas the longest wavelength band in the C_{2v}-isomer appears at ~700 nm with the end-absorption tailing to ~900 nm, the D_3-isomer shows a set of distinct bands between 700 and 850 nm.

Why do the experimental results differ from the calculations by Manolopoulos and Fowler, who predicted the preferential formation of the D_{3h}-isomer (18), and why do

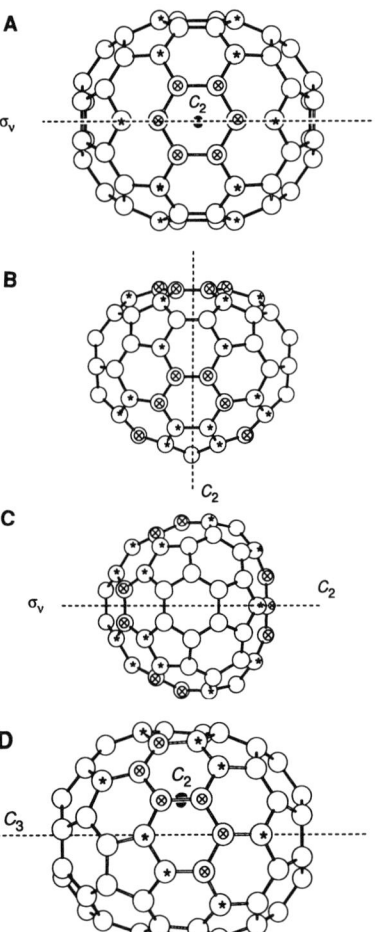

Fig. 3. (**A** to **C**) Views of C_{2v}-C_{78} (A) onto the flat face shaped by a coronene subunit, (B) perpendicular to the symmetry plane, and (C) onto one of the two caps. The starred C atoms are corannulene-type sites, and the C atoms marked by a circled × are pyrene-type sites. (**D**) View of D_3=C_{78} (A) onto one of the three symmetry-related, smoothly curved faces that shows corannulene-type sites as starred atoms and pyrene-type sites marked by a circled ×.

Fig. 2. Expanded view of the fullerene region in the ^{13}C NMR spectrum of 1.6 mg of C_{2v}-C_{78} in CS_2, taken at 125.6 MHz, with acetone as an internal lock. In addition to the fullerene resonances, the full spectrum from 0 to 220 ppm only shows the acetone peaks near 30 and 206 ppm and the solvent peak near 190 ppm.

Table 2. Line positions in the ^{13}C NMR spectra of C_{2v}-C_{78} and D_3-C_{78} in CS_2. Centers of gravity in the spectra: C_{60}, 143.2; C_{70}, 145.0; C_{76}, 142.7; C_{2v}-C_{78}, 141.9; and D_3-C_{78}, 141.1.

	C_{2v}-C_{78}		D_3-C_{78}*
δ (ppm)	Integrated intensity	Carbons (no.)	δ (ppm)
132.31	4.1	4	132.18
132.63	2.3	2	132.90
133.98	4.2	4	134.85
136.49	3.9	4	139.55
137.67	3.9	4	140.45
138.08	3.9	4	140.82
138.39	6.7†	4	140.91
138.41		4	141.83
142.19	1.7	2	142.88
143.34	4.2	4	144.53
143.89	3.9	4	145.45
144.49	4.5	4	148.14
144.72	4.5	4	149.45
144.89	4.1	4	
144.99	1.9	2	
145.96	3.8	4	
146.04	4.7	4	
146.12	4.2	4	
146.99	4.1	4	
147.56	4.6	4	
147.62	4.7	4	

*The integrated intensities are not inconsistent with 13 × 6 C signals; high uncertainties of individual values are due to a poor signal-to-noise ratio of the small resonances of D_3-C_{78} in the spectrum. †Two peaks.

we observe the formation specifically of the C_{2v}- and D_3-isomers? The latter question of isomeric preference was until now irrelevant in fullerene chemistry, because only one IPR-satisfying structure exists for C_{60} and C_{70}. The prediction of D_{3h}-C_{78} as the most stable structure was predominantly based on the qualitative MO theory calculation of a large gap between the highest occupied MO and the lowest unoccupied MO; it allows no comparative estimate of the strain among the IPR-satisfying isomers. We performed MM3 force-field calculations (21, 22) on the five IPR-satisfying C_{78} fullerene structures and found that the two isolated isomers are much more

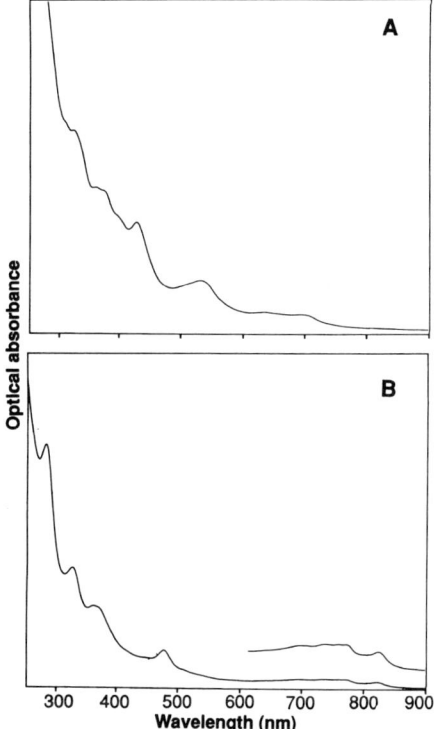

Fig. 4. Optical spectra recorded in CH_2Cl_2 for (**A**) C_{2v}-C_{78} and (**B**) D_3-C_{78} (path length of 1 cm). Absorption maxima λ (nanometers) and absolute extinction coefficients ε (liter mol^{-1} cm^{-1}) for C_{2v}-C_{78}: λ 696 (ε 3,400); 638 (4,000); 528 (11,500); 424 (23,800); 390 shoulder (sh) (25,500); 368 sh (30,300); 359 (30,800); 325 (42,700); and 308 sh (44,400). Absorption maxima of D_3-C_{78} (λ) 820, 768, 757, 734, 700, 472, 362 sh, 356, 322, and 276. The concentration of the D_3-C_{78} sample is unknown.

stable than the one predicted by qualitative MO theory (Table 1). Also, model building clearly shows that the D_{3h}-isomer with three flat coronene faces separated by rather sharp edges is much more strained than the C_{2v}-isomer, which has only one quasi-planar coronene face (Fig. 3) and a more uniform curvature over two thirds of the helmet-shaped ball surface. Strain is better delocalized in the C_{2v}- than in the D_{3h}-isomer.

Beyond energetics and stability, an explanation for the formation specifically of C_{2v}- and D_3-C_{78} became apparent when we explored possible mechanisms of interconversion among the five C_{78} fullerene structures. The predicted D_{3h}- and the observed dominant C_{2v}-isomers are closely related: They are interchangeable by 90° rotation of a single C_2-unit in the pyracylene rearrangement (Fig. 5), which was first suggested by Stone and Wales (23) for C_{60} and subsequently proposed as the first step in the degradation of C_{60} to smaller fullerenes through consecutive C_2-losses (24). The pyracylene rearrangement is a general interconversion route on the C_{78} fullerene hypersurface; as many as four isomers of C_{78} are interchangeable in single-step rearrangements (Eq. 1):

$$D_{3h} \rightleftharpoons C_{2v} \rightleftharpoons C_{2v}' \rightleftharpoons D_{3h}' \qquad (1)$$

In the D_{3h}-structure, the orientation of all three central pyracylene subunits (Fig. 5) generates a central circular acene of nine fused benzene rings. A single pyracylene rearrangement leads to the C_{2v}-structure (Fig. 5), a second to the C_{2v}'-structure, and a third to the D_{3h}'-structure. In the latter isomer, all three central pyracylenes now take the orientation shown in Fig. 5 for the C_{2v}-structure. The caps in all four fullerenes are structurally identical (Fig. 3C) and remain unaffected by the rearrangement. In contrast, no similar simple interconversion is possible between any of the four isomers in Eq. 1 and the chiral D_3-isomer that we isolated as the minor component. It becomes clear that two fullerene C_{78}-forming channels are active in the production process. One leads into the manifold of four isomers shown in Eq. 1 and ultimately, under thermodynamic control during the "cooling period" of the production process

(10), to exclusively the most stable C_{2v}-isomer (25). A second product channel leads to the minor D_3-product, which can be isolated because there exists no good mechanism for conversion into the more stable C_{2v}-isomer. We believe that we have discovered a general selection rule for predicting which isomers of the higher fullerenes have a high probability for isolation. This rule should facilitate their structural characterization, particularly since the number of IPR-satisfying fullerene isomers increases rapidly with larger molecular size. For C_{84}, Manolopoulos and Fowler calculated 24 isomers, and this number increases to 46 for C_{90} (26). As illustrated above for C_{78}, molecular mechanics calculations in combination with the analysis of isomer interconversions should allow us to predict how many product channels exist for a higher fullerene and which isomers one can expect to isolate.

REFERENCES AND NOTES

1. H. W. Kroto, J. R. Heath, S. C. O'Brien, R. F. Curl, R. E. Smalley, *Nature* **318**, 162 (1985).
2. W. Krätschmer, L. D. Lamb, K. Fostiropoulos, D. R. Huffman, *ibid.* **347**, 354 (1990).
3. H. Ajie *et al.*, *J. Phys. Chem.* **94**, 8630 (1990).
4. R. Taylor *et al.*, *J. Chem. Soc. Chem. Commun.* **1990**, 1423 (1990).
5. R. D. Johnson, G. Meijer, D. S. Bethune, *J. Am. Chem. Soc.* **112**, 8983 (1990).
6. R. E. Haufler *et al.*, *J. Phys. Chem.* **94**, 8634 (1990); J. M. Hawkins *et al.*, *J. Org. Chem.* **55**, 6250 (1990).
7. D. M. Cox *et al.*, *J. Am. Chem. Soc.* **113**, 2940 (1991).
8. R. Tycko *et al.*, *J. Phys. Chem.* **95**, 518 (1991).
9. P.-M. Allemand *et al.*, *J. Am. Chem. Soc.* **113**, 1050 (1991).
10. R. L. Whetten *et al.*, *Mater. Res. Soc. Symp. Proc.* **206**, 639 (1991).
11. F. Diederich *et al.*, *Science* **252**, 548 (1991).
12. D. H. Parker *et al.*, *J. Am. Chem. Soc.* **113**, 7499 (1991).
13. J. C. Fetzer and E. J. Gallegos, *J. Polycycl. Aromat. Compd.*, in press.
14. K. Kikuchi *et al.*, *Chem. Lett.* **1991**, 1607 (1991).
15. R. Ettl *et al.*, *Nature* **353**, 149 (1991).
16. D. E. Manolopoulos, *J. Chem. Soc. Faraday Trans.* **87**, 2861 (1991).
17. _____, J. C. May, S. E. Down, *Chem. Phys. Lett.* **181**, 105 (1991).
18. P. W. Fowler, D. E. Manolopoulos, R. C. Batten, *J. Chem. Soc. Faraday Trans.* **87**, 3103 (1991).
19. Vydac C_{18} reversed-phase column, The Separations Group, Hesperia, California.
20. R. D. Johnson, G. Meijer, J. R. Salem, D. S. Bethune, *J. Am. Chem. Soc.* **113**, 3619 (1991).
21. N. L. Allinger *et al.*, *ibid.* **111**, 8551 (1989).
22. M. Saunders, *Science* **253**, 330 (1991).
23. A. J. Stone and D. J. Wales, *Chem. Phys. Lett.* **128**, 501 (1986).
24. S. C. O'Brien, J. R. Heath, R. F. Curl, R. E. Smalley, *J. Chem. Phys.* **88**, 220 (1988).
25. Although the large difference in MM3 heat of formations, $\Delta(\Delta H_f°) = 6.6$ kcal mol^{-1}, between the C_{2v}- and D_{3h}-isomers is a significant value, the calculated difference of 1.4 kcal mol^{-1} between the C_{2v}-isomers is not very large, and the stability sequence among the two C_{2v}-isomers might actually be reversed.
26. D. E. Manolopoulos and P. W. Fowler, *Chem. Phys. Lett.* **88**, 220 (1988).
27. Supported by the Office of Naval Research (F.D. and R.L.W.), Nato Fellowships (R.E. and C.T.), and the Grand-Duchy of Luxembourg (C.T.). We thank N. L. Allinger for valuable discussions.

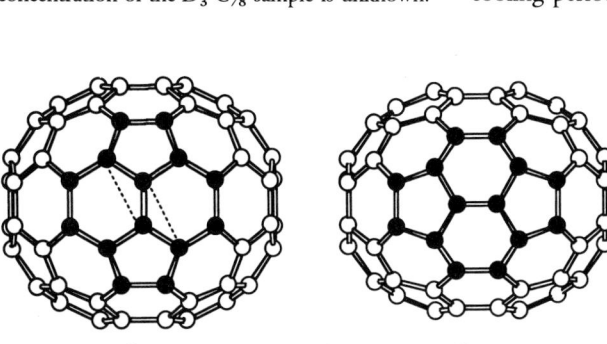

Fig. 5. Interconversion between the C_{2v}- and D_{3h}-isomers of C_{78} through the 4e$^-$ pyracylene rearrangement.

Reprinted with permission from Nature
Vol. 357, No. 6374, pp. 142–145, 14 May 1992
© 1992 Macmillan Magazines Limited

NMR characterization of isomers of C_{78}, C_{82} and C_{84} fullerenes

Koichi Kikuchi*, Nobuo Nakahara,
Tomonari Wakabayashi, Shinzo Suzuki,
Haruo Shiromaru, Yoko Miyake, Kazuya Saito,
Isao Ikemoto, Masatsune Kainosho & Yohji Achiba*

Department of Chemistry, Tokyo Metropolitan University, Hachioji,
Tokyo 192-03, Japan

* To whom correspondence should be addressed.

FOLLOWING the development of a method for bulk synthesis of C_{60} and other fullerenes[1], the isolation of higher fullerenes ranging from C_{76} to C_{96} has been achieved using chromatographic techniques[2–5]. Whereas C_{60} and C_{70} have unique, high-symmetry structures[6], theoretical calculations for fullerenes larger than C_{76} have suggested that each may exist in at least two isomeric forms[7]. For C_{84}, 24 isomers have been postulated[7], and for C_{96} calculations have yielded 196 distinct isomers[8]. Diederich et al.[9] have used liquid chromatography and ^{13}C NMR to identify two isomers of C_{78}, but previous experimental studies of other higher fullerenes[2] have produced ambiguous results. Here we use ^{13}C NMR to determine the structures of some principal isomers of C_{78}, C_{82} and C_{84}. We find a third isomer of C_{78}, which was not reported in ref. 9. Characterization of the structures of these larger fullerenes should provide new understanding of the factors determining the stability of hollow carbon clusters.

We isolated C_{78}, C_{82} and C_{84} by high-performance liquid chromatography (the preparation is described in detail elsewhere[4,5]). Laser-desorption time-of-flight mass spectra were obtained to confirm the purity of the isolated samples. We used an ArF (193 nm) laser as the desorption light source[5]. Mass spectra for the samples of C_{78}, C_{82} and C_{84} are shown as inserts to Fig. 1a–c. We measured ^{13}C NMR spectra of the higher fullerenes using CS_2 as the solvent with $Cr(C_5H_7O_2)_3$ as a relaxant.

C_{78}. Figure 1a displays the ^{13}C-NMR spectrum of C_{78}, showing 56 distinct lines. From a recent paper by Diederich et al.[9], one can easily identify the 21 lines (marked * in Table 1) due to one of the two C_{2v}-C_{78} fullerenes out of the five fullerenes that satisfy the 'isolated pentagon rule' (IPR)[10], which are slightly solvent-shifted by 0.17 p.p.m. Likewise one can easily identify the 13 lines (marked † in Table 1) that correspond to D_3-C_{78} (solvent-shifted by ~0.09 p.p.m.). This leaves 22 peaks to be accounted for, 17 of which have relative intensity 3.9–6.0 and five have relative intensity 1.6–2.4 (Table 1). The other of the two C_{2v}-C_{78} candidates, corresponding to C'_{2v} in ref. 9, possesses 22 NMR lines with numbers of independent resonances $5 \times 2C$ and $17 \times 4C$. This situation is consistent with the present observation. Therefore, we can conclude that C_{78} consists of three isomers with symmetry C'_{2v}, C_{2v} and D_3, in the ratio 5:2:2. The structures of the three isomers determined here are shown in Fig. 2a. Our conclusion is, however, inconsistent with that of Diederich et al.[9], who showed the formation of two isomers for C_{78}, C_{2v} and D_3, in the ratio 5:1.

C_{82}. The ^{13}C-NMR spectrum of C_{82} (Fig. 1b) consists of 41 strong lines, in addition to 29 lines of moderate intensity (the line marked 47, 48 looks like a single strong line, but in fact we believe it to be a superposition of two weaker lines; see Fig. 1b). Positions and intensities of the strong and moderate lines of C_{82} are summarized in Table 1. In Fig. 1b, we can also see at least 48 weak lines and more than 80 very weak lines. Here we note that the positions of some very weak lines coincide with the lines due to C_{84} described later. This evidence is consistent with the mass spectrum of C_{82} which shows it to be contaminated by a small amount of C_{84}.

TABLE 1   Line positions and relative intensities of the ^{13}C NMR spectra of C_{78}, C_{82} and C_{84} fullerenes

No.	C_{78} Shift (p.p.m.)	C_{78} Relative intensity	C_{82} Shift (p.p.m.)	C_{82} Relative intensity	C_{84} Shift (p.p.m.)	C_{84} Relative intensity
1	132.28†	2.3	128.20	1.0	133.81	1.7
2	132.48*	2.6	129.36	0.9	134.97	1.8
3	132.76	5.3	131.74	2.3	135.48	1.9
4	132.80*	1.1	131.93	0.7	137.39	1.6
5	133.04†	1.7	131.94	0.6	137.50	1.8
6	134.15*	2.8	132.00	0.7	137.94	1.6
7	134.88†§	1.9	132.78	2.2	138.46	1.4
8	135.36	4.9	133.10	2.1	138.57	1.7
9	136.34	5.0	133.60	1.9	138.88	2.0
10	136.66*	2.3	133.83	0.7	138.89	2.0
11	137.08	4.6	134.52	2.0	139.63	1.5
12	137.20	4.6	134.61	1.2	139.70	1.7
13	137.40	4.9	135.12	1.9	139.75	1.8
14	137.83*	2.6	136.68	2.0	139.77	1.5
15	138.23*	2.8	136.71	2.2	139.82	1.0
16	138.46	3.9	137.13	2.0	139.98	1.7
17	138.56*	2.2	137.34	2.2	140.27	1.6
18	138.58*	2.9	137.91	0.8	140.33	1.7
19	138.60	4.3	138.18	2.0	140.49	1.9
20	139.64†	1.6	138.27	1.8	140.61	1.3
21	139.71	4.6	138.33	2.0	140.99	1.5
22	140.55†	1.8	138.96	1.9	141.33	1.7
23	140.90†	1.6	139.19	2.1	141.58	1.7
24	141.00†	2.2	139.25	1.8	142.12	1.9
25	141.81§	2.1	139.32	2.4	142.55	1.8
26	141.92†	2.0	139.58	2.1	142.88	2.0
27	142.34*	1.5	139.72	1.9	143.77	1.8
28	142.82	4.6	139.88	0.6	143.80	1.7
29	142.97†	2.2	139.98	2.1	143.98	1.9
30	143.49*	3.0	140.15	2.1	144.48	1.9
31	143.86	4.9	140.56	2.3	144.58	2.2
32	144.06*	2.6	140.65	2.0	144.60	2.2
33	144.41	5.7	140.86	0.6		
34	144.43	6.0	141.02	0.6		
35	144.61†	2.0	141.06	0.7		
36	144.66*	2.7	141.14	0.7		
37	144.88*	3.0	141.45	2.3		
38	145.06*	2.5	141.50	0.7		
39	145.12	4.5	141.70	0.6		
40	145.17*	0.9	141.85	2.0		
41	145.39	1.6	141.88	1.8		
42	145.59†	1.6	142.17	0.8		
43	146.12*	2.6	142.25	2.2		
44	146.20*	2.8	142.38	0.9		
45	146.29*	2.5	142.41	1.9		
46	146.36	5.5	142.45	0.7		
47	146.40	2.4	143.23	0.7‡		
48	146.91	5.1	143.23	0.7‡		
49	147.02	5.0	143.29	2.0		
50	147.14*	3.0	143.36	2.0		
51	147.42	2.1	143.67	0.8		
52	147.51	1.6	143.69	0.8		
53	147.74*	2.8	143.85	2.2		
54	147.78*	2.5	143.97	2.0		
55	148.23†	2.3	144.02	1.9		
56	149.54†	2.0	144.65	2.5		
57			144.66	2.5		
58			145.03	1.1		
59			145.18	0.6		
60			145.27	0.8		
61			145.28	2.2		
62			145.52	2.3		
63			145.61	2.2		
64			145.83	2.1		
65			146.10	1.1		
66			149.04	2.4		
67			150.35	2.1		
68			150.56	0.6		
69			151.41	0.6		
70			151.47	0.7		

* Attributed to D_3-C_{78}. † Attributed to C_{2v}-C_{78}.

‡ The intensity of the line 47, 48 was divided into two for each line.

§ An alternative explanation, suggested by R. Taylor (personal communication) is that these lines are due to C'_{2v} and D_3, respectively.

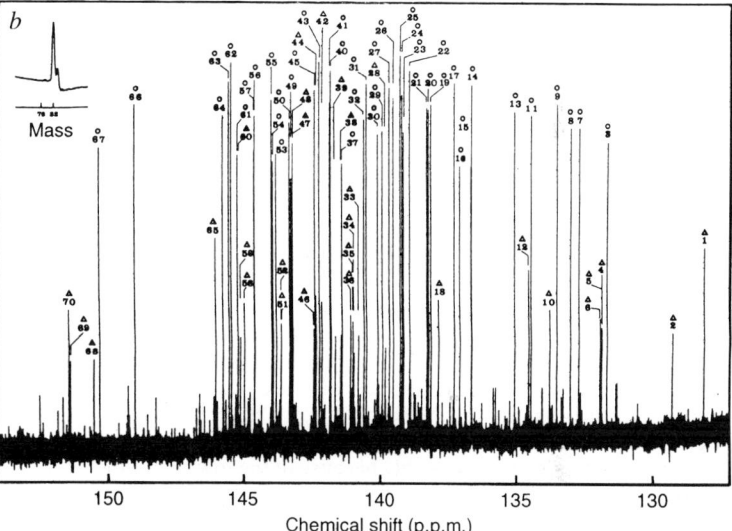

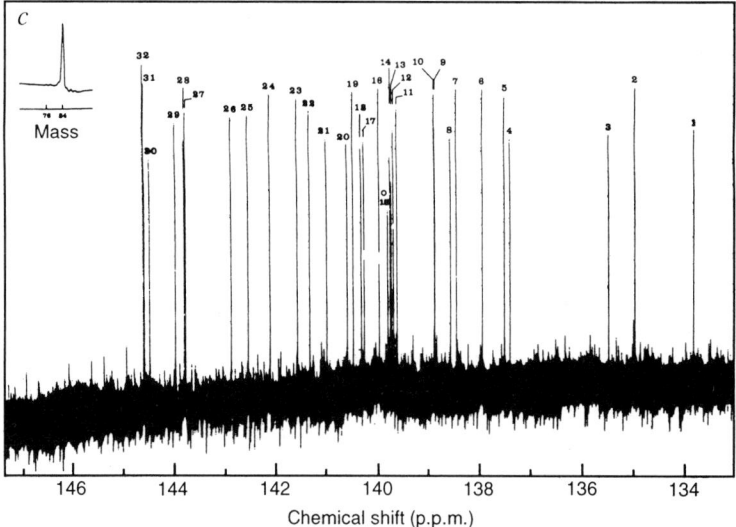

FIG. 1 ^{13}C-NMR spectra of fullerenes. a, C_{78} after 23,000 scans; b, C_{82} after 35,000 scans and C_{84} after 90,000 scans, in CS_2 solution, taken at 125.6 MHz, with acetone as an internal lock; ○, strong lines, △, moderate lines. The region of lines 51 and 52 in a is expanded so that the linewidths can be compared; the different widths suggest differences in the spin-lattice relaxation rates of the carbon nuclei. Laser-desorption time-of-flight mass spectra are also shown for samples of C_{78}, C_{82} and C_{84} isolated by high-performance liquid chromatography. The line denoted 47 and 48 in b seems to be a single line with almost the same intensity as those of the strong lines (○). But throughout our two independent preparations and measurements of C_{82}, it was strongly suggested that the line arises from the overlap of two kinds of C_{82} isomers. One of its components was very sensitive to the experimental conditions. In one case (not shown here), the intensity of the line decreased to about half the intensity shown here. This reduction of intensity was accompanied by a decrease in eleven other lines, 1, 2, 5, 12, 18, 51, 52, 60, 65, 68 and 69. From the numbers of these NMR lines, it may be deduced that the component that is sensitive to the sample preparation is due to the isomer C_{3v}-C_{82}.

According to Manolopoulos and Fowler[7], C_{82} has nine IPR-satisfying isomers $(2 \times C_{3v}, C_{2v}, 3 \times C_2$ and $3 \times C_s)$, out of which the three isomers with C_2 symmetry give 41 NMR lines with nearly equal intensity. The 41 strong lines with intensities of 2.3-1.8 (Table 1) can therefore be easily assigned to the three C_2-C_{82}. Three structural candidates for C_2-C_{82} are shown in Fig. 2b.

On the other hand, C_{2v}-C_{82} and C_{3v}-C_{82} are expected to have combinations of $(17 \times 4C, 7 \times 2C)$ and $(12 \times 6C, 3 \times 3C, 1 \times 1C$ or $11 \times 6C, 5 \times 3C, 1 \times 1C)$ lines, respectively[7]. Out of these 40 lines, 17 lines for C_{2v} and 12 lines (or 11 lines) for C_{3v} should be twice or six times as strong as the other lines. Therefore, if we assume that C_{2v}-C_{82} and one of the C_{3v}-C_{82} isomers are produced in almost equal amounts, the 29 observed lines with

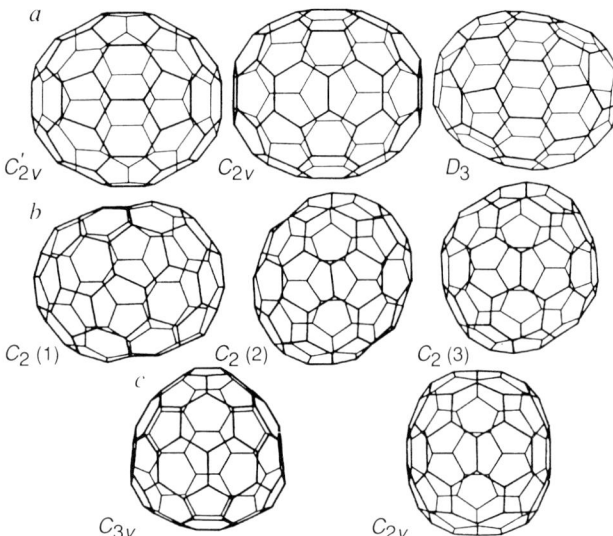

FIG. 2 The structures of fullerene isomers suggested by the ^{13}C-NMR measurements. a, Three isomers of C'_{2v}-, C_{2v}- and D_3-C_{78}. b, Three structural candidates for C_{82} fullerene with C_2 symmetry. c, Structures of C_{2v}- and C_{3v}-C_{82} isomers suggested by the ^{13}C-NMR measurements.

Received 6 February; accepted 14 April 1992.

1. Krätschmer, W., Lamb, L. D., Fostiropoulos, K. & Huffman, D. R. Nature **347**, 354–358 (1990).
2. Diederich, F. et al. Science **252**, 548–551 (1991).
3. Ettl, F. et al. Nature **352**, 149–153 (1991).
4. Kikuchi, K. et al. Chem. Lett. 1607–1610 (1991).
5. Kikuchi, K. et al. Chem. Phys. Lett. **188**, 177–180 (1991).
6. Kroto, H. W., Heath, J., O'Brien, S. C., Curl, R. F. & Smalley, R. E. Nature **318**, 162–164 (1985).
7. Manolopoulos, D. E. & Fowler, P. W. Chem. Phys. Lett. **187**, 1–7 (1991).
8. Manolopoulos, D. E. & Fowler, P. W. J. chem. Phys. (submitted).
9. Diederich, F. et al. Science **254**, 1768–1770 (1991).
10. Fowler, P. W., Batten, R. C. & Manolopoulos, D. E. J. chem. Soc. Faraday Trans. **87**, 3103–3104 (1991).
11. Diederich, F. & Whetten, R. B. Accts chem. Res. (in the press).

ACKNOWLEDGEMENTS. This work was supported by the Ministry of Education, Science and Culture of Japan.

moderate intensity can readily be rationalized. The remaining nine weaker lines associated with these cages are considered to be hidden in the weak or very weak lines. In fact, we found that no other combination of C_{82} isomers could produce 29 NMR lines of nearly equal intensity. Quantitative analysis for the NMR lines with strong and moderate intensities indicates that the production ratios of C_{82} are 8 : 1 : 1 for C_2-, C_{2v}- and C_{3v}-C_{82}, respectively. The numbers of unidentified weak and very weak lines, on the other hand, suggest the existence of at least three more C_2- and/or C_s-C_{82} isomers.

C_{84}. The first attempt to determine the structures of C_{84} was made by Diederich et al.[2]. Possibly because of the low signal-to-noise ratios, their determination was not conclusive. The NMR spectrum of C_{84} obtained here is shown in Fig. 1c. Despite C_{84} being larger than the former two fullerenes, the NMR spectrum shows a much simpler profile, consisting of 31 strong lines and one additional line at half the intensity. Considering the predicted numbers and intensities of the NMR lines for C_{84} isomers[8], we conclude that C_{84} has two major isomers. The most abundant isomer is one of the four possible D_2 cages and the other is the D_{2d}-C_{84}, which uniquely corresponds to the structure numbered 23 in Fig. 4 of ref. 8. The ratio of D_2 to D_{2d} is 2 : 1. This conclusion is surprising, because out of 24 candidates, only two isomers with relatively low symmetries are preferentially formed as a stable C_{84} fullerene.

These results provide some general insight into the formation and stability of higher fullerenes. Among many IPR-satisfying candidates, only limited numbers of isomers with, in general, low symmetry were found to be favoured in the carbon soot generated by arc-heating of graphite. Furthermore, comparison of the present results with those previously reported strongly indicates that unidentified experimental factors in the sample preparation influence the stabilization of certain kinds of isomers as well as fullerenes. All of this experimental evidence suggests that growth processes may have an essential role in determining the structure of fullerene cages, rather than the structures being determined simply by thermodynamical stability considerations.

After this work was completed, we learned that Diederich and Whetten had reported 30 NMR lines at almost the same positions as ours[11]. The formation of C_{82} has also been suggested to be sensitive to experimental conditions; no formation of C_{82} was seen by Diederich and Whetten[11] (see also ref. 2). □

SOLID C$_{60}$

We have already seen a first rough description of solid C$_{60}$ as a close-packed array of spheres in the pioneering paper of Krätschmer *et al.* in Chapter I of this volume.[Kr90*] As sample preparation techniques have progressed, a very detailed understanding of solid C$_{60}$ has emerged. In this chapter, we discuss some of the highlights.

The initial x-ray investigation was performed on a sample of mixed C$_{60}$ and C$_{70}$ that was crystallized from a benzene solution. It was described as a hexagonal-close-packed (hcp) structure, with significant background intensity, indicative of imperfect long range order. However, all of the observed lines are also present in a face-centered-cubic (fcc) lattice of the same intermolecular separation. Fleming *et al.* grew solvent-free single crystals of chromatographically purified C$_{60}$, and found them to be unambiguously fcc, with a high degree of orientational disorder.[Fl91a] The lattice constant was 14.198 Å, leading to a closest interfullerene distance of 10.04 Å.

It is worth noting that several subsequent experiments have found structural modifications when C$_{60}$ co-crystallizes with a solvent. A clathrate structure with *n*-pentane, having an unusual pseudo-tenfold symmetry, has been described.[Fl91b, Pe92] Another example is afforded by C$_{60}$ crystals grown from CS$_2$ solution, which have an orthorhombic lattice, 25.01 × 25.58 × 10.00 Å.[Ar92]

Early NMR experiments directly gave information about the nature of the molecular disorder in solid C$_{60}$.[Ty91a*, Ya91a] By observing a single sharp line at room temperature in solid C$_{60}$, it could be deduced that the molecule was rotating within the lattice, thereby averaging the chemical shift anisotropy. As the sample was cooled, the NMR spectrum broadened, indicating that the characteristic time for molecular rotation had become slower than about 50 μsec at 100 K.

Despite the molecular disorder, high energy neutron scattering experiments are able to determine bond lengths from the (instantaneous) radial distribution function.[Li91] Given the known molecular topology, the intramolecular bond lengths were found to be 1.39 Å for the double bonds (shared by two hexagons) and 1.46 Å for the single bonds (on a pentagon), in satisfactory agreement with NMR and gas-phase electron diffraction measurements.[He91c, Ya91b]

Recently, Chow *et al.* have performed very detailed measurements of the structure of solid C$_{60}$ at room temperature by synchrotron x-ray crystallography.[Ch92a] They found that the charge distribution of each fullerene is nearly spherical (averaged over time) due to its tumbling motion. There is a 26% variation from the lowest to the highest charge density around the spherical surface.

The electronic structure of C$_{60}$ was determined by photoemission experiments.[We91*] Such experiments measure the kinetic energy of an electron ejected from a solid film of C$_{60}$ when excited by soft x-rays. Furthermore, some information about the symmetry of the electronic state is revealed in the energy dependence of the shape of the spectrum. In that work, good agreement was found between the experimental spectra and theoretical (local density approximation) predictions for an isolated C$_{60}$ molecule.[We91*] Independent calculations also agreed with these results.[Zh91, Sa91b] The latter two groups explicitly considered the modifications to the free C$_{60}$ molecule imposed by the crystalline environment, and found it to be quite weak. Specifically, the molecules are bound into a lattice by their van der Waals attraction. Further experimental and theoretical work on the distribution of electronic states is reviewed by Weaver.[We92a, We92b]

The picture of van der Waals bonding between fullerene molecules in the crystalline state is reinforced by measurements of the compressibility of solid C$_{60}$.[Fi91*] The linear compressibility was found to be the same as the interlayer compressibility of graphite, within 10% experimental error. These authors also suggested that the difference between the apparent van der Waals diameter of carbon inferred from the interfullerene separation in solid C$_{60}$ (2.92 Å [Kr90*]) *vs.* that of graphite

(3.35 Å) was a consequence of the splaying of the π orbitals on the curved C_{60} molecule. This paper also describes briefly the steps required to prepare high purity ($> 99.5\%$), solvent-free powders of C_{60}.

A structural phase transition from the high-temperature fcc phase of solid C_{60} was found in low-temperature x-ray diffraction experiments by Heiney et al.[He91a*] Below 249 K, extra x-ray reflections corresponding to a simple cubic lattice appear. This group found that the low-temperature phase was still a fcc lattice of fullerenes, but that the four molecules in a cubic unit cell became orientationally inequivalent. Subsequent measurements by that and other groups have reported a range of transition temperatures, averaging around 255 K. Developments are thoroughly reviewed by Heiney.[He92]

The paper describing the appearance of the orientationally ordered phase suggested one possible structure for the low temperature phase (space group Pn3), but it did not agree satisfactorily with the experimental x-ray diffraction spectrum. Soon thereafter, a different pattern of orientations within the unit cell was suggested (space group Pa3), which gave a much better fit.[Sa91a*, He91b] The reader is referred to papers reprinted in this volume and to the review by Heiney for a description of the fullerene orientation in these structures.[He91a*, Sa91a*, He92] The experimentally determined Pa3 structure is illustrated on the cover of this book. An exhaustive discussion of the possible symmetries of the low-temperature phase has been given by [Ha92].

Structural refinements of the low temperature phase quickly showed that the Pa3 model was correct.[Da91] Furthermore, they gave an explanation for the curious pattern of fullerene orientations within the lattice. The chosen orientation optimizes the electrostatic interaction between fullerenes, by placing electron-poor pentagons in opposition to electron rich double bonds.

A detailed crystallographic analysis, such as that of [Da91], probably gives the most accurate measurement of molecular bond lengths. That experiment gave an average 6:6 (double) bond length of 1.39 Å, and 6:5 (single) bond length of 1.45 Å, in excellent agreement with room temperature neutron diffraction,[Li91] low temperature NMR, [Ya91b] and gas-phase electron diffraction measurements.[He91c]

In a more detailed study of the temperature dependence of the structure, it was found that a significant number of the fullerene molecules adopt a different, presumably higher energy, orientation relative to their neighbors.[Da92*] The fraction of these alternate orientations decreases with decreasing temperature to about 90 K, at which it freezes out at around 18%. Below that temperature, the structure may correspond to an orientational glass.[Da92*] There is also a cusp in the lattice constant at this temperature. "Misplaced" molecules are also seen in radial distribution functions from high energy neutron scattering.[Hu92]

The first theoretical predictions of the ground-state structure of solid C_{60} were based on Lennard-Jones interactions between carbon atoms on different fullerenes.[Gu91, Ch92b] The fcc phase was found to be more stable than hcp, although the calculations did not give the correct, observed (simple cubic) structures. Lu, Li, and Martin independently predicted the importance of bond charge in the attraction of double bonds and pentagons on adjacent fullerenes.[Lu92*] This study also found many local minima separated by large potential barriers in the orientational potential, helping to explain the glassy structure and features of the thermal conductivity and sound attenuation.

A different low temperature phase, with even lower symmetry, has been seen in electron diffraction measurements.[Va92] They propose a fcc structure with double the lattice constant of the room temperature form, containing 32 C_{60} molecules. This structure is supported by optically excited electron paramagnetic resonance measurements.[Gr92] However, it has been argued that this structure is inconsistent with existing powder diffraction data.[Ha93] Subsequent experiments have suggested that small quantities of impurities in the solid phase produce similar superlattice phases.[Fi93]

Moving from structure to dynamics, we recall the NMR results, [Ty91*, Ya91a] which indicate that the molecules in solid C_{60} tumble rapidly at room temperature but freeze at lower temperatures. Considering the structural data reviewed above, one would expect a more-or-less continuous rotational diffusion in the room temperature fcc phase, and either small oscillations or molecular jumps between favorable orientations in the low temperature phase. This difference was very clearly seen in the overall spin-lattice relaxation time near the orientational phase transition.[Ty91b] It was also found that the molecules undergo rapid reorientations well below the phase transition temperature.[Ty91b] The molecular reorientation time was determined directly from the magnetic field dependence of the spin-lattice relaxation time.[Jo92] The rotational motion was found to be thermally activated in both phases.

Neutron scattering has also been used to probe the dynamics of solid C_{60}. The rotational diffusion constant in the orientationally disordered fcc phase agrees with the NMR result.[Ne91, Jo92] At low temperature, well-defined librational excitations are observed, which soften and broaden as the temperature increases.[Ne92] The neutron scattering experiments are reviewed by Copley et al.[Co92] and by Prassides et al.[Pr92]

Intramolecular vibrations in solid C_{60} have been extensively probed by Raman scattering. Eklund et al. give a useful, critical review.[Ek92]

Dynamics of C_{60} molecules in the solid state are indirectly probed by a number of other techniques. Glassy dynamics at low temperatures are suggested by measurements of sound attenuation [Sh92] and thermal conductivity [Yu92]. The specific heat of solid C_{60} is dominated by a linear term below 3 K, suggestive of two-level tunneling centers commonly observed in glasses.[Be92]

Recently, Eklund's group has reported that solid C_{60} transforms from a van der Waals solid into a covalently bonded network upon irradiation with visible or ultraviolet light.[Ra93] The transformed phase is resistant to toluene, a good solvent for C_{60}. Significant changes in the infrared, Raman, UV–vis, and x-ray diffraction spectra were noted. However, the spectra are sufficiently similar to untransformed C_{60} that it appears that the fullerene framework is largely intact in the phototransformed phase.

Bibliography

References marked with * are reprinted in this volume.

Ar92   T. Arai et al., "Irreversible Structural Transition of Orthorhombic C_{60} Single Crystal to Face-Centered Cubic Phase," J. Phys. Soc. Japan 61, 1821–1822 (1992).

Be92   W. P. Beyermann, M. F. Hundley, J. D. Thompson, F. N. Diederich and G. Grüner, "Low-Temperature Specific Heat of C_{60}," Phys. Rev. Lett. 68, 2046–2049 (1992).

Ch92a   P. C. Chow et al., "Synchrotron X-ray Study of Orientational Order in Single Crystal C_{60} at Room Temperature," Phys. Rev. Lett. 69, 2943–2946 (1992), Erratum ibid. 3591 (1992).

Ch92b   A. Cheng and M. L. Klein, "Molecular-Dynamics Investigation of Orientational Freezing in Solid C_{60}," Phys. Rev. B45, 1889–1895 (1992).

Co92   J. R. D. Copley, D. A. Neumann, R. L. Cappelletti and W. A. Kamitakahara, "Neutron Scattering Studies of C_{60} and its Compounds," J. Phys. Chem. Solids 53, 1353–1371 (1992).

Da91   W. I. F. David et al., "Crystal Structure and Bonding of Ordered C_{60}," Nature 353, 147–149 (1991).

Da92*   W. I. F. David et al., "Structural Phase Transitions in the Fullerene C_{60}," Europhys. Lett. 18, 219–225 (1992).

Ek92   P. C. Eklund, P. Zhou, K. A. Wang, G. Dresselhaus and M. S. Dresselhaus, "Optical Phonon Modes in Solid and Doped C_{60}," J. Phys. Chem. Solids 53, 1391–1413 (1992).

Fi91*   J. E. Fischer et al., "Compressibility of Solid C_{60}," Science 252, 1288–1290 (1991).

Fi93   J. E. Fischer et al., "Existence of High-Order Superlattices in Orientationally Ordered C_{60}," Phys. Rev. B47, 14614–14617 (1993).

Fl91a   R. M. Fleming et al., "Diffraction Symmetry in Crystalline, Close-Packed C_{60}," in Clusters and Cluster-Assembled Materials, edited by R. S. Averback, D. L. Nelson and J. Bernholc, Mat. Res. Soc. Symp. Proc. 206, 691–695 (Materials Research Society, Pittsburgh, 1991).

Fl91b   R. M. Fleming et al., "Pseudotenfold Symmetry in Pentane-Solvated C_{60} and C_{70}," Phys. Rev. B44, 888–891 (1991).

Gr92   E. J. J. Groenen et al., "Triplet Excitation of C_{60} and the Structure of the Crystal at 1.2 K," Chem. Phys. Lett. 197, 314–318 (1992).

Gu91   Y. Guo, N. Karasawa and W. A. Goddard III, "Prediction of Fullerene Packing in C_{60} and C_{70} Crystals," Nature 351, 464–467 (1991).

Ha92   A. B. Harris and R. Sachidanandam, "Orientational Ordering of Icosahedra in Solid C_{60}," Phys. Rev. B46, 4944–4957 (1992).

Ha93   A. B. Harris and R. Sachidanandam, "Comment on 'New Orientationally Ordered Low-Temperature Superstructure in High-Purity C_{60}'," Phys. Rev. Lett. 70, 102 (1993).

He91a* P. A. Heiney *et al.*, "Orientational Ordering Transition in Solid C_{60}," *Phys. Rev. Lett.* **66**, 2911–2914 (1991).

He91b P. A. Heiney *et al.*, "Heiney *et al.* Reply [to a Comment by Sachidanandam and Harris]," *Phys. Rev. Lett.* **67**, 1468 (1991).

He91c K. Hedberg *et al.*, "Bond Lengths in Free Molecules of Buckminsterfullerene, C_{60} from Gas-Phase Electron Diffraction," *Science* **254**, 410–412 (1991).

He92 P. A. Heiney, "Structure, Dynamics and Ordering Transition of Solid C_{60}," *J. Phys. Chem. Solids* **53**, 1333–1352 (1992).

Hu92 R. Hu, T. Egami, F. Li and J. S. Lannin, "Local Intermolecular Correlations in C_{60}," *Phys. Rev.* **B45**, 9517–9520 (1992).

Jo92 R. D. Johnson, C. S. Yannoni, H. C. Dorn, J. R. Salem and D. S. Bethune, "C_{60} Rotation in the Solid State: Dynamics of a Faceted Spherical Top," *Science* **255**, 1235–1238 (1992).

Kr90* W. Krätschmer, L. D. Lamb, K. Fostiropoulos and D. R. Huffman, "Solid C_{60}: A New Form of Carbon," *Nature* **347**, 354–358 (1990).

Li91 F. Li, D. Ramage, J. S. Lannin and J. Conceicao, "Radial Distribution Function of C_{60}: Structure of Fullerene," *Phys. Rev.* **B44**, 13167–13170 (1991).

Lu92* J. P. Lu, X.-P. Li and R. M. Martin, "Ground State and Phase Transitions in Solid C_{60}," *Phys. Rev. Lett.* **68**, 1551–1554 (1992).

Ne91 D. A. Neumann *et al.*, "Coherent Quasielastic Neutron Scattering Study of the Rotational Dynamics of C_{60} in the Orientationally Disordered Phase," *Phys. Rev. Lett.* **67**, 3808–3811 (1991).

Ne92 D. A. Neumann *et al.*, "Rotational Dynamics and Orientational Melting of C_{60}: A Neutron Scattering Study," *J. Chem. Phys.* **96**, 8631–8633 (1992).

Pe92 S. Pekker *et al.*, "Structure and Stability of Crystalline C_{60} · *n*-pentane Clathrate," *Solid State Commun.* **83**, 423–426 (1992).

Pr92 K. Prassides *et al.*, "Fullerenes and Fullerides in the Solid State: Neutron Scattering Studies," *Carbon* **30**, 1277–1286 (1992).

Ra93 A. M. Rao *et al.*, "Photoinduced Polymerization of Solid C_{60} Films," *Science* **259**, 955–957 (1993).

Sa91a* R. Sachidanandam and A. B. Harris, "Comment on 'Orientational Ordering Transition in Solid C_{60}'," *Phys. Rev. Lett.* **67**, 1467 (1991).

Sa91b S. Saito and A. Oshiyama, "Cohesive Mechanism and Energy Bands of Solid C_{60}," *Phys. Rev. Lett.* **66**, 2637–2640 (1991).

Sh92 X. D. Shi *et al.*, "Sound Velocity and Attenuation in Single-Crystal C_{60}," *Phys. Rev. Lett.* **68**, 827–830 (1992).

Ty91a* R. Tycko *et al.*, "Solid State Magnetic Resonance Spectroscopy of Fullerenes," *J. Phys. Chem.* **95**, 518–520 (1991).

Ty91b R. Tycko *et al.*, "Molecular Dynamics and the Phase Transition in Solid C_{60}," *Phys. Rev. Lett.* **67**, 1886–1889 (1991).

Va92 G. Van Tendeloo, S. Amelinckx, M. A. Verheijen, P. H. M. van Loosdrecht and G. Meijer, "New Orientationally Ordered Low-Temperature Superstructure in High-Purity C_{60}," *Phys. Rev. Lett.* **69**, 1065–1068 (1992).

We91* J. H. Weaver *et al.*, "Electronic Structure of Solid C_{60}: Experiment and Theory," *Phys. Rev. Lett.* **66**, 1741–1744 (1991).

We92a J. H. Weaver, "Fullerenes and Fullerides: Photoemission and Scanning Tunnelling Microscopy Studies," *Acc. Chem. Res.* **25**, 143–149 (1992).

We92b J. H. Weaver, "Electronic Structures of C_{60}, C_{70}, and the Fullerides: Photoemission and Inverse Photoemission Studies," *J. Phys. Chem. Solids* **53**, 1433–1447 (1992).

Yu92 R. C. Yu, N. Tea, M. B. Salamon, D. Lorents and R. Malhotra, "Thermal Conductivity of Single Crystal C_{60}," *Phys. Rev. Lett.* **68**, 2050–2053 (1992).

Ya91a C. S. Yannoni, R. D. Johnson, G. Meijer, D. S. Bethune and J. R. Salem, "^{13}C NMR Study of the C_{60} Cluster in the Solid State: Molecular Motion and Carbon Chemical Shift Anisotropy," *J. Phys. Chem.* **95**, 9–10 (1991).

Ya91b C. S. Yannoni, P. P. Bernier, D. S. Bethune, G. Meijer and J. R. Salem, "NMR Determination of Bond Lengths in C_{60}," *J. Am. Chem. Soc.* **113**, 3190–3192 (1992).

Zh91 Q.-M. Zhang, J.-Y. Yi and J. Bernholc, "Structure and Dynamics of Solid C_{60}," *Phys. Rev. Lett.* **66**, 2633–2636 (1991).

Reprinted with permission from The Journal of Physical Chemistry
Vol. 95, No. 2, pp. 518–520, 1991

Solid-State Magnetic Resonance Spectroscopy of Fullerenes

R. Tycko,* R. C. Haddon, G. Dabbagh, S. H. Glarum, D. C. Douglass, and A. M. Mujsce

AT&T Bell Laboratories, 600 Mountain Avenue, Murray Hill, New Jersey 07974
(Received: November 2, 1990)

We report solid-state ^{13}C NMR measurements on powder samples of C_{60} and of a mixture of C_{60} and C_{70}. The NMR results show that, at 296 K, C_{60} molecules rotate rapidly and isotropically in the solid state, while C_{70} molecules rotate somewhat more anisotropically. These results are consistent with the proposed spherical geometry of C_{60} and prolate spheroidal geometry of C_{70}. The rotational correlation time of C_{60} molecules in the solid state becomes greater than 50 μs at about 100 K.

Introduction

Interest in the structures and properties of fullerenes[1] has received new impetus from the recent discovery that the molecules C_{60} and C_{70} can be prepared in large quantities by comparatively simple procedures.[2] The ready availability of solid samples of C_{60} and C_{70} now permits their characterization by a variety of physical methods. In this paper, we report the results of solid-state ^{13}C nuclear magnetic resonance (NMR) measurements on powder samples of C_{60} and of a mixture of C_{60} and C_{70}. Our NMR results indicate that C_{60} rotates rapidly and nearly isotropically in the solid state at 296 K and that C_{70} also rotates at 296 K, although somewhat anisotropically. The rotation of C_{60} molecules becomes slow on the time scale of our measurements at about 100 K.

Materials and Methods

Powder samples containing a mixture of C_{60} and C_{70} were prepared as described previously.[2,3] The ^{13}C NMR spectrum of the dissolved powder in C_6D_6 shows the line at 143.2 ppm characteristic of C_{60}. Solid-state magic angle spinning (MAS) ^{13}C NMR spectra (see below) show five additional lines from C_{70} and indicate a $C_{60}:C_{70}$ ratio of $3.9 \pm 0.6:1$. Comparison of the integrated ^{13}C MAS NMR signal from the fullerene powder with the ^{13}C NMR signal from a known weight of isopropyl alcohol indicates that the entire mass of the powder is accounted for in the ^{13}C MAS spectrum, within the accuracy of the measurement (estimated to be 10%). Mass spectrometry shows no species other than C_{60} and C_{70} in the mass range 700–900 amu. Samples of pure C_{60} were prepared by column chromatography on neutral alumina, using hexane as the solvent.

Solid-state ^{13}C NMR spectra at 100.48 MHz were obtained on a Chemagnetics CMX spectrometer, using home-built static and MAS probes. Approximately 70 mg of powder were used for the NMR measurements. Frequency scales in the figures are in ppm with respect to TMS or in kilohertz with respect to the carrier frequency.

Solid-state electron spin resonance (ESR) spectra at 9.25 GHz were also measured, using microwave power levels <1 mW and conventional field-modulation techniques.

Results and Discussion

Figure 1a shows the solid-state ^{13}C NMR spectrum of the C_{60} powder at 296 K, without sample spinning. A single line is observed with a width of about 5 ppm. For comparison, Figure 1c shows the ^{13}C NMR spectrum of polycrystalline pentamethylbenzene, also without sample spinning. The broad powder pattern line shape in Figure 1c, with a width of about 200 ppm, is typical of aromatic carbons in the solid state, resulting from the large chemical shift anisotropies (CSA) of aromatic carbons.[4] The fact that the spectrum of C_{60} shows a line at a frequency characteristic of aromatic carbons but with a greatly reduced line width, indicates that *the C_{60} molecules rotate rapidly and nearly isotropically in the solid state*, thus averaging out the CSA. The rotational correlation time must be short compared to the inverse of the CSA width, in this case about 50 μs. This result is consistent with the spherical geometry proposed for C_{60} and with packing of C_{60} molecules in the solid that does not strongly favor particular relative orientations of neighboring molecules. Figure 1b shows the spectrum of C_{60} at 100 K. At this temperature, most of the spectral intensity is in a broad powder pattern component, although a sharp peak at the isotropic frequency remains, demonstrating that the rotational correlation time is on the order of or greater than 50 μs.[5,6]

(1) Kroto, H. W.; Heath, J. R.; O'Brien, S. C.; Curl, R. F.; Smalley, R. E. *Nature* **1985**, *318*, 162.

(2) Kratschmer, W.; Lamb, L. D.; Fostiropoulos, K.; Huffman, D. R. *Nature* **1990**, *347*, 354.

(3) Haufler, R. E.; Conceicao, J.; Chibante, L. P. F.; et al. *J. Phys. Chem.* **1990**, *94*, 8634.

(4) Duncan, T. M. *A Compilation of Chemical Shift Anisotropies*; Farragut Press: Chicago, 1990.

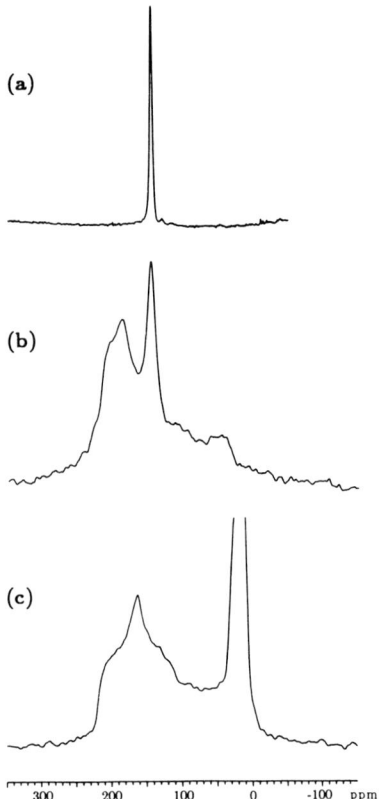

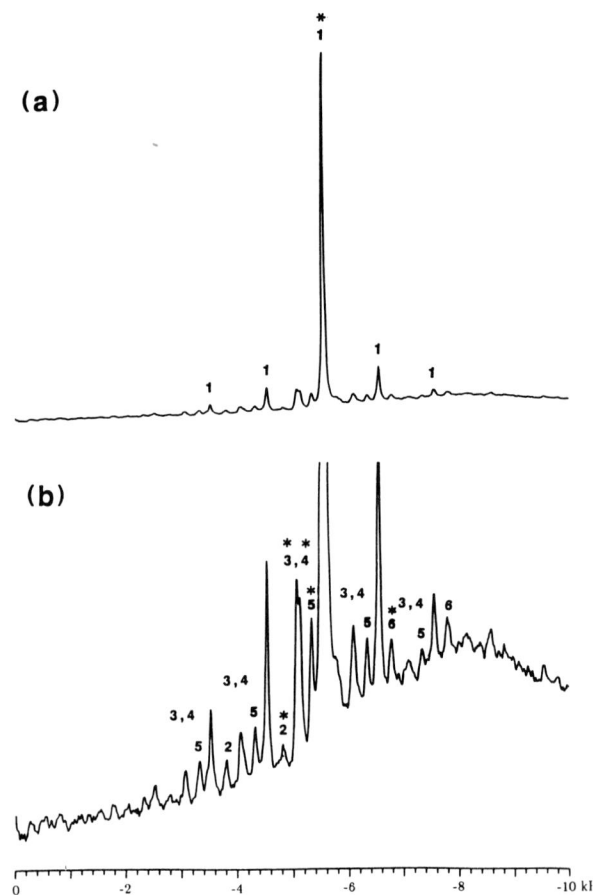

Figure 1. (a) ^{13}C NMR spectrum of C_{60} powder at 296 K, without sample spinning. The narrowness of the line indicates rapid rotation of C_{60} molecules in the solid state. (b) Spectrum of C_{60} at 100 K. The increase in the molecular rotational correlation time with decreasing temperature causes the line to broaden into a powder pattern shape. (c) ^{13}C NMR spectrum of polycrystalline pentamethylbenzene, shown for the sake of comparison. The line between 10 and 40 ppm is due to methyl carbons; the broad powder pattern extending to 220 ppm is due to aromatic carbons.

Figure 3. MAS spectrum at ν_R = 1.01 kHz, shown on two different vertical scales. Spinning sidebands are labeled according to their corresponding centerbands. Centerbands are indicated by asterisks. Baseline curvature is due to probe background. The ratios of sideband to centerband intensities are greater for C_{70} lines than for C_{60} lines, indicating that the rotation of C_{70} molecules is more anisotropic.

at 151.4, 148.8, 148.1, 146.2, and 131.7 ppm are resolved. To obtain approximate measurements of residual CSA, which results from incomplete averaging due to anisotropy of the molecular rotational motion, we acquire MAS spectra at a reduced spinning frequency. Figure 3 shows the MAS spectrum at ν_R = 1.01 kHz. Spinning sidebands are clearly seen, spaced at multiples of ν_R from the center bands. Spinning sidebands are generally observed in MAS spectra when ν_R does not greatly exceed the CSA width. Sidebands and centerbands arising from the same ^{13}C resonance are indicated by the same number (1–6) in Figures 2 and 3. The centerbands are identified by asterisks in Figure 3. The striking feature in Figure 3 is that the ratios of sideband to centerband intensities are much greater for the C_{70} peaks than for the C_{60} peaks. Assuming that the CSA values for nonrotating C_{60} and C_{70} molecules are approximately equal, this observation suggests that *the rotation of C_{70} molecules in the solid state is more anisotropic than that of C_{60} molecules* over a time scale of at least 50 μs. The more anisotropic motion of C_{70} is consistent with its proposed prolate spheroidal geometry.

The ^{13}C lines are all downfield of the resonances usually observed in planar aromatic molecules. This is consistent with the rehybridization expected in the fullerene structures.[7] The C_{70} resonance at 131.7 ppm in the solid may then be due to carbon atoms in the belt of the prolate spheroid, which require the least rehybridization.[8]

We have also measured ^{13}C spin–lattice relaxation times (T_1) in the mixed powder in room-temperature MAS experiments, using the saturation–recovery technique. T_1 for C_{60} is 28 ± 5 s, while

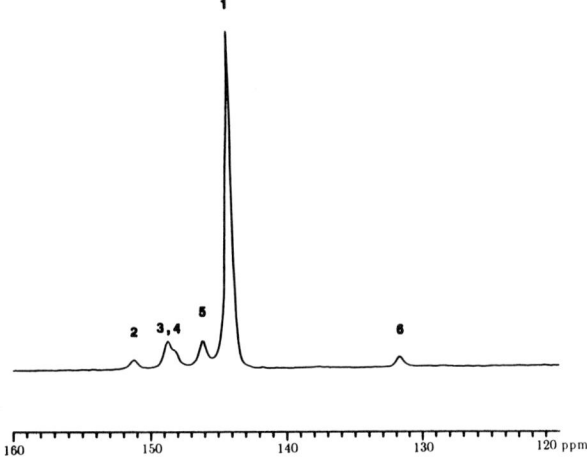

Figure 2. (a) ^{13}C MAS NMR spectrum of a powder containing a mixture of C_{60} and C_{70}, ν_R = 5.00 kHz. The C_{60} line is labeled 1. The C_{70} lines are labeled 2–6.

Figure 2 shows the ^{13}C MAS spectrum of a mixture of C_{60} and C_{70} at 296 K, at a spinning frequency ν_R = 5.00 kHz. The C_{60} line at 144.1 ppm is now less than 0.5 ppm wide, and C_{70} lines

(5) Mehring, M. *Principles of High Resolution NMR in Solids*; Springer-Verlag: New York, 1983.

(6) Yannoni, C. S.; Johnson, R. D.; Meijer, G.; Bethune, D. S.; Salem, J. R. Submitted for publication. These authors show that the isotropic component of the spectrum is no longer present at 77 K. Slight differences between their results and ours for the temperature dependence of the ^{13}C NMR spectrum may be due to differences in sample purity and in the external magnetic field strength used in the NMR measurements.

(7) Haddon, R. C. *Acc. Chem. Res.* **1988**, *21*, 243.

(8) Haddon, R. C.; Brus, L. E.; Raghavachari, K. *Chem. Phys. Lett.* **1986**, *131*, 165.

T_1 for the C_{70} lines is shorter. Generally speaking, the relaxation times are short compared with those of typical rigid, diamagnetic, insulating organic compounds in the solid state, supporting the idea that there is substantial molecular motion in the solid fullerenes.

Finally, we have detected an ESR signal with $g = 2.0019$ and a derivative peak-to-peak line width of 1.9 G at 296 K in solid samples before chromatography. Integrated absorption plots indicate about 1 electron spin/1000 C_{60} units. This signal may be due to a carbon spheroid composed of an odd number of carbon atoms.

Acknowledgment. We thank R. E. Smalley, R. E. Haufler, Y. Chai, L. P. F. Chibante, and J. Conceicao for generously providing the sample of C_{60} and C_{70} used in our initial experiments. We are grateful to C. S. Yannoni and R. D. Johnson for discussing their NMR results with us before submitting them for publication.

Reprinted with permission from Physical Review Letters
Vol. 66, No. 13, pp. 1741–1744, 1 April 1991
© 1991 The American Physical Society

Electronic Structure of Solid C$_{60}$: Experiment and Theory

J. H. Weaver, José Luís Martins, T. Komeda, Y. Chen, T. R. Ohno, G. H. Kroll, and N. Troullier

Department of Materials Science and Chemical Engineering, University of Minnesota,
Minneapolis, Minnesota 55455

R. E. Haufler and R. E. Smalley

Rice Quantum Institute and Departments of Chemistry and Physics, Rice University, Houston, Texas 77251
(Received 20 December 1990)

Synchrotron-radiation and x-ray photoemission studies of the valence states of condensed phase-pure C$_{60}$ showed seventeen distinct molecular features extending ~23 eV below the highest occupied molecular states with intensity variations due to matrix-element effects involving both cluster and free-electron-like final states. Pseudopotential calculations established the origin of these features, and comparison with experiment was excellent. The sharp C $1s$ main line indicated a single species, and the nine satellite structures were due to shakeup and plasmon features. The 1.9-eV feature reflected transitions to the lowest unoccupied molecular level of the excited state.

PACS numbers: 79.60.−i, 31.20.Sy, 33.60.−q

The structure of C$_{60}$ has been proposed to be a truncated icosahedron with twenty six-membered rings and twelve five-membered rings.[1] In this allotrope, the atoms are equivalent, giving a closed-shell electronic structure and a molecule that is unique in nature. While the existence of C$_{60}$ has been known for several years, moderately large-scale production and phase separation of C$_{60}$ and other fullerenes were not possible before the work of Krätschmer et al.[2] That breakthrough virtually assured rapid development in understanding the properties of these novel forms of matter.

This paper focuses on the electronic states of C$_{60}$. The experimental results, obtained with photoemission, show seventeen resolvable features within ~23 eV of the highest occupied electronic state. Many of these structures are remarkably sharp, rising from near-zero backgrounds. Indeed, the spectra resemble those of simple molecules rather than other forms of pure carbon, while exhibiting an overall bandwidth and a distribution of σ and π character that is the same as for graphite and diamond.[3,4] These results provide a long awaited test for calculations of the electronic configuration of C$_{60}$ and related fundamental properties. The theoretical results presented here were obtained with a pseudopotential-based local-density-approximation calculation of the truncated icosahedral structure. Comparison with experiment shows agreement for all of the features, with a maximum error of 0.5–0.7 eV in the p_σ levels ~10 eV below the highest occupied states. These results demonstrate that the properties of this symmetric stable structure of carbon can be described quantitatively to a high level of accuracy, in contrast to what has been found for clusters with substantial numbers of atoms but less definite structure.

The fullerenes were formed by the contact arc method,[5] with subsequent separation by solution with toluene. Phase-pure C$_{60}$ was obtained by a liquid chroma-tography process on alumina diluted with mixtures of hexanes. The resulting C$_{60}$ was then rinsed in methanol, dried, and placed in Ta boats that were ~10 cm from the substrate onto which C$_{60}$ was to be condensed. Degassing to ~475 °C desorbed the toluene, as well as H$_2$O and CO. Evaporations were performed at ~550 °C. The uppermost experimental features were very sharp, and the emission intensity between them was essentially "base line," indicating the phase-pure character of films grown in this manner. Substrates were prepared by cleaving GaAs(110) and InP(110) in situ at $< 10^{-10}$ Torr, and C$_{60}$ was condensed onto these mirror-like surfaces at 300 K. The synchrotron-radiation studies exploited the tunability of the source to obtain high-resolution valence-band spectra for $40 \le h\nu \le 200$ eV, as well as C $1s$ spectra at higher energy. The x-ray photoemission measurements used a monochromatized Al $K\alpha$ source ($h\nu = 1486.6$ eV) to examine C $1s$–derived structure and the valence bands. The results presented here are for relatively thick films that attenuated the substrate emission completely. For thinner films, the persistence of the substrate core-level features provided an independent energy reference.

To our knowledge, these results are the first to reveal the full valence-band region and structure associated with the C $1s$ level. However, Lichtenberger et al.[6] have obtained photoemission results with $h\nu = 21.2$ and 40.8 eV, and comparison shows good agreement in the region of overlap, although our results show additional structure. They did not detect shakeup or plasmon features associated with the C $1s$ emission, but we show that there are nine such structures above the main line. The only previous studies[7] used ultraviolet photoemission to investigate the gap between the highest occupied and lowest empty molecular orbitals for gas-phase C$_{60}^{-}$.

The calculations were performed using the pseudopotential plane-wave formalism and the local-density ap-

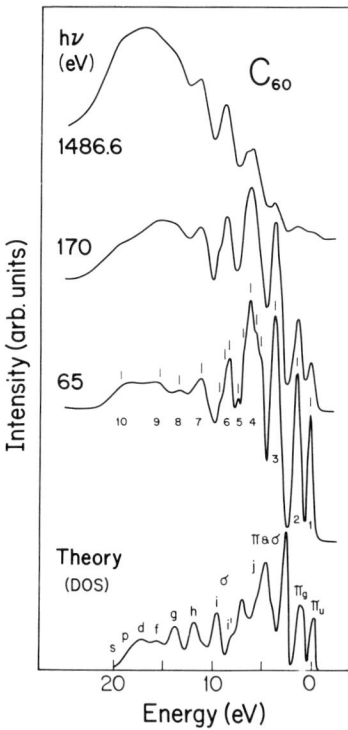

FIG. 1. Representative photoemission EDCs for condensed C$_{60}$ showing the full valence band and modulation with $h\nu$ of the cluster features. Those within ~5 eV of the highest occupied level are p_π derived, those between ~5 and ~12 eV are hybrids of s-p_σ character, and features below ~12 eV are primarily s derived. The full bandwidth is the same as graphite and diamond, but only C$_{60}$ has the richness in structure. The bottom curve is the density of states (DOS) calculated with the pseudopotential local-density method. The numbers and vertical lines associate experimental and theoretical features.

proximation.[8] The atomic positions for C$_{60}$ were determined from the minimization of the total energy, yielding C-C bond lengths of 1.382 and 1.444 Å. We assumed that the solid has a fcc lattice with a tetrahedral space group. Plane waves with an energy up to a cutoff of 49 Ry were included in the basis set (~28 000 plane waves). With this energy cutoff and using new "soft" pseudopotentials,[9] the total energy was converged to within 50 meV per atom. The one-electron Schrödinger equation was solved with an iterative method.[10] The self-consistent potential was calculated using the single Γ point for the Brillouin-zone integration. Two special $\bar{k}$ points[11] were used to calculate the density of states.

Figure 1 shows representative energy distribution curves (EDCs) for C$_{60}$ taken at $h\nu = 65$, 170, and 1486.6 eV; additional spectra acquired from 20 to 200 eV in 2-eV increments will be discussed elsewhere.[12] Figure 2 shows an EDC acquired at 50 eV with an experimental resolution of 0.2 eV. The zero of energy is the emission maximum of the highest occupied feature. Calculations for neutral C$_{60}$, C$_{60}^-$, and C$_{60}^+$ indicate that the removal or addition of one electron would displace these levels rigidly.[13] With account of the position the Fermi level

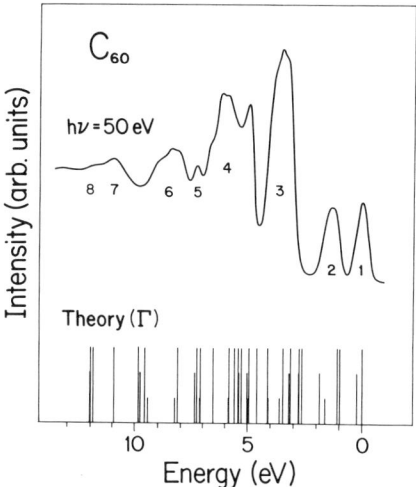

FIG. 2. High-resolution EDC showing seventeen resolvable experimental features in the photoemission spectrum of C$_{60}$. The theoretical lines show the electron levels at the point Γ of the Brillouin zone. Their height is proportional to the degeneracy.

E_F and the electron affinity of the GaAs(110) substrate, the highest occupied feature of photoexcited solid C$_{60}$ would be 7.3 eV below the vacuum level, assuming alignment of the vacuum level of C$_{60}$ with the substrate. An independent check of this energy was obtained by x-ray-photoemission-spectroscopy studies of C$_{60}$ on InP(110) where a value of 7.4 eV was obtained. Lichtenberger et al.[6] reported a value of 7.6 ± 0.2 eV.

The results of Figs. 1 and 2 show dramatic variations in the line shapes as a function of photon energy. The lower-photon-energy results show dominant emission from the upper half of the valence bands, especially those features labeled 1–4. With increasing $h\nu$, the intensity of features 1–4 diminished relative to the deeper structures, notably 9–10. Features 6 and 7 were well defined at all energies. These changes reflect the photoionization cross sections of the s- and p-derived states. In particular, the cross section of C $2p$–derived states is large at low energy, is comparable to that of the s states at $h\nu = 125$ eV, and is about 13 times smaller than the s states for energies of about 1486 eV.[14] The $2s$ enhancement relative to the $2p$ character reflects the nodal structure of the former and the larger matrix elements for coupling to plane-wave states at ~1475 eV. McFeely et al.[3] and Bianconi, Hagström, and Bachrach[4] used this property to identify the character of the states in graphite and diamond. Both groups concluded that $2p_\pi$ levels fell within ~5 eV of E_F, that $2p_\sigma$ levels were concentrated between ~5 and ~10 eV of E_F, and that deeper states were derived from s-like σ bands. From Figs. 1 and 2, we see that C$_{60}$ exhibits an analogous behavior, despite differences in bond angles and structure, with π character in the leading structures and s character at the bottom. Significantly, the overall bandwidths of C$_{60}$, graphite, and diamond are essentially the same, namely,

~23 eV. Such similarities notwithstanding, C_{60} exhibits much more detailed structure, and these features give insight into the electronic structure.[12]

Analysis of the variations in relative intensities of the peaks identified in Figs. 1 and 2 for the cluster levels shows that the high-lying final states of the cluster play an important role in the photoexcitation probability. From Figs. 1 and 2, and especially from more closely spaced EDCs,[12] it is clear that the intensities of features 1–3 oscillate until about $h\nu = 140$ eV. For example, the peak-height intensity ratio I_1/I_2 was 0.88 at $h\nu = 40$ eV, 1.48 at 45 eV, 1.06 at 50 eV, and 0.78 at 65 eV. Such changes reflect excitations to states in the continuum which are not free-electron-like, with differences due to the fact that feature 1 has odd symmetry while feature 2 has even symmetry. The importance of direct transitions and selection rules to ~140 eV stands in contrast to results for crystalline graphite where the free-electron-like final-state regime was reached by ~90 eV.[4] For C_{60}, the spectra replicate the state density by $h\nu \cong 170$ eV, with fivefold degeneracy of feature 1 and ninefold degeneracy of feature 2.

Our results also show interesting matrix-element effects for features 4 and 5. In particular, the shallowest component of peak 4 at 6 eV is almost resolution limited at $h\nu = 50$ eV (Fig. 2) but is only a weak shoulder at $h\nu = 65$ eV (Fig. 1). Feature 5 is clear from $h\nu = 50$ to 80 eV but was not discernible at other energies, despite the fact that the experimental resolution was not varied. Again, these effects demonstrate the importance of dipole selection rules for these highly symmetric molecules. In contrast, solids rarely exhibit modulation in structure at such high photon energies except for transitions related to critical points, i.e., primary Mahan cone emission.[15]

Examination of the high-resolution EDC of Fig. 2 shows seventeen distinct molecular levels associated with ten spectral features. The width of feature 2 almost certainly indicates that it is derived from at least two lines or a broadened band, but we have not yet been able to resolve them at 300 K. We know of no cluster containing sixty atoms that retains levels so well resolved as those in Figs. 1 and 2. While studies of small molecules have shown very detailed structure with vibrational splittings,[16] such effects are lost for clusters with such large numbers of metal or semiconductor atoms and the accompanying thermal broadening. C_{60}, with its high symmetry and high molecular level degeneracy, is a beautiful counterexample. One can only speculate that such detail will be found in larger fullerenes.

In Fig. 2 we show the predicted energy levels at Γ for the ground state of C_{60}, aligned to the highest occupied level. The degeneracy is indicated by the height of the lines. The first feature (derived from the fivefold-degenerate h_u state of the isolated molecule) appears as a pair of levels split by the tetrahedral crystal field with threefold and twofold degeneracy, while feature 2 has four distinct eigenvalues derived from h_g and g_g states. The density of states of solid C_{60} calculated using two special $\bar{k}$ points is shown in the bottom of Fig. 1. The theoretical spectrum was convoluted with a Gaussian of width 0.23 eV$+0.2|\Delta E|$ to simulate experimental resolution and lifetime effects. ΔE is the binding energy with respect to the highest occupied level. The agreement between theory and experiment is very good, with discrepancies of ~0.5 and ~0.7 eV in the positions of features 6 and 7.

The quasispherical shape of the C_{60} molecule suggested that we should analyze the wave functions in terms of their expansion in spherical harmonics. We found that they were dominated by a single l component and we used it to label the lowest peaks of the density of states in Fig. 1 $(s,p,d,\ldots)$. The highest peaks also have well-defined angular character, but they have significant splittings and overlaps that will be described in detail elsewhere. Features 1 and 2 in the spectrum, for example, correspond to π states with $l = 5$ and 4 angular character, respectively.

Several authors[16] have reported calculations of the electronic states of isolated C_{60}, and their analyses show p_z-derived eigenstates within ~4 eV of the highest molecular level arising from orbitals directed radially from the cluster center with h_u, g_g, h_g, and g_u symmetry. Our experimental and theoretical results are in qualitative agreement with them, supporting their predictions that the levels were narrow with little orbital mixing. Comparison of the predictions of Saito[13] and Bernholc[17] to the experimental results shows good agreement. Note that all of the calculations were done without knowledge of the experimental spectra, and peak associations may change as refinements are added.

Figure 3 shows the C $1s$ spectrum measured with $h\nu = 1486.6$ eV. Of the ten structures that can be readily resolved, the first corresponds to the main line itself, centered at 282.9 eV relative to the center of the leading valence-band feature. Its FWHM, 0.65 eV, is probably resolution limited by the spectrometer. The main line is symmetric and offers no indication of more than a single C species. Features 2–9 probably represent π to π^* or σ^* transitions, with linewidths comparable to those observed in small molecules such as benzene. Following the discussion of Bigelow and Freund,[18] we can assume that the energy levels of the π^* configuration that result from the sudden creation of the core hole are displaced rigidly from those of the ground state. In this simple approximation, the first feature at 1.9 eV could be attributed to a shakeup process involving direct excitation across the gap. Such transitions involve states of the same symmetry and are dictated by monopole selection rules.

The satellite features of Fig. 3 for C_{60} resemble those of small molecules at low energy but also mimic the behavior of graphite, glassy carbon, and diamond in the region of the plasmon losses, ~30 eV.[3] Here we see a broad feature centered ~28 eV below the main line.

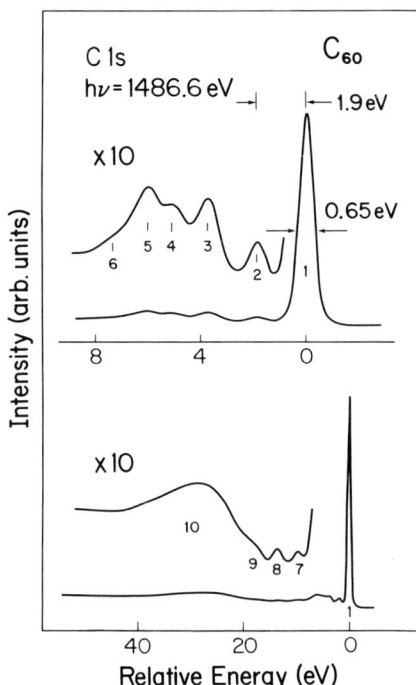

FIG. 3. C 1s features for C_{60} referenced to the main line 282.9 eV below the center of the highest-occupied-state feature. Features 2–9 reflect shakeup structures of the form π to π^* or σ^* akin to those in small C-based molecules. Feature 2 probably reflects excitation across the gap of the excited state. Feature 10 reflects a plasmon loss due to excitation of a collective mode of the cluster or the condensed solid.

The origin of this feature is particularly intriguing because equivalent structures are not observed in small molecules. We can speculate that it is due to collective excitations of the electrons of the cluster itself, representing the normal modes of a spherical shell of charge of average radius ~3.5 Å with a charge-free central region. At the same time, the loss structure might be due to a bulk plasmon associated with solid C_{60}. Inverse-photoemission studies and investigations of C_{60} condensed in a dispersed form should make it possible to choose one or the other possibility, and such studies are presently underway.[19]

This study has shown that seventeen experimental features can be identified in the high-resolution photoemission spectrum of C_{60} and that their origins can be determined from *ab initio* calculations of the eigenstate spectrum. The sharpness and number of such levels is reminiscent of the molecular orbits of smaller molecules. There appears to be no fundamental limitation in our ability to fully explore these levels and their implication on the physical and chemical properties of C_{60}. At the same time, it is remarkable that a structure containing so many electrons can have such a sharp distribution, a fact that no doubt reflects the high symmetry of the molecule. In the near future, it can be anticipated that the preparation of phase-pure C_{70} and other fullerenes will allow detailed comparisons of their orbital structures. In the meantime, we regard the agreement of theory based on a truncated icosahedron with the experiment as strong evidence that the structure is correct.

The work was supported by the National Science Foundation (University of Minnesota and Rice University), the Minnesota Supercomputer Institute, and the Robert A. Welch Foundation (Rice University). The synchrotron-radiation studies were done at Aladdin, a user facility supported by the National Science Foundation, and the assistance of its staff is gratefully acknowledged. Special thanks are due to M. B. Jost and D. M. Poirier for stimulating discussions and to S. Saito and J. Bernholc for sharing their calculations prior to publication.

[1] H. W. Kroto, J. R. Heath, S. C. O'Brien, R. F. Curl, and R. E. Smalley, Nature (London) **318**, 162 (1985).

[2] W. Krätschmer, L. D. Lamb, K. Fostiropoulos, and D. R. Huffman, Nature (London) **347**, 354 (1990).

[3] F. R. McFeeley, S. P. Kowalczyk, L. Ley, R. G. Cavell, R. A. Pollak, and D. A. Shirley, Phys. Rev. B **9**, 5268 (1974).

[4] A. Bianconi, S. B. M. Hagström, and R. Z. Bachrach, Phys. Rev. B **16**, 5543 (1977).

[5] R. E. Haufler *et al.*, J. Chem. Phys. **94**, 8634 (1990); R. E. Haufler, Y. Chai, L. P. F. Chibante, J. Conceicao, C.-M. Jin, L.-S. Wang, S. Maruyama, and R. E. Smalley, Mater. Res. Soc. Symp. Proc. **206** (to be published).

[6] D. L. Lichtenberger, K. W. Nebesny, C. D. Ray, D. R. Huffman, and L. D. Lamb, Chem. Phys. Lett. (to be published).

[7] S. H. Yang, C. L. Pettiette, J. Conceicao, O. Cheshnovsky, and R. E. Smalley, Chem. Phys. Lett. **139**, 233 (1987); R. F. Curl and R. E. Smalley, Science **242**, 1017 (1988).

[8] For a review, see W. E. Pickett, Comput. Phys. Rep. **9**, 115 (1989).

[9] N. Troullier and J. L. Martins, Solid State Commun. **74**, 613 (1990); Phys. Rev. B **43**, 1993 (1991).

[10] J. L. Martins and M. L. Cohen, Phys. Rev. B **37**, 6134 (1988).

[11] D. J. Chadi and M. L. Cohen, Phys. Rev. B **8**, 5747 (1973).

[12] P. J. Penning, D. M. Poirier, N. Troullier, J. L. Martins, J. H. Weaver, R. E. Haufler, and R. E. Smalley (to be published).

[13] S. Saito, Mater. Res. Soc. Symp. Proc. **206** (to be published).

[14] U. Gelius, in *Electron Spectroscopy*, edited by D. A. Shirley (North-Holland, Amsterdam, 1972), p. 311.

[15] F. J. Himpsel, Adv. Phys. **32**, 1 (1983); E. W. Plummer and W. Eberhardt, Adv. Chem. Phys. **49**, 533 (1982).

[16] S. Satpathy, Chem. Phys. Lett. **130**, 545 (1986); R. L. Disch and J. M. Schulman, Chem. Phys. Lett. **125**, 465 (1986); H. P. Lüth and J. Almlöf, Chem. Phys. Lett. **135**, 357 (1987); P. D. Hale, J. Am. Chem. Soc. **108**, 6087 (1986); S. Larsson, A. Volosov, and A. Rosén, Chem. Phys. Lett. **137**, 501 (1987).

[17] J. Bernholc (private communication).

[18] R. W. Bigelow and H.-J. Freund, J. Chem. Phys. **77**, 5552 (1982).

[19] M. B. Jost, N. Troullier, D. M. Poirier, J. L. Martins, J. H. Weaver, L. P. F. Chibante, and R. E. Smalley (to be published).

Reprinted with permission from Science
Vol. 252, pp. 1288–1290, 31 May 1991
© 1991 The American Association for the Advancement of Science

Compressibility of Solid C_{60}

John E. Fischer,* Paul A. Heiney, Andrew R. McGhie,
William J. Romanow, Arnold M. Denenstein,
John P. McCauley, Jr., Amos B. Smith III

Room-temperature powder x-ray diffraction profiles have been obtained at hydrostatic pressures $P = 0$ and 1.2 gigapascals on the solid phase of cubic C_{60} ("fullerite"). Within experimental error, the linear compressibility $d(\ln a)/dP$ is the same as the interlayer compressibility $d(\ln c)/dP$ of hexagonal graphite, consistent with van der Waals intermolecular bonding. The volume compressibility $-d(\ln V)/dP$ is $7.0 \pm 1 \times 10^{-12}$ square centimeter per dyne, 3 and 40 times the values for graphite and diamond, respectively.

THE RECENT DISCOVERY OF AN EFFIcient synthesis of C_{60} and C_{70} has, among other things, facilitated the study of a new class of molecular crystals ("fullerites") based on these molecules ("fullerenes") (1). The first x-ray powder diffraction profile of solid C_{60} was analyzed in terms of a faulted hexagonal close-packed (hcp) lattice (1), consistent with close packing of spherical molecules but with weak second-neighbor intermolecular interactions. A more recent single-crystal study shows that the molecules are actually centered on sites of an unfaulted face-centered cubic (fcc) Bravais lattice but with a high degree of rotational disorder (2). With ei-

J. E. Fischer, Laboratory for Research on the Structure of Matter and Department of Materials Science and Engineering, University of Pennsylvania, Philadelphia, PA 19104.
P. A. Heiney and A. M. Denenstein, Laboratory for Research on the Structure of Matter and Department of Physics, University of Pennsylvania, Philadelphia, PA 19104.
A. R. McGhie and W. J. Romanow, Laboratory for Research on the Structure of Matter, University of Pennsylvania, Philadelphia, PA 19104.
J. P. McCauley, Jr., and A. B. Smith III, Laboratory for Research on the Structure of Matter and Department of Chemistry, University of Pennsylvania, Philadelphia, PA 19104.

*To whom correspondence should be addressed at the Department of Materials Science and Engineering.

ther indexing, the center-to-center distance between neighboring molecules is 10.0(2) Å, implying a van der Waals (VDW) separation of 2.9 Å for a calculated C_{60} diameter of 7.1 Å (3). Nuclear magnetic resonance (NMR) spectroscopy clearly indicates the existence of dynamical disorder (presumably free rotation) which decreases with decreasing temperature (4, 5). More work is needed to reconcile conclusions pertaining to dynamic effects on the different time scales of x-ray and NMR experiments.

The nature of intermolecular bonding is of considerable interest, both in its own right and as a clue to the potential electronic properties of fullerites and their derivatives. One expects a priori that the bonding would consist primarily of VDW interactions, analogous to interlayer bonding in graphite. Isothermal compressibility is a sensitive probe of interatomic-intermolecular bonding in all forms of condensed matter and also provides a check on the potentials used in molecular dynamics simulations. We have performed such an experiment on pure solid C_{60}, using standard diamond anvil techniques and powder x-ray diffraction. We find that, within experimental error, the linear compressibility $d(\ln a)/dP$ of cubic C_{60} is the same as the c-axis or interlayer com-

pressibility $d(\ln c)/dP$ of graphite. This indicates that the functional relations between energy and close-packed layer separation are similar in the two solids.

We used standard techniques to prepare our powder sample: soot production by "burning" graphite rods in 300-torr He, Soxhlet extraction in boiling toluene, and liquid chromatography in hexanes on neutral alumina (6). High-performance liquid chromatography (HPLC) with a Pirkle column (7) showed >99.5% pure C_{60}. The resulting powder was dried in flowing N_2 at 400°C to drive off all traces of solvents. Preliminary powder diffraction profiles were consistent with an fcc cell, $a = 14.1$ Å, with no detectable peaks from other phases. A small amount of powder was packed into a hole 1 mm in diameter in a stainless steel gasket 0.6 mm thick, located between the anvils of a standard Merrill-Bassett diamond anvil cell (DAC). A hydrostatic environment was assured by filling the remaining volume with a 50:50 mixture of ethanol: methanol. Attempts to incorporate a small amount of powdered solid with known compressibility along with the fullerite were unsuccessful; after several tries, we were unable to produce a usable combination of relative scattering intensities, clearly defined peaks, and measurable pressure shifts using any of the usual standards. Thus we resorted to a secondary pressure scale based on the torque of the three screws that compress the liquid and deform the gasket (as established in previous experiments). After measuring the fullerite at "high" pressure, we reloaded the cell with graphite powder, using an identical gasket, and measured the position of the graphite (002) reflection at the same torque settings used for the fullerite measurements. This procedure established that the torque-pressure relation conformed to the secondary standard to within 10%.

We measured powder profiles in the DAC using a two-axis diffractometer consisting of a Mo rotating anode operating at 7.5 kW, a focusing graphite (004) monochromator, a Soller slit after the sample, and a NaI scintillation detector. The longitudinal resolution ΔQ was about 0.05 Å^{-1} full width at half maximum. The positions of the three strongest peaks of solid C_{60} [the (111), (220), and (311) reflections], recorded at 0 and 1.2 GPa, were fitted to the predicted values for an fcc lattice with a as an adjustable parameter. The results are shown in Fig. 1.

Two differences are noticeable between the profiles at 0 and 1.2 GPa. First, at higher pressure all the peaks shift to higher Q, indicating a reduction in lattice constant. Furthermore, the relative intensities of (111), (220), and (311) reflections exhibit significant pressure dependence. This can be understood in terms of the unusual structure factor (described below) and a decrease in $\bar{a}$ with fixed C_{60} diameter. A similar change in relative intensity occurs at low temperature and $P = 0$ GPa due to thermal contraction (8). Scans at intermediate torque values indicate that both of these changes are monotonic with increasing pressure and that the P-induced shift is reversible upon release of the torque.

The profile fits in Fig. 1 show that a has decreased by 0.4 Å between 0 and 1.2 GPa.

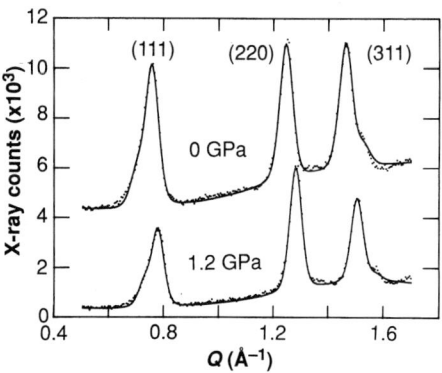

Fig. 1. Powder x-ray profiles of solid C_{60} at atmospheric pressure (top) and 1.2-GPa hydrostatic pressure (bottom). Dots are experimental points (approximately 70 per point), and the solid curves are least-squares fits to an fcc structure with adjustable lattice constant a. The fitted relative intensities have no physical significance in this simple model. The scattered wave vector $Q = 4\pi\sin\theta/\lambda$, where θ is the Bragg angle; for these profiles wavelength $\lambda = 0.71$ Å. Indexing of the strongest peaks is indicated. The high-Q shoulder on the (311) is the weak (222) reflection; the low-Q shoulder on the (111), observed to some extent in all our nominally pure C_{60} samples, is presently unidentified. The variable intensity of this shoulder has little effect on the lattice constant of a particular sample, so we can safely conclude that it has no effect on the compressibility derived from the present data.

If there is no change in molecular radius, this corresponds to a reduction in intermolecular spacing from 2.9 to 2.5 Å. The a-axis compressibility $-d(\ln a)/dP$ is 2.3×10^{-12} cm^2/dyne, essentially the same as the interlayer compressibility $-d(\ln c)/dP$ of graphite within our combined 10% experimental error on P and a. The system retains the fcc structure up to at least 1.2 GPa, so it is reasonable to assume that all the volume reduction is accommodated by decreasing the VDW separation between molecules rather than by compressing or deforming the spheres. This is also consistent with the fact that a pressure of 1.2 GPa has no measurable effect on the in-plane lattice constant of graphite (9).

At both atmospheric pressure and 1.2 GPa, the (111), (220), and (311) reflections are observed with comparable intensities but there is no detectable (200) intensity (expected near 0.9 Å^{-1}). This is unusual for fcc structures but can be understood in terms of the x-ray form factor. If we assume, as suggested by NMR results, that the C_{60} molecule is freely rotating at room temperature, then its ensemble-averaged charge density is that of a spherical shell of charge. The Fourier transform of a uniform shell of radius R_0 is $j_0(QR_0) = \sin(QR_0)/QR_0$, where j_0 is the zero-order spherical Bessel function. This ad hoc form factor in fact describes the data very well. Figure 2 shows a powder profile recorded over a wider range of scattering angles and with higher resolution than the DAC profiles (10) (Fig. 1), along with a fit to a model consisting of shells with 3.5 Å radius centered on fcc sites but neglecting translational thermal disorder. The excellent agreement indicates that

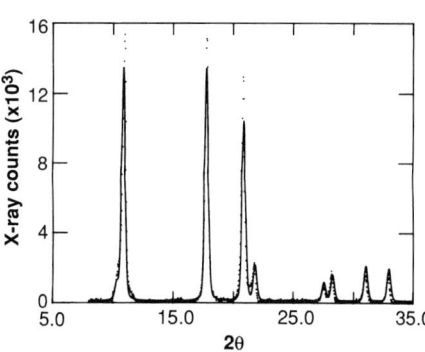

Fig. 2. Powder profile of solid C_{60} at atmospheric pressure, measured on a diffractometer equipped with a position-sensitive detector (8) and a 1.5-kW sealed Cu source monochromatized by the (002) reflection of graphite ($\lambda = 1.54$ Å). The powder sample was contained in a Lindemann capillary tube (0.7 mm in diameter). Dots are the measured points (2 hours accumulation), and the solid curve is a least-squares fit to an fcc lattice of uniform spherical shells. The best-fit parameters are $a = 14.11$ Å and shell radius $R_0 = 3.5$ Å. This sample exhibits much less intensity in the low-angle shoulder of the (111) reflection.

the C_{60} molecules exhibit little or no orientational order at 300 K, 1 atm. In principle, the disorder could be either dynamic or static; the molecules could be spinning rapidly, as inferred from NMR measurements (4, 5), or the symmetry axes of the icosahedral molecules could exhibit no site-to-site correlation in their directions. Either conjecture is consistent with the fact that a single-crystal refinement at 300 K fails to localize the polar and azimuthal angles of individual C atoms (2). The absence of detectable (200) intensity (and, indeed of any ($h00$) peaks with h even) is entirely due to the fact that $j_0(QR_0)$ has minima at the corresponding Q values if $R_0 = 3.5$ Å and $a = 14.11$ Å.

Kratschmer *et al.* noted (1) that the inferred VDW C diameter in solid C_{60} (2.9 to 3.0 Å) is considerably less than the 3.3 Å value characteristic of planar aromatic molecules and graphite. The observation of different diameters yet similar compressibilities can be rationalized qualitatively as follows. Intermolecular or interlayer separations in C_{60} and graphite, respectively, are determined by the balance between attractive and repulsive energies, whereas the corresponding compressibilities are defined by gradients of these energies. It is easy to show that the number of VDW bonds per unit area parallel to a close-packed layer is only 1/7 as large in C_{60} as in graphite (if neighboring C_{60} molecules orient with adjacent hexagonal faces in opposition). This implies a large difference in the total energies of the two solids but does not directly account for the reduced VDW separation; if the close-packed layers were very stiff, the equilibrium spacing would be independent of bonds per area. However, the nature of the bonds is qualitatively different. The lobes of p_z charge in fullerene are normal to the spherical surface and probably remain nearly so in the solid. This orbital structure permits a closer approach of neighboring C_{60} molecules as compared to the spacing in graphite because the lobes extending into the intermolecular gap in C_{60} are "splayed out" with respect to a normal to the close-packed plane rather than strictly normal to the plane as in graphite. Simple trigonometry shows that this effect alone can account for more than one half the reduction in the VDW separation.

The p_z splaying also adds a new repulsive contribution to the total energy relative to that of planar systems. In graphite, a small reduction in interlayer spacing compresses the π charge; in fullerite there would also be a tendency to increase the splay angle, which requires an energetically unfavorable hybridization between pi and sigma orbitals. The VDW "bonds" in fullerite are thus stiffer and shorter, as well as less dense, than in graphite. We propose that the reduced gap is due to the combination of shorter bonds and reduced areal density, whereas the comparable compressibilities result from cancellation between different bonds per area and bond strengths. A direct test of this proposal can be performed if these experiments are extended to higher pressure; if the C_{60} bonds are indeed stiffer than in graphite, the compressibility should decrease rapidly with increasing P.

Isothermal volume compressibilities $-1/V (dV/dP)$ are 6.9×10^{-12}, 2.7×10^{-12}, and 0.18×10^{-12} cm^2/dyne for solid C_{60}, graphite, and diamond, respectively. Clearly, fullerite is the softest all-C solid currently known. The linear compressibility normal to the close-packed planes is nearly equal for solid C_{60} and graphite (11). This suggests that there should be no elastic impediments to the formation of "intercalated" (12) fullerites.

REFERENCES AND NOTES

1. W. Kratschmer, L. D. Lamb, K. Fostiropoulos, D. R. Huffman, *Nature* **347**, 354 (1990).
2. R. M. Fleming *et al.*, *Mater. Res. Soc. Symp. Proc.*, in press.
3. H. Kroto, *Science* **242**, 1139 (1988).
4. C. S. Yannoni, R. D. Johnson, G. Meijer, D. S. Bethune, J. R. Salem, *J. Phys. Chem.* **95**, 9 (1991).
5. R. Tycko *et al.*, *ibid.*, p. 518.
6. H. Ajie *et al.*, *ibid.* **94**, 8630 (1990).
7. J. M. Hawkins *et al.*, *J. Org. Chem.* **55**, 6250 (1990).
8. P. A. Heiney *et al.*, in preparation.
9. R. Clarke and C. Uher, *Adv. Phys.* **33**, 469 (1984).
10. Manufactured by INEL, Les Ulis, France.
11. The effect of hydrostatic pressure on a cubic crystal is to reduce all lengths proportionally regardless of direction with respect to the crystal axes. Our results therefore show that the compressibility of close-packed (111) planes in cubic C_{60} is the same as that of the honeycomb (002) planes in hexagonal graphite.
12. J. E. Fischer and T. E. Thompson, *Phys. Today* **31** (no. 7), 36 (1978).
13. We acknowledge the technical assistance of O. Zhou and V. B. Cajipe. We are grateful to R. M. Fleming and E. J. Mele for helpful discussions and to F. A. Davis for the use of his HPLC column. This work was supported under National Science Foundation Materials Research Laboratory Program grant DMR88-19885 and under Department of Energy grants DE-FC02-86ER45254 and DE-FG05-90ER75596.

Reprinted with permission from Physical Review Letters
Vol. 66, No. 22, pp. 2911–2914, 3 June 1991

Orientational Ordering Transition in Solid C$_{60}$

Paul A. Heiney, [a] John E. Fischer, [b] Andrew R. McGhie, William J. Romanow,
Arnold M. Denenstein, [a] John P. McCauley, Jr., [c] and Amos B. Smith, III [c]
Laboratory for Research on the Structure of Matter, University of Pennsylvania,
Philadelphia, Pennsylvania 19104

David E. Cox
Physics Department, Brookhaven National Laboratory, Upon, New York 11973
(Received 8 April 1991)

Synchrotron-x-ray powder-diffraction and differential-scanning-calorimetry measurements on solid C$_{60}$ reveal a first-order phase transition from a low-temperature simple-cubic structure with a four-molecule basis to a face-centered-cubic structure at 249 K. The free-energy change at the transition is approximately 6.7 J/g. Model fits to the diffraction intensities are consistent with complete orientational disorder at room temperature, and with the development of orientational order rather than molecular displacements or distortions at low temperature.

PACS numbers: 61.50.−f, 35.20.Bm, 64.70.Kb

The recent discovery[1] of an efficient synthesis of C$_{60}$ has spurred intense interest in the chemical, electronic, and physical properties of these highly symmetric molecules, which take the form of truncated icosahedra.[2,3] At room temperature, single-crystal x-ray diffraction shows[4] that the molecules are centered on sites of a face-centered-cubic (fcc) Bravais lattice, $a_0 = 14.2$ Å, with a high degree of rotational disorder. The center-to-center distance between neighboring molecules is 10.0 Å; the calculated[2] C$_{60}$ diameter is 7.1 Å. The crystal is quite soft, with the compressibility at low pressure along *any* axis being comparable to that of graphite along its *c* axis.[5,6] Nuclear magnetic resonance (NMR) clearly indicates the existence of dynamical disorder (presumably free rotation) which decreases with decreasing temperature.[7,8] This implies several possible low-temperature structures: icosahedral glass, quasicrystal, or an orientationally ordered crystalline phase. In this Letter we present x-ray data and analyses which are consistent with only the last option. We also show that the ordered phase is stable to 249 K which is a surprisingly high temperature.

We used high-resolution synchrotron-x-ray powder diffraction to study solid C$_{60}$ between 300 and 11 K. A powder sample was prepared and purified using standard techniques as previously detailed,[5] resulting in better than 99.5%-pure C$_{60}$ and no detectable solvent. About 1 mg was loaded into a 0.7-mm-diam glass capillary tube. Measurements were performed at beam line X7/A of the National Synchrotron Light Source. A wavelength of $\lambda = 1.1992$ Å was selected by a channel-cut Si(111) monochromator. Measurements of the strongest peak with a Ge(220) crystal analyzer gave $\Delta(2\theta) = 0.07°$ full width at half maximum (FWHM), implying a positional correlation length $\xi \geq 1000$ Å at 300 K. In order to obtain adequate counting rates, subsequent measurements were made with a narrow receiving slit (0.5 mm), in-

stead of a crystal analyzer, in front of a Kevex detector, resulting in an instrumental resolution of $\Delta(2\theta) \sim 0.10°$ FWHM.

Figure 1 shows full diffraction profiles measured at 300 and 11 K. All of the peaks at 300 K can be indexed[9] as fcc, with $a_0 = 14.17 \pm 0.01$ Å, as previously reported.[4,5] Note that *no* $h00$ peaks are present in this profile, despite the fact that $h00$ is allowed for h even. At 11 K, a_0 has decreased to 14.04 ± 0.01 Å and many new peaks have appeared. The new peaks can all be indexed as simple-cubic (sc) reflections with mixed odd and even indices (i.e., forbidden fcc reflections). The crystal has therefore undergone a transition to a simple-cubic structure, but since the cube edge has not changed appreciably the basis must still consist of four molecules per unit cell, which were equivalent in the fcc structure but which somehow become inequivalent at low temperature.

Figure 2, top, shows the *T*-dependent integrated intensity of the 451 peak. This peak is fcc forbidden, and its intensity is therefore proportional to the square of the sc order parameter. Abut half the 0-K intensity is lost in a continuous precursor extending from 0 to T_c; the remainder disappears much more abruptly at $T_c = 249 \pm 1$ K. There is no measurable hysteresis, and no measurable cell volume change at T_c. These observations are all consistent with weakly first-order behavior. Scans of the range 2.7 Å$^{-1} < q < 4.5$ Å^{-1} at 230 and 255 K show that all of the sc peaks have the same qualitative *T* dependence.

Figure 2, bottom, shows differential-scanning-calorimetry (DSC) measurements on a 12-mg sample (N$_2$ flow, 20 °C/min heating). The DSC endothermal also shows a broad precursor terminating in an abrupt transition whose onset appears a few degrees above T_c due to the more rapid heating rate. The area under the precursor and the sharp endotherm is about 6.7 J/g. Attribut-

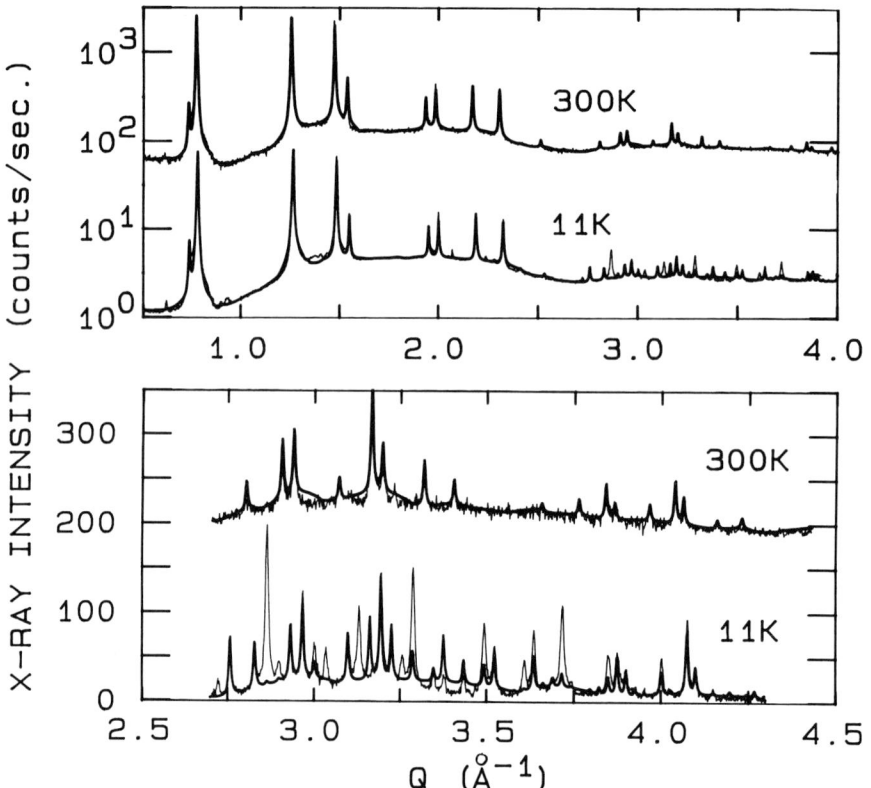

FIG. 1. Total x-ray scattering intensity (powder sample plus capillary) of pure C_{60} at 300 and 11 K. Light and heavy curves are data and model fits, respectively. Intensities are normalized to counts/sec at a synchrotron ring current of 100 mA; the data were typically collected for 5 sec/point at a current of 150 mA. Top panel shows the entire profile on a semilogarithmic scale, and bottom panel shows a blowup of the same data in the region $2.5 \leq q < 4.5$ Å^{-1} on a linear scale. All profiles have been offset for clarity.

ing this entirely to a change in configurational entropy gives $\Delta S = R \ln(8)$ per C_{60}, but this attribution underestimates the correct value since the DSC scan misses some of the precursor, and overestimates it to the extent that there could be a small energy contribution as well.

We now discuss the analysis of the x-ray intensities. The atoms of the C_{60} molecule are placed at the vertices of a truncated icosahedron.[2,10] The x-ray structure factor is given by the Fourier transform of the electronic charge density; this can be factored into an atomic carbon form factor times the Fourier transform of a thin shell of radius R modulated by the angular distribution of the atoms. For a molecule with icosahedral symmetry, the leading terms in a spherical-harmonic expansion of the charge density are $Y_{00}(\Omega)$ (the spherically symmetric contribution) and $Y_{6m}(\Omega)$, where Ω denotes polar and azimuthal coordinates. The corresponding terms in the molecular form factor are proportional to $S_0^{\text{mol}}(q) \propto j_0(qR) \equiv \sin(qR)/qR$ and

$$S_6^{\text{mol}}(Q) \propto j_6(qR) \sum_m A_{6m} Y_{6m}(\Omega_q),$$

where j_0 and j_6 are spherical Bessel functions, the constants A_{6m} are derived from the details of the charge distribution, and Ω_q, is the orientation of the momentum transfer $\mathbf{q}$ in the reference frame of the icosahedron.

For the high-temperature fcc structure of C_{60}, both NMR[7,8] and previous structural studies[4,5] indicate a high degree of orientational disorder. Accordingly, it is reasonable to truncate the spherical-harmonic expansion at the first term, thus approximating the molecule by a spherical shell of charge as appropriate for a freely rotating molecule. The solid curve shown with the 300-K data in Fig. 1 is the best fit to the measured integrated intensities with this model. The integrated intensity at Bragg vector $\mathbf{q} = \mathbf{G}_{hkl}$ is calculated as

$$I(\mathbf{G}_{hkl}) = I_0 |j_0(G_{hkl}R) f_C(G_{hkl})|^2 \times L_P \times M_{hkl},$$

where f_C is the carbon atomic form factor, L_P is the Lorentz-polarization factor, and M_{hkl} is the multiplicity. The only adjustable parameters are I_0, R, and the cube edge length a_0. The best fit gave $R = 3.52 \pm 0.01$ Å and $a_0 = 14.17$ Å. Incorporating a Debye-Waller term $\exp(-G^2 u^2)$ did not improve the fits, indicating that the thermal disorder is primarily rotational rather than translational. The fit clearly captures most of the important features of the data, including the absence of $h00$ peaks which fortuitously coincide with zeros of j_0 for even h. This gives strong additional evidence for a high degree of molecular orientational disorder at 300 K. The data are equally well described on the basis of a static

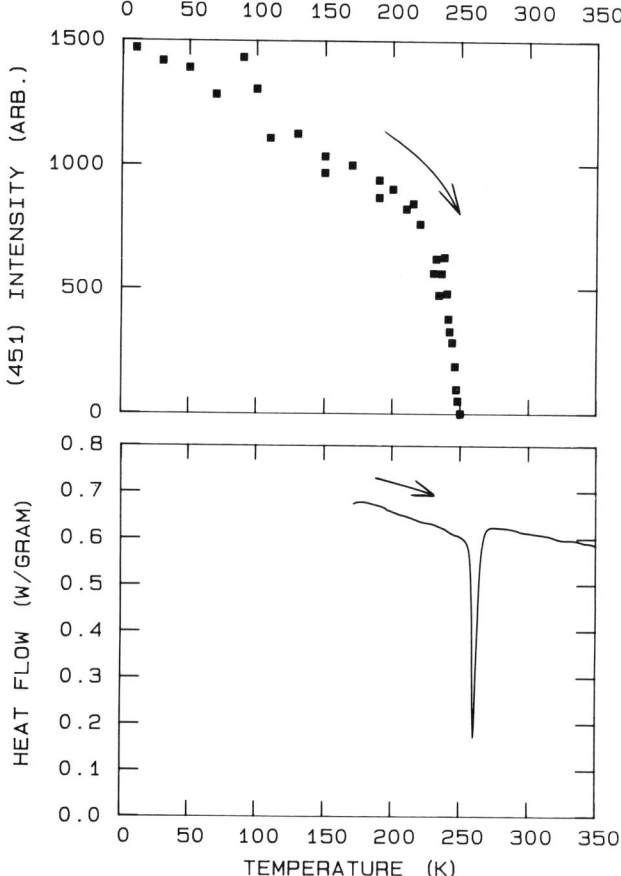

FIG. 2. Top: Integrated intensity of the C_{60} 451 peak as a function of temperature. This peak is forbidden for a fcc lattice. Bottom: Results of a differential-scanning-calorimetry measurement on a powder sample of C_{60}.

served. Indeed, fits to the entire spectrum allowing these symmetries yielded zero amplitudes for Y_{1m} and Y_{2m}. A displacive transition can also be ruled out on physical grounds, given the proximity of neighboring molecules. Molecular-dynamics simulations[11] do show football-shaped instantaneous distortions at very high T, but a static (or average) distortion of this kind at low T is certainly unfavorable energetically.

By contrast, the development of orientational order is to be expected at low temperature. Such order could be represented in $S(q)$ by $j_6(qR)$ and higher-order terms; the first maximum of j_6 in the units of Fig. 1 is at $q = 2.3$ Å^{-1}, consistent with the observation that the strongest sc-allowed peaks are observed between 2.5 and 4.5 Å^{-1}. This gives us confidence that the fcc → sc transition is due to the existence of inequivalent molecular orientations on the four fcc Bravais sites at low T. The transition is analogous to the orientational ordering transitions observed in CD_4 and related systems,[12] and in particular to solid H_2, in which the long axes of the molecule are oriented along different $\langle 111 \rangle$ cubic axes.[13]

The low-T sc lattice imposes severe constraints on possible models, since the equivalence of the x, y, and z axes and the corresponding threefold rotation axes must be maintained. One model which satisfies these criteria and gives reasonable agreement with the data is as follows. Four molecules, centered on the fcc Bravais lattice sites, are oriented such that one of the ten threefold icosahedral axes (normal to the pseudohexagonal faces) is aligned with one of the four $\langle 111 \rangle$ directions, and three mutually orthogonal twofold molecular axes are aligned with $\langle 100 \rangle$ directions.[3] It follows[14] that three *other* molecular threefold axes are *also* aligned with the three remaining $\langle 111 \rangle$ crystal axes. At this point there are no remaining rotational degrees of freedom; all molecules are equivalent and the structure is still fcc. The equivalence is now broken by rotating the four molecules through the same angle Γ but about *different* $\langle 111 \rangle$ axes: the molecule at (000) about [111], that at $(\frac{1}{2} 0 \frac{1}{2})$ about $[\bar{1} 1 \bar{1}]$, that at $(\frac{1}{2} \frac{1}{2} 0)$ about $[\bar{1} \bar{1} 1]$, and that at $(0 \frac{1}{2} \frac{1}{2})$ about $[1 \bar{1} \bar{1}]$. This choice preserves the cubic symmetry, although the twofold axes are no longer aligned along $\langle 100 \rangle$ directions.

The above-described structure could be modeled by a sufficient number of terms in the Y_{1m} expansion. We chose instead to embed sixty *discrete* atoms in a smooth spherical shell of charge, with atomic angular coordinates taken from an equilibrium structure derived from *ab initio* Hartree-Fock calculations of Scuseria,[10] and with molecular rotations as described above. This model requires the addition of only two parameters to the 300-K model (the rotation angle Γ and a fraction α of spherically symmetric j_0 amplitude). The best fit to the measured intensities with the above model is shown in Fig. 1; this fit gave $R = 3.54 \pm 0.01$ Å, $\Gamma = 22° \pm 5°$, and $\alpha = 0.5 \pm 0.3$. The model captures some but not all features of

model similar to that proposed by Fleming *et al.*,[4] i.e., with inequivalent carbon atoms $C(1)$ in 48(h) sites at 0, y, z, and $C(2)$ and $C(3)$ in 96(i) positions at x, y, z in space group $Fm3$. This model requires an additional seven adjustable parameters and also gives a much larger value of $u \sim 0.3$ Å, indicative of pronounced translational thermal fluctuations. We accordingly believe that the free-rotation model is more appealing physically.

We can imagine several models for the onset of simple-cubic order below T_c. The molecules at the corner and three face-centered sites must be made inequivalent. This could be accomplished, for example, by displacements away from fcc Bravais lattice sites, quadrupolar distortions into a "football" shape, or development of orientational order. The first two mechanisms correspond, respectively, to finite Y_{1m} or Y_{2m} spherical-harmonic components, resulting in terms in the structure factor proportional to $j_1(qR)$ or $j_2(qR)$. However, maxima in j_1 and j_2 occur at small arguments (corresponding to $q = 0.59$ and 0.93 Å^{-1} in the units of Fig. 1 if $R = 3.52$ Å), so these two mechanisms would predict detectable 100 and 200 intensities which are not obthe data. It incorporates the cubic symmetry, and correctly predicts that the new peaks should only have appreciable intensity above 2.5 Å^{-1}. Out of sixty allowed peaks in the 2.5–4.5-Å^{-1} range, the model seriously underestimates eight intensities, overestimates one, and reproduces the rest to within a factor of 2. Work is in progress on more sophisticated models involving correlated or anisotropic rotational disorder.

Note that the low-temperature structure is neither a quasicrystal nor an icosahedral glass. Given the molecular symmetry, one might have predicted the latter, whereas icosahedral quasicrystals require two distinct structural units to satisfy space filling. While the precise energetic requirements for crystal versus icosahedral glass formation are not understood, it seems likely that the structural order at low T is driven both by a preference for close packing and by local orientational order.

The x-ray results presented here show both consistencies and discrepancies with NMR observations. The most serious discrepancy is the implied coexistence of static and mobile C nuclei well below our T_c, deduced from the NMR observation of superposed motionally narrowed and powder pattern signals at temperatures as low as 140 K.[7,8] On the other hand, a minimum in T_1 at 233 K is observed in one NMR experiment.[7] In fact, the two techniques probe different aspects of the structure. NMR experiments to date cannot distinguish between free rotation and jump rotational diffusion between symmetry-equivalent orientations. X-ray diffraction is sensitive to orientational order (as a canonical average of snapshots) even in the presence of substantial thermal disorder, as long as one set of orientations is statistically preferred and the orientational order is long range. Indeed, our measurements indicate that much of the sc order is reduced by orientational fluctuations at T_c.

An orientational ordering transition temperature of 249 K is unusually high compared with, for example, the value of 20.4 K measured[12] for CD_4. However, C_{60} is a large molecule, with a rotational inertia several orders of magnitude larger than that of CD_4. This means that its motion will be much closer to the classical limit, and quantum tunneling will be substantially suppressed. A quantitative calculation of T_c would incorporate the angular dependence of the two-molecule C_{60} pair potential, and also the high symmetry of the C_{60} molecule. The large number of equivalent orientations of a single molecule implies that the decrease in entropy per molecule from free rotation to fixed orientation is relatively small, resulting in an increased value of T_c. Our estimate $\Delta S \sim R \ln(8)$ should provide some guidance in this regard.

We have shown that orientational ordering below 249 K is the only reasonable explanation for the low-T sc structure, and that this occurs by breaking the equivalence of the four sites per fcc cell in the high-T

phase. Similar phenomena are expected to occur in solid C_{70}, the smallest nonspherical fullerene.[2]

We acknowledge helpful conversations with A. Cheng, R. M. Fleming, A. B. Harris, E. J. Mele, and R. Tycko. This work was supported by the National Science Foundation Materials Research Laboratories Program, Grant No. DMR88-19885, by NSF Grant No. DMR89-01219 (P.A.H., J.P.M., and A.B.S.), and by the Department of Energy, Grant No. DEFC02-86ER45254 (J.E.F.). The work at Brookhaven was supported by DOE, Division of Materials Science, under Contract No. DEAC02-76CH00016.

(a)Also at Department of Physics, University of Pennsylvania, Philadelphia, PA 19104.

(b)Also at Department of Materials Science and Engineering, University of Pennsylvania, Philadelphia, PA 19104.

(c)Also at Department of Chemistry, University of Pennsylvania, Philadelphia, PA 19104.

[1]W. Kratschmer, L. D. Lamb, K. Fostiropoulos, and D. R. Huffman, Nature (London) **347**, 354 (1990).

[2]H. W. Kroto, J. R. Heath, S. C. O'Brien, R. F. Curl, and R. E. Smalley, Nature (London) **318**, 162 (1985); H. W. Kroto, Science **242**, 1139 (1988); R. F. Curl and R. E. Smalley, *ibid*. **242**, 1017 (1988).

[3]An icosahedron possesses twelve fivefold vertices, twenty triangular faces, and thirty edges with twofold rotation symmetry. In the truncated icosahedron, the fivefold vertices are replaced by pentagons with atoms at the vertices, and the triangular faces are replaced by pseudohexagonal faces.

[4]R. M. Fleming, T. Siegrist, P. M. March, B. Hessen, A. R. Kortan, D. W. Murphy, R. C. Haddon, R. Tycko, G. Dabbagh, A. M. Mujsce, M. L. Kaplan, and S. M. Zahurak, Mater. Res. Soc. Symp. Proc. (to be published).

[5]J. E. Fischer, P. A. Heiney, A. R. McGhie, W. J. Romanow, A. M. Denenstein, J. P. McCauley, Jr., and A. B. Smith, III, Science (to be published).

[6]S. J. Duclos (private communication).

[7]C. S. Yannoni, R. D. Johnson, G. Meijer, D. S. Bethune, and J. R. Salem, J. Phys. Chem. **95**, 9 (1991).

[8]R. Tycko, R. C. Haddon, G. Dabbagh, S. H. Glarum, D. C. Douglass, and A. M. Mujsce, J. Phys. Chem. **95**, 518 (1991).

[9]There are two exceptions to the cubic indexing. A low-angle shoulder on the 111 peak has been identified from electron-microscopy studies as a truncation rod arising from planar defects [D. E. Luzzi (private communication)]. The very weak unindexed peak at 0.61 Å^{-1} is attributed to an unknown impurity.

[10]G. E. Scuseria, Chem. Phys. Lett. **176**, 423 (1991).

[11]Q. M. Zhang, Jae-Yel Yi, and J. Bernholc, Phys. Rev. Lett. **66**, 2633 (1991).

[12]W. Press and A. Kollmar, Solid State Commun. **17**, 405 (1975).

[13]I. F. Silvera, Rev. Mod. Phys. **52**, 393 (1980).

[14]See, e.g., J. W. Cahn, D. Shechtman, and D. Gratias, J. Mater. Res. **1**, 13 (1986), for a discussion of Cartesian representations of the icosahedron.

Reprinted with permission from Physical Review Letters
Vol. 67, No. 11, p. 1467, 9 September 1991

Comment on "Orientational Ordering Transition in Solid C_{60}"

In an interesting Letter, Heiney *et al.* [1] study the x-ray scattering from powdered solid C_{60}. For temperatures above 249 K they find an orientationally disordered phase in which the freely rotating molecules form an fcc lattice. At lower temperatures their data indicate an orientationally ordered structure (OOS), which, within their resolution, occurs without motion of the centers of the molecules away from their fcc positions. All the powder-diffraction peaks could be indexed according to a simple-cubic unit cell containing four C_{60} molecules. However, since their fit to the intensities showed some discrepancies, we considered other structures for oriented icosahedra on an fcc lattice which would be consistent with simple-cubic indexing. We determined that most of the fifteen simple-cubic space groups [2] are excluded. For instance, when the icosahedra are on an fcc lattice their symmetry is higher (containing an inversion element) than permitted for space groups 195 and 198. In fact, there are two main possibilities [3]: the one actually used in Ref. [1], $P2/n\bar{3}$ (or $Pn3$), and that [4] of solid H_2, $P2_1/a\bar{3}$ (or $Pa3$), which Heiney *et al.* mentioned but did not actually use. These differ only in the way the molecular threefold axes are distributed over various $\langle 111\rangle$ directions.

The most general OOS for space group $P2_1/a\bar{3}$ is obtained as follows. Start from a fcc crystal in which all molecules have their twofold axes aligned along the three $(1,0,0)$ directions. Then the molecules centered at $(0,0,0)$, $(\frac{1}{2},\frac{1}{2},0)$, $(\frac{1}{2},0,\frac{1}{2})$, and $(0,\frac{1}{2},\frac{1}{2})$ are rotated through an angle $\phi\neq0$ about the $(1,1,1)$, $(1,\bar{1},\bar{1})$, $(\bar{1},\bar{1},1)$, and $(\bar{1},1,\bar{1})$ directions, respectively, so that the 240 atoms in the unit cell occupy ten sets of "d" sites [2]. Even after this rotation, these directions remain threefold axes for each of the four sites in the unit cell.

The calculated powder-diffraction intensity for the $Pa3$ structure was optimized with respect to the angle ϕ and the authors of Ref. [1] have kindly supplied the results shown in Fig. 1, which was obtained, following the procedures of Ref. [1], taking the lattice parameter to be

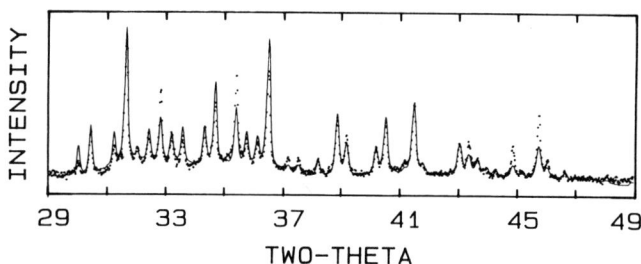

FIG. 1. Dots: A portion of the experimental 11-K x-ray profile from Ref. [1] in which most of the reflections are associated with the OOS. Wavelength 1.1992 Å. Continuous line: Model profile as in Ref. 1 but based on $Pa3$ rather than $Pn3$ space group.

$a=14.05$ A and $\phi=26°$ to best fit the 11-K data. This fit is significantly better than that found in Ref. [1]. We thus conclude that the OOS structure of solid C_{60} is $P2_1/a\bar{3}$. The structural parameters and symmetry will be discussed elsewhere [3].

We thank P. A. Heiney for many helpful discussions and the National Science Foundation MRL Program, Grant No. DMR88-19885, for partial support. We thank the authors of Ref. [1] for supplying us with the fit shown in Fig. 1.

Ravi Sachidanandam and A. B. Harris
    Department of Physics
    University of Pennsylvania
    Philadelphia, Pennsylvania 19104

Received 21 June 1991
PACS numbers: 61.50.−f, 35.20.Bm

[1] P. A Heiney, J. E. Fischer, A. R. McGhie, W. J. Romanow, A. M. Denenstein, J. P. McCauley, Jr., A. B. Smith, III, and D. E. Cox, Phys. Rev. Lett. **66**, 2911 (1991).

[2] *International Tables for Crystallography,* edited by Theo Hahn (Reidel, Boston, 1983), Vol. 4.

[3] A. B. Harris and R. Sachidanandam (to be published).

[4] A. B. Harris, S. Washburn, and H. Meyer, J. Low Temp. Phys. **50**, 151 (1983); see Sec. 3.1.

Reprinted with permission from Europhysics Letters
Vol. 18, No. 3, pp. 219–225, 21 February 1992
© 1992 Europhysics Physical Society

Structural Phase Transitions in the Fullerene C_{60}.

W. I. F. David(*), R. M. Ibberson(*), T. J. S. Dennis(**)
J. P. Hare(**) and K. Prassides(**)

(*) ISIS Science Division, Rutherford Appleton Laboratory
Chilton, Didcot, Oxon OX11 0QX, UK
(**) School of Chemistry and Molecular Sciences, University of Sussex
Brighton BN1 9QJ, UK

(received 4 November 1991; accepted in final form 16 January 1992)

PACS. 61.12 – Neutron determination of structures.
PACS. 64.70K – Solid-solid transitions.
PACS. 35.20B – General molecular conformation and symmetry; stereochemistry.

Abstract. – High-resolution powder neutron diffraction has been used to study the crystal structure of the fullerene C_{60} in the temperature range 5 K to 320 K. Solid C_{60} adopts a cubic structure at all temperatures. The experimental data provide clear evidence of a continuous phase transition at ca. 90 K and confirm the existence of a first-order phase transition at 260 K. In the high-temperature face-centred-cubic phase ($T > 260$ K), the C_{60} molecules are completely orientationally disordered, undergoing continuous reorientation. Below 260 K, interpretation of the diffraction data is consistent with uniaxial jump reorientation principally about a single $\langle 111 \rangle$ direction. In the lowest-temperature phase ($T < 90$ K), rotational motion is frozen although a small amount of static disorder still persists.

1. Introduction.

C_{60} buckminsterfullerene [1-3] is the most stable member of the whole family of closed carbon cage molecules—the fullerenes [4,5]. Its extraction and purification from arc-processed carbon [2,3] have not only enabled the original structural proposal [1] to be confirmed [2,3,6] but have also led to numerous experiments that are revealing many novel physical and chemical properties [7]. At room temperature, crystalline C_{60} adopts a face-centred cubic crystal structure in which each of the C_{60} molecules is orientationally disordered [8,9]. The structure may be regarded as cubic-closed-packed in which rotations and orientations of individual C_{60} molecules are uncorrelated with their neighbours [9]. ^{13}C NMR [10-12] and quasi-elastic neutron scattering [13] measurements confirm this rapid isotropic reorientation at room temperature that results in time-averaging of the truncated icosahedron to spherical symmetry. Perhaps surprisingly in view of the almost spherical symmetry, this orientational disorder does not persist to low temperatures. Differential scanning calorimetry (DSC) [14] and X-ray diffraction [9] measurements established the existence of a first-order phase transition near 250 K. More recent work [9,15,16] has confirmed an ordered simple cubic crystal structure for C_{60} at low temperatures. The reason for the orientational order has been discussed in terms of van der Waals bonding and electrostatic repulsion that results in the facing of the most electron-poor regions (the pentagonal faces) and the most electron-rich regions (the higher bond-order inter-pentagon bonds) of adjacent molecules [15].

We have undertaken a high-resolution powder neutron diffraction study of the temperature dependence of the crystal structure of C_{60}. We confirm the first-order phase transition at 260 K and, for the first time, provide clear evidence of the presence of a second phase transition at ca. 90 K. There is also weak evidence for a small structural anomaly at 155 K. We conclude that all phases possess cubic symmetry in strong disagreement with theoretical calculations that predicted tetragonal and orthorhombic symmetries [17] for ordered C_{60}: deviations from cubic symmetry could not be stabilised in our Rietveld profile refinement analyses. In discussing the configuration of C_{60}, we use the terminology of David *et al.* [15]. The truncated icosahedron consists of 12 pentagons and 20 hexagons; we denote the bonds fusing two hexagons and a hexagon and pentagon as 6:6 and 6:5 bonds, respectively. Ideal C_{60} molecular symmetry is consistent with space group $Fm\bar{3}$ with atoms located at $a(0, \pm 1/2, \pm 3\tau/2)\curvearrowright$, $a(\pm \tau/2, \pm 1, \pm 1/2(2\tau + 1))\curvearrowright$, $a(\pm \tau, \pm 1/2, \pm 1/2(2 + \tau))\curvearrowright$, where $\curvearrowright$ implies cyclic permutation of coordinates, τ is the golden ratio $1.61803 = (1 + \sqrt{5})/2$ and a is the ratio of the average C-C distance to cubic lattice constant. Although C_{60} does not adopt space group $Fm\bar{3}$ at any temperature, the $Fm\bar{3}$ configuration serves as a reference for C_{60} molecular orientation in $Pa\bar{3}$; the angle ϕ used in this paper denotes an anticlockwise rotation about the [111] direction of this ideal $Fm\bar{3}$ configuration.

2. Experimental.

Pure C_{60} was prepared and purified using standard procedures [18]. Traces of solvent were removed by heating at 170 °C under reduced pressure (0.5 mm Hg) for a few hours. The sample was characterised by X-ray powder diffraction, infrared, ^{13}C NMR and inelastic neutron scattering [19] spectroscopy. The diffraction profiles of C_{60} were recorded on the high-resolution powder diffractometer (HRPD) at ISIS, Rutherford Appleton Laboratory, U.K. with the sample at the low resolution $(\Delta d/d = 8 \cdot 10^{-4})$ high-flux position. Two extended runs at 5 and 115 K were obtained over a time-of-flight range $(30 \div 230)$ ms, equivalent to a d-spacing range of $(0.6 \div 3.2)$ Å. A larger set of shorter, 30 minute runs over a smaller time-of-flight range $((40 \div 115)$ ms; $286 \geqslant h^2 + k^2 + l^2 \geqslant 35)$ were also recorded between 5 and 320 K in 10 K steps. Data analysis was performed using the ISIS powder diffraction software package [20].

3. Results.

The first stage of data analysis, the extraction of reliable lattice constants at temperatures between 5 and 320 K, was performed using the Pawley method. This procedure is unbiased by a particular structural model since the only parameters that are refined are cell constants, integrated intensities (in the form of unnormalised structure factors $|F^2|$) and peak width parameters. Figure 1 shows the temperature evolution of the lattice constant a. The most striking feature is the discontinuity apparent at 260 K; the lattice constant at this temperature jumps from 14.1501(9) Å to 14.1015(6) Å, a change of 0.0486(11) Å, corresponding to a 0.344(8)% lattice contraction which is the signature of the orientational ordering transition identified before by DSC [14], NMR [10,12] and X-ray diffraction [9]. At temperatures above 260 K, the structure conforms to face-centred-cubic (f.c.c.) crystal symmetry (space group $Fm\bar{3}m$). Below 260 K, reflections with mixed odd and even indices have appeared and simple cubic (s.c.) crystal symmetry (space group $Pa\bar{3}$ [15,21]) is appropriate. Both the f.c.c. orientationally disordered and s.c. orientationally ordered phases co-exist at 260 K. Lattice con-

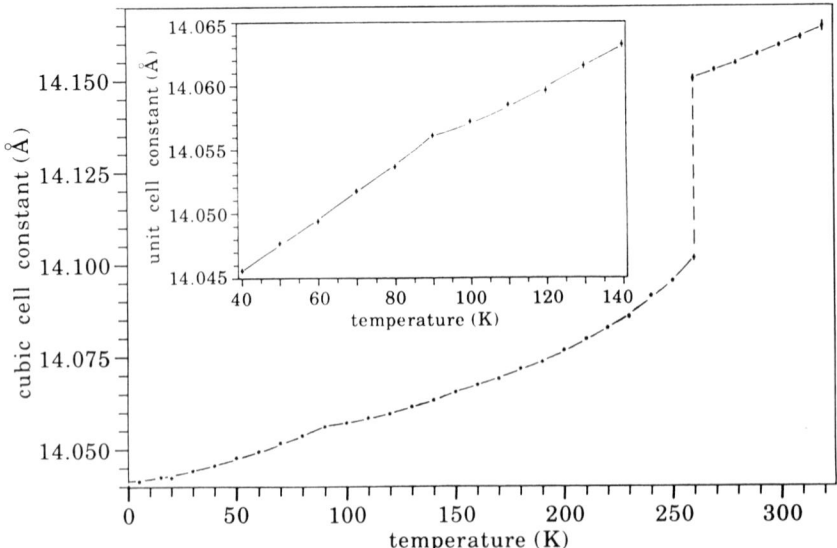

Fig. 1. – Temperature variation of the cubic lattice constant of C_{60}. The solid lines indicate the three temperature regions $(5 \div 90)$ K (phase I) $(100 \div 260)$ K (phase II) and $(260 \div 320)$ K (phase III). The temperature range 40 K to 140 K shown in the inset highlights the phase transition at 90 K.

stants $a_{f.c.c.}$ and $a_{s.c.}$ extracted at this temperature extrapolate from the single-phase lines of fig. 1, confirming that temperature equilibration has occurred and, as a consequence, phases I and II coexist in thermodynamic equilibrium at 260 K. At temperatures below 260 K, the cubic lattice constant varies in a smooth monotonic manner with a distinct curvature down to 90 K. At this temperature, there is a well-defined cusp, followed by a further smooth monotonic decrease down to 5 K (inset, fig. 1). Our data thus provide compelling evidence for the existence of a second-order phase transition at 90 K.

Refinements of the structural model based on bonding and packing considerations [15] were attempted in order to improve the fit to the low-temperature phase (phase III). In the space group $Pa\overline{3}$ the four C_{60} molecular units in the unit cell have become inequivalent because of anticlockwise rotations of 98° about various ⟨111⟩ directions. The result is that the three axes [110], [101] and [011] are almost, but not exactly, situated at the centres of three of the pentagonal faces of each C_{60} molecule; the ideal directions are ⟨11δ⟩ where $\delta = -0.05218$ [15]. We investigated the possibility that the onset of a small rhombohedral distortion at low temperatures may result in more efficient packing, distorting the unit cell in such a way that the value of the deviation δ was closer to zero. However, Rietveld refinements based on the space group $R\overline{3}$ led to no statistically significant improvement. Indeed it proved impossible in refinements to stabilise the rhombohedral distortion. Hence the various phase transitions in solid C_{60} must be discussed solely within the context of cubic symmetry and must involve detailed consideration of the reorientational dynamic behaviour of C_{60} molecules and their intermolecular interactions.

The structure of ordered C_{60} at 5 K [15] is characterised by optimisation of the intermolecular interactions and lack of bonding frustration, as the short electron-rich inter-pentagon bonds face the electron-poor pentagon faces of neighbouring C_{60} molecules. The rotation angle of $\phi = 98°$ about [111] (fig. 2a)) ensures that all twelve nearest-neighbour interactions are optimised in this way. However, facile van der Waals calculations, using a Lennard-Jones «6-12» potential indicate that the solely van der Waals contribution to the free energy predicts a different optimal structure situated at $\phi = 33.5°$, roughly midway between the sym-

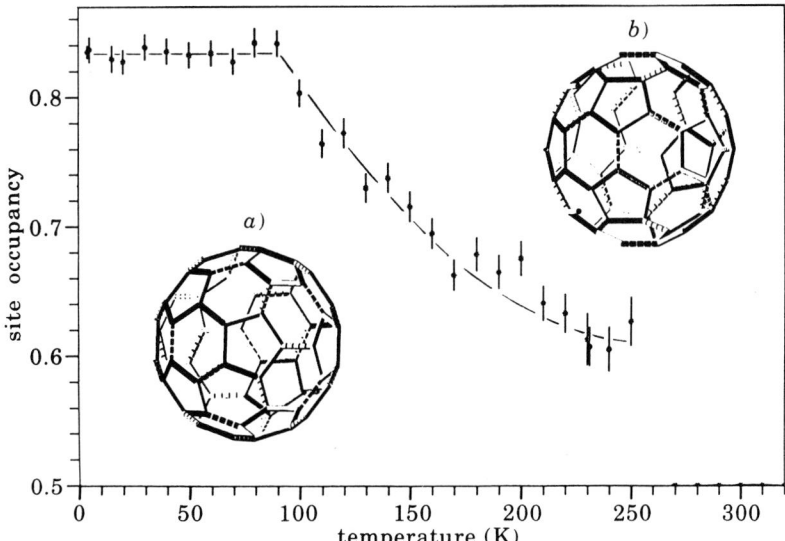

Fig. 2. – Temperature variation of the fraction of C_{60} molecules locked into the $\phi = 98°$ anticlockwise rotation around the [111] direction from the ideal $Fm\overline{3}$ configuration. The occupancies of the $\phi = 98°$ and $\phi = 38°$ configurations add to unity. Inset a): projection of two neighbouring C_{60} molecules along their centre-to-centre direction for the 98° configuration; b) as in a), but for the 38° configuration.

metry-equivalent ordered structures at $\phi = -22°$ and $\phi = 98°$. The packing motif associated with this new configuration is illustrated in fig. 2b); a hexagon now faces the 6:6 bond of the neighbouring C_{60} molecule. However, the broad similarities between the two configurations led us to surmise that i) although the global minimum energy configuration occurs at $\phi = 98°$, a local minimum will occur at $\phi \simeq 38°$ with only marginally higher energy, ii) the potential barrier for rotation about the [111] direction is lower than that for rotations about other directions and, as a consequence, uniaxial reorientation about [111] may preferentially occur, and iii) this uniaxial reorientation will take the form of a fast 60° hop with relatively long residence times at the sites corresponding to $\phi \approx (98 + 60n)°$ ($n = 0, 1, 2, ...$). This model of uniaxial reorientation in phase II between 90 and 260 K provides an explanation for the apparent paradox of the diffraction data [9, 15, 16, 21] that clearly indicate primitive cubic symmetry and ^{13}C NMR spectroscopy [10-12] that shows motional narrowing of lines implying rapid C_{60} rotation at temperatures well below 260 K.

In the phase III region ((5 ÷ 90) K), we thus assumed that the major fraction of the C_{60} molecules is frozen into the optimal configuration ($\phi = 98°$), while a minority assumes an orientation close to the «optimal van der Waals» configuration at $\phi = 38°$. Refinement of the 5 K data in space group $Pa\overline{3}$ using this model and incorporating refinement of the relative occupancies of the carbon atoms in the two configurations, led to an improved fit and excellent agreement with the experimental data ($R_{wp} = 2.7\%$, $R_E = 1.9\%$, $\chi^2 = 2.0$). Surprisingly, the final occupancies did not refine to full population of the $\phi = 98°$ configuration. Instead for $\phi = 98°$ and $\phi = 38°$ the occupancies were 0.835(4) and 0.165(4), respectively, a ratio of approximately $p = 5.1(1):1$. The ratio p remained virtually unchanged throughout the phase III region (fig. 2). Interestingly, these occupancies correspond, to within experimental error, to two of the twelve nearest neighbours adopting the «wrong» configuration. This may represent the mechanism for relieving the small degree of frustration associated with the fact that the middle of the pentagon faces cannot align precisely along ⟨110⟩ directions. A short-range ordered superstructure may result. However, no evidence of this was detected in the diffraction data. A more straightforward explanation, however, is that this ratio may depend on cooling rate and that the structure is frozen into an orientational glass.

The same structural model was used successfully in the refinements of the data sets in the phase II region $((100 \div 260)\,K)$. The major difference with the phase III region is that the preference for the $\phi = 98°$ configuration is reduced as the temperature increases and the molecules can execute rotational jumps between the two energetically preferred configurations; the ratio p smoothly decreases as the temperature increases towards the critical temperature of the first-order phase transition (fig. 2) and reaches a minimum value of 1.7(1), corresponding to relative occupancies of 0.626(18) and 0.374(18) for $\phi = 98°$ and $\phi = 38°$ configurations, respectively.

4. Discussion and conclusions.

Using high-resolution powder neutron diffraction, we have followed the temperature dependence of the structure of C_{60} between 5 and 320 K. This temperature range may be divided into three regions which we identify as phase I $((260 \div 320)\,K)$, phase II $((90 \div 260)\,K)$ and phase III $((5 \div 90)\,K)$. In the phase I region, our results are in agreement with earlier X-ray [9], NMR [10-12] and quasi-elastic scattering [13] measurements; the C_{60} molecules rotate freely, effectively randomly and independently of each other. The powder neutron diffraction data are consistent with the f.c.c. space group $Fm\bar{3}m$, with all four spherical C_{60} units being symmetry equivalent. The transition between f.c.c. (phase I) and s.c. crystal symmetry (phase II) occurs abruptly at 260 K for our sample. Both phases co-exist at 260 K. The phase transition is accompanied by an abrupt lattice contraction and is first-order in nature, as expected from symmetry considerations; $Pa\bar{3}$ is not a direct subgroup of $Fm\bar{3}m$—two irreducible representations are required to describe the order parameter associated with the phase transition [22].

Our analysis has also clarified the nature of the orientationally ordered C_{60} structure in the low-temperature $(5 \div 90)\,K$ region (phase III). Phase III is characterised by lack of rotational motion as the individual C_{60} molecules are locked principally into their orientationally optimal configuration although a small amount of static disorder [16] is still present. We have previously identified the structural motif [15] that results from the combination of optimum local orientational order and close-packing requirements. Electron-deficient regions (pentagons) face electron-rich regions (inter-pentagon bonds) of neighbouring C_{60} units (fig. 2a)), leading to optimal molecular electrostatic interactions and a reduction in local symmetry from $I_h (m\bar{5}m)$ to $S_6 (\bar{3})$. Each C_{60} unit has six of its twelve pentagonal faces and six of its thirty inter-pentagon bonds facing its 12 nearest neighbours. This bonding interaction arises through the rigid anticlockwise rotation of the C_{60} unit at the origin of the unit cell around the [111] axis by 98° (fig. 2a)). The other three C_{60} molecules in the unit cell, centred on the cube faces, are similarly rotated through the same angle but about the other three $\langle 111 \rangle$ axes. Although the fit to the experimental data is very good, small discrepancies between the observed and calculated intensities led us to re-examine a diffraction data set which was collected at 5 K over a longer counting time (12 hours). Retaining the same space group $(Pa\bar{3})$ and introducing a second configuration of C_{60} units (fig. 2b)), characterised by a rotation angle $\phi = 38°$, related to the optimal one by 60°, led to improved agreement between the observed and calculated profiles. The new structural motif (fig. 2b)) corresponds to an energetically favourable intermolecular interaction; this time the short electron-rich inter-pentagon bonds face hexagonal faces of neighbouring C_{60} molecules. Such a configuration does not optimise as much the nearest-neighbour molecular electrostatic interactions if compared to the 98° configuration,

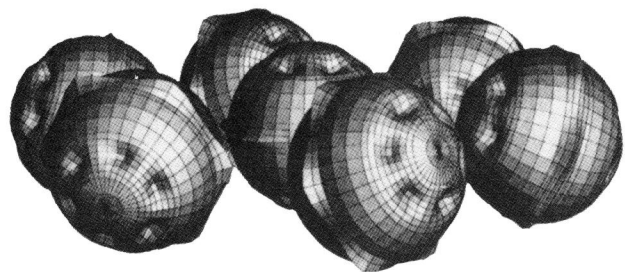

Fig. 3. – Close-packed (111) plane of C_{60} molecules depicting uniaxial reorientation about $\langle 111 \rangle$ directions (phase II, (90 ÷ 260) K). The «dimples» correspond to the centres of electron-poor hexagonal/pentagonal faces; the equatorial protrusions depict electron-rich 6:6 bonds. The constraints demanded by $Pa3$ symmetry imply that the electron-rich and electron-poor regions have well-defined loci corresponding to constant latitudes relative to [111]. These electrostatic considerations are consistent with the assumption that the preferred easy reorientation direction is [111].

but leads to an improved van der Waals interaction. The relative proportion of the molecules between the 98° and the 38° configurations is 5.1(1):1 at 5 K and remains essentially constant in the 5 to 90 K range; this implies that for every C_{60} unit, two of its twelve nearest neighbours are locked into the 38° configuration.

The intermediate phase II ((90 ÷ 260) K) region is more difficult to model precisely as subtle dynamic effects may only be inferred from diffraction data. A least-squares polynomial fit to the lattice constant data (fig. 1) is not as good as for phases I and III and there is weak evidence of a possible change in slope in the vicinity of 155 K (fig. 1 and 2) that may indicate the presence of a further phase change in this temperature region. (This anomaly may separate phases involving uniaxial rotation about a single $\langle 111 \rangle$ direction ($T < 155$ K) and uniaxial rotations about a number of distinct, presumably $\langle 111 \rangle$, directions.) Rietveld refinement of the relative proportions of the two configurations reveal that the fraction of the 38° configuration is continuously increasing as the temperature increases at the expense of the 98° configuration (fig. 2). We interpret this as resulting from orientational fluctuations about the $\bar{3}$ axis; the molecules reorientate between a number of different sites but principally perform 60° jumps about [111] residing in the two configurations corresponding to 38° and 98° rotations that appear to be energetically preferred (fig. 3). Thus, although the orientational disorder of the C_{60} molecules is not completely frozen until much below the first-order phase transition temperature, the nature of the disorder persisting below 260 K is fundamentally different from that of phase I. The NMR studies of Tycko et al. [10] and Johnson et al. [12] have identified the presence of large-amplitude reorientations well below 260 K that resulted in motional narrowing of the single NMR line, approximated by thermally activated jumping of C_{60} molecules among symmetry-equivalent orientations. Our diffraction results are compatible with these NMR works and offer a detailed explanation of the nature of the molecular motion in solid C_{60}. Although the data are not of sufficient quality to be able to exclude reorientations about alternative directions, on energetic and symmetry grounds, 60° reorientational jumps about the [111] direction appear to dominate with the two configurations of fig. 2 statistically preferred. We note that a similar sequence of phase transition behaviour, namely orientational ordering ⇒ uniaxial reorientation ⇒ isotropic reorientation, is observed in carboranes [23].

In conclusion, we have shown that our diffraction data are consistent with rapid reorientational disorder above 260 K, shuffling of the C_{60} molecules between positions related by 60° rigid reorientational jumps about the $\langle 111 \rangle$ direction in the temperature range 90 to 260 K and freezing of the reorientational motion below 90 K, with some static disorder still persisting to 5 K.

* * *

We are indebted to H. KROTO, R. TAYLOR and D. WALTON for their support and to J. P. LU for useful discussions. We thank the SERC for financial support and the Rutherford Appleton Laboratory for provision of neutron beam facilities.

REFERENCES

[1] KROTO H. W. et al., Nature, 318 (1985) 162.
[2] KRATSCHMER W., LAMB L. D., FOSTIROPOULOS K. and HUFFMAN D. R., Nature, 347 (1990) 354.
[3] TAYLOR R. et al., J. Chem. Soc. Chem. Commun. (1990) 1423.
[4] KROTO H. W., Nature, 329 (1987) 529.
[5] SCHMALZ T. G., SEITZ W. A., KLEIN D. J. and HITE G. E., J. Am. Chem. Soc., 110 (1988) 1113.
[6] HAWKINS J. M. et al., Science, 252 (1991) 312.
[7] KROTO H. W., ALLAF A. W. and BALM S. P., Chem. Rev., 91 (1991) 1213.
[8] FLEMING R. M. et al., Proceedings of the Materials Research Society, to be published.
[9] HEINEY P. A. et al., Phys. Rev. Lett., 66 (1991) 2911.
[10] TYCKO R. et al., Phys. Rev. Lett., 67 (1991) 1886.
[11] YANNONI C. S. et al., J. Phys. Chem., 95 (1991) 9.
[12] JOHNSON R. D. et al., Science, in press.
[13] NEUMANN D. A. et al., Phys. Rev. Lett., 67 (1991) 3808.
[14] DWORKIN A. et al., C. R. Acad. Sci. Paris, Serie II, 312 (1991) 979.
[15] DAVID W. I. F. et al., Nature, 353 (1991) 147.
[16] COPLEY J. R. D. et al., Physica B, in press.
[17] GUO Y., KARASAWA N. and GODDARD III W. A., Nature, 351 (1991) 464; LU J. P. et al., Phys. Rev. Lett., in press.
[18] HARE J. P., KROTO H. W. and TAYLOR R., Chem. Phys. Lett., 177 (1991) 394.
[19] PRASSIDES K. et al., Chem. Phys. Lett., 187 (1991) 455.
[20] DAVID W. I. F. et al., Rutherford Appleton Laboratory, Report RAL-88-103 (1988).
[21] SACHIDANANDAM R. and HARRIS A. B., Phys. Rev. Lett., 67 (1991) 1467.
[22] TOLEDANO J.-C. and TOLEDANO P., The Landau Theory of Phase Transitions (World Scient. Publ.) 1987, p. 194.
[23] REYNHARDT E. C. and FRONEMAN S., Mol. Phys., 74 (1991) 61.

Reprinted with permission from Physical Review Letters
Vol. 68, No. 10, pp. 1551–1554, 9 March 1992

Ground State and Phase Transitions in Solid C_{60}

Jian Ping Lu, X.-P. Li, and Richard M. Martin

Department of Physics and Material Research Laboratory, University of Illinois at Urbana-Champaign, Urbana, Illinois 61801
(Received 19 August 1991; revised manuscript received 9 December 1991)

A simple model is developed to describe the intermolecular interactions in solid C_{60}. The model predicts correctly the observed ground-state structure $Pa\bar{3}$ and the first-order transition to the high-temperature fcc phase. The calculated transition temperature $T_c \sim 270$ K and its pressure dependence $dT_c/dP = 11.5$ K/kbar agree very well with recent experiments. Below T_c, there exist nearly degenerate orientations which are separated by potential barriers of order ~ 300 meV, leading to a glassy behavior with $T_g \sim 90$–130 K. It is suggested that similar orientational disorder exists in K_3C_{60} and other fullerides.

PACS numbers: 61.50.-f, 61.55.-x, 64.70.Kb

The recent discoveries of an efficient synthesis [1] of C_{60} and superconductivity in K_3C_{60} and Rb_3C_{60} [2] have generated great interest in the structural and electronic properties of these materials. It is known that at room temperature C_{60} molecules are centered at fcc Bravais lattice sites in both solid C_{60} and K_3C_{60} [3]. Recently it was reported [4] that solid C_{60} undergoes a structural phase transition around 250 K from fcc to the simple cubic $Pa\bar{3}$ structure at low temperature. Both x-ray and NMR measurements [5] indicate that this transition is related to the orientational order of the C_{60} molecules.

We have developed a model to study the basic structural properties of solid C_{60}. The model consists of two distinct types of intermolecular interactions. The dominant one is the van der Waals–type interactions between carbon atoms on different C_{60} molecules. A secondary short-range Coulomb interaction is modeled by a small charge transfer between the two types of bonds in the C_{60} molecule. In contrast to early calculations [6] which include the van der Waals interactions only, our model predicts correctly the observed cubic ground-state structure $Pa\bar{3}$. Many structural properties calculated, such as the compressibility, cohesive energy, and specific heat, are in good agreement with experiments [7].

Most importantly, the model enables us to examine the possible structural phase transitions and orientational order in bulk C_{60}. In this Letter we describe our model and summarize some of our basic results which include the following: (1) The orientationally ordered ground state is simple cubic with symmetry $Pa\bar{3}$ as observed in experiments [8]. (2) At T_c the structure undergoes a first-order transition from $Pa\bar{3}$ to the high-temperature fcc phase where each molecule can freely rotate. The estimated transition temperature is $T_c \sim 270$ K, in agreement with experiments [4,9]. (3) The pressure dependence of the T_c is calculated to be $dT_c/dP = 11.5$ K/kbar which agrees very well with the observed values of 11.7 and 10.4 K/kbar [10]. (4) Below T_c, there are many nearly degenerate orientations for each C_{60} which are separated by potential barriers of order ~ 300 meV. This leads to novel frequency-dependent relaxation dynamics. (5) We predict that around $T_g \sim 90$–130 K the system undergoes a glassy transition when the equilibrium relaxation time

exceeds the laboratory time scale. This explains the existence of disorder at low temperature as observed in the neutron-scattering experiments [9,11]. (6) It is suggested that similar disorder exists in fullerides such as K_3C_{60} and Rb_3C_{60}.

The intermolecular interactions and the ground state.— As a result of the strong intra-C_{60} covalent bonds one expects that the structural properties of bulk C_{60} at low temperature are dominated by the weak inter-C_{60} interactions. This is supported by the fact that the internal molecular structure is unchanged in the solid state. Since the minimum distance between two C atoms on different C_{60} molecules (3 Å) is much larger than the covalent bond length, the interaction is primarily van der Waals in nature and can be described by the Lennard-Jones potential. As the minimum distance is comparable to the interlayer spacing in graphite (3.5 Å), it is reasonable to assume that the van der Waals interaction is similar to that in graphite. There are two parameters in the standard Lennard-Jones potential:

$$U(R) = 4\epsilon[(\sigma/R)^{12} - (\sigma/R)^6]. \qquad (1)$$

In the case of graphite it is easy to see that lattice constant c and the modulus of elasticity c_{33} are solely determined by the above potential. From the well-documented graphite data [12], $c = 6.708$ Å and $c_{33} = 0.408 \times 10^{12}$ dyn/cm^2, it is found that $\epsilon = 2.964$ meV and $\sigma = 3.407$ Å.

If Eq. (1) is the only interaction between two C_{60} molecules, then the ground state is found to be orthorhombic [6], in contradiction to the observed simple cubic structure $Pa\bar{3}$. On the other hand, the cohesive energy calculated is in fair agreement with experiment. This demonstrates that there exists a secondary interaction which makes the $Pa\bar{3}$ structure more stable. Indeed there is one crucial difference between a sheet of graphite and the C_{60} molecule— the former contains only one type of covalent bond while there are two in the latter. Because of the differences in the bond length in C_{60}, 1.45 Å for "single" bonds and 1.40 Å for "double" bonds, one expects that there will be a deficiency of electrons in the single bonds and an excess in the double bonds. Such a charge transfer will naturally lead to short-ranged intermolecular Coulomb interactions. (Because the C_{60} molecule is charge neutral and has a high symmetry, the interaction decays as $1/R^{13}$ at long distance.) This interaction will favor a nearest-neighbor configuration in which a pentagon (consisting of five single bonds) of a C_{60} molecule faces an oppositely charged double bond of the other molecule. Indeed this has been found in experiments [8].

To model the secondary interaction we introduce an effective charge q on single bonds (by the charge neutrality the effective charge on double bonds is $-2q$). Thus the total interaction between two C_{60} molecules is

$$V_{12} = \sum_{i,j=1}^{60} 4\epsilon \left[\left(\frac{\sigma}{|\mathbf{r}_{1i} - \mathbf{r}_{2j}|} \right)^{12} - \left(\frac{\sigma}{|\mathbf{r}_{1i} - \mathbf{r}_{2j}|} \right)^{6} \right]$$
$$+ \sum_{m,n=1}^{90} \frac{q_m q_n}{|\mathbf{b}_{1m} - \mathbf{b}_{2n}|}, \tag{2}$$

where $\mathbf{r}_{1i,2j}$ are coordinates of C atoms, $\mathbf{b}_{1m,2n}$ are coordinates of bond centers, and $q_{m,n}$ are the effective bond charges. The above interaction not only depends on the relative distance between the two molecules, but also on the orientations of two molecules, thus giving rise to interesting orientational dynamics in solid C_{60}.

Using the above interaction a search for the ground state was carried out by minimizing the total energy using the steepest descent method with four independent C_{60} per unit cell. It is found that for $q > 0.21e$, where e is the electron charge, the lowest energy state is always cubic with the $Pa\bar{3}$ symmetry. The ground-state orientation is described by rotating four molecules through the same angle $\phi = 21.3°$ (clockwise) but about different $\langle 111 \rangle$ axes: the molecule at $(0,0,0)$ about $[111]$, that at $(0,\frac{1}{2},\frac{1}{2})$ about $[\bar{1}1\bar{1}]$, that at $(\frac{1}{2},0,\frac{1}{2})$ about $[\bar{1}\,\bar{1}1]$, and that at $(\frac{1}{2},\frac{1}{2},0)$ about $[1\bar{1}\,\bar{1}]$ [8]. These rotations result in twelve identical nearest neighbors for each C_{60} with six pentagons almost directly facing the double bonds of neighboring molecules, thus maximizing the short-range Coulomb interactions. Experimentally the rotational angle was found to be 22° and 26° [8], in good agreement with our calculation.

Since the only free parameter in our model is the effective charge q, we fixed it by fitting the low-temperature lattice constant. From $a = 14.041$ Å [9] one obtains $q = 0.27e$, which is reasonable as one expects that the charge transfer should be a fraction of a single electron charge. Once the effective charge is known, the intermolecular interactions, and hence the low-temperature structural properties, are completely determined. Elsewhere we summarize elastic properties, phonon and libron spectra, and specific heat [7]. These results are in close agreement with experiments where data are available. The calculated cohesive energy is -1.990 eV/C_{60}, of which 90% comes from the Lennard-Jones interaction and only 10% is due to the Coulomb part.

The sc-fcc first-order transition. — As the temperature increases, C_{60} molecules can be thermally activated into other orientations. Therefore above a certain temperature T_c one expects that they can rotate freely and the average structure will be fcc. Indeed at room temperature this is the observed experimental structure.

To estimate the transition temperature we employ a mean-field argument by calculating the free energies. For the fcc phase we approximate each C_{60} as a three-dimensional free rotor. The free energy is the sum of the static energy E_{fcc} and the free energy of three-dimensional rotors at temperature T, $F_{\text{fcc}}(T) = N[E_{\text{fcc}} - k_B T \times \ln 8\pi^2 (Ik_B T/2\pi\hbar^2)^{3/2}]$, where $I = 1.0 \times 10^{-43}$ kg m^2 is the moment of inertia for the C_{60} molecule. The static energy E_{fcc} is calculated by averaging over random configurations. In the $Pa\bar{3}$ phase each molecule vibrates around its optimal orientation. So, the free energy is the sum of the static energy $E_{Pa\bar{3}}$ and that of three-dimensional harmonic oscillators, $F_{Pa\bar{3}}(T) = N[E_{Pa\bar{3}} + 3k_B T \ln(1 - e^{-\hbar\omega_0/k_B T})]$, where $\omega_0 = 1.86 \times 10^{12}$ s^{-1} is the average frequency of librons [7]. The transition temperature is identified as the point where the two free energies are equal. Using the lattice constant of $a = 14.16$ Å [9] we find $E_{\text{fcc}} = -1.772$ eV/C_{60} and $E_{Pa\bar{3}} = -1.968$ eV/C_{60}, from which we obtain a first-order transition at $T_c \sim 270$ K. The change of entropy at the transition temperature is estimated to be $\Delta S \sim 6R$ from the calculated T_c. Experimentally it is found that there is a first-order transition at 260 K [9]. The measured entropy discontinuity ranges from $2.1R$ [4] to $3.7R$ [13]. We expect that our mean-field estimation of the entropy discontinuity is very crude; a Monte Carlo study [14] is being carried out to calculate thermodynamic properties near T_c.

Using the above criteria we also calculated the pressure dependence of the transition temperature. Elsewhere [7]

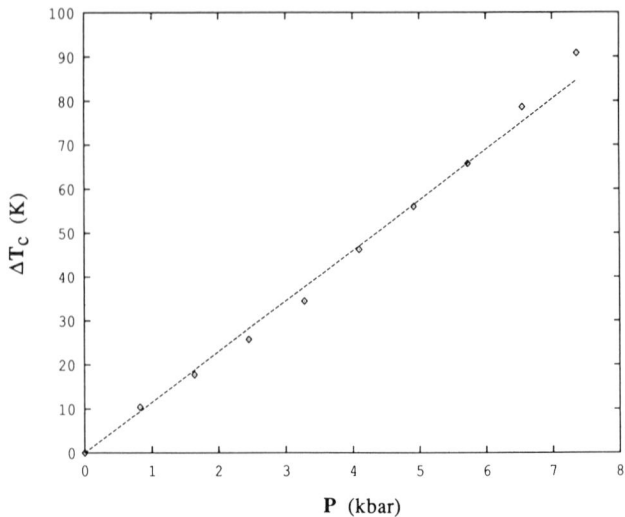

FIG. 1. The increment of the first-order transition temperature T_c as a function of the pressure. The calculated T_c at zero pressure is 270 K. A linear fitting through the calculated points gives $dT_c/dP = 11.5 \pm 0.7$ K/kbar. This is in good agreement with recent experimental results of 11.7 and 10.4 K/kbar [10].

we have calculated the compressibility of the ground state to be $K_0 = 5.175$ Mbar^{-1} which leads to $da_0/dP = 0.024$ Å/kbar. Assuming that the only effect of the pressure is to change the lattice constant we calculated the static energy $E_{Pa\bar{3}}$ and E_{fcc} as functions of pressure from which the transition temperature is obtained. Figure 1 shows the calculated increment of T_c as a function of the pressure. The linear fitting gives a slope of $dT_c/dP = 11.5 \pm 0.7$ K/kbar. This result agrees very well with recent experimental results of 11.7 and 10.4 K/kbar [10].

The glassy transition at low temperature.— As described in the previous section, in the ideal ground-state structure each C_{60} molecule has a unique orientation. However, there are many local minima with energy close to that of the ground state. These local minima are separated by large potential barriers of 300 meV. An example is shown in Fig. 2 where the potential energy for the C_{60} at $(0,0,0)$ is shown as a function of the rotation angle along three high-symmetry axes. As one can see there are many local minima; particularly a rotation of 60° along the threefold axis leads to a deep local minimum. Near T_c one expects there will be a thermal distribution of orientations among different local minima, achieved by thermal activation of molecules from one local minimum to another. As temperature decreases, the time needed to reach equilibrium increases exponentially, similar to what occurs in the orientational glasses such as $(KBr)_{1-x}(KCN)_x$ [15]. Thus we expect that at a temperature for which the equilibrium time exceeds the experimental time scale, the system freezes with a finite amount of disorder which persists down to zero temperature. This explains the neutron-scattering results which indicate that even at 14 K there is still about 20%–30% of disorder [9,11].

The freezing temperature T_g can be roughly estimated by considering the transition rate $1/\tau$ for a molecule to be thermally activated between local minima,

$$1/\tau = v e^{-E_b/k_B T}, \qquad (3)$$

where $v \sim 10^{12}$ Hz is the libration frequency and $E_b \sim 300$ meV is the typical potential barrier. τ should serve as a time scale needed for the system to reach equilibrium. From Eq. (3) one concludes that τ changes from 1 s to 1 day between 130 and 90 K. Therefore we expect that within this temperature range the system undergoes a glassy transition below which disorder becomes frozen. The amount of disorder will depend on the experimental details such as the cooling rate.

Equation (3) has several other important implications which can be directly confirmed by finite-frequency probes. One example is the motion-narrowing effect in NMR experiments which is expected to disappear when $1/\tau$ is below the chemical-shift-anisotropy (CSA) width. Indeed the NMR results of Tycko *et al.* [16] indicate that for a CSA width of 18.2 kHz the line broadens below 190 K and develops a powder pattern at lower temperature. This is in fair agreement with the 200 K calculated from Eq. (3). They also concluded that the thermal activation energy is around 260 meV below T_c, again close to the values we calculated. The glassy dynamics can be probed by other experiments such as sound attenuation, microwave absorption, and thermal conductivity. In particular the characteristic temperature will depend on probe frequency. Such studies are essential to fully understand the low-temperature orientational dynamics.

Finally, we comment on the existence of similar disorder in doped fullerides. For the metallic K_3C_{60} it has been shown that at room temperature the structure is fcc with C_{60} centered on the fcc Bravais lattice site and three K occupying the tetrahedral and octahedral interstitial sites [3]. It is also known that there are only two possible orientations for each molecule, corresponding to two possible ways of lining up the three orthogonal twofold axes of the icosahedron with cubic lattice axes. At high temperature, one expects that each C_{60} molecule can randomly flip between the two orientations. Because of the presence of ionic charge, one expects that the potential barrier between the two orientations is substantially larger than that in pure C_{60}. Hence the freezing temperature could be higher than that estimated for the bulk C_{60}. Whether it is higher than the characteristic temperature when K ions start to be mobile is an interesting question. At the moment, we are not aware of any experiments which probe this orientational disorder. It is expected that if such a glassy transition exists, it could have a strong effect on the normal-state transport properties.

In conclusion, we have constructed a model for the intermolecular interactions in solid C_{60}. The basic structural properties calculated are in good agreement with experiments. It is shown that around $T_c \sim 270$ K the structure undergoes a first-order transition from the high-

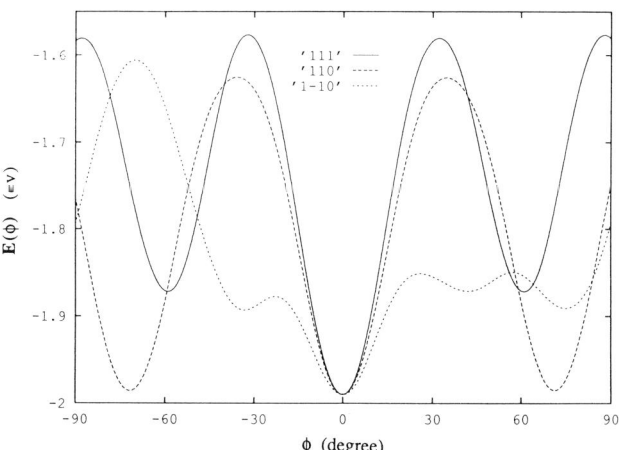

FIG. 2. The potential energy for the C_{60} sitting at $(0,0,0)$ as a function of rotation angle away from its equilibrium orientations. "111," "110," and "1-10" correspond, respectively, to threefold, fivefold, and twofold rotation axes of the molecule.

temperature fcc phase to the ordered $Pa\bar{3}$ structure. The pressure dependence of the transition temperature is calculated to be $dT_c/dP = 11.5$ K/kbar which agrees very well with recent experiments. We predict that between 90 and 130 K the system should undergo a glassy transition below which there is a finite amount of frozen disorder. This is due to the existence of many local minima which are separated by potential barriers of order 300 meV. The glassy behavior should manifest interesting frequency-dependent relaxation dynamics [17]. Finally, we expect that similar disorder exists in K_3C_{60} and other fullerides.

We appreciate helpful discussions with M. Gelfand, K. Ghiron, N. Goldenfeld, M. Grumbach, Y. Kwon, R. Liu, M. Salamon, K. Sorn, N. Tea, J. Yu, and R. Yu. We are grateful to K. Prassides and P. Heiney for communicating their experimental results. The research at the University of Illinois is supported by the National Science Foundation (Grant No. NMR 88-09854) through the Science and Technology Center for Superconductivity and the Department of Energy (Grant No. DEFG02-91ER45439).

[1] W. Kratschmer et al., Nature (London) 347, 354 (1990).

[2] H. Hebard et al., Nature (London) 350, 600 (1991); M. Rosseinsky et al., Phys. Rev. Lett. 66, 2830 (1991); K. Holczer et al., Science 252, 1154 (1991).

[3] P. Stephens et al., Nature (London) 351, 632 (1991); O. Zhou et al., Nature (London) 351, 462 (1991).

[4] P. Heiney et al., Phys. Rev. Lett. 66, 2911 (1991).

[5] R. Tycko et al., J. Phys. Chem. 95, 518 (1991); C. Yannoni et al., J. Phys. Chem. 95, 9 (1991).

[6] Y. Guo et al., Nature (London) 351, 464 (1991); J. P. Lu et al. (unpublished).

[7] X.-P. Li et al. (to be published).

[8] W. David et al., Nature (London) 353, 147 (1991); R. Schidanandam and A. Harris, Phys. Rev. Lett. 67, 146 (1991).

[9] W. David et al. (to be published); P. Heiney et al. (to be published).

[10] G. Kriza et al., J. Phys. I (France) 1, 1361 (1991); G. Samara et al., Phys. Rev. Lett. 67, 3136 (1991).

[11] J. Copley et al. (to be published).

[12] W. Gauster and I. Fritz, J. Appl. Phys. 45, 3309 (1991); Encyclopedia of X-Rays and Gamma Rays, edited by G. Clark (Reinhold, New York, 1963).

[13] T. Atake (to be published).

[14] Y. Kwon et al. (unpublished).

[15] J. De Yoreo et al., Phys. Rev. B 34, 8828 (1988); E. Grannan et al., Phys. Rev. Lett. 60, 1402 (1988).

[16] R. Tycko et al., Phys. Rev. Lett. 67, 1886 (1991).

[17] After we finished the manuscript we learned of recent experiments by X. Shi et al. on sound attenuation (to be published) and by R. Yu et al. on thermal conductivity (private communication). Their results indicate that there is a frequency-dependent glassy behavior at low temperature, in agreement with our predictions.

SUPERCONDUCTIVITY AND MAGNETISM IN FULLERIDES

As other research avenues on fullerenes flourished, a startling discovery came from a group at AT&T Bell Labs: C_{60} films "doped" with alkali metals became conducting,[Ha91a*] and, indeed, superconducting, with an onset temperature of 18 K.[He91*] This immediately generated new excitement throughout the solid-state physics and materials research community. Within two months of the first announcement of metallic conductivity, two groups had published transition temperatures of 29 ± 1 K for Rb_xC_{60}.[Ro91*, Ho91a*] Matters quickly warmed up to 33 K,[Ta91] at which they are holding steady, at least for the time being. (A report of superconductivity at 45 K has been withdrawn.[Iq91, Iq92]) An entirely new family of (moderately) high temperature superconductors does not come along every day! The only known superconductors with higher transition temperatures than the alkali fullerides are those based on copper-oxide planes.

Several detailed reviews of superconducting fullerides have appeared in the literature. Especially useful is the overview by Haddon [Ha92], reviews of the synthesis and structures by Murphy *et al.*[Mu92] and by Zhou and Cox [Zh92b], superconducting and normal state properties by Holczer and Whetten [Ho92], and electronic structure, by Weaver [We92a, We92b].

The first experiments were carried out on films of solid C_{60}, 100 to 1000 Å thick, exposed to alkali metal vapors.[Ha91a*] It was observed that the conductivity increased by more than seven orders of magnitude, to 500 $(\Omega \text{ cm})^{-1}$. This is still a much lower conductivity than any metallic element, but on the order of such organic systems as doped polyacetylene. Haddon *et al.* made the plausible suggestion, subsequently proved, that the smaller alkali ions were intercalated into the voids between the much larger fullerenes, and donated their charge to the unoccupied fullerene t_{1u} molecular orbital. The ratio of alkali atoms to fullerenes in the (super) conducting phase was not then known, nor was it known whether the structure was based on the

face-centered-cubic structure of pure C_{60} or upon some rearrangement of the fullerene lattice. As the alkali content was increased, the film again became insulating. We will return to these normal-state transport properties below.

It is worthwhile to point out that the alkali fullerides are extremely air-sensitive. All sample preparations with alkali metals must be carried out in a glove box with oxygen levels on the order of one part per million. However, prepared samples can survive some exposure to air.[Ho91a*]

There are two definitive tests for superconductivity: zero electrical resistance and expulsion of magnetic flux. Both were demonstrated in the first K_xC_{60} experiments, in thin films and powder samples, respectively.[He91*] Initial efforts to observe superconductivity in Rb-doped films were unsuccessful, but the exposure of C_{60} powder to Rb vapor, followed by appropriate heat treatment, did produce superconducting samples with a transition temperature of 28 K.[Ro91*, Ho91a*]

Rosseinsky *et al.* also suggested a simple reason for the large increase in transition temperature when Rb is substituted for K. The most elementary, widely understood theory of superconductivity, the weak-coupling Bardeen-Cooper-Schrieffer (BCS) model,[Ba57] predicts a simple connection between transition temperature T_c, phonon energy ω, electron-phonon coupling strength V, and the density of electron states at the Fermi surface $N(E_F)$,

$$kT_c = \hbar\omega \exp \frac{-1}{N(E_F)V}. \qquad (1)$$

It is plausible to assume that ω and V are independent of the identity of the alkali ion, and that the lattice is expanded by the substitution of the larger Rb ion. As the lattice expands, the electronic wave functions on neighboring fullerenes overlap less strongly, narrowing the energy band, leading to an increase of $N(E_F)$ and thereby, an increase of T_c.

In these initial experiments, only a small fraction of the sample was superconducting. This hindered any determination of the composition and the structure of the superconducting phase. A significant advance occurred when Holczer et al. reported techniques for preparing relatively homogeneous powder samples with shielding diamagnetism fractions of at least 67%.[Ho91a*] Such a high fraction in a powder sample indicates that the superconducting phase is the definite majority, with any impurity phase a relatively minor component. The highest diamagnetic fraction was obtained for a composition of K_3C_{60}, implying that this was the composition of the superconducting phase. They also independently discovered superconductivity in Rb_3C_{60} at an onset temperature of 30 K, but with a significantly lower total shielding diamagnetism than their K_3C_{60} samples[Ho91a*].

With homogeneous samples available, the detailed characterization of the superconducting phase could proceed. X-ray diffraction experiments gave a structure consisting of a face-centered cubic lattice of C_{60}, with K ions in all of the octahedral and tetrahedral voids, confirming the composition K_3C_{60}.[St91*] From these x-ray measurements, it could be determined that the fullerene molecules were not rotating freely, in contrast to the case of solid C_{60} at room temperature. Indeed, there would not be enough space for the cations if the C_{60} were spherically disordered; each C_{60} is oriented so that eight of its twenty hexagonal faces form nests for the tetrahedral cations. There are two such possible orientations for the C_{60} within the lattice, and they appear to be randomly distributed.[St91*] The molecular orientational disorder persists to low temperatures.[Stephens et al., unpublished] Subsequent model calculations have indicated that the orientational disorder has a significant effect on the band structure and the conduction electron scattering rate.[Ge92, Ma93a]

X-ray experiments show that Rb_3C_{60} also has the same fcc structure.[Fl91b*, St92a, Zh91b] However, that conclusion seems to be inconsistent with recent [87]Rb nuclear magnetic resonance experiments, which indicate that the tetrahedral cations are not all equivalent.[Wa93]

Magnetization measurements on the bulk powders of K_3C_{60} allowed the determination of additional parameters of the superconducting state. Upper and lower critical fields were analyzed to determine a penetration depth of $\lambda = 2400$ Å, and a superconducting coherence length of $\xi =$

26 Å.[Ho91b*] Muon spin resonance experiments gave a somewhat different value, $\lambda = 4800$ Å, and provided evidence that the superconducting gap was spherically symmetric, implying singlet, s-wave pairing.[Ue91]

Earlier photoemission experiments had studied the modifications of the electronic structure of a C_{60} film upon exposure to potassium vapor.[Be91, We91] These experiments found that the potassium donates an electron to the unoccupied $C_{60}t_{1u}$ molecular orbital, creating a partially filled (metallic) band. Further photoemission and inverse photoemission experiments have been reviewed in detail by Weaver.[We92a, We92b] These experimental results may be compared with electronic band structure calculations, such as [Sa91], [Er91], and [Ha91b], reviewed in detail in [Os92b].

More recently, this interpretation of the electronic band structure has been challenged by Lof et al.[Lo92] In rough agreement with earlier work, they find a band gap of 2.3 eV. Comparing the photoelectron (one hole) with the Auger electron (two holes) spectra, they directly measured an on-site hole-hole repulsion of 1.6 eV, in agreement with earlier estimates. Because the electron correlation energy is greater than the widths of the bands derived from the HOMO and LUMO, Lof et al. argue that K_3C_{60} should be regarded as a strongly correlated system. If that were the case, K_3C_{60} could actually be a Mott-Hubbard insulator, with metallic behavior occurring only for off-stoichiometric concentrations.[Lo92]

More insights into the superconducting state were provided by measurements of the transition temperature at elevated pressures. It was found that the superconducting transition temperature T_c was lowered by pressure, with $(dT_c/dP)_{P=0} = -7.8$ K/GPa (-0.78 K/kbar), among the strongest depressions of T_c reported for any superconductor.[Sp91] As these authors pointed out, this fits nicely into the picture that the lattice constant has a strong effect on the transition temperature through the conduction electron density of states. On the other hand, as discussed below, other pictures of the superconductivity are also able to account for this observation.

Various information about the superconducting K_3C_{60} material was revealed by ^{13}C nuclear magnetic resonance (NMR) experiments.[Ty91*] From the observation of a single sharp NMR line at room temperature, it could be deduced that the C_{60} molecules were reorienting rapidly with respect to the NMR chemical shift anisotropy time scale of 50 μsec. In view of the x-ray results [St91*], the C_{60} motion must be a jump among the 120 discrete orientations (for an isotopically labeled fullerene) allowed by the structure. The NMR linewidth and relaxation time increase below room temperature, indicating that the C_{60} reorientation time slows to less than the NMR time scale. This NMR experiment also found that there were no stable phases of K_xC_{60} for $0 < x < 3$.

One of the most important experimental results to emerge in the early days of fullerene superconductivity was the simple correlation between the lattice constant a and the transition temperature T_c. This was most clearly expressed in experiments which compared fullerides made with various mixtures of K, Rb, and Cs,[Ch91a, Fl91b*] and in the pressure dependence discussed earlier. According to x-ray measurements, all have the same structure, but the larger cations push a to larger values.[Fl91b*] A plot of T_c vs. a shows a steady increase of about 55 K/Å. Unfortunately, this steady increase of transition temperature appears to stop at Rb_2CsC_{60};[Ta91, Fl91b*] efforts to synthesize Cs_3C_{60} have not been successful. More important than the quest for higher transition temperatures is the understanding afforded by the T_c vs. a data. It provides a quantitative connection between the parameters in Eq. (1), leading to an improved understanding of the phenomenon of fullerene superconductivity.

During this period, the theoretical understanding began to advance beyond hand-waving arguments based upon Eq. (1). The principal challenge is to determine whether transition temperatures as high as those observed can be understood within the context of conventional electron-phonon BCS theory, or if some novel mechanism is involved. It is safe to say that there is no widely-accepted theory of superconductivity for the layered cuprate "high-T_c" materials, and so the highest transition temperature that may be claimed to be understood is 23 K for Nb_3Ge. (Actually, $Ba_{1-x}K_xBiO_{3-y}$ is in a situation very similar to A_3C_{60}, a rather poor metal with superconducting T_c near 30 K, almost understood through the electron-phonon interaction.[Sh89, Lo89, Ta92c]) If an electron-phonon interaction is responsible for binding the electrons into Cooper pairs, one hopes to determine which phonons are responsible for the attractive interaction between electrons. Based on Eq. (1), we must be dealing with higher excitation energies ω, narrower bands

[large $N(E_F)$], or stronger electron-phonon interactions V than are observed in other materials.

Two early calculations of electron-phonon pairing strengths reached similar conclusions.[Va91*, Sc92a*] Both groups performed calculations showing that particular intra-fullerene vibrational modes produced a strong attractive electron pairing interaction (dynamical Jahn-Teller interaction). The relevant vibrations have relatively high energy ω. At the same time, the electronic band width is predicted to be very narrow because it is derived from a single molecular orbital on the bare fullerene molecule, broadened only by the weak quantum mechanical tunneling between fullerenes. Both papers argued that the transition temperatures observed were consistent with their calculations, as was the observation that T_c falls rapidly with decreasing lattice constant. Further details are given in [Sc92c]. A subsequent paper treated the intramolecular electron-phonon interaction more rigorously, and gave quantitative postdictions for T_c in reasonable agreement with experiment.[Ma92]

An alternative theoretical approach focuses on direct electronic correlation effects within a single fullerene molecule, without reference to vibrations.[Ch91b*] These authors argue that correlations among the sixty π electrons on a fullerene are so strong that they cannot be treated in an effective single particle picture of only the highest occupied molecular orbital (as is done in the electron-phonon theories discussed above). In this picture, the electrostatic repulsion between each pair of electrons on a given fullerene can actually produce attraction in the multi-electron system. Because of correlations, the energy to add two electrons to a given fullerene may be less that twice the energy to add one electron. If the attractive electron-electron interaction on each fullerene is less than the electronic band width, it produces a weak pairing force between conduction electrons, which can be treated in a framework very similar to BCS theory [Eq. (1)]. In particular, the pressure dependence of the transition temperature is also consistent with this theory. There are other predictions of this model, which will be discussed in the context of further experimental results.

As one advances beyond the simplest ideas about superconductivity, more parameters of the theory must be included. In particular, the effective attraction between electrons is diminished by their electrostatic repulsion. There has been a lively debate between the two schools of thought described above about plausible estimates for this "Coulomb pseudopotential".[Ch92a, Sc92b]

Returning to Rb_3C_{60}, magnetization measurements gave values for the London penetration depth and superconducting coherence length of $\lambda = 2470$ Å, and $\xi = 20$ Å, respectively.[Sp92a*] In the same work, the pressure dependence of the transition temperature in Rb_3C_{60} was measured and compared to that of K_3C_{60}. The zero-pressure slope $(dT_c/dP)_{P=0}$ was -9.7 K/GPa, even higher than that of K_3C_{60}. The analysis in that paper concluded that the differences in T_c, dT_c/dP, and ξ between the K and Rb fullerides were all consistent with a 15% difference in the conduction electron density of states.

The conduction electron density of states can in principle be measured through the nuclear spin-lattice relaxation time T_1, although there are many complexities.[Wo92, Ho92] Comparisons of the T_1 data gave a 30% higher density of states for Rb_3C_{60} than K_3C_{60}.[Ty92*] This further validates the model surrounding Eq. (1), although it is not in quantitative agreement with the 15% discussed above. At the same time, electronic structure calculations weigh in with a 20% change in the density of states between those two materials.[Os92a]

It was also found that the plot of T_c vs. P for the two materials coincided, if the pressure scale was translated by 1.06 GPa.[Sp92*] This would match the T_c vs. lattice constant a data of Fleming et al. if the linear compressibility of K_3C_{60} were 1.21×10^{-2} GPa^{-1}.[Sp92*] This is almost identical to the experimental value for the compressibility measured contemporaneously.[Zh92a*] Indeed, with measurements of K_3C_{60} and Rb_3C_{60} lattice constants vs. pressure, the previous data sets of T_c vs. P [Sp91, Sp92a*] could be compared directly to the T_c vs. a data [Fl91b*] for various compounds $(K{:}Rb{:}Cs)_3C_{60}$ discussed above. At a first glance, all of the data fall on the same universal curve, indicating that, for this family of materials, the superconducting transition temperature depends only on the lattice constant. This implies, within the weak coupling BCS theory of Eq. (1), that the relevant phonon energy ω, the electron-phonon coupling strength V, and the density of states $N(E_F)$ depend principally on the lattice constant a, with at most a weaker dependence on cation mass. Taking a closer look, there are differences as much as 4 K in T_c at the same a. It is difficult (though obviously important) to judge whether this is a significant variation or undetected error.

A comparison of fullerene vibrations between C_{60} and A_3C_{60} (A = K or Rb) is a significant issue, in view of the importance of intra-fullerene vibrational modes in certain models of superconductivity. Direct observations were made by Raman spectroscopy,[Du91*] and inelastic neutron scattering.[Pr91*] The neutron scattering experiments have been further reviewed by Prassides et al.[Pr92] In both experiments, it was found that the higher modes of H_g symmetry became very broad and/or weak. Those are precisely the vibrations which the electron-phonon models had identified as significant for superconductivity,[Va91*, Sc92a*] and it is plausible that they have become strongly coupled to the conduction electrons. There has been some theoretical debate on the precise interpretation of these results.[Ch92a, Sc92b] Overall, it is safe to say that the Raman and neutron data support the idea that the H_g fullerene vibrations are important modes for superconductivity.

Another experimental result which might elucidate directly the mechanism of superconductivity is the shift of transition temperature T_c with isotopic substitution. This phenomenon had provided the first suggestion for the role of phonons in superconductivity well before the birth of the BCS theory. If the mass M of all of the atoms in a sample is increased, the lattice vibration frequencies will all decrease as $\omega \sim M^{-1/2}$. Looking back to Eq. (1), this predicts that $T_c \sim M^{-1/2}$, if all other parameters remain the same. In practice, measured values of the exponent α defined as $-(\partial \ln T_c / \partial \ln M)$ are frequently somewhat smaller than $1/2$, due to a mass dependence of the coupling strength V or other electron-electron interactions. Some puzzles on the isotope shift in other superconductors are reviewed by Allen.[Al88]

Probably the cleanest measurement of an isotope effect in a fulleride superconductor was made on 99% enriched $K_3{}^{13}C_{60}$, giving an exponent of $\alpha = 0.3 \pm 0.06$.[Ch92b*] Other works on both K and Rb have given a bewildering array of results, tabulated below.

Table 1 Published Isotope Shift Experiments on Superconducting Fullerides

Cation	^{13}C Fraction	α	Reference
Rb	33%	1.6 ± 0.5	Eb92
Rb	$(75 \pm 5)\%$	0.37 ± 0.05	Ra92
K	99%	0.3 ± 0.06	Ch92b*
K, Rb	60%	1.2–2.25	Za92

There has been considerable discussion among these groups and others about the large differences in the reported values of α, without any clear resolution to date. One possible variation among experiments is in the distribution of isotopes in samples of less than 100% ^{13}C enrichment. A most interesting recent result is that inter-fullerene isotopic disorder ($\mathrm{Rb_3\,^{13}C_{60_{0.5}}\,^{12}C_{60_{0.5}}}$) reduces the transition temperature more (larger measured α) than does intrafullerene disorder ($\mathrm{Rb_3(^{13}C_{0.5}\,^{12}C_{0.5})_{60}}$).[Ch93] It is clear that rigorous comparisons of sample purity via infrared and mass-spectrometric techniques, as well as careful, parallel preparation of standard comparison samples will be required to clarify the experimental differences.

Interpreting these widely different results may be dangerous. It is, however, unambiguously clear that there is a significant isotope shift, with an exponent α of at least 0.3. While this would immediately suggest that phonons are implicated in the pairing mechanism, one must proceed with some caution. First, it is difficult to see how electron-phonon coupling could lead to values of α larger than 1/2, as some of the experiments have given. Second, there is a possible mechanism for an isotope dependence in a purely electronic pairing mechanism. Substitution of ^{13}C could reduce the size of each fullerene through the quantum-mechanical zero-point motion of the carbon-carbon bonds, leading to an estimated suppression of T_c consistent with the experimental results.[Ch92c]

Tunneling spectroscopy is a widely used technique for investigation and characterization of superconductivity. Briefly, it provides a direct measure of the superconducting energy gap Δ. Measurements in both $\mathrm{K_3C_{60}}$ and $\mathrm{Rb_3C_{60}}$ gave a reduced energy gap of $2\Delta/kT_c = 5.3$.[Zh91c*] Conventional weak coupling BCS theory predicts that the electronic energy gap should be related to the transition temperature T_c as $2\Delta/kT_c = 3.53$. The higher value observed indicates strong coupling, necessitating a more complicated theoretical treatment of the electron pairing interaction. The tunneling observation of a clean energy gap is not a trivial one; it has not been seen in the layered cuprate (high T_c) superconductors. The same value for the gap is also revealed in infrared measurements.[Ro92b, De92]

We return to the normal state of the (super)conducting fullerides, first with two papers on transport properties in thin films.[Ko92a*, Pa92*] Recalling that the metallic phase occurs at a composition of $\mathrm{K_3C_{60}}$, and that distinct phases of $\mathrm{K_xC_{60}}$ at $x = 0$, 3, 4, and 6 are known, it is of interest to determine the state of a polycrystalline film of $\mathrm{K_xC_{60}}$ as the potassium dose x is increased. Such measurements show a distinct minimum in the resistivity at $x = 3$.[Ko92a*] The temperature dependence of the resistivity is activated for $x \neq 3$, leading to an interpretation of hopping conductivity between grains. The lowest resistivity observed for thick films is very high for a metal, 2200 $\mu\Omega$ cm. Combined with the known carrier density (one electron per K ion), this leads to an unphysically small electronic mean free path of 2.3 Å. That result implies that the film is not homogeneous, and indeed a model of intergranular transport gave a satisfactory fit to the data. The composition at the resistivity minimum was determined by Rutherford backscattering to be $x = 3.00 \pm 0.05$.

The temperature dependence of the resistivity and the Hall coefficient were studied in a complementary paper.[Pa92*] The resistivity increased 20% as the sample cooled from room temperature to the superconducting transition, due to the granular nature of the film. Resistivity measurements near the superconducting transition of a $\mathrm{K_3C_{60}}$ film also indicated granular conduction, with a grain size of 70 Å.[Pa92*] Upper critical field measurements on these film samples led to a coherence length of $\xi = 26$ Å, in good agreement with earlier data on powders.[Ho91b*] The authors argued that this small coherence length was an effect of the grain size, and estimated that, in the absence of sample granularity, ξ would increase to 150 Å. This larger value for the coherence length is also favored by theory.[Ma92] However, it is not clear that sample granularity could have suppressed ξ in the powder measurements, because x-ray experiments have given grain sizes of at least 500 Å.[St91*]

Single crystals of superconducting $\mathrm{K_3C_{60}}$ were first synthesized by Maruyama et al., [Ma91] by exposing a solvated crystal of $\mathrm{C_{60}:CS_2}$ to K vapor. K stoichiometry was controlled by monitoring the sample resistance. They observed that the resistivity decreased with decreasing temperature, as expected for a metal. Furthermore, the single crystal had a superconductive transition less than 1 K wide from onset to zero resistance, much narrower than had been measured on any other sample. This technique was refined by Xiang et al.,[Xi92*] by starting with $\mathrm{C_{60}}$ crystals grown by vapor transport, and repeated cycles of annealing during the K dosing process. This yielded samples with a room temperature resistivity of 5000 $\mu\Omega$ cm ($\pm$50%), clearly metallic transport with a resistivity ratio $\rho(300\ \text{K})/\rho(20\ \text{K}) = 2$ and $\rho(T)$ curving upward, and a superconducting transition width of less than 0.2 K. Fluctuation conductivity above the critical temperature was subsequently measured by the same group.[Xi93]

We have been concentrating exclusively on the superconducting A_3C_{60} phases, but there are other alkali fullerides. K_6C_{60}, Rb_6C_{60}, and Cs_6C_{60} all have a body-centered cubic structure, with alkali ions in distorted tetrahedral interstices.[Zh91a] In that structure, the fullerenes are all fixed in the same orientation, in contrast to the disorder and motion observed in K_3C_{60}.[St91*, Ty91*] The A_6C_{60} phases are insulators, as would be expected from the fact that six donated electrons will completely fill the bands created by the fullerene t_{1u} and t_{1g} molecular orbitals.

The transfer of six electrons to each fullerene in A_6C_{60} has a significant effect on the vibrational spectrum, shifting the infrared absorption frequencies by as much as 11%, and increasing their oscillator strength by factors as large as 100.[Fu92, Ma93b] This has been interpreted as the coupling of a molecular vibration to a virtual intramolecular electronic transition.[Ri92] It may be possible to use this effect to determine directly the electron-phonon coupling constant in the superconducing A_3C_{60} phases, at least to IR active modes.

There are also body-centered tetragonal phases A_4C_{60} for A = K, Rb, and Cs.[Fl91a] ^{13}C NMR relaxation time measurements indicate that Rb_4C_{60} is an insulator,[Mu92] which is slightly surprising in view of the fact that four donated electrons should leave the band derived from the C_{60} t_{1g} orbital partially occupied. The tetragonal symmetry may be enough to split that band, or it may be an effect of electron correlations. While single-phase powders have been produced, it appears that alkali-dosed C_{60} films do not exhibit this phase, growing as a two-phase mixture of K_3C_{60} and K_6C_{60}.[Ko92a*, Mu92]

Sodium fullerides present some interesting puzzles. Na_3C_{60} fails to superconduct, even though its room-temperature lattice constant would lead one to predict a transition temperature of ~ 15 K.[Ro92a] Na_3C_{60} appears to disproportionate into insulating Na_2C_{60} and Na_6C_{60} as it is cooled below room temperature.[Gu92, Ro92a] Some mixed $(Li, Na)_x(K, Rb, Cs)_{3-x}C_{60}$ phases exhibit superconductivity, but a plot of their T_c vs. lattice constant differs from that of fullerides of only K, Rb, and Cs.[Ta92b] This departure from the

previous relationship, through the same range of lattice constants, suggests that the motion of the smallest alkali ions in the fullerite interstices is somehow inimical to superconductivity.

Unlike the larger alkalis, Na has stable face-centered cubic fullerides Na_xC_{60} for x as large as 10, with a small body-centered cube of nine Na atoms (ions?) occupying the octahedral hole.[Yi92] In that phase, the nearest Na–Na distance is a remarkably small 2.8 Å, much less than the 3.67 Å in Na metal.

Recently, it has been discovered that superconductivity in the fullerides is not limited to alkali cations. Ca_5C_{50} is a superconductor below $T_c = 8.4$ K.[Ko92b] Through a range of Ca compositions up to Ca_6C_{60}, the structure is face-centered cubic, similar to K_3C_{60}. However, weak simple-cubic reflections are seen at Ca_5C_{60}, possibly indicating some structural ordering in the three-fold occupied octahedral holes. Photoemission spectra indicate that the superconducting phase has a fully occupied t_{1u} band, and a partially filled t_{1g} band.[We92c] The (super)conductivity is associated with the latter, in contrast to the alkali A_3C_{60} phases. Photoemission experiments have also been performed on other alkaline earth fullerides.[Ch92d] Sr:C_{60} and Ba:C_{60} have metallic phases, but Mg:C_{60} does not.

Nor is superconductivity limited to (approximately) fcc phases. Body-centered-cubic Ba_6C_{60} is superconducting below a transition temperature of 7 K.[Ko92c]

We have been discussing C_{60} salts with electron donors; it is interesting to consider the possibility of a halogen compound of C_{60}. The solution electrochemistry of C_{60} is not encouraging, inasmuch as oxidation waves have not been detected. A well defined phase of I_4C_{60} has been characterized, with an interfullerene distance comparable to those of the (superconducting) alkali fullerides.[Zh92c] However, this phase is an insulator at room temperature, and shows no signs of superconductivity to 4 K.

An entirely different ordered state is produced when fullerenes are combined with strong organic donors. Other organic charge-transfer salts comprise several interesting families of materials, exhibiting superconductivity, spin density waves, and other electronic phenomena. When C_{60} was mixed with the strong organic donor TDAE (tetrakis(dimethylamino)ethylene, $C_2N_4(CH_3)_8$), it crystalized into a material which exhibited a huge magnetic susceptibility below 16 K.[Al91*] This is by far the highest magnetic ordering temperature observed for any compound made up of only first-row elements. One may wish to be careful about calling this phenomenon ferromagnetism, because the material does not exhibit any remanent magnetization or hysteresis. However, the susceptibility is large enough to rule out such phenomena as superparamagnetism.

The reason for the different behaviors of fullerides produced from organic *vs.* alkali or alkaline-earth donors was clarified by the determination of the lattice structure of TDAE–C_{60}.[St92b*] It was found that the solid has a significantly lower crystal symmetry, which cannot really be viewed as a close-packing of fullerenes with donors in the interstices. Structurally, this comes about because the donor molecule has a size in the order of that of the fullerene. The fullerenes are arrayed in chains, with a nearest neighbor distance along each chain of 9.98 Å, and an intra-chain separation of 10.25 Å. These two distances span the range of nearest neighbor distances in the superconducting alkali fullerides. One therefore expects that the intermolecular overlap is sufficient to create a partially filled electronic band, whose structure would be significantly anisotropic. However, there is still no detailed calculation of the electronic structure of this compound.

In contrast, TDAE–C_{70} does not show magnetic order. The ESR and magnetization properties of TDAE–C_{60} and TDAE–C_{70} were compared by Tanaka *et al.*, who reproduced the magnetic transition in TDAE–C_{60}, but found only Curie paramagnetism in TDAE–C_{70}.[Ta92a] It is not yet known whether this difference is due to the difference in the electronic structure of C_{60} *vs.* C_{70}, or if TDAE–C_{70} adopts some different crystal structure.

Another useful insight into the magnetism of TDAE–C_{60} is afforded by the observation that the magnetism is suppressed very rapidly with applied pressure; no magnetic transition is visible at a pressure of 1.6 kbar.[Sp92b] Interestingly, application of pressure reduces the magnitude of the moment more strongly than it does the transition temperature. These results are discussed further, and placed in the overall context of organic ferromagnetism in a review by Wudl and Thompson.[Wu92]

Bibliography

References marked with * are reprinted in this volume.

Al88   P. B. Allen, "Isotope Shift Controversies," *Nature* **335**, 396–397 (1988).

Al91*   P.-M. Allemand *et al.*, "Organic Molecular Soft Ferromagnetism in a Fullerene C_{60}," *Science* **253**, 301–303 (1991).

Ba57   J. Bardeen, L. N. Cooper and J. R. Schrieffer, "Microscopic Theory of Superconductivity," *Phys. Rev.* **106**, 162–164 (1957).

Be91   P. J. Benning, J. L. Martins, J. H. Weaver, L. P. F. Chibante and R. E. Smalley, "Electronic States of $K_x C_{60}$: Insulating, Metallic, and Superconducting Character," *Science* **252**, 1417–1419 (1991).

Ch91a   C. C. Chen, S. P. Kelty and C. M. Lieber, "$(Rb_x K_{1-x})_3 C_{60}$ Superconductors: Formation of a Continuous Series of Solid Solutions," *Science* **253**, 886–888 (1991).

Ch91b*   S. Chakravarty, M. P. Gelfand and S. Kivelson, "Electronic Correlation Effects and Superconductivity in Doped Fullerenes," *Science* **254**, 970–974 (1991).

Ch92a   S. Chakravarty, S. Khlebnikov and S. Kivelson, "Comment on 'Electron-Phonon Coupling and Superconductivity in Alkali-Intercalated C_{60} Solid'," *Phys. Rev. Lett.* **69**, 212 (1992).

Ch92b*   C. C. Chen and C.M. Lieber, "Synthesis of Pure $^{13}C_{60}$ and Determination of the Isotope Effect for Fullerene Superconductors," *J. Am. Chem. Soc.* **114**, 3141–3142 (1992).

Ch92c   S. Chakravarty, S. A. Kivelson, M. I. Salkola and S. Tewari, "Istotope Effect in Superconducting Fullerenes," *Science* **256**, 1306–1308 (1992).

Ch92d   Y. Chen, F. Stepniak, J. H. Weaver, L. P. F. Chibante and R. E. Smalley, "Fullerides of Alkaline Earth Metals," *Phys. Rev.* **B45**, 8845–8848 (1992).

Ch93   C. C. Chen and C. M. Lieber, "Isotope Effect and Superconductivity in Metal-doped C_{60}," *Science* **259**, 655–658 (1993).

De92   L. Degiorgi *et al.*, "Optical Response of the Superconducting State of $K_3 C_{60}$ and $Rb_3 C_{60}$," *Phys. Rev. Lett.* **69**, 2987–2990 (1992).

Du91*   S. J. Duclos, R. C. Haddon, S. Glarum, A. F. Hebard and K. B. Lyons, "Raman Studies of Alkali-Metal Doped $A_x C_{60}$ Films (A = Na, K, Rb, and Cs; x = 0, 3, and 6)," *Science* **254**, 1625–1627 (1991).

Eb92   T. W. Ebbesen *et al.*, "Isotope Effect on Superconductivity in $Rb_3 C_{60}$," *Nature* **355**, 620–622 (1992).

Er91   S. C. Erwin and W. E. Pickett, "Theoretical Fermi-Surface Properties and Superconducting Parameters for $K_3 C_{60}$," *Science* **254**, 842–845 (1991).

Fl91a   R. M. Fleming *et al.*, "Preparation and Structure of the Alkali-Metal Fulleride A_4C_{60}," *Nature* **352**, 701–703 (1991); *Erratum ibid.* **353**, 868 (1991).

Fl91b*   R. M. Fleming *et al.*, "Relation of Structure and Superconducting Transition Temperatures in A_3C_{60}," *Nature* **352**, 787–788 (1991).

Fu92   K. J. Fu *et al.*, "Giant Vibrational Resonances in A_6C_{60} Compounds," *Phys. Rev.* **B46**, 1937–1940 (1992).

Ge92   M. P. Gelfand and J. P. Lu, "Orientational Disorder and Electronic States in C_{60} and A_3C_{60}, where A is an Alkali Metal," *Phys. Rev. Lett.* **68**, 1050–1053 (1992).

Gu92   G. Gu *et al.*, "Metallic and Insulating Phases of Li_xC_{60}, Na_xC_{60}, and Rb_xC_{60}," *Phys. Rev.* **B45**, 6348–6351 (1992).

Ha91a*   R. C. Haddon *et al.*, "Conducting Films of C_{60} and C_{70} by Alkali-Metal Doping," *Nature* **350**, 320–322 (1991).

Ha91b   N. Hamada, S. Saito, Y. Miyamoto and A. Oshiyama, "Fermi Surfaces of Alkali-Metal-doped C_{60} Solid," *Japan. J. of Appl. Phys.* **30**, L2036–L2038 (1991).

Ha92   R. C. Haddon, "Electronic Structure, Conductivity, and Superconductivity of Alkali Metal Doped C_{60}," *Acc. Chem. Res.* **25**, 127–133 (1992).

He91*   A. F. Hebard *et al.*, "Superconductivity at 18 K in Potassium-doped C_{60}," *Nature* **350**, 600–601 (1991).

Ho91a*   K. Holczer *et al.*, "Alkali-Fulleride Superconductors: Synthesis, Composition, and Diamagnetic Shielding," *Science* **252**, 1154–1157 (1991).

Ho91b*   K. Holczer *et al.*, "Critical Magnetic Fields in the Superconducting State of K_3C_{60}," *Phys. Rev. Lett.* **67**, 271–274 (1991).

Ho92   K. Holczer and R. L. Whetten, "Superconducting and Normal State Properties of the A_3C_{60} Compounds," *Carbon* **30**, 1261–1276 (1992).

Iq91   Z. Iqbal *et al.*, "Superconductivity at 45 K in Rb/Tl Codoped C_{60} and C_{60}/C_{70} Mixtures," *Science* **254**, 826–829 (1991).

Iq92   Z. Iqbal *et al.*, "Superconducting Transition Temperature of Doped C_{60}: Retraction," *Science* **256**, 950–951 (1992).

Ko92a*   G. P. Kochanski, A. F. Hebard, R. C. Haddon and A. T. Fiory, "Electrical Resistivity and Stoichiometry of K_xC_{60} Films," *Science* **255**, 184–186 (1992).

Ko92b   A. R. Kortan *et al.*, "Superconductivity at 8.4 K in Calcium-doped C_{60}," *Nature* **355**, 529–532 (1992).

Ko92c   A. R. Kortan *et al.*, "Superconductivity in Barium Fulleride," *Nature* **360**, 566–568 (1992).

Lo89   C. K. Loong *et al.*, "High-Energy Oxygen Phonon Modes and Superconductivity in $Ba_{1-x}K_xBiO_3$: An Inelastic-Neutron-Scattering Experiment and Molecular-Dynamics Simulation," *Phys. Rev. Lett.* **62**, 2628–2631 (1989).

Lo92   R. W. Lof, M. A. van Veenendaal, B. Koopmans, H. T. Jonkman and G. A. Sawatzky, "Band Gap, Excitons, and Coulomb Interaction in Solid C_{60}," *Phys. Rev. Lett.* **68**, 3924–3927 (1992).

Ma91   Y. Maruyama *et al.*, "Observation of Metallic Conductivity and Sharp Superconducting Transition at 19 K in Potassium-Doped Fulleride, C_{60}, Single Crystal," *Chemistry Letters*, 1849–1852 (1991).

Ma92   I. I. Mazin *et al.*, "Quantitative Theory of Superconductivity in Doped C_{60}," *Phys. Rev.* **B45**, 5114–5117 (1992).

Ma93a   I. I. Mazin *et al.*, "Orientational Order in A_3C_{60}: Antiferromagnetic Ising Model for the fcc Lattice," *Phys. Rev. Lett.* **70**, 4142–4145 (1993).

Ma93b   M. C. Martin, D. Koller and L. Mihaly, "*In Situ* Infrared Transmission Study of Rb- and K-Doped Fullerenes," *Phys. Rev.* **B47**, 14607–14610 (1993).

Mu92   D. W. Murphy *et al.*, "Synthesis and Characterization of Alkali Metal Fullerides: A_xC_{60}," *J. Phys. Chem. Solids* **53**, 1321–1332 (1992).

Os92a   A. Oshiyama and S. Saito, "Linear Dependence of Superconducting Transition Temperature on Fermi-Level Density-of-States in Alkali-Doped C_{60}," *Solid State Commun.* **82**, 41–45 (1992).

Os92b   A. Oshiyama, S. Saito, N. Hamada and Y. Miyamoto, "Electronic Structures of C_{60} Fullerides and Related Materials," *J. Phys. Chem. Solids* **53**, 1457–1471 (1992).

Pa92*   T. T. M. Palstra, R. C. Haddon, A. F. Hebard and J. Zaanen, "Electronic Transport Properties of K_3C_{60} Films," *Phys. Rev. Lett.* **68**, 1054–1057 (1992).

Pr91*   K. Prassides *et al.*, "Vibrational Spectroscopy of Superconducting K_3C_{60} by Inelastic Neutron Scattering," *Nature* **354**, 462–463 (1991).

Pr92   K. Prassides *et al.*, "Fullerenes and Fullerides in the Solid State: Neutron Scattering Studies," *Carbon* **30**, 1277–1286 (1992).

Ra92   A. P. Ramirez *et al.*, "Isotope Effect in Superconducting Rb_3C_{60}," *Phys. Rev. Lett.* **68**, 1058–1060 (1992).

Ri92   M. J. Rice and H. Y. Choi, "Charged-Phonon Absorption in Doped C_{60}," *Phys. Rev.* **B45**, 10173–10176 (1992).

Ro91* M. J. Rosseinsky *et al.*, "Superconductivity at 28 K in Rb_xC_{60}," *Phys. Rev. Lett.* **66**, 2830–2832 (1991).

Ro92a M. J. Rosseinsky *et al.*, "Structural and Electronic Properties of Sodium-Intercalated C_{60}," *Nature* **356**, 416–418 (1992).

Ro92b L. D. Rotter *et al.*, "Infrared Reflectivity Measurements of a Superconducting Energy Scale in Rb_3C_{60}," *Nature* **355**, 532–534 (1992).

Sa91   S. Saito and A. Oshiyama, "Ionic Metal K_xC_{60}: Cohesion and Energy Bands," *Phys. Rev.* **B44**, 11536–11539 (1991).

Sc92a* M. Schlüter, M. Lannoo, M. Needels, G. A. Baraff and D. Tománek, "Electron-Phonon Coupling and Superconductivity in Alkali-Intercalated C_{60} Solid," *Phys. Rev. Lett.* **68**, 526–529 (1992).

Sc92b M. A. Schlüter, M. Lannoo, M. F. Needels, G. A. Baraff and D. Tománek, "Schlüter *et al.* Reply [to a comment by S. Chakravarty *et al.*]," *Phys. Rev. Lett.* **69**, 213 (1992).

Sc92c M. Schlüter, M. Lannoo, M. Needels, G. A. Baraff and D. Tománek, "Superconductivity in Alkali Intercalated C_{60}," *J. Phys. Chem. Solids* **53**, 1473–1485 (1992).

Sh89   M. Shirai, N. Suzuki and K. Motizuki, "Superconductivity in $BaPb_{1-x}Bi_xO_3$ and $Ba_xK_{1-x}BiO_3$," *J. Phys. Condens. Matter* **1**, 2939–2943 (1989).

Sp91   G. Sparn *et al.*, "Pressure Dependence of Superconductivity in Single-Phase K_3C_{60}," *Science* **252**, 1829–1831 (1991).

Sp92a* G. Sparn *et al.*, "Pressure and Field Dependence of Superconductivity in Rb_3C_{60}," *Phys. Rev. Lett.* **68**, 1228–1231 (1992).

Sp92b G. Sparn *et al.*, "Pressure Dependence of Magnetism in C_{60}TDAE," *Solid State Commun.* **82**, 779–782 (1992).

St91* P. W. Stephens *et al.*, "Structure of Single-Phase Superconducting K_3C_{60}," *Nature* **351**, 632–634 (1991).

St92a P. W. Stephens *et al.*, "Structure of $Rb:C_{60}$ Compounds," *Phys. Rev.* **B45**, 543–546 (1992).

St92b* P. W. Stephens *et al.*, "Lattice Structure of the Fullerene Ferromagnet $TDAE–C_{60}$," *Nature* **355**, 331–332 (1992).

Ta91   K. Tanigaki *et al.*, "Superconductivity at 33 K in $Cs_xRb_yC_{60}$," *Nature* **352**, 222–223 (1991).

Ta92a K. Tanaka *et al.*, "Magnetic Properties of $TDAE–C_{60}$ and $TDAE–C_{70}$. A Comparative Study," *Phys. Lett. A* **164**, 221–226 (1992).

Ta92b K. Tanigaki *et al.*, "Superconductivity in Sodium- and Lithium-Containing Alkali-Metal Fullerides," *Nature* **356**, 419–421 (1992).

Ta92c S. Tajima, M. Yoshida, N. Koshizuka, H. Sato and S. Uchida, "Raman-Scattering Study of the Metal-Insulator Transition in $Ba_{1-x}K_xBiO_3$," *Phys. Rev.* **B46**, 1232–1235 (1992).

Ty91* R. Tycko *et al.*, "^{13}C NMR Spectroscopy of K_xC_{60}: Phase Separation, Molecular Dynamics, and Metallic Properties," *Science* **253**, 884–886 (1991).

Ty92* R. Tycko *et al.*, "Electronic Properties of Normal and Superconducting Alkali Fullerides Probed by [13]C Nuclear Magnetic Resonance," *Phys. Rev. Lett.* **68**, 1912–1915 (1992).

Ue91   Y. J. Uemura *et al.*, "Magnetic-Field Penetration Depth in K_3C_{60} Measured by Muon Spin Relaxation," *Nature* **352**, 605–607 (1991).

Va91* C. M. Varma, J. Zaanen and K. Raghavachari, "Superconductivity in the Fullerenes," *Science* **254**, 989–992 (1991).

Wa93   R. E. Walstedt, D. W. Murphy and M. Rosseinsky, "Structural Distortion in Rb_3C_{60} Revealed by [87]Rb NMR," *Nature* **362**, 611–613 (1993).

We91   G. K. Wertheim *et al.*, "Photoemission Spectra and Electronic Properties of K_xC_{60}," *Science* **252**, 1419–1421 (1991).

We92a J. H. Weaver, "Fullerenes and Fullerides: Photoemission and Scanning Tunneling Microscopy Studies," *Acc. Chem. Res.* **25**, 143–149 (1992).

We92b J. H. Weaver, "Electronic structures of C_{60}, C_{70}, and the Fullerides: Photoemission and Inverse Photoemission Studies," *J. Phys. Chem. Solids* **53**, 1433–1447 (1992).

We92c G. K. Wertheim, D. N. E. Buchanan and J. E. Rowe, "Charge Donation by Calcium into the t_{1g} Band of C_{60}," *Science* **258**, 1638–1640 (1992).

Wo92   W. H. Wong *et al.*, "Normal-State Magnetic Properties of K_3C_{60}," *Europhys. Lett.* **18**, 79–84 (1992).

Wu92 F. Wudl and J. D. Thompson, "Buckminster-fullerene C_{60} and Organic Ferromagnetism," *J. Phys. Chem. Solids* **53**, 1449–1455 (1992).

Xi92* X. D. Xiang *et al.*, "Synthesis and Electronic Transport of Single Crystal K_3C_{60}," *Science* **256**, 1190–1191 (1992).

Xi93 X. D. Xiang, J. G. Hou, V. H. Crespi, A. Zettl and M. L. Cohen, "Three-dimensional Fluctuation Conductivity in Superconducting Single Crystal K_3C_{60} and Rb_3C_{60}," *Nature* **361**, 54–56 (1993).

Yi92 T. Yilidrim *et al.*, "Intercalation of Sodium Hetero-Clusters into the C_{60} Lattice," *Nature* **360**, 568–571 (1992).

Za92 A. A. Zakhidov *et al.*, "Enhanced Isotope Effect in ^{13}C-Rich Superconducting M_xC_{60} ($M = K, Rb$): Support for Vibronic Pairing," *Physics Letters*, **A164**, 355–361 (1992).

Zh91a O. Zhou *et al.*, "Structure and Bonding in Alkali-Metal-doped C_{60}," *Nature* **351**, 462–464 (1991).

Zh91b Q. Zhu *et al.*, "X-Ray Diffraction Evidence for Nonstoichiometric Rubidium–C_{60} Intercalation Compounds," *Science* **254**, 545–548 (1991).

Zh91c* Z. Zhang, C.-C. Chen and C. M. Lieber, "Tunneling Spectroscopy of M_3C_{60} Superconductors: The Energy Gap, Strong Coupling, and Superconductivity," *Science* **254**, 1619–1621 (1991).

Zh92a* O. Zhou *et al.*, "Compressibility of M_3C_{60} Fullerene Superconductors: Relation Between T_c and Lattice Parameter," *Science* **255**, 833–835 (1992).

Zh92b O. Zhou and D. E. Cox, "Structures of C_{60} Intercalation Compounds," *J. Phys. Chem. Solids* **53**, 1373–1390 (1992).

Zh92c Q. Zhu *et al.*, "Intercalation of Solid C_{60} with Iodine," *Nature* **355**, 712–714 (1992).

Reprinted with permission from Nature
Vol. 350, No. 6316, pp. 320-322, 28 March 1991
© 1991 Macmillan Magazines Limited

Conducting films of C$_{60}$ and C$_{70}$ by alkali-metal doping

R. C. Haddon, A. F. Hebard, M. J. Rosseinsky,
D. W. Murphy, S. J. Duclos, K. B. Lyons, B. Miller,
J. M. Rosamilia, R. M. Fleming, A. R. Kortan,
S. H. Glarum, A. V. Makhija, A. J. Muller, R. H. Eick,
S. M. Zahurak, R. Tycko, G. Dabbagh & F. A. Thiel

AT&T Bell Laboratories, Murray Hill, New Jersey 07974-2070, USA

THE recent syntheses[1,2] of macroscopic quantities of C$_{60}$ have suggested possible applications in host–guest and organic chemistry, tribology, electrochemistry and semiconductor technology. Here we report the preparation of alkali-metal-doped films of C$_{60}$ and C$_{70}$ which have electrical conductivities at room temperature that are comparable to those attained by n-type doped polyacetylene. The highest conductivities observed in the doped films are: $4\,\mathrm{S\,cm}^{-1}$ (Cs/C$_{60}$), 100 (Rb/C$_{60}$), 500 (K/C$_{60}$), 20 (Na/C$_{60}$), 10 (Li/C$_{60}$), 2 (K/C$_{70}$). The doping process is reversed on exposure of the films to the atmosphere. At high doping levels, the films become more resistive. We attribute the conductivity induced in these films to the formation of energy bands from the π orbitals of C$_{60}$ or C$_{70}$, which become partially filled with carriers on doping. The smaller alkali metal ions should be able to fit into the interstices in the lattice without disrupting the network of contacts between the carbon spheroids. In the case of C$_{60}$, this would allow the development of an isotropic band structure, and we therefore propose that these materials may constitute the first three-dimensional 'organic' conductors.

Previous highly conducting organic charge-transfer systems have had low-dimensional (quasi-one-dimensional or -two-dimensional) electronic structures. In extended systems, the dimensionality is imposed by the host—for example, polymers (one-dimensional)[3,4] and graphite (two-dimensional)[5]—whereas in molecular systems the dimensionality is governed by the counter-ion, as in κ-ET$_2$Cu(NCS)$_2$ (refs 4, 6), or by crystal packing as in TTF–TCNQ[4,6]. The high symmetry and electron affinity[7-10] of the carbon spheroids (fullerenes)[1,11] provides a unique opportunity to circumvent these limitations. It should be possible to intercalate small dopant ions into the fullerene crystal without disrupting the network of contacts between the spheroids, thereby generating the first three-dimensional isotropic organic conductor. The structure of C$_{60}$ is comprised of rigid spheres of radius 4.98 Å with face-centred cubic packing ($a = 14.1$ Å)[12], leaving two vacant tetrahedral sites and one octahedral site per C$_{60}$ molecule of sufficient size to accommodate spheres of radius 1.12 Å and 2.06 Å respectively. The former sites are close to the size of Na$^+$ and K$^+$ ions. Furthermore, changes in the packing of the C$_{60}$ lattice could provide a variety of sites for other alkali-metal cations. Occupation of some of these sites and exclusion of extraneous co-intercalants should produce charge transfer from the metals to C$_{60}$ without disrupting the direct contact between C$_{60}$ molecules. Electrochemical experiments[9,10] have demonstrated that C$_{60}$

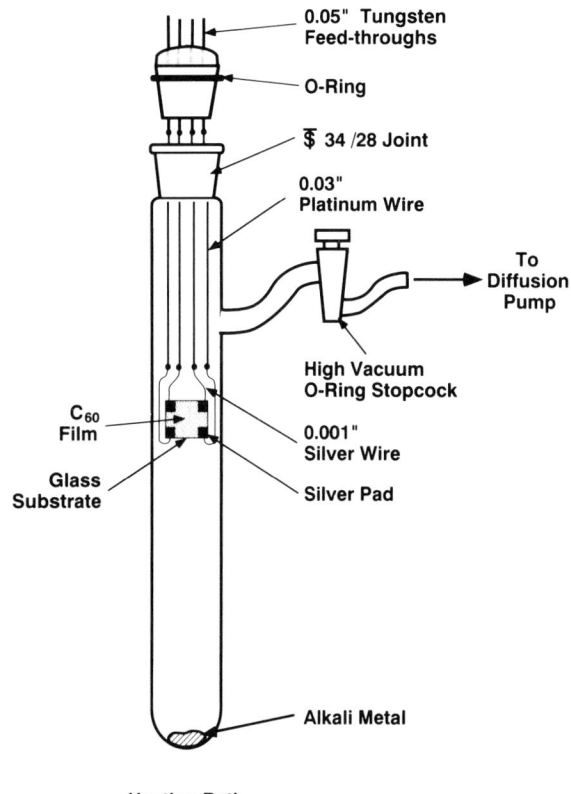

FIG. 1 Apparatus for vapour-phase doping of C$_{60}$ and C$_{70}$ films with alkali metals.

undergoes three reversible one-electron reductions, and theoretical considerations indicate that highly reduced states should be accessible under appropriate circumstances[7]. The charge-transfer salt (Ph$_4$P$^+$C$_{60}^-$) · 2(Ph$_4$P$^+$Cl$^-$) has been isolated by electrochemical crystal-growth techniques[13], but has a conductivity of less than 10^{-5} S cm^{-1}. The low conductivity of this compound is presumably a consequence of the bulky Ph$_4$P$^+$ units incorporated in the lattice, which inhibit close contact between the C$_{60}^-$ units. Furthermore, the fullerenes show a marked tendency to crystallize with solvent molecules trapped in the lattice. These considerations, together with the high electron affinity[7-10] of the fullerenes, led us to attempt to dope thin films of C$_{60}$ and C$_{70}$ with alkali metals. As we wished to obtain highly reduced states of these molecules, we constructed a high-vacuum apparatus which would allow us to measure conductivities *in situ* (Fig. 1).

We deposited thin films of C$_{60}$ and C$_{70}$ on a variety of substrates. The C$_{60}$ and C$_{70}$ samples used as source material in film growth were obtained by reversed-phase column chromatography of fullerite[1] on octadecylsilanized silica using 40:60 toluene/isopropanol as the eluant. The purities of the C$_{60}$ and C$_{70}$ were checked by high-performance liquid chromatography using ultraviolet detection. High-quality films of C$_{60}$ (C$_{70}$) of thickness 100–1,000 Å were deposited by high-vacuum sublimation from an alumina crucible regulated at 300 °C (400 °C) at a pressure of 1.5×10^{-6} torr. The C$_{60}$ (C$_{70}$) films appeared smooth and pale yellow (pink) to the eye and were poorly crystalline as evidenced by X-ray diffraction. Infrared spectra of the C$_{60}$ films deposited on KBr substrates showed the four characteristic absorption peaks of C$_{60}$ (ref. 1) with no evidence of contaminants.

For the conductivity measurements, the films were deposited on glass slides which had been precoated with 1,000-Å-thick stripes or pads of evaporated silver metal. The deposited silver metal was bonded to silver wires with silver epoxy which were in turn connected to tungsten connectors by platinum wires (Fig. 1). A high-vacuum glass apparatus was constructed with which

TABLE 1 Conductivities of alkali-metal-doped films of C$_{60}$ and C$_{70}$

Film	Dopant	Bath temperature (°C)	Maximum conductivity (S cm^{-1})
C$_{60}$	Li	*	10
C$_{60}$	Na	180	20
C$_{60}$	K	130	500
C$_{60}$	Rb	120	100
C$_{60}$	Cs	40	4
C$_{70}$	K	120	2

* Lithium metal in contact with Kovar container and flame-heated.

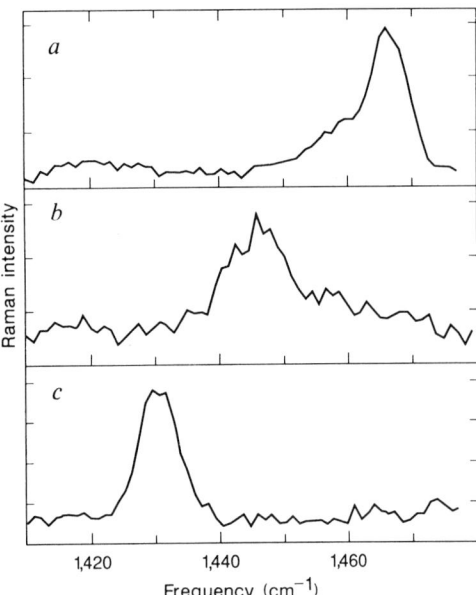

FIG. 2 *In situ* Raman spectra of a C_{60} film taken during potassium doping. *a*, Initial (pristine) film; *b*, Conducting film; *c*, Insulating film (see text).

we could monitor continuously the film conductivity in a variety of electrical configurations. The measured two-probe resistance in the pristine films was greater than 10^{10} Ω, which, for a film surface area of 1×1 cm and thickness 400 Å, implied a conductivity of less than 10^{-5} S cm^{-1}. Here we describe conductivities below this value as insulating.

The conductivity apparatus was loaded with the alkali-metal dopant in a dry box, before evacuation to a pressure at the pump of about 10^{-5} torr. Initial experiments were conducted with a two-probe apparatus, and higher conductivities confirmed in a four-probe van der Pauw configuration. The bottom of the apparatus was immersed in an oil bath so that the oil level was 5–8 cm below the film, and the temperature raised slowly until conductivity in the film could be detected. All of the doping experiments showed qualitatively the same behaviour—the conductivity first increased by several orders of magnitude and then decreased, usually to a point below our threshold of detection. Using caesium as dopant and with the heating bath at 40 °C, we detected conductivity in a C_{60} film after 2 h. On continued heating, the conductivity increased to a maximum value of 4 S cm^{-1} over an additional period of 2 h before decreasing to a final value of 0.05 S cm^{-1} over a further period of about 4 h. At this point the film had darkened and the initially smooth surface was distinctly granular and rough. The disruption of the C_{60} film is presumably a reflection of the difficulty in incorporating the large Cs^+ counter-ion (radius 1.67 Å) in the lattice. Doping of the C_{60} films with the other alkali metals occurred on similar timescales and led to magenta films with fairly good surface quality, which remained specular. The C_{70} film showed little colour change on doping. The results for the combinations tested are summarized in Table 1.

Of the C_{60} dopants, potassium gave rise to the highest conductivities, and we therefore selected this system for further study. During the doping process, this film maintained excellent surface quality. We measured the Raman spectra of this film *in situ* (Fig. 2). The pristine film (Fig. 2*a*), shows the expected feature[14] at 1,467 cm^{-1}, which has been assigned to an A_g mode of C_{60}. In the highly conducting state (Fig. 2*b*), this line shifts to 1,445 cm^{-1}. In the highly doped, insulating state the spectral feature moves to 1,430 cm^{-1} (Fig. 2*c*). We assign these two latter vibrations to the corresponding A_g mode in reduced forms of C_{60}. Addition of electrons to the C_{60} framework is expected to soften the bond stretching modes as the added electrons enter

antibonding molecular orbitals[7]. To provide a calibration of the frequency shifts, we measured the Raman spectrum of $(Ph_4As^+C_{60}^-) \cdot 2(Ph_4As^+Cl^-)$. This salt is grown by solution electrochemistry and is known to contain the C_{60} mono-anion[13]. The Raman peak occurred at 1,461 cm^{-1} in this salt—a shift of 6 cm^{-1} relative to the neutral film. The lowest unoccupied molecular orbital of C_{60} is the triply degenerate t_{1u} level[7], and the shift of 37 cm^{-1} in the Raman line of the insulating C_{60} film (Fig. 2*c*) therefore suggests that this state might correspond to C_{60}^{6-}, in which the t_{1u} level is completely filled. The conducting state may then be associated with energy bands derived from the partially filled t_{1u} level. A comparison may be made with graphite, in which the intralayer Raman mode at 1,582 cm^{-1} is shifted to 1,547 cm^{-1} in the intercalation compound KC_8 (ref. 15). Shifts in the Raman bands of C_{60} adsorbed on gold have also been reported[16]. The doped C_{60} film rapidly returned to its pale yellow colour on exposure to the atmosphere, and a Raman spectrum of this film showed that the 1,467-cm^{-1} line from neutral C_{60} had been restored. These observations suggest that the molecular integrity of C_{60} is maintained on doping and that the process is chemically reversible.

To ensure that the enhanced conductivity was not due to the deposition of an alkali-metal film on the surface of the C_{60} and C_{70} films, we ran a blank using two substrates each prepared with two silver contact stripes. One substrate was used as prepared, whereas the other had a C_{60} film deposited onto it as before. The two slides were placed side by side in the same apparatus with potassium as dopant. On heating, the C_{60} film first became conducting and then insulating as before, whereas the blank remained insulating throughout.

It should be noted that the materials are synthesized under non-equilibrium conditions as the experiments are performed in a dynamic vacuum, and the local vapour pressure of the alkali metal is unknown. The rate and extent of reaction will depend on the nature of the alkali metal, the temperature of the film and the presence of residual ambient gas impurities, which are not controlled in these preliminary experiments. At present, the effect of these variables on the conductivities cannot be assessed. Synthesis in a closed system will be required to determine the relevant thermodynamic parameters.

We attribute the conductivity of the doped films to population of energy bands formed from the C_{60} π-orbitals[17]. As these bands become filled at high doping levels the material becomes insulating. Based on the energy-level structure of C_{60}, it is likely that the conductor–insulator transition occurs at the point where each molecule has accepted about six electrons, thereby filling the t_{1u} level[7]. We intend to carry out *in situ* X-ray, Hall effect, reflectivity and low-temperature studies on the intercalated films, as well as studying ways in which to measure and optimize the doping level and to improve the film quality. Encapsulation of an appropriate dopant within the fullerene cage could lead to increased chemical stability at the optimal doping level. □

Received 12 February; accepted 5 March 1991.

1. Kratschmer, W., Lamb, L. D., Fostiropoulos, K. & Huffman, D. R. *Nature* **347**, 354–358 (1990).
2. Meijer, G. & Bethune, D. S. *J. chem. Phys.* **93**, 7800–7802 (1990).
3. Skotheim, T. A. (ed.) *Handbook of Conductive Polymers* (Marcel Dekker, New York, 1986).
4. Ferraro, J. R. & Williams, J. M. *Introduction to Synthetic Electrical Conductors* (Academic, London, 1987).
5. Legrand, A. P. & Flandrois, S. (eds) *Chemical Physics of Intercalation* (Plenum, New York, 1987).
6. Ishiguro, T. & Yamaji, K. *Organic Superconductors* (Springer, Berlin, 1990).
7. Haddon, R. C., Brus, L. E. & Raghavachari, K. *Chem. Phys. Lett.* **125**, 459–464 (1986).
8. Curl, R. F. & Smalley, R. E. *Science* **242**, 1017–1022 (1988).
9. Hauffler, R. E. *et al. J. phys. Chem.* **94**, 8634–8636 (1990).
10. Allemand, P.-M. *et al. J. Am. chem. Soc.* **113**, 1050–1051 (1991).
11. Kroto, H. W., Heath, J. R., O'Brien, S. C., Curl, R. F & Smalley, R. E. *Nature* **318**, 162–164 (1985).
12. Fleming, R. M. *et al. Proc. Mat. Res. Soc. Symp. G* Boston, 1990 (in the press).
13. Allemand, P.-M. *et al. J. Am. chem. Soc.* (in the press).
14. Bethune, D. S., Meijer, G., Tang, W. C. & Rosen, H. J. *Chem. Phys. Lett.* **174**, 219–222 (1990).
15. Nemanich, R. J., Solin, S. A. & Guerard, D. *Phys. Rev. B* **16**, 2965–2972 (1977).
16. Garrell, R. L. *et al. J. Am. chem. Soc.* (submitted).
17. Haddon, R. C. *Acc. Chem. Res.* **21**, 243–249 (1988).

Reprinted with permission from Nature
Vol. 350, pp. 600–601, 18 April 1991
© 1991 Macmillan Magazines Limited

Superconductivity at 18 K in potassium-doped C$_{60}$

A. F. Hebard, M. J. Rosseinsky, R. C. Haddon, D. W. Murphy, S. H. Glarum, T. T. M. Palstra, A. P. Ramirez & A. R. Kortan

AT&T Bell Laboratories, Murray Hill, New Jersey 07974-2070, USA

THE synthesis of macroscopic amounts of C$_{60}$ and C$_{70}$ (fullerenes)[1] has stimulated a variety of studies on their chemical and physical properties[2,3]. We recently demonstrated that C$_{60}$ and C$_{70}$ become conductive when doped with alkali metals[4]. Here we describe low-temperature studies of potassium-doped C$_{60}$ both as films and bulk samples, and demonstrate that this material becomes superconducting. Superconductivity is demonstrated by microwave, resistivity and Meissner-effect measurements. Both polycrystalline powders and thin-film samples were studied. A thin film showed a resistance transition with an onset temperature of 16 K and essentially zero resistance near 5 K. Bulk samples showed a well-defined Meissner effect and magnetic-field-dependent microwave absorption beginning at 18 K. The onset of superconductivity at 18 K is the highest yet observed for a molecular superconductor.

The sensitivity to air of alkali-metal-doped fullerenes (A$_x$C$_n$) limits the choice of sample preparation and characterization techniques. To avoid sample degradation, we carried out reactions with the alkali metal vapour and C$_{60}$ in sealed tubes either in high vacuum or under a partial pressure of helium. The C$_{60}$ was purified by chromatography[1] of fullerite[2] and was heated at 160 °C under vacuum to remove solvents.

Small amounts of the individual fullerenes (~0.5 mg) were placed in quartz tubes with alkali metals and sealed under vacuum. These samples were subjected to a series of heat treatments and tests for superconductivity by 9-GHz microwave-loss experiments[5]. Preliminary tests indicated that only the K-doped C$_{60}$ showed a response consistent with a superconducting transition (Fig. 1). For this reason, together with the fact that K$_x$C$_{60}$ showed the highest film conductivity[1], we focused our studies on the K-doped compound.

The conductivity measurements were performed on potassium-doped films of C$_{60}$ that were prepared in a one-piece all-glass version of the apparatus described previously[4]. This reaction vessel was sealed under a partial pressure of helium

before reaction. This configuration allowed both *in situ* doping and low-temperature studies of thin films. All measurements were made in a four-terminal Van der Pauw configuration using a 3-μA a.c. current at 17 Hz. Figure 2 shows the temperature dependence of the resistivity of a 960-Å-thick K$_x$C$_{60}$ film. The film was doped with potassium until the resistivity had fallen to 5×10^{-3} Ω cm. The resistivity increases by a factor of two on cooling the sample to near 20 K. Below 16 K, the resistivity starts to decrease; zero resistivity ($<10^{-4}$ of the normal state) is obtained below 5 K. The 10–90% width of the transition is 4.6 K. At 4 K we measured the lower bound to the critical current to be 40 A cm^{-2}.

A bulk polycrystalline sample of nominal composition K$_3$C$_{60}$ was prepared by reaction of 29.5 mg of C$_{60}$ with 4.8 mg potassium. The amount of potassium was controlled volumetrically by using potassium-filled pyrex capillary tubing cut to size in a dry box. The reaction was run with the C$_{60}$ in a 5-mm fused silica tube joined to a larger tube in which the potassium-containing capillary was placed. The tube was sealed after being evacuated and refilled with 10^{-2} torr of helium to serve later as a thermal-exchange gas for low-temperature measurements. With the C$_{60}$-containing end of the tube at room temperature,

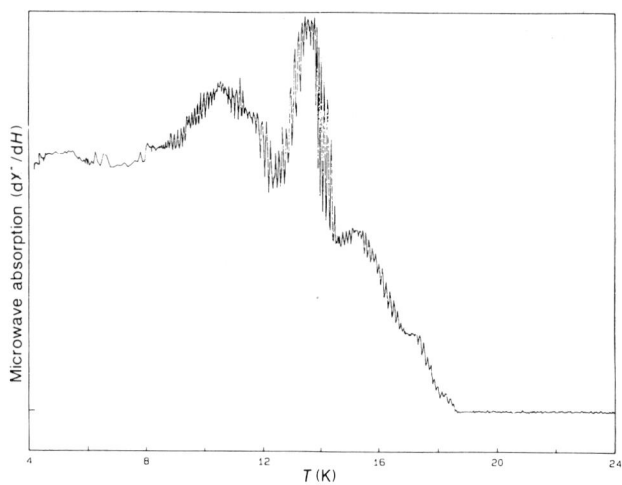

FIG. 1 Microwave loss as a function of temperature for K$_x$C$_{60}$ in a static field of 20 Oe.

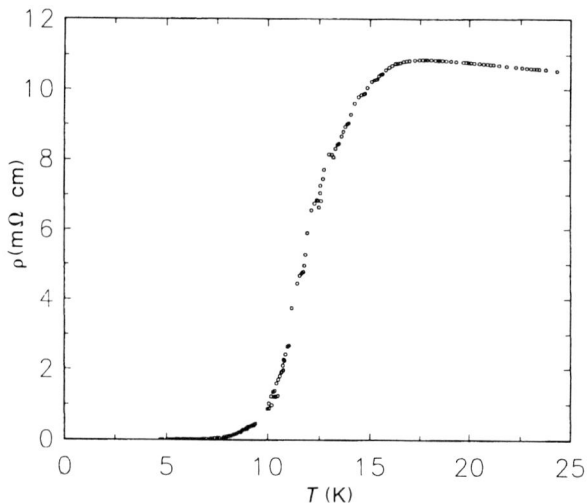

FIG. 2 Temperature dependence of the electrical resistivity of a 960-Å-thick film of K_xC_{60}.

the potassium was distilled from the capillary in a furnace at 200 °C. Some reaction of the potassium with the quartz tube, visible as a dark brown discoloration, was observed at this temperature. Unreacted potassium was observed after this period. Following distillation of the potassium to the C_{60} end, the tube was shortened by sealing to about 8 cm and heated to 200 °C for 36 h. Finally, the tube was resealed to a length of about 4 cm for magnetic measurements.

The temperature dependence of the d.c. magnetization of the sample with nominal composition K_3C_{60} was measured in a SQUID magnetometer (Fig. 3). On zero-field cooling the sample to 2 K, a magnetic field of 50 Oe was applied. On warming, this field is excluded by the sample to 18 K; this verifies the presence of a superconducting phase. The bulk nature of superconductivity in the sample is demonstrated unambiguously by cooling in a field of 50 Oe. A well defined Meissner effect (flux expulsion) develops below 18 K. The shape of the magnetization curve, in particular the temperature-independent signal at low temperature, indicates good superconducting properties for this sample. Also noteworthy is the relatively narrow transition width. The magnitude of the flux exclusion for the zero-field-cooled curve corresponds to 1% volume fraction. This small fraction is possibly due to non-optimal doping or the granular nature of the sample. The large value of the Meissner effect for the field-cooled curve relative to the total exclusion, however, indicates bulk superconductivity in the electrically connected regions.

The universally accepted tests for superconductivity, namely a transition to zero resistance and a Meissner effect showing the expulsion of magnetic field, demonstrate unequivocally the existence of superconductivity in K_xC_{60}. The 18-K transition temperature is the highest yet reported for a molecular superconductor. This may be compared with the previously reported occurrence of superconductivity at 0.55 K in potassium-intercalated graphite[6]. We expect that optimization of composition and crystallinity will lead to further improvement in the superconducting properties. □

Received 26 March; accepted 4 April 1991.

1. Kroto, H. W., Heath, J. R., O'Brien, S. C., Curl, R. F. & Smalley, R. E. Nature **318**, 162–164 (1985).
2. Kratschmer, W., Lamb, L. D., Fostiropoulos, K. & Huffman, D. R. Nature **347**, 354–358 (1990).
3. Meijer, G. & Bethune, D. S. J. chem. Phys. **93**, 7800–7802 (1990).
4. Haddon, R. C. et al. Nature **350**, 320–322 (1991).
5. Haddon, R. C., Glarum, S. H., Chichester, S. V., Ramirez, A. P. & Zimmerman, N. M. Phys. Rev. B **43**, 2642–2647 (1991).
6. Hannay, N. B. et al. Phys. Rev. Lett. **14**, 225–226 (1965).

ACKNOWLEDGEMENTS. We thank G. Dabbagh, S. J. Duclos, R. H. Eick, A. T. Fiory, R. M. Fleming, M. L. Kaplan, K. B. Lyons, A. V. Makhija, B. Miller, A. J. Muller, K. Raghavachari, J. M. Rosamilia, L. F. Schneemeyer, F. A. Thiel, J. C. Tully, R. Tycko, R. B. Van Dover, J. V. Waszczak, W. L. Wilson and S. M. Zahurak for valuable contributions to this work.

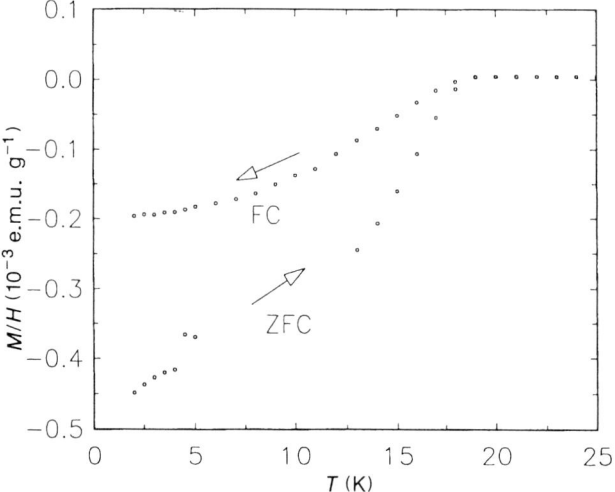

FIG. 3 Temperature dependence of the magnetization of a K_xC_{60} crystalline sample. The direction of temperature sweep in the field-cooled (FC) and the zero-field-cooled (ZFC) curves is indicated by the arrows.

Reprinted with permission from Physical Review Letters
Vol. 66, No. 21, pp. 2830–2832, 27 May 1991

Superconductivity at 28 K in Rb$_x$C$_{60}$

M. J. Rosseinsky, A. P. Ramirez, S. H. Glarum, D. W. Murphy, R. C. Haddon, A. F. Hebard,
T. T. M. Palstra, A. R. Kortan, S. M. Zahurak, and A. V. Makhija

AT&T Bell Laboratories, Murray Hill, New Jersey 07974-2070
(Received 26 April 1991)

Meissner-effect and microwave-absorption measurements on bulk samples show that Rb$_x$C$_{60}$ is superconducting with a maximum transition temperature of 28 K. This is a 10-K (60%) increase over the K-doped material. Only Ba$_{0.6}$K$_{0.4}$BiO$_3$ and the cuprate superconductors have higher transition temperatures.

PACS numbers: 74.10.+v, 61.55.−x

The observation of superconductivity at 18 K in K$_x$C$_{60}$ (Ref. 1) demonstrated the possibility of obtaining high superconducting transition temperatures in the doped fullerenes. In this paper we report the observation of superconductivity at 28 K on doping of C$_{60}$ with rubidium. This transition temperature is dramatically higher than previously observed in any molecular, elemental, or intermetallic superconductor, and is surpassed only by Ba$_{0.6}$K$_{0.4}$BiO$_3$ and the cuprate superconductors.

Previous work on K$_x$C$_{60}$ showed the need for a variety of characterization techniques for these highly air sensitive materials.[1,2] These included *in situ* conductivity, Raman spectroscopy, microwave absorption, and dc susceptibility. In this study, as well, all doping reactions and measurements were carried out using sealed tubes. C$_{60}$ was purified by chromatography of fullerite[3] on octadecylsilanised silica with toluene-isopropanol eluent, and dried at 160 °C under vacuum.

Room-temperature conductivities for alkali-metal-doped C$_{60}$,[1] A$_x$C$_{60}$, ranged from 4 to 500 S cm^{-1}, with the highest value for $A=$K. These conductivities were dependent on doping level and a maximum conductivity was obtained for each alkali metal, beyond which the conductivity decreased. Our initial screening for superconductivity in A$_x$C$_{60}$ ($A=$alkali metal) by conductivity measurements in thin films (at the conductivity maximum) and by magnetic-field-dependent microwave absorption was positive only for potassium.[1] Subsequently, microwave absorption indicating superconductivity was observed with rubidium following appropriate heat treatments.

A bulk polycrystalline sample of nominal composition Rb$_3$C$_{60}$ was prepared by reaction of 30 mg of C$_{60}$ with 10.7 mg of Rb. The quantity of rubidium was controlled by cutting Rb-filled Pyrex capillary tubing to the required length in a dry box, to avoid overdoping the material.[1,2] The C$_{60}$ was loaded into a 5-mm Pyrex tube sealed to a larger-diameter tube containing the Rb capillary, and the tube was sealed under 10^{-2} torr of helium. The tube was then heated for 24 h with the Rb end at 450 °C and the C$_{60}$ end at 400 °C (corresponding to the natural gradient of our furnace). Some discoloration of

the tube due to reaction with the glass was visible, but no unreacted rubidium was observed. Following this initial reaction, the tube was resealed close to the C$_{60}$ end for final reaction (450 °C, 24 h) and characterization.

The dc magnetization [$M(T)$] of the Rb$_x$C$_{60}$ sample, shown in Fig. 1, was measured in a SQUID magnetometer in an applied field of 2 Oe. The sample was first cooled to 4.2 K in zero field and data were taken in a 2-Oe field on warming the sample to 40 K [zero-field-cooled (ZFC) data]. The sample was then cooled in the same field down to the base temperature [field-cooled (FC) data]. Data qualitatively similar to those in Fig. 1 were obtained at field values up to 50 Oe. The Pyrex sample container was paramagnetic and contributed less than 1% to the observed signal at 5 K. The overall behavior of $M(T)$ demonstrates superconductivity with an onset T_c of 28 K. In particular, the negative sign of the magnetization for $T \ll T_c$ and the Meissner-Ochsenfeld effect of flux expulsion, namely, the FC data of equal or

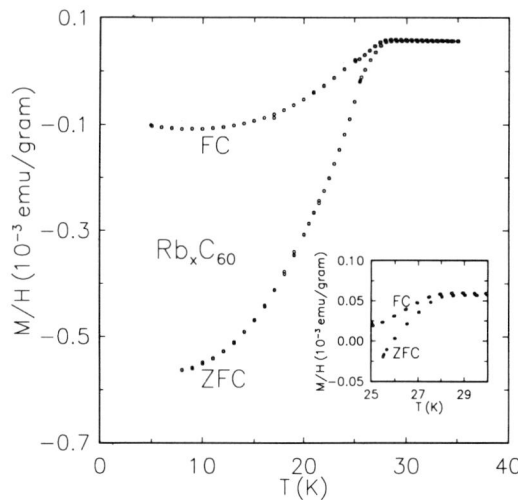

FIG. 1. Magnetization of a sample of nominal composition Rb$_3$C$_{60}$. The data labeled ZFC were obtained upon warming in a field of 2 Oe, after cooling the sample in zero applied field. The FC data were obtained by cooling the sample in 2 Oe, illustrating flux expulsion.

lower magnitude than the ZFC data, cannot be explained by any known phenomena other than superconductivity. The magnitude of the ZFC shielding signal is 1% of that expected for a homogeneous superconductor. This value is similar to that found in the K-doped samples, and is most likely due to sample inhomogeneity, a feature that is also reflected in the broad width of the superconducting transition.

Superconductivity was also demonstrated on the same sample by microwave-absorption measurements, shown in Fig. 2. These data, taken at a frequency of 9 GHz in an applied field of 100 Oe, show a dramatic rise in the absorption coincident with the onset of diamagnetism at 28 K. This absorption is attributed to motion of flux vortices and has become a well-documented test for superconductivity.[4] Absorption well below T_c is common in superconductors with weak links between superconducting regions.

Thin films of Rb_xC_{60} were prepared by a modification of the procedure reported previously.[2] Doping was achieved by heating Rb metal at the bottom of the ampoule or by using a commercial SAES Rb source in combination with a Zr getter. Both methods yielded room-temperature resistivities of about 10 mΩ cm. Cooling to liquid-nitrogen temperature increased the sheet resistance by more than 4 orders of magnitude. Some samples showed resistance drops at ≈ 5 K, but zero resistance was not observed. Examination of the various electrode configurations in the van der Pauw geometry indicated inhomogeneous doping. In addition, high contract resistances were noted. It should be noted that these preparative conditions are quite different from those used for the bulk sample, and, therefore, the observations are not inconsistent with the data on the bulk sample.

In the case of K_xC_{60}, superconductivity was readily

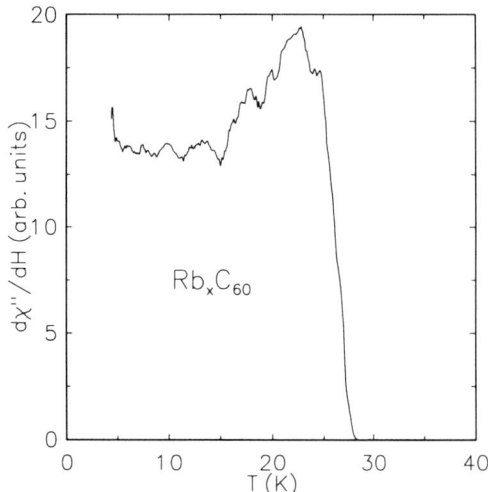

FIG. 2. Microwave loss vs temperature for a sample of nominal composition Rb_3C_{60} in a static field of 100 Oe.

obtained and resistivity, microwave absorption, and dc susceptibility all indicated a maximum T_c above 18 K.[1] With Rb, higher reaction temperatures appear to be required and the conditions are not readily duplicated for the thin films. For both K and Rb, optimization of synthetic conditions is still required to obtain homogeneous, single-phase samples.

Although the intercalated C_{60}'s are molecular systems, they have marked similarities to extended solids. The salient characteristics of molecular systems include (1) discrete molecular units and (2) formation of compounds with fixed stoichiometries. Extended solids in contrast (1) have no discrete molecular entities, but rather repeat units, and (2) form phases with either fixed or variable stoichiometry (solid solutions). The doped C_{60} may well have solid-solution regions, miscibility gaps, and phase diagrams more typical of extended solids. All of the classes of superconducting materials exhibit strong structure-bonding-property relationships, with T_c being optimized by control of both electron count and relevant interatomic distances. The fixed stoichiometries of other known molecular superconductors diminish the role of electron count as an experimentally variable parameter. However, a marked dependence of T_c on counterion size in molecular organic systems has been noted.[5] In extended systems, such as the copper oxides, the size and electron count are both variable and the range of properties is accordingly more extensive. In A_xC_{60}, several experiments sensitive to doping level suggest that variable stoichiometry is possible, e.g., the existence of a minimum in the resistivity, the shift in the 1467-cm^{-1} Raman mode,[2] and the continuous filling of the t_{1u} lowest-unoccupied-molecular-orbital-derived band seen in photoemission.[6] The possibility of varying the electron count in this case is unique for a molecular system.

Some comparisons with other systems are appropriate. The T_c's of graphite intercalation compounds of the same elements are 0.128–0.55 K for C_8K and 0.03 K for C_8Rb.[7,8] The ambient-pressure molecular organic superconductors, $(BEDT-TTF)_2X$, have fixed composition, with T_c varying with the size of the anion X.[5,9] However, the Chevrel phases[10] may provide the closest analogy with C_{60} superconductors. In these compounds, Mo_6-$(S,Se)_8$ clusters condense to form an extended solid (the clusters are covalently linked and exist as molecules only as chemically modified derivatives). The parent compounds $Mo_6(S,Se)_8$ can be prepared, as well as ternary phases $MMo_6(S,Se)_8$. Several of the ternaries superconduct with T_c strongly influenced by the ternary atom [e.g., $BaMo_6Se_8$, $T_c = 2.7$ K; $PbMo_6S_8$, $T_c = 15$ K (Ref. 10)], similar to the large difference in T_c between K_xC_{60} and Rb_xC_{60}.

Several explanations are conceivable for the dramatic variation in T_c with alkali-metal dopant. Because of the small superconducting fraction, neither the composition nor structure of the superconducting phase has been identified for K or Rb. Assuming similar structure and composition of the superconducting phase in both cases, the changes in T_c may be discussed using a BCS-type relation,

$$T_c \propto E \exp[-1/VN(E_F)] = E \exp(-1/\lambda) . \quad (1)$$

Here E is the frequency of the pairing-mediating excitation, $N(E_F)$ the density of states at the Fermi level, and V the electron-excitation coupling strength. In view of the high transition temperature of 28 K and the polarizable π system of the C_{60} molecule, excitonic mechanisms of the type envisaged by Little,[11] where E is an electronic excitation energy, may be of relevance.

If we limit further discussion to phonon-mediated pairing, then $E = \omega_{ph}$, the average energy of the most strongly coupled phonon modes. In C_{60}, these modes possess characteristic energies of 1000–2000 K (intramolecular),[2] 80 K (intermolecular),[12] and ≤ 80 K (rotational). Doping will affect the modes in different ways. Population of the conduction band, derived from t_{1u} molecular orbitals (which are antibonding with respect to the C_{60} molecules), will decrease the *intramolecular* phonon frequencies.[2] The Raman frequencies for fully doped $A_x C_{60}$ ($A =$ Na, K, Rb, Cs) films are very similar, implying little change in intramolecular phonon modes between different alkali metals.[13] The *intermolecular* vibrations will harden on doping because the lower part of the conduction band is slightly bonding between C_{60} molecules. The rotational modes should also stiffen due to this increased intermolecular coupling.

In view of the high transition temperatures, we consider coupling to the highest-frequency phonons in these materials, i.e., the intramolecular vibrations. These frequencies are significantly larger than found in the intermetallic superconductors. Assuming a relevant phonon frequency of $\omega_{ph} = 2000$ K, the observed $T_c = 28$ K requires a coupling constant $\lambda = 0.35$, for which the BCS treatment is valid. The 10-K difference in T_c between K and Rb requires a change in λ of only 10%. In general, one expects that on expansion V will decrease due to the nonlinearity in the dependence of orbital overlap on ion separation, and $N(E_F)$ will increase due to band narrowing. If it is assumed that the C_{60} lattice responds elastically to the differing alkali-metal atom sizes ($r_K = 1.32$ Å, $r_{Rb} = 1.49$ Å), then one is led to the conclusion that T_c is primarily determined by modulation of $N(E_F)$. This shows that a variation in $N(E_F)$ is a physically reasonable explanation for the difference in T_c.

It is also conceivable that the high transition temperatures are related to the lower-frequency modes, i.e., the intermolecular and rotational phonons. A T_c of 28 K with a relevant phonon frequency of 80 K requires a coupling constant $\lambda = 5$ using strong-coupling theory.[14] However, in this case λ would have to change by as much as a factor of 2 to explain the different T_c's for Rb and K doping. If, however, weak coupling ($\lambda \leq 0.5$) is valid, then the phonons would have to stiffen by a factor of more than 2 to obtain a T_c of 28 K, with a smaller change in λ between K and Rb than in the strong-coupling limit.

In summary, superconductivity in intercalated C_{60} is shown to occur for Rb as well as K. The 28-K transition temperature observed by dc susceptibility is dramatically higher than observed in any molecular, elemental, or intermetallic superconductor, and is surpassed only by $Ba_{0.6}K_{0.4}BiO_3$ and the cuprate superconductors.

We wish to thank M. Schluter for valuable discussions.

[1]A. F. Hebard, M. J. Rosseinsky, R. C. Haddon, D. W. Murphy, S. H. Glarum, T. T. M. Palstra, A. P. Ramirez, and A. R. Kortan, Nature (London) **350**, 600 (1991).

[2]R. C. Haddon, A. F. Hebard, M. J. Rosseinsky, D. W. Murphy, S. J. Duclos, K. B. Lyons, B. Miller, J. M. Rosamilia, R. M. Fleming, A. R. Kortan, S. H. Glarum, A. V. Makhija, A. J. Muller, R. H. Eick, S. M. Zahurak, R. Tycko, G. Dabbagh, and F. A. Thiel, Nature (London) **350**, 321–322 (1991).

[3]W. Kratschmer, L. D. Lamb, K. Fostiropoulos, and D. R. Huffman, Nature (London) **347**, 354–358 (1990).

[4]S. H. Glarum, J. H. Marshall, and L. F. Schneemeyer, Phys. Rev. B **37**, 7491 (1988).

[5]J. R. Ferraro and J. M. Williams, *Introduction to Synthetic Electrical Conductors* (Academic, London, 1987).

[6]J. H. Weaver *et al.* (to be published); G. K. Wertheim, J. E. Rowe, D. N. E. Buchanan, E. E. Chaban, A. F. Hebard, A. R. Kortan, A. V. Makhija, and R. C. Haddon (to be published).

[7]N. B. Hannay, T. H. Geballe, B. T. Matthias, K. Andres, P. Schmidt, and D. Macnair, Phys. Rev. Lett. **14**, 225 (1965).

[8]Y. Koike, K. Suematsu, K. Higuchi, and S. Tanuma, Solid State Commun. **27**, 623 (1978).

[9]H. Urayama, H. Yamochi, G. Saito, K. Nozawa, T. Sugano, M. Kinoshita, S. Sato, K. Oshima, A. Kawamoto, and J. Tanaka, Chem. Lett. **55** (1988).

[10]K. Yvon, Curr. Top. Mater. Sci. **3**, 53 (1979).

[11]W. A. Little, Phys. Rev. **134**, 1416 (1964).

[12]Y. Wang, D. Tomanek, and G. F. Bertsch (to be published).

[13]S. J. Duclos, R. C. Haddon, and S. H. Glarum (to be published).

[14]V. Z. Kresin, Phys. Lett. A **122**, 434 (1987).

Reprinted with permission from Science
Vol. 252, pp. 1154–1157, 24 May 1991
© 1991 The American Association for The Advancement of Science

Alkali-Fulleride Superconductors: Synthesis, Composition, and Diamagnetic Shielding

Károly Holczer, Oliviér Klein, Shiou-Mei Huang,
Richard B. Kaner, Ke-Jian Fu, Robert L. Whetten,
Francois Diederich

The recent report of a superconductivity onset near the critical temperature $T_c = 18$ K in potassium-doped C_{60} raises questions concerning the composition and stability of the superconducting phase. The effects of mixing and heat treatment of $K_x C_{60}$ samples prepared over a wide range of initial compositions on the superconducting transition was determined from shielding diamagnetism measurements. A single superconducting phase ($T_c = 19.3$ K) occurs for which the composition is K_3C_{60}. The shielding reaches a maximum of greater than 40 percent of the perfect diamagnetism, a high value for a powder sample, in samples prepared from 3:1 mixtures. A Rb_xC_{60} sample prepared and analyzed in an analogous way exhibited evidence for superconductivity with $T_c = 30$ K and a diamagnetic shielding of 7 percent could be obtained.

THE RECENT DISCOVERY (1) AND separation (2, 3) of molecular forms of solid carbon has made possible the formation of new semiconducting (4) and conducting (5) ²anionic charge-transfer compounds of C_{60} (and C_{70}) with molecular and alkali counterions, respectively. From the latter studies, Hebard et al. (6) have provided unambiguous evidence for superconductivity in solids composed of icosahedral C_{60} molecules "doped" with potassium (K_xC_{60}) with an onset near $T_c = 18$ K. They combined three experiments on two different morphologies (microwave absorption and magnetic susceptibility χ of powders, and dc resistivity of films) to assert superconductivity under unusually difficult, poorly controlled chemical conditions. As the powder is prepared by a solid-state reaction, only the initial composition is known, so no claim could be made about the

homogeneity of the end product. Therefore, the actual composition of the superconducting phase was not directly determined, and the lower bound of 1% superconducting phase, established by shielding diamagnetism measurements, leaves open questions about the compositional stability of this phase within the K-C_{60} phase diagram.

We report measurements of the shielding diamagnetism curves (χ versus T) conducted over a range of compositions and treatments with the aim of separating and identifying the superconducting phase. In an attempt to narrow the composition range of the superconducting material, the initial composition was systematically varied to locate the maximum fraction of shielding diamagnetism, which is a measure of the actual quantity of superconducting material. This maximum, found at a composition K_3C_{60}, is >40%, a high value for a powder sample. This phase appears to be stable, that is, it is present after indefinitely long heating and mixing, for all nominal sample compositions, x <6. When this same procedure

Departments of Physics and of Chemistry and Biochemistry and Solid State Science Center, University of California, Los Angeles, CA 90024–1569.

was applied to Rb-doped C_{60}, a 7% shielding diamagnetism was measured on a sample, starting at an apparent $T_c = 30$ K.

The molecular carbon sample was prepared following the resistive-heating procedure of Krätschmer et al. (1) as described by Ajie et al. (2) followed by solvent extraction, chromatographic separation of higher fullerenes, and several hours drying of solvent under vacuum. The potassium (Alfa, 99.95%) and rubidium (Mackay, high purity) were used as obtained. All manipulations were performed in a dry box under He atmosphere, with an O_2 background level <1 ppm. Our procedure for reacting the solids and for obtaining homogeneous samples differs from the previous report in that it consists of three heat treatment stages, as follows:

1) In the mixing stage, a given quantity of K was sealed under vacuum with C_{60} powder (typically 6 to 25 mg) in a 5-mm or 8-mm diameter Pyrex tube. This tube was placed in a furnace at 200°C for 20 to 24 hours, during which time the K appears to be completely absorbed by the C_{60} powder, which has negligible vapor pressure at this temperature. Residual unreacted K was evident only for mixtures of 6:1 composition and greater. The material was collected, put in a capillary tube (1.5 mm diameter), and sealed under (1 bar) He for initial shielding diamagnetism measurements.

2) In the diffusion stage, the samples from stage 1, still in the capillary under 1 bar He, were again heated to 200°C for ~22 hours, after which time the shielding diamagnetism measurement was repeated.

3) In the relaxation stage, we checked that an equilibrium had been achieved by

Fig. 2. The shielding diamagnetism, $\chi_{dia}(T)$, of a series of samples plotted versus nominal sample composition x at the three stages of treatment.

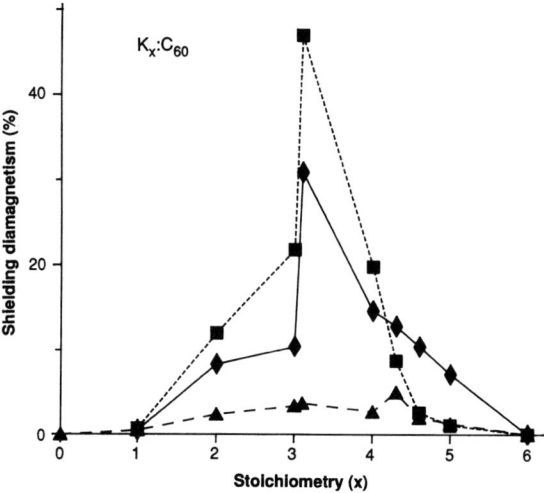

heating the samples from stage 2 again but at 250°C for six or more hours, and additional changes in the shielding diamagnetism were measured. Further heat treatment on a limited number of samples revealed no further changes. Samples appear to be completely stable for at least 2 weeks.

Shielding diamagnetism curves, the primary quantity used here to assess the amount (volume fraction) of the superconducting phase and its homogeneity, were obtained with a standard magnetic inductance coil bridge, operated at 100 kHz with phase-sensitive detection (7). The amplitude of the ac susceptibility was typically measured over the range from 4 to 100 K. The diamagnetic contribution, defined here as $\chi_{dia} = \chi_{normal} - \chi(4.2 \text{ K})$, is expressed as a fraction of a Nb standard. No correction has been applied to take into account the degree of compactness of the powder nor the mag-

netic field distribution inside the capillary. These effects both tend to reduce the observed superconducting fraction, so that the estimates serve as conservative lower bounds.

Three characteristic diamagnetic shielding curves representative of those obtained on 15 samples of ten different nominal compositions at the varying stages (1, 2, and 3) are shown in Fig. 1. Curves A and B are of the shape expected for a single phase transition occurring at 19.3 K. The transition is sharp; the 10 to 50% full width of the diamagnetism transition is <1 K. These represent samples of x:1 nominal composition measured at stages 2 and 3, respectively. Curve C is typical of the measurement at the mixing stage on samples of nominal composition ≥ 3:1. These curves exhibit a more complicated structure. Their unusual form, somewhat similar to that of Hebbard et al. (6), might indicate the presence of more than one phase or a continuum of phases, or even grains with superconducting surface and normal interiors. After subsequent treatment (diffusion and relaxation stages), these samples also show curves of the same form as A and B, indicative of a single superconducting phase.

The different forms of the diamagnetic shielding curve $\chi(T)$ found after the mixing stage give the impression that, despite the apparently complete macroscopic mixing, there exist many different compositions (or a continuum) of $K_x C_{60}$. None of these exhibit diamagnetism above 20 K. Among these, the curves like C resemble most the one shown in (6). Second, after stage 2, the various mixtures all have curves for which the amplitudes can be scaled to match one another, with an always sharp, identical transition temperature of $T_c = 19.3$ K, which implies that there is only one stable superconducting phase of $K_x C_{60}$. Thus the other phase (or phases) that account for the

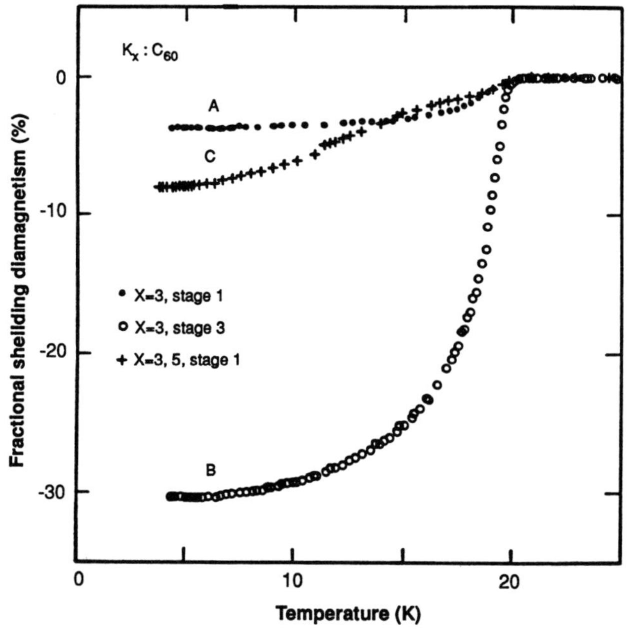

Fig. 1. Magnetic susceptibility curves $\chi(T)$ for three samples of $K_x C_{60}$ powder exhibiting shielding diamagnetism expressed as a percentage of a Nb standard measured at 4.2 K. Curves A and B represent the same sample (nominal composition 3:1) after the mixing and final stages, respectively. Curve C is a composition 3.5:1 sample after the mixing stage.

bulk of the material in most samples are *not* superconducting above 1.5 K or failed to segregate under the preparation conditions used.

In an attempt to determine the actual composition of this superconducting phase, we examined how the volume percent of diamagnetic shielding varied with composition at the mixing stage and at nominal equilibrium. In Fig. 2 the diamagnetic shielding, as defined above (χ_{dia}), is plotted against nominal composition (Table 1). At the first stage (mixing), the maximum in the χ_{dia} versus x curve is definitely at $x > 3$. At later stages, the χ_{dia} increases strongly for small $x \leq 3$, but decreases for most of the samples with $x > 4$. At the final stage, the $x = 3$ sample exhibits the largest shielding observed, 40% or more, a high value for a powder sample. As the typical grain size of the powder determined with an optical microscope is ~1 μm, this is necessarily a lower bound on the fraction of the sample that is in the superconducting phase; it could be that this sample attains a high degree of homogeneity. Despite their reputed air sensitivity (1) a powder sample exposed to air after 3 hours continued to exhibit a Meissner effect of one-half its original magnitude, and showed no diamagnetic shielding after 3 days.

We conclude that the single superconducting phase has a composition close to, or slightly greater than 3:1, and definitely less than 4:1. However, the surprising stability and reproducibility of T_c favors a stoichiometric composition, hence K_3C_{60}, which confirms the hypothesis of the AT&T group (5, 6). After completing this systematic study on relatively small amounts of materi-

Table 1. Variation of fractional shielding diamagnetism as a function of composition x during different stages of preparation. The relaxation stage values were reproducible to ±5% (determined for two or more samples at several compositions).

x	Mixing	Diffusion	Relaxation
1.0	0.5	0.5	0.7
2.0	2.4	8.3	12.0
3.0	3.4	10.4	21.8
3.1	3.7	30.9	47.0
4.0	2.7	14.6	19.8
4.3	5.0	12.8	8.7
4.6	2.0	10.4	2.6
5.0	1.3	7.1	1.1
6.0	0.0	0.0	0.0

al, we prepared several larger quantities (20 to 30 mg each) with the initial composition 3.1:1 and subjected them to extended heat treatments. These resulted in reproducible shielding diamagnetism fractions ranging from 40 to 67%. Keeping in mind that this measurement results in an underestimation of the actual superconducting fraction, it becomes clear that we have found the means to produce samples in which the K_3C_{60} phase is in the definite majority, with "impurity" phases as low as a few percent remaining to be removed, rather than the opposite situation in which the superconducting phase is a very minor component.

The mixing and heat treatment procedure developed on the K_xC_{60} system was used to explore other members of the alkali-doped fullerene class of materials, namely Rb and Cs. Although experiments on Cs_xC_{60} are so far inconclusive in showing signs of a phase transition, a Rb sample prepared from a 3:1

mixture exhibited a clear transition at $T_c = 30$ K with a diamagnetic shielding not less than 7% (Fig. 3). The sharpness (10 to 50% width of ~2 K) and distinctness (no other features) of this transition indicate that a well-defined composition is responsible (8).

We now consider the kinetics of doping, both of the solid-gas reaction and diffusion in the solid. Besides the maximum at $x = 3$ (Fig. 2), there are two surprising aspects. For nominal compositions $x < 3$, and even at $x = 1$, observable shielding diamagnetism is present and increases monotonically with x toward the maximum, suggesting that there are macroscopic precipitates of the superconducting phase to provide a continuous pathway for shielding-current loops. For $x > 3$, the diamagnetic shielding increases during the diffusion stage, but then decreases as the final homogenization is assumed to occur, so that it achieves a maximum at an intermediate stage of the diffusion. All of these observations are consistent with the following simple picture of the doping process:

1) Mixing stage. In vacuum, the entire quantity of alkali metal is vaporized and adsorbed (irreversibly under the conditions used, $T < 300°C$) onto the C_{60} powder granules, which we suppose produces a uniform surface concentration (coverage) of all granules. Accompanying the adsorption, a limited amount of inward diffusion occurs, leading to a steep K concentration gradient toward the grain interiors. Even if at some radial distance the right composition is reached, no shielding would be observed unless the thickness of this shell of superconducting material is comparable to the penetration depth of the superconductor. Separate measurements (9) of the critical-field curves have evaluated this quantity to be $\lambda = 2500$ Å, or about one-fourth of the mean grain size. In fact, the criterion that the grain as a whole is shielded is that it has a closed shell of superconducting material of thickness greater than λ located somewhere between the surface and the grain center.

2) Diffusion and equilibration processes. The effective irreversibility of the adsorption process implies that the quantity of K adsorbed onto a C_{60} grain is proportional to its surface area, rather than to its volume (or quantity of C_{60}). Therefore the smaller than average grains take up more K per volume than average, and the larger grains have less K. In the second stage, under the He atmosphere, diffusion of dopant within a particle takes place. We believe that the small shielding diamagnetism observed at one-third of the optimum composition, that is, nominally 1:1 in Fig. 2, is produced in this way, by generating a 3:1 composition in small grains, which have a surface-to-volume ratio

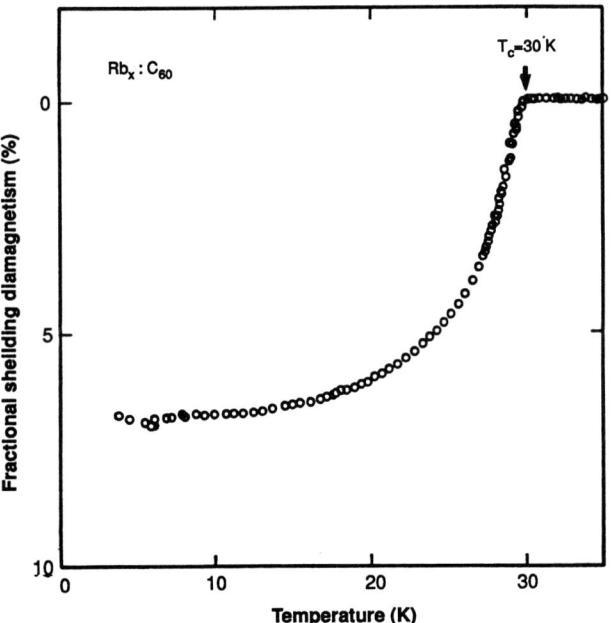

Fig. 3. Magnetic susceptibility curves $\chi(T)$, as in Fig. 1, for a Rb_xC_{60} sample. A single transition is observed at an apparent $T_c = 30$ K.

such that after the first stage they have the right quantity to yield a sufficiently homogeneous particle of 3:1 composition. This process provides the continuous pathway for shielding currents. At the other extreme of nominal composition, the smaller grains become superconducting during the diffusion stage, but upon homogenization this superconductivity is destroyed by the excess K, much as the normal state conductivity is destroyed by extended doping (5), as supported by photoemission results (10) and orbital-filling arguments (5). On the other hand, the larger than average grains, at equilibrium, can turn out to have the right concentration to be superconducting. This accounts for the nearly symmetrical curve of shielding diamagnetism curve versus nominal composition (Fig. 2) and also for the rise and fall of shielding diamagnetism as a function of heat treatment in the high-x samples. The sample with the observed maximum shielding diamagnetism corresponds to transforming to the right concentration the average grain size and hence the majority of grains. We believe this conclusion reflects the inherent limitations of the gas-solid reaction procedure. Different techniques, such as precipitation from organic solution, could result in higher yields. This is why we view the >40% value as a very high yield from these powder samples.

The actual quality of the material obtained here has provided the capability to measure critical field curves $H_{c1}(T)$ and $H_{c2}(T)$ (9). The zero-temperature extrapolated values are H_{c1}=132 G and H_{c2}=49 T. These values allow one to evaluate the penetration depth $\lambda = 2400$ Å and the coherence length $\xi = 26$ Å, and hence a Ginzburg-Landau parameter $\kappa = \lambda/\xi$ around 100, an extremely high value approaching that of the high-T_c copper oxides.

We find that it is the K_3C_{60} composition that is the superconducting phase, and neither under- nor overdoped phases (relative to the correct one) are superconducting. The discovery of the $T_c = 30$ K superconducting phase of Rb_xC_{60} should stringently test the different theoretical possibilities, as it offers the possibility of comparing the dependence of T_c on slight chemical modifications. Prerequisite to this is a completion of the systematic study of the phase diagram of Rb and other alkali-C_{60} mixtures.

Note added in proof: During these studies we learned that Rosseinsky et al. (11) found a superconducting transition at 28 K in a Rb-doped C_{60} sample.

REFERENCES AND NOTES

1. W. Krätschmer, L. D. Lamb, K. Fostiropoulos, D. R. Huffman, *Nature* **347**, 354 (1990).
2. H. Aije et al., *J. Phys. Chem.* **94**, 8630 (1990).
3. R. Taylor, J. P. Hare, A. K. Abdul-Sada, H. W. Kroto, *J. Chem. Soc. Chem. Commun.* **1990**, 1423 (1990).
4. P.-M. Allemand et al., *J. Am. Chem. Soc.* **113**, 2780 (1991).
5. R. C. Haddon et al., *Nature* **350**, 320 (1991).
6. A. F. Hebard et al., *ibid.*, 600.
7. M. Tinkham, *Introduction to Superconductivity* (McGraw-Hill, New York, 1975).
8. K. Holczer et al., unpublished results.
9. K. Holczer et al., unpublished results.
10. P. J. Benning, J. L. Martins, J. H. Weaver, L. P. F. Chibante, R. E. Smalley, *Science*, in press.
11. M. J. Rosseinsky et al., *Phys. Rev. Lett.*, in press.
12. S. J. Anz and F. Ettl produced and separated the molecular carbon samples used in this research. Supported by NSF grants (to R.L.W. and F.N.D. and to R.B.K.) and by the David and Lucille Packard Foundation (R.L.W. and R.B.K.).

8 May 1991; accepted 10 May 1991

Reprinted with permission from Physical Review Letters
Vol. 67, No. 2, pp. 271–274, 8 July 1991
© 1991 The American Physical Society

Critical Magnetic Fields in the Superconducting State of K_3C_{60}

K. Holczer,[a] O. Klein, and G. Grüner

Department of Physics and Solid State Science Center, University of California, Los Angeles, California 90024

J. D. Thompson

Los Alamos National Laboratory, Los Alamos, New Mexico 87545

F. Diederich and R. L. Whetten

Department of Chemistry and Solid State Science Center, University of California, Los Angeles, California 90024
(Received 7 May 1991)

We have measured the temperature dependence of the lower and upper critical fields in superconducting K_3C_{60}. From the measurements, we have evaluated the penetration depth ($\lambda = 2400$ Å) and superconducting coherence length ($\xi = 26$ Å). The parameters are in agreement with a superconducting state formed by a narrow band.

PACS numbers: 74.40.+k, 74.30.Ci

The material C_{60}, also called buckminsterfullerene, upon reacting with donors, like alkali metals, forms a new class of molecular solids. Several of them are conductors, and superconductivity was repored for K_3C_{60} with a critical temperature [1] $T_c = 18$ K and Rb_3C_{60} with $T_c = 30$ K [2]. So far, practically nothing is known about the normal and superconducting states of these materials, although the first findings are suggestive of type-II superconductivity. In this paper we report on our measurements of the critical magnetic fields of the K_3C_{60} compound, and evaluate the penetration depth λ and coherence length ξ. We also briefly discuss the implications of our results.

The samples were prepared from solid-phase reaction of high-purity C_{60} powder with potassium in a way similar to that reported originally by the AT&T group [1]. A wide range of composition was explored and different heat treatments were applied to maximize the superconducting fraction of the resulting material. In the case of the sample used in this study, a starting composition of K_4C_{60} was heated in a vacuum at 200°C for 22 h. The material was subsequently annealed in He atmosphere at 200°C for 24 h, followed by another annealing at 250°C for another 6 h. This procedure leads to a maximum shielding fraction of about 15% of the perfect diamagnetism. Throughout the whole preparation, extreme care has been taken to avoid oxygen contamination. A detailed study of the reaction kinetics and of the determination of the stoichiometry of the superconducting compound is published elsewhere [2]. We believe that we have measured K_3C_{60}.

dc magnetization was measured with a Quantum Design SQUID magnetometer in fields up to 5 T and temperatures down to 2 K. The pickup coils were arranged in a second gradiometer geometry in order to eliminate any spurious gradient of the applied magnetic field. The powder of K_3C_{60} was sealed in a Pyrex capillary under 1 atm of helium as an exchange gas. The

powder can be approximated by a set of independent spheres. In this case the magnetic field inside the sample is related to the externally applied magnetic field by

$$H_{in} = \frac{H_{ext}}{1-n}, \quad M = -\frac{1}{4\pi}\frac{H_{ext}}{1-n}, \quad (1)$$

where n is the demagnetization factor, a geometrical constant equal to $\frac{1}{3}$ for a sphere. In all the measurements, care was taken to ensure that the sample was not exposed to a field gradient larger than $\pm 0.03\%$ of the applied field.

We have evaluated the lower critical field from the magnetic-field dependence of the zero-field-cooled dc magnetization. From the normal phase ($T = 55$ K), the sample was cooled down to the superconducting phase ($T < T_c$) in zero field (magnetic field below 1 Oe). Once the desired temperature was reached, the dc magnetization was measured as a function of increasing applied field. The lower critical field $H_{c1}(T)$ was defined as $1/(1-n)$ times the lowest value of the applied field which leads to a departure from a linear behavior in the dc magnetization. The accuracy of the determination of the lower critical field was found to be around ± 5 Oe at low temperature and ± 10 Oe near T_c (the linear regime narrows and thus becomes difficult to evaluate).

The upper critical field $H_{c2}(T)$ was evaluated from the temperature dependence of the field-cooled magnetization. From the normal phase, the sample was cooled to the superconducting state in an external magnetic field (2 kOe $< H_{ext} <$ 50 kOe). We monitored the magnetization while heating the sample until the normal state is reached, with the applied field held constant. The intercept of a linear extrapolation of the magnetization in the superconducting state with the normal-state base line defines the transition temperature T, and the upper critical field $H_{c2}(T)$ is equal to the applied field.

In Fig. 1 we display the zero-field-cooled (ZFC) and field-cooled (FC) temperature dependence of the dc mag-

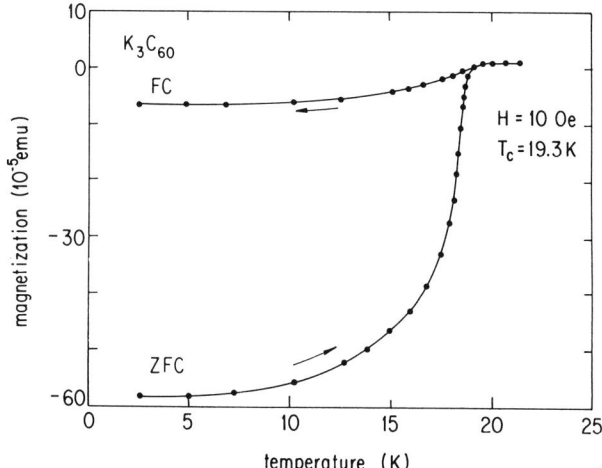

FIG. 1. Temperature dependence of the zero-field-cooled (ZFC) and field-cooled (FC) dc magnetization with a fixed applied field of 10 Oe. The mass of the sample was $m = 0.013$ g.

netization at a fixed field of 10 Oe. The protocol was the following: From its normal state ($T = 55$ K) the sample was first cooled to 2 K with no field applied; we then monitored the dc magnetization as the sample was heated above the critical point under an applied field of 10 Oe (ZFC). We then cooled the sample to $T = 2$ K under the same field, 10 Oe (FC). The onset of a diamagnetic transition occurs at $T = 19.3$ K. The magnetic transition (10%–50% diamagnetic shielding) is narrow, less than 1 K, suggesting that the superconducting transition is relatively homogeneous. From the ZFC curve one can compute the diamagnetic shielding fraction. Assuming the powder is made of independent spheres, this leads to a shielding fraction of 15% of the perfect diamagnetism, a much higher figure than the one first reported [1]. Hysteresis between the FC and ZFC measurement indicates that flux is trapped in the sample as it is cooled below the transition temperature.

In Fig. 2 we display the magnetic-field dependence of the dc magnetization at a fixed temperature $T = 5$ K. From the normal state, the sample was cooled to $T = 5$ K in zero field ($H_{ext} < 1$ Oe) and the dc magnetization was monitored as the field was increased up to 40 kOe and then back to zero. The hysteresis found suggests the presence of substantial flux pinning, i.e., critical current density, that decreases strongly with applied field. One can get an order of magnitude for the critical-current density using the hysteresis curve in Fig. 2. If we assume that we have a cylindrical specimen of radius R, then we can use the formula [3]

$$J_c(H)(\text{A/cm}^2) = 15 \frac{M_+ - M_- (\text{emu/cm}^3)}{R(\text{cm})}, \quad (2)$$

where M_+ and M_- are the hysteretic magnetizations at a given field H. Using a typical grain size of the powder, determined using an optical microscope, $R = 1$ μm, we evaluate $J_c(H = 10 \text{ kOe}) = 1.2 \times 10^5$ A/cm^2.

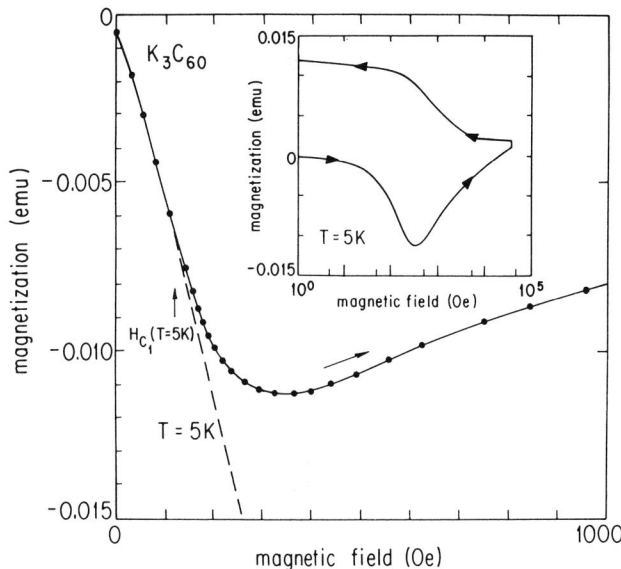

FIG. 2. Magnetic-field dependence of the magnetization at a fixed temperature $T = 5$ K. The zero offset is due to the Pyrex capillary in which the sample was sealed.

In Fig. 3 we have plotted the temperature dependence of the upper and lower critical fields, determined as described earlier. For the upper critical field we observe that, except in the vicinity of the critical temperature

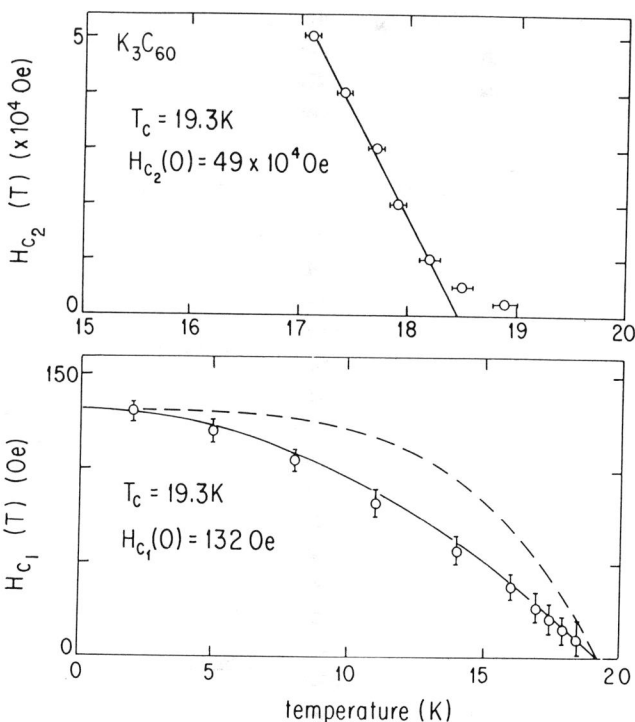

FIG. 3. Temperature dependence of the upper and lower critical fields. For the upper panel, the solid line is a linear fit of the temperature dependence; the slope of the fit is -3.73 T/K. For the lower panel, we compare the temperature dependence of H_{c1} to $H_{c1}(T)/H_{c1}(0) = 1 - (T/T_c)^n$, where $n = 2$ corresponds to the empirical law (solid line) and $n = 4$ is appropriate for the two-fluid model (dashed line).

$(T_c = 19.3$ K), a linear dependence adequately fits the temperature dependence of H_{c2}. The critical-field slope from the linear fit is -3.73 T/K. The critical field extrapolated to $T = 0$ K using the Werthamer-Helfand-Hohenberg (WHH) formula [4]

$$H_{c2}(0) = 0.69 \left| \frac{\partial H_{c2}}{\partial T} \right|_{T_c} T_c \tag{3}$$

is $H_{c2}(0) = 49^{+5}_{-10}$ T. This large value exceeds the Pauli limit [5] of 35 T assuming no electron-phonon enhancement of the Pauli field, H_P [6]. Rigorously, Eq. (3) is only valid in the dirty limit. In the clean limit, the value of $H_{c2}(0)$ computed from the WHH formula is higher by a few percent than the one in the dirty limit [7]. If one includes effects of Pauli spin paramagnetism [3], the value of $H_{c2}(0)$, extrapolated from the slope near T_c (WHH formula), will be greatly reduced. Direct measurement of $H_{c2}(0)$ would be particularly informative in this regard. Using the relation [8]

$$H_{c2}(0) = \phi_0 / 2\pi\xi^2 , \tag{4}$$

this corresponds to a zero-temperature coherence length of $\xi = 26^{+4}_{-2}$ Å for $H_{c2}(0) = 49$ T. If the applied field is below 5 kOe, then $\partial H_{c2} / \partial T$ is smaller than the slope expected from the linear fit shown in Fig. 3. Such behavior is found commonly in conventional superconductors and attributed to slight variations in the local T_c.

The temperature dependence of the lower critical field follows within a few percent the empirical law [8]

$$H_{c1}(T) = H_{c1}(0)[1 - (T/T_c)^2] . \tag{5}$$

From the fitting we extrapolate the lower critical field at zero temperature $H_{c1}(0) = 132^{+10}_{-20}$ Oe and using the formula [8]

$$H_{c1}(0) = (\phi_0 / 4\pi\lambda^2) \ln(\lambda/\xi) , \tag{6}$$

we estimate $\lambda_L = 2400^{+100}_{-200}$ Å. For comparison we have also plotted in Fig. 3 the temperature dependence of H_{c1} expected by assuming that λ/ξ is independent of temperature and that $H_{c1}(T) \propto \lambda(T)^{-2}$. Using the two-fluid temperature dependence for $\lambda(T)$ gives

$$H_{c1}(T) = H_{c1}(0)[1 - (T/T_c)^4] . \tag{7}$$

Shown in Fig. 4 are two sets of raw data used to determine H_{c2}. Below T_c, the slopes are $\partial M/\partial T = 0.063$ and 0.089 G/K for the $H = 5$ and 2 kOe data sets, respectively. In calculating these slopes, we used the sample mass $m = 0.013$ g and the x-ray density [9] of K_3C_{60}, $\rho = 1.91$ g cm^{-3}, to arrive at the conversion $M(G) = 4\pi(\rho/m)M$ (emu). These slopes can be compared to that calculated from [10]

$$\frac{\partial M}{\partial T} = - \left[\frac{1}{4.64\pi(2\kappa^2 - 1)} \right] \frac{\partial H_{c2}}{\partial T} , \tag{8}$$

where $\kappa = \lambda/\xi = 92$ and $\partial H_{c2}/\partial T = -1.6$ T/K from Fig.

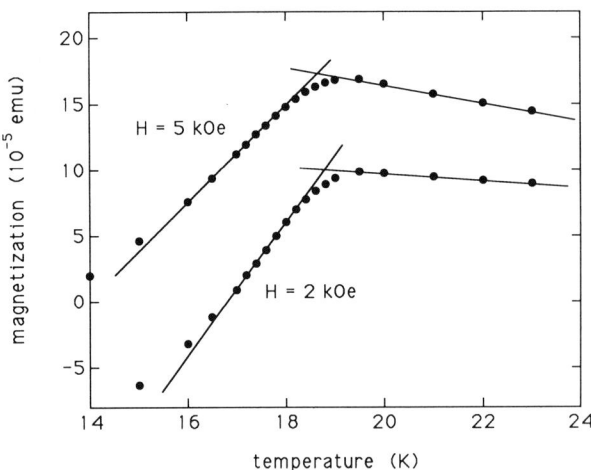

FIG. 4. Magnetization vs temperature measured near T_c in applied fields of 5 and 2 kOe. The intersection of the linear extrapolations made both below and above T_c defines $H_{c2}(T)$.

3. Equation (8) gives $\partial M/\partial T = 0.063$ G/K $\pm 10\%$, with the assumption that κ is temperature independent and the Pyrex capillary contributes negligibly to the slope at these low fields. The agreement is rather good, confirming consistency in our estimate of H_{c2}. At higher fields, the latter assumption is no longer valid and direct comparison is not possible.

The electronic structure of C_{60} can be described crudely by a tight-binding approach, and the overlap of the neighboring C_{60} units can be accounted for by an overlap integral t_1. The fact that the coherence length exceeds the diameter d ($\simeq 10$ Å) of the C_{60} spheres suggests that the appropriate starting point for understanding the superconducting state is not that of a weakly Josephson coupled unit but that of a conventional superconducting state where the relevant normal-state parameter is the transfer integral t_1. The coherence length can be written as [11]

$$\xi/d = D/\Delta , \tag{9}$$

where the relevant bandwidth D equals $8t_1$ in three dimensions. The measured coherence length, together with a weak-coupling expression for the superconducting gap $\Delta = 3.52k_BT_c$, leads to a bandwidth $D = 200$ K. The clean-limit expression of the penetration depth [8]

$$\lambda = \frac{c}{\omega_p}, \quad \omega_p = \left[\frac{4\pi n e^2}{m_{eff}} \right]^{1/2} , \tag{10}$$

where $n = 1.9 \times 10^{21}/\text{cm}^3$ is the number of carriers (assuming one conduction electron per C_{60}) and m_{eff} is the effective band mass, gives $m_{eff} = 4m_e$, i.e., a slightly enhanced effective mass, a situation similar to that encountered in high-temperature superconductors, and in full agreement with our previous conclusion on the bandwidth D. Taking one conduction electron per C_{60} is equivalent to assuming that the threefold-degenerate Fermi level of the pure C_{60} splits with doping. In the case where degeneracy holds, then we have to take a density of carriers 3 times larger, leading to an effective mass $m_{eff} = 12m_e$.

In conclusion, our experiments on the critical magnetic fields in K_3C_{60} show a strongly type-II superconducting state with the temperature dependence of the parameters well described by mean-field theory. The coherence length and the penetration depth are in agreement with the picture of a superconducting state that develops within a relatively narrow band, which we believe is controlled by the overlap of the wave functions of the neighboring C_{60} units. The $T = 0$ upper critical field, extrapolated from the WHH model, exceeds the Clogston limit, suggesting that $H_{c2}(0)$ is Pauli paramagnetically limited as might be expected for a material with a modestly large density of states at the Fermi level. We note that the superconducting properties of K_3C_{60} are rather similar to those of conventional A15 superconductors [6] (e.g., Nb_3Sn). Also throughout this paper we have assumed that the electronic structure is isotropic, and both the normal and superconducting states have a three-dimensional character. The extent to which K_3C_{60} deviates from three-dimensional character and influences our conclusions remains an open question.

We thank S. M. Huang, R. B. Kaner, S. J. Anz, and F. Ettl for their help in the sample preparation, also S. Chakravarty and S. Kivelson for useful discussions. This research was supported by the INCOR program of the University of California. Work at Los Alamos was performed under the auspices of the U.S. Department of Energy.

---

[a] Permanent address: Central Research Institute for Physics, P.O. Box 49, H1525 Budapest, Hungary.

[1] A. F. Hebard, M. J. Rosseinsky, R. C. Haddon, D. W. Murphy, S. H. Glarum, T. T. M. Palstra, A. P. Ramirez, and A. R. Kortan, Nature (London) **350**, 600 (1991).

[2] K. Holczer, O. Klein, G. Grüner, S.-M. Huang, R. B. Kaner, K.-J. Fu, R. L. Whetten, and F. Diederich, Science **252**, 1154 (1991).

[3] W. A. Fietz and W. W. Webb, Phys. Rev. **178**, 657 (1969).

[4] N. R. Werthamer, E. Helfand, and P. C. Hohenberg, Phys. Rev. **147**, 295 (1966).

[5] A. M. Clogston, Phys. Rev. Lett. **9**, 266 (1962).

[6] T. P. Orlando, E. J. McNiff, S. Foner, and M. R. Beasley, Phys. Rev. B **19**, 4545 (1979).

[7] E. Helfand and N. R. Werthamer, Phys. Rev. **147**, 288 (1966).

[8] M. Tinkham, *Introduction to Superconductivity* (McGraw-Hill, New York, 1975).

[9] P. W. Stephens, L. Mihàly, P. L. Lee, R. L. Whetten, S. M. Huang, R. Kaner, F. Diederich, and K. Holczer (to be published).

[10] U. Welp, W. K. Kwok, G. W. Crabtree, K. G. Vandervoort, and J. Z. Liu, Phys. Rev. Lett. **62**, 1908 (1989).

[11] J. M. Ziman, *Principles of the Theory of Solids* (Cambridge Univ. Press, Cambridge, 1986).

Reprinted with permission from Nature
Vol. 351, pp. 632–634, 20 June 1991
© 1991 Macmillan Magazines Limited

Structure of single-phase superconducting K_3C_{60}

Peter W. Stephens*, Laszlo Mihaly*, Peter L. Lee†,
Robert L. Whetten‡, Shiou-Mei Huang‡,
Richard Kaner‡, François Deiderich‡
& Karoly Holczer‡

* Department of Physics, State University of New York, Stony Brook,
New York, 11794, USA
† New York State Institute on Superconductivity and Department of
Chemistry, State University of New York, Buffalo, New York 14214, USA
‡ Department of Physics, Department of Chemistry and Biochemistry, and
Solid State Science Center, University of California, Los Angeles,
California 90024–1569, USA

RECENT reports[1-3] of superconductivity in alkali-metal-doped compounds of the icosahedral C_{60} (buckminsterfullerene) molecule have attracted great experimental and theoretical interest. Superconductivity was originally discovered in samples prepared from gas–solid reactions[1], which made it impossible to determine the composition or structure of the superconducting phase. Holczer et al.[4] demonstrated that potassium-doped C_{60} has only a single stable superconducting phase, K_3C_{60}, with a transition temperature of 19.3 K. Improvements have since resulted in the preparation of 100% bulk superconductors[3]. Because of the absence of impurity phases, we have been able to perform accurate Rietveld analysis of X-ray diffraction data from the superconducting phase. Here we report our results for the crystal structure of K_3C_{60}, determining that this superconducting compound has a face-centred cubic structure with a well defined stoichiometry. These results should open the way to rigorous description of the normal and superconducting properties of this compound.

The structures that have been determined so far include those of the pure, electrically insulating C_{60} (ref. 5) and the heavily doped K_6C_{60} (ref. 6). At room temperature, solid C_{60} forms a face-centred cubic (f.c.c.) lattice with 10.0 Å intercluster separation[5]. K_6C_{60} has a body-centred cubic structure, with K atoms in distorted tetrahedral sites[6]. No superconductivity was observed in that material. Cs_6C_{60} has the same structure[6]. We have found that the superconductor K_3C_{60} has a f.c.c. structure derived from that of C_{60} by incorporating K ions into all of the octahedral and tetrahedral interstices of the host lattice.

Preparation of these powder samples has been described previously[2]. Briefly, the reaction between stoichiometric

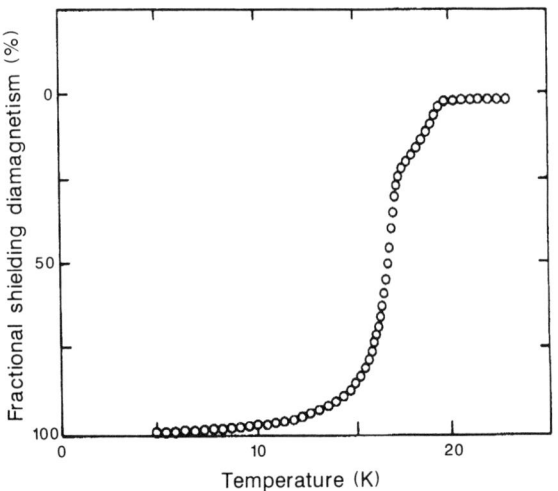

FIG. 1 Diamagnetic shielding signal against temperature for the sample of K_3C_{60} used in the detailed X-ray study.

TABLE 1 Structural parameters from Rietveld analysis for K_3C_{60}

	Site		x	y	z	Occupancy	B (Å^2)
C1	96	j	0	0.046	0.245	50%	1.2 ± 0.5
C2	192	l	0.213	0.084	0.096	50%	1.2 ± 0.5
C3	192	l	0.184	0.160	0.051	50%	1.2 ± 0.5
K1	8	c	1/4	1/4	1/4	100% ± 20%	6.5 ± 2
K2	4	b	1/2	1/2	1/2	100% ± 10%	16 ± 6

Crystallographic space group is $Fm\bar{3}m$ (O_h^5). Cubic lattice parameter $a = 14.24 \pm 0.01$ Å. R values for fit are $R_{wp} = 8\%$, $R_i = 17\%$.

amounts of solid C_{60} and K vapour proceeds in several stages, until a material that is very rich in the superconducting phase is achieved. Powder samples of 50% or higher diamagnetic shielding at 4.2 K are produced. The superconducting ceramic pellets are made by one or more cycles of pressing, sintering and grinding[4].

Figure 1 shows the temperature dependence of the diamagnetic shielding. Data were taken using a standard magnetic inductance coil bridge, operated at 100 kHz; the diamagnetic fraction is expressed relative to a niobium standard. Within the precision of the instrument, 100% diamagnetism is achieved for the ceramic sample at temperatures below 10 K, indicating that the magnetic field is completely expelled from the bulk of the sample. Apart from the onset temperature at 19.3 K, the curve shows a kink and a sharp decline below 17.5 K. As discussed elsewhere[4], such features may be interpreted in terms of weak links between superconducting grains.

A number of pressed samples, of volume ~1 mm^3, were cut from pellets and sealed in pyrex capillaries for powder diffraction measurements. Before the X-ray studies, the diamagnetic fraction was measured and found to be close to 100%. The samples were never removed from the capillaries, so we can be confident that they did not degrade from exposure to air. In addition to the detailed spectrum presented below, we observed essentially identical diffraction patterns from a total of five samples from three separate batches.

Powder diffraction spectra were collected at the State University of New York X3A beamline at the National Synchrotron Light Source. X-rays of wavelength $\lambda = 0.800$ Å were produced by a double Si(111) monochromator, and the diffracted beam was analysed with a Ge(111) crystal. The resulting resolution was 0.05° full width at half maximum. The room-temperature diffraction pattern is shown in Fig. 2. For comparison, we also show the powder diffraction pattern of pure C_{60} taken at $\lambda = 1.500$ Å. The C_{60} spectrum matches previous results[5].

Both the C_{60} and K_3C_{60} patterns correspond to f.c.c. lattices of essentially identical unit-cell size: 14.11 Å for C_{60} and 14.24 Å for K_3C_{60}. The full width at half maximum of the strongest reflections in the K_3C_{60} sample is 0.1°, corresponding to a grain size of at least 500 Å. The higher background scattering in this sample is a consequence of its heavier capillary; any amorphous background scattering is no stronger than for the pure C_{60} material.

As both samples have a f.c.c. structure of equal unit-cell dimensions, it is apparent that the K_3C_{60} structure is based on a f.c.c. lattice of C_{60} molecules, with K atoms inserted into empty spaces. The possibility of incorporating alkali ions into these sites was noted previously by Haddon et al.[7]. The clearest difference between the C_{60} and the K_3C_{60} diffraction patterns occurs in the relative intensities of the strongest few peaks, most notably (220) and (311). It is therefore evident that the sites occupied by K interfere destructively and constructively with the C atoms, respectively, for these two reflections. This immediately suggests that the K occupy sites such as $(\frac{1}{4}, \frac{1}{4}, \frac{1}{4})$ or $(\frac{1}{2}, 0, 0)$, shown in Fig. 3. The fact that the stoichiometry requires three K atoms for each C_{60} molecule implies that K atoms occupy both of these inequivalent sites. The structure is essentially the well known cryolite structure, which applies to a large class of ionic solids[8].

An important issue in describing this structure is the orientational order of the C_{60} molecules. Consideration of the size of the constituents severely limits the possible choices. Free rotation of the molecules, as is observed in pure C_{60}, can be immediately ruled out. The shortest distance between a K atom at the $(\frac{1}{4}, \frac{1}{4}, \frac{1}{4})$ position and C atom on the freely rotating C_{60} would be 2.66 Å, much smaller than the 3.06 Å measured in intercalated graphite[9]. The estimate by Haddon et al.[7] also suggests that the molecules are locked into a fixed position: a rotating (spherical) C_{60} unit leaves a tetrahedral site of radius 1.12 Å, whereas the radius of a K^+ ion is 1.33 Å. In fact, there are only two possible orientations of the C_{60} molecule that provide sufficient space for all eight nearest-neighbour alkali-metal atoms. In both cases the C_{60} molecule is oriented with eight of its twenty hexagonal faces along the cubic $\langle 111 \rangle$ directions. The two orientations are related by a 90° rotation about the (100) direction, producing different positions for the pentagonal faces, as shown in Fig. 3.

We performed a Rietveld refinement of the K_3C_{60} powder diffraction pattern for two structures allowed by these size restrictions. In the first case, we assumed that all C_{60} units have the same orientation. The space group is then $Fm\bar{3}$, and we find that the refinement always yields a significant peak at 3.3 Å^{-1}, which is not seen in the experiment. In the second case, the molecules are disordered between the two possible orientations, as illustrated in Fig. 3. The fit shown in Fig. 2 is in excellent agreement with the data. Table 1 summarizes the fitting parameters. The space group is $Fm\bar{3}m$, which is a higher symmetry

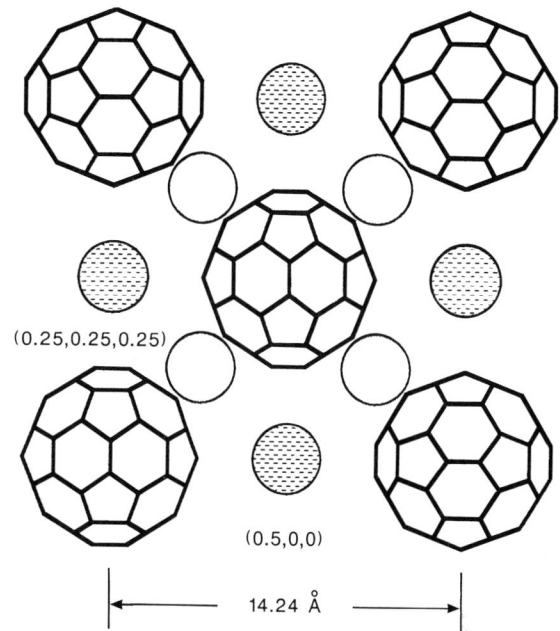

(0.25,0.25,0.25)

(0.5,0,0)

14.24 Å

FIG. 3 The structure of K_3C_{60}. The open and hatched spheres represent the potassium at the tetrahedral and octahedral sites, respectively.

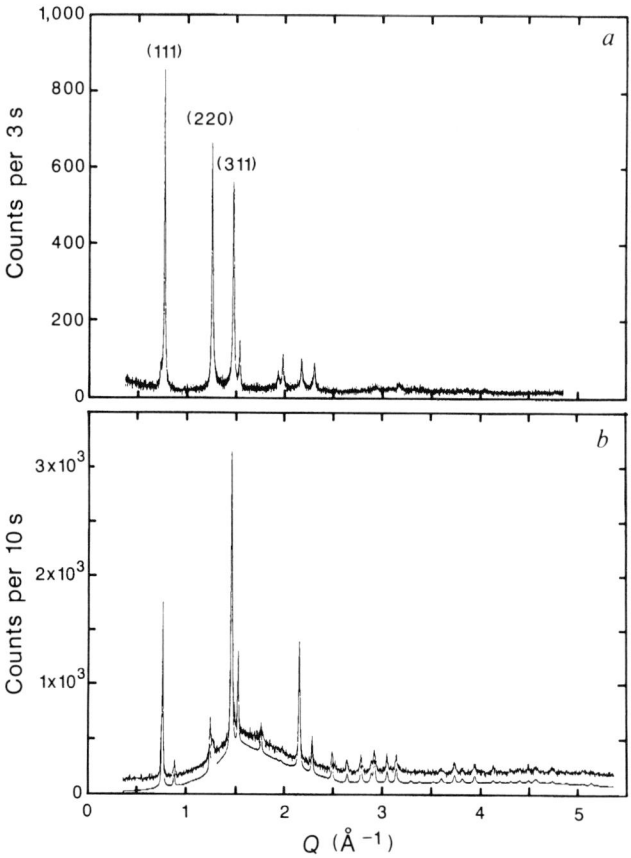

1,000

800

600

400

200

0

(111)

(220)

(311)

a

Counts per 3 s

3×10^3

2×10^3

1×10^3

0

b

Counts per 10 s

0   1   2   3   4   5

Q (Å^{-1})

FIG. 2 X-ray diffractogram of a, the pure C_{60} and b, the K_3C_{60} compounds. The Rietveld fit, shown for the K_3C_{60}, is shifted for clarity.

than allowed for the icosahedral molecule, but can be interpreted as a 50% occupancy of the two orientations. X-ray diffraction cannot distinguish between dynamic or frozen disorder.

We find the radius of the C_{60} molecule to be 3.54 Å, in satisfactory agreement with values derived from undoped C_{60}. The carbon atoms move from their ideal positions by at most 0.02 Å, a small but significant distortion. The temperature factors are strongly correlated with the background correction, but two points can be made: first, B factors for the carbon atoms are small enough to confirm that there is little rotation of the molecules, and second, the large thermal factor for the octahedral cation reflects the larger space available for it.

The nearest-neighbour C-K distance in this structure is 3.27 Å, separating the potassium atoms at the $(\frac{1}{4}, \frac{1}{4}, \frac{1}{4})$ tetrahedral site from the six-membered carbon rings directly facing them in each of four C_{60} molecules. The distance from the octahedral K site at $(\frac{1}{2}, 0, 0)$ to the nearest C atom (at $(0, 0.0501, 0.243)$) is 3.69 Å. The closest K-K distance is 6.17 Å, much larger than in metallic potassium.

The determination of the structure is of great interest in connection with other known properties of superconducting K_3C_{60}. Our observation of orientational disorder and the large thermal factor for K suggests some interesting possibilities. The most important feature of the structure determined here is that it is so similar to that of the undoped solid. Pure C_{60} is a molecular insulator with no partially filled electronic bands, whereas the doped compound is a metal. As there is no significant change in the lattice spacing, it is possible to assume that the doping changes the filling of the band while leaving the band structure unchanged. Because the bands are formed by weak overlap of molecular orbitals, they are expected to be narrow[10]. Magnetization measurements on a nearly optimized powder sample with $T_c = 19.3$ K indicate that K_3C_{60} is an extreme type II superconductor with a London penetration depth of 2,400 Å and a coherence length of 26 Å (ref. 11). These parameters indicate that a high density of states builds up in a narrow energy range around the Fermi level. The observed strong decrease of T_c with pressure (near -0.8 K kbar^{-1}) further confirms this picture[4].

The successful Rietveld fit of the data, in which the K ions occupy all of the available sites in a virtually unchanged C_{60} lattice, is a direct experimental indication of the microscopic homogeneity of a material that has a stoichiometric $K:C_{60}$ ratio of $3:1$. The presence of two crystallographically different K sites suggests that it may be possible to derive other stable compounds. It remains to be seen if these possibilities can be realized by doping with potassium or with other elements, such as rubidium, known to lead to superconductivity at even higher temperatures[2,3]. □

Received 31 May; accepted 5 June 1991.

1. Hebard, A. F. et al. Nature **350**, 600–601 (1991).
2. Holczer, K. et al. Science **252**, 1154 (1991).
3. Rosseinsky, M. J. et al. Phys. Rev. Lett. **66**, 2830–2832 (1991).
4. Sparn, G. et al. Science (submitted).
5. Heiney, P. A. et al. Phys. Rev. Lett. **66**, 2911–2914 (1991).
6. Zhou, O. et al. Nature **351**, 462–464 (1991).
7. Haddon, R. C. et al. Nature **350**, 320–323 (1991).
8. Bode, H. & Voss, E. Z. Anorg. Chem. **290**, 1 (1957).
9. Rudorff, W. Advances in Inorganic Chemistry and Radiochemistry Vol. 1 (eds Emelius, H. J. & Sharp, A. G.) 225 (Academic Press, New York, 1959).
10. Weaver, J. H. et al. Phys. Rev. Lett. **13**, 1741–1744 (1991).
11. Holczer, K. et al. Phys. Rev. Lett. (submitted).

ACKNOWLEDGEMENTS. S. J. Anz and F. Ettl produced and separated the molecular carbon samples used in this research. We thank G. Gruner for sharing the resources of his laboratory and his NSF grant. Funding was provided by the NSF (P.W.S., L.M., R.L.W., F.N.D. and R.B.K.) and by the David and Lucille Packard Foundation (R.L.W. and R.B.K.). The SUNY beamline at NSLS is supported by the Department of Energy (P.W.S. and P.L.). Part of the research was carried out at the National Synchrotron Light Source, Brookhaven National Laboratory, which is supported by the US Department of Energy, Division of Materials Sciences and Division of Chemical Sciences.

Reprinted with permission from Science
Vol. 253, pp. 884–886, 23 August 1991

^{13}C NMR Spectroscopy of K_xC_{60}: Phase Separation, Molecular Dynamics, and Metallic Properties

R. Tycko, G. Dabbagh, M. J. Rosseinsky, D. W. Murphy,
R. M. Fleming, A. P. Ramirez, J. C. Tully

The results of ^{13}C nuclear magnetic resonance (NMR) measurements on alkali fullerides K_xC_{60} are reported. The NMR spectra demonstrate that material with $0 < x < 3$ is in fact a two-phase system at equilibrium, with $x = 0$ and $x = 3$. NMR lineshapes indicate that C_{60}^{3-} ions rotate rapidly in the K_3C_{60} phase at 300 K, while C_{60}^{6-} ions in the insulating K_6C_{60} phase are static on the time scale of the lineshape measurement. The temperature dependence of the ^{13}C spin-lattice relaxation rate in the normal state of K_3C_{60} is found to be characteristic of a metal, indicating the important role of the C_{60}^{3-} ions in the conductivity. From the relaxation measurements, an estimate of the density of electronic states at the Fermi level is derived.

THE RECENT DISCOVERY THAT ELEC-trically conducting (1) and superconducting (2, 3) alkali fullerides with general formula M_xC_{60} (M = K, Rb, Cs) can be prepared by reacting solid C_{60} [buckminsterfullerene (4, 5)] with alkali metals has stimulated widespread interest in the properties of these materials. Crystal structures of K_3C_{60} (6) and K_6C_{60} (7) have been reported. It is generally believed (1–3, 8), that the K_3C_{60} phase is responsible for superconductivity in K_xC_{60}, while the K_6C_{60} phase is an insulator. At present, substantial questions remain concerning the structure, stability, and electronic properties of intermediate phases, that is, phases with $x \neq 0, 3, 6$. In this paper, we report ^{13}C nuclear magnetic resonance (NMR) measurements on alkali fullerides with nominal stoichiometry K_xC_{60}. We find distinctive spectral features for the K_3C_{60} and K_6C_{60} phases. Our NMR spectra demonstrate that, for $x \leq 3$, only the C_{60} and K_3C_{60} phases are stable. We show that this finding is consistent with the calculated electrostatic energies of the $x = 1, 2,$ and 3 phases. The NMR lineshapes indicate that C_{60}^{3-} ions rotate rapidly in K_3C_{60} at room temperature and that C_{60}^{6-} ions are static in K_6C_{60} at room temperature. Finally, we present measurements of the ^{13}C spin-lattice relaxation time (T_1) in K_3C_{60} in the temperature range from 133 K to 363 K that reveal behavior characteristic of a metal and provide an experimental determination of the local density of electronic states at the Fermi level.

Powder samples of K_xC_{60} were prepared as described previously (3, 5, 8). Briefly, the appropriate amounts of C_{60} powder and K were sealed in pyrex tubes in vacuo. The tubes were heated at 225°C for 3 to 5 days to form the potassium fulleride compounds. The samples were then resealed in ≤20 torr $He_{(g)}$ and annealed at temperatures between 250° and 300°C for periods ranging from 10 to 25 days. NMR measurements are reported for four samples, with nominal stoichiometries $K_{1.5}C_{60}$ (sample I), $K_{2.0}C_{60}$ (sample II), $K_{3.1}C_{60}$ (sample III), and $K_{6.0}C_{60}$ (sample IV). Superconducting fractions, as determined by magnetic flux expulsion measurements, were 16, 33, 32, and 0% for samples I to IV, respectively. In all cases, a single superconductivity transition at 19.0 ± 0.5 K was observed. NMR spectra were obtained at a ^{13}C frequency of 100.5 MHz (9.39-T field). Frequencies are reported in parts per million with respect to tetramethylsilane. The total mass of carbon that contributes to the NMR spectrum was determined for each K_xC_{60} sample by comparing the initial amplitude of the free induction decay (FID) signal with that of a known weight of $CH_3OH_{(l)}$. In all cases, the entire mass of the sample was accounted for in the NMR spectrum to within the accuracy of the calibration (±15%). T_1 values were determined using the saturation-recovery technique. Recovery curves were well fit by a single exponential.

NMR spectra were taken of pure solid C_{60} and of samples I through IV at room temperature after the above preparation (Fig. 1). As reported earlier (9–11), the spectrum of C_{60} (Fig. 1a) shows a single line at 143 ppm, with a width of 2.5 ppm. The

spectrum of sample III (Fig. 1d) also shows a single line, but centered at 186 ppm and with a width of 15 ppm. We assign this line to C_{60}^{3-} in the K_3C_{60} phase. The spectra of samples I and II show only the K_3C_{60} and C_{60} lines. We see no evidence for resonances from intermediate phases. The ratio of the areas of the two lines in the spectrum of sample I is 1:1, precisely what one would predict if the intermediate phases were un-

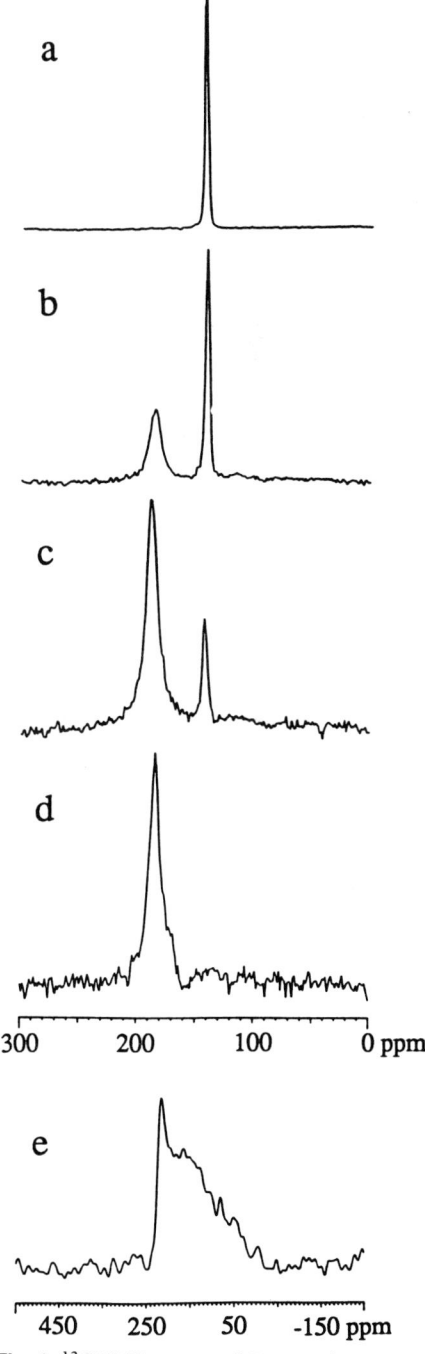

Fig. 1. ^{13}C NMR spectra of C_{60} powder (**a**) and K_xC_{60} with nominal values of $x = 1.5$ (**b**), 2.0 (**c**), 3.1 (**d**), and 6.0 (**e**). $T = 300$ K; sample weights ≈ 30 mg; number of shots = 16, 256, 256, 128, 780 (a–e); delay between shots = 300, 180, 180, 30, 120 s (a–e).

AT&T Bell Laboratories, 600 Mountain Avenue, Murray Hill, NJ 07974.

stable with respect to disproportionation into K_3C_{60} and C_{60} phases. The area ratio is 5:1 in the spectrum of sample II, indicating an actual stoichiometry $K_{2.5}C_{60}$.

X-ray powder diffraction patterns were also acquired for C_{60} and for samples I, III, and IV (Fig. 2). The similarity of the patterns for samples I and III is attributable to the fact that there is virtually no change in the lattice parameters or symmetry between C_{60} and K_3C_{60} (6, 12). Thus, x-ray powder diffraction without profile analysis is of marginal utility in distinguishing between phase separation and the formation of a solid solution for $0 < x < 3$. The diffraction pattern for sample IV is in good agreement with the x-ray data for K_6C_{60} reported previously (7).

To check that equilibrium was established in our K_xC_{60} samples, we examined the NMR spectrum of sample I after further heating. Spectra were taken after initial preparation (Fig. 3a, same as Fig. 1b), after an additional 24 hours at 350°C (Fig. 3b), and after an additional 72 hours at 400°C (Fig. 3c). Although the width of the K_3C_{60} line changes upon heating, the ratio of the areas of the K_3C_{60} and the C_{60} lines remains roughly 1:1. In addition, we see the development of a broad, low-intensity feature in the spectrum. X-ray powder diffraction at this point shows a substantial broadening and weakening of the diffraction peaks. We attribute these observations to sample decomposition. We conclude that the spectra in Fig. 1 are those of the equilibrium phases.

The narrowness of the K_3C_{60} line in Figs. 1 and 3 indicates that C_{60}^{3-} ions rotate rapidly in K_3C_{60} at room temperature, thereby averaging out the ^{13}C chemical shift anisotropy (CSA) as previously observed for C_{60} molecules in the undoped solid (9, 10, 13). In contrast, the spectrum of sample IV

(Fig. 1e) shows a broad, poorly defined CSA powder pattern lineshape. The fact that the ^{13}C CSA is not averaged out in K_6C_{60} indicates that large amplitude reorientations of C_{60}^{6-} ions in K_6C_{60} at room temperature are infrequent on the time scale of the inverse of the CSA width (roughly 50 μs). Once the C_{60}^{6-} ions are static, the individual carbon nuclei are no longer all magnetically equivalent. The NMR lineshape in Fig. 1e is then a superposition of CSA powder patterns from inequivalent, but similar, carbon sites. Refinement of the x-ray diffraction data for K_3C_{60} and K_6C_{60} indicates that the data is better fit in both cases by structures in which C_{60} ions are orientationally ordered, rather than rotationally averaged (6, 7). Our NMR lineshapes then imply that the dynamics of C_{60}^{3-} ions are better described by jump motion among symmetry-equivalent orientations, as observed in the orientationally ordered phase of undoped C_{60} (13, 14), rather than by continuous rotational diffusion.

The two-phase behavior of K_xC_{60} with $0 < x < 3$ can be understood in terms of electrostatic forces. Pure, undoped C_{60} adopts a structure in which C_{60} molecules are positioned with their centers of mass on a face-centered-cubic (fcc) lattice (12). In K_3C_{60}, K^+ ions occupy the one octahedral

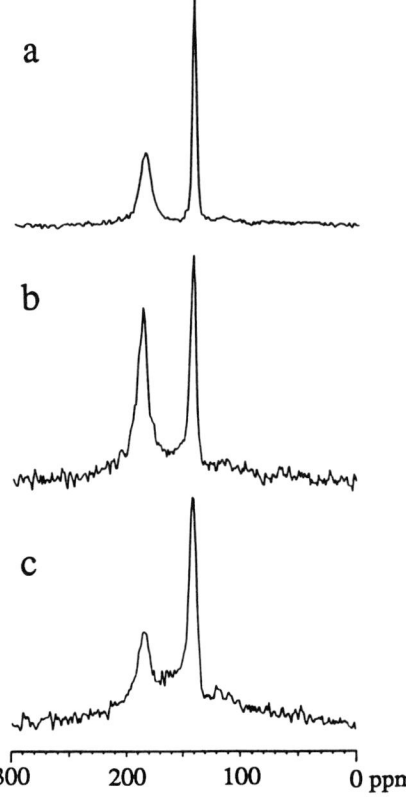

Fig. 3. ^{13}C NMR spectra of $K_{1.5}C_{60}$ after initial preparation (**a**), an additional 24 hours of annealing at 350°C (**b**), and an additional 72 hours at 400°C (**c**).

and the two tetrahedral holes per C_{60}^{3-} ion in the fcc lattice (6). Assuming that the ions act as point charges, we calculate a total electrostatic (Madelung) energy ME_3 of −22.4 eV per K_3C_{60} unit. The electrostatic energies of a putative K_1C_{60} structure, in which K^+ ions occupy only the octahedral holes and only C_{60}^- ions are present, and a putative K_2C_{60} structure, in which K^+ ions occupy only the tetrahedral holes and only C_{60}^{2-} ions are present, are $ME_1 = -3.5$ eV and $ME_2 = -11.8$ eV per K_1C_{60} and K_2C_{60}, respectively. In addition to the Madelung energies, we must take into account the electron-electron repulsions on individual C_{60} ions. We take the electron affinity of an isolated C_{60} neutral molecule to be $EA = 2.6$ eV, as determined experimentally (15). We assume that a second electron is bound by $EA - RE_2$, where RE_2 is the electron-electron repulsive energy per C_{60}^{2-} ion. Taking for RE_2 the minimum energy of two point charges on a sphere, RE_2 has the value $e^2/2R = 2.1$ eV ($R = 3.5$ Å, the radius of a C_{60} molecule). This gives a net binding energy of 0.5 eV for the second electron. Similarly, the interaction energy of a third electron is assumed to be $-EA + RE_3 - RE_2$, where RE_3 is the repulsive energy per C_{60}^{3-} ion. We assume that RE_3 has the value $\sqrt{3}e^2/R = 7.1$ eV, corresponding to the minimum energy of three point charges on a sphere. With this assumption, addition of a third electron is energetically unfavorable by 2.5 eV. Even with this strongly repulsive value for the energy of placing a third electron on C_{60}, electrostatic energies favor the K_3C_{60} phase. The electrostatic energy of disproportionation of K_1C_{60} into one-third K_3C_{60} and two-thirds C_{60} is $(ME_3 - 3ME_1 + RE_3)/3 = -1.6$ eV. The energy of disproportionation of K_2C_{60} into two-thirds K_3C_{60} and one-third C_{60} is $(2ME_3 - 3ME_2 + 2RE_3 - 3RE_2)/3 = -0.4$ eV.

Relaxation time (T_1) measurements were made on the K_3C_{60} line in sample II between 133 K and 363 K (Fig. 4). For $T \leq 273$ K, the T_1 data can be fit to the form $T_1T = \kappa$, as expected for a metal (16), with $\kappa = 140$ K-s. This result provides experimental evidence that molecular orbitals on C_{60} molecules contribute to the conduction band in K_3C_{60}, as anticipated, and that the normal state of K_3C_{60} is a Fermi liquid. Above 273 K, T_1T is reduced. We also see an increase in the NMR linewidth, from 15 ppm to 100 ppm, with decreasing temperature in the range from 300 K to 220 K. We therefore attribute the reduction in T_1T above 273 K to a contribution to the spin-lattice relaxation rate from rotation of C_{60}^{3-} ions (9, 13), which vanishes when the rotational motion freezes out and the line broadens. Using $\kappa = 140$ K-s and the standard

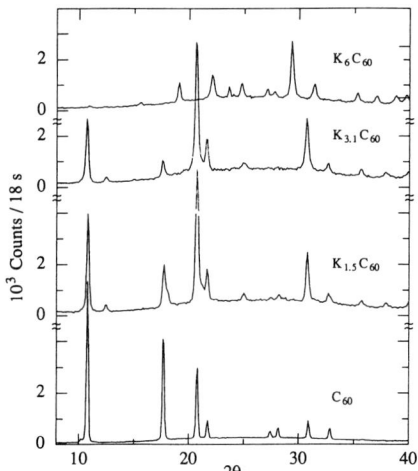

Fig. 2. X-ray powder diffraction patterns (Cu K_α radiation) for K_xC_{60} with indicated nominal stoichiometries.

886

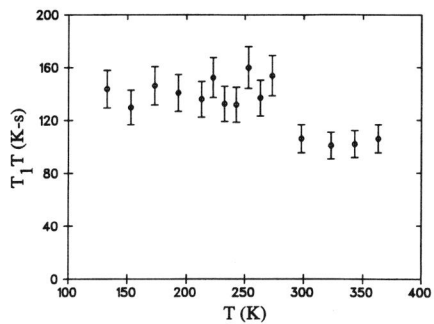

Fig. 4. Temperature dependence of the ^{13}C spin-lattice relaxation time T_1 in K_3C_{60}, plotted as T_1T versus T. Error bars are standard deviations estimated from five T_2 measurements at 298 K.

expression for spin-lattice relaxation by a contact hyperfine interaction (16),

$$\frac{1}{\kappa} = \frac{64}{9}\pi^3 k\hbar^3 \gamma_e^2 \gamma_n^2 |\Psi(0)|^4 \rho^2(E_f) \quad (1)$$

where γ_e and γ_n are the gyromagnetic ratios of the electron and the ^{13}C nucleus, we calculate the local electron density of states at the Fermi level and at a carbon nucleus in the normal state of K_3C_{60} to be $|\Psi(0)|^2\rho(E_f) = 7.6 \times 10^{24}$ eV^{-1} cm^{-3} per C_{60}^{3-} ion. We estimate $|\Psi(0)|^2$, the average electron density at a carbon nucleus for an orbital at the Fermi level (normalized to 1 for each C_{60} molecule), to be of order 4×10^{23} cm^{-3}. This value is determined from unrestricted Hartree-Fock calculations of the ratio of the unpaired spin density in carbon $2s$ orbitals in C_{60}^{3-} to that in planar methyl radical (17) and from the measured isotropic ^{13}C hyperfine coupling in methyl radical (18). $2s$-$2p$ hybridization (19) associated with the nonplanarity of C_{60} and core polarization make comparable contributions to $|\Psi(0)|^2$. We then derive a density of states at the Fermi level $\rho(E_f) = 20$ eV^{-1} per C_{60}^{3-} ion.

We interpret the 43 ppm downfield shift of the K_3C_{60} ^{13}C NMR resonance relative to the C_{60} resonance to be primarily a Knight shift attributable to hyperfine coupling between ^{13}C nuclei and conduction electron spins. A downfield shift of only 14 ppm relative to neutral C_{60} has been observed for a diamagnetic C_{60} anion of unknown charge in solution (20), suggesting a small orbital contribution to the shift in K_3C_{60}. With $\kappa = 140$ K-s, the Korringa relation for a Fermi gas of noninteracting electrons predicts a ^{13}C Knight shift of 170 ppm. The discrepancy between the observed and predicted shifts raises the possibility that orbital couplings (21) make a substantial contribution to the relaxation rate.

In conclusion, our ^{13}C NMR spectra of K_xC_{60} provide strong evidence for phase separation in material with $0 < x < 3$. The NMR lineshapes demonstrate the presence of rapid, large amplitude molecular reorien-

tations at room temperature in K_3C_{60} and the absence of such reorientations in K_6C_{60}. NMR relaxation measurements indicate the central role played by C_{60}^{3-} ions in the conductivity of K_3C_{60} and suggest that ^{13}C NMR will be an important probe of the superconducting state of K_3C_{60} and other alkali fullerides. Further low temperature measurements are in progress.

Note added in proof: We find the ^{13}C NMR T_1T to be 69 ± 8 K-s in Rb_3C_{60} between 213 K and 344 K, indicating that $\rho(E_f)$ is 40% larger than K_3C_{60}.

REFERENCES AND NOTES

1. R. C. Haddon et al., Nature **350**, 320 (1991).
2. A. F. Hebard et al., ibid., p. 600.
3. M. J. Rosseinsky et al., Phys. Rev. Lett. **66**, 2830 (1991).
4. H. W. Kroto, J. R. Heath, S. C. O'Brien, R. F. Curl, R. E. Smalley, Nature **318**, 162 (1985).
5. W. Krätschmer, L. D. Lamb, K. Fostiropoulos, D. R. Huffman, ibid. **347**, 354 (1990).
6. P. W. Stephens, et al., ibid. **351**, 632 (1991).
7. O. Zhou et al., ibid., p. 462.
8. K. Holczer, Science **252**, 1154 (1991).
9. C. S. Yannoni, R. D. Johnson, G. Meijer, D. S. Bethune, J. R. Salem, J. Phys. Chem. **95**, 9 (1991).
10. R. Tycko et al., ibid., p. 95.
11. R. Taylor, J. P. Hare, A. K. Abdul-Sada, H. W. Kroto, J. Chem. Soc. Chem. Comm. **1990**, 1423 (1990).
12. R. M. Fleming et al., Mat. Res. Soc. Proc., in press.
13. R. Tycko et al., Phys. Rev. Lett., in press.
14. P. A. Heiney et al., ibid. **66**, 2911 (1991).
15. R. E. Haufler et al., Chem. Phys. Lett. **179**, 449 (1991).
16. C. P. Slichter, Principles of Magnetic Resonance (Springer-Verlag, New York, ed. 3, 1990).
17. K. Raghavachari, unpublished results.
18. R. W. Fessenden, J. Phys. Chem. **71**, 74 (1967).
19. R. C. Haddon, J. Am. Chem. Soc. **109**, 1676 (1987).
20. J. W. Bausch et al., ibid. **113**, 3205 (1991).
21. J. Winter, Magnetic Resonance in Metals (Oxford Univ. Press, London, 1971).
22. We thank S. M. Zahurak, A. V. Makhija, and R. C. Haddon for supplying purified C_{60}, M. F. Needels for providing a program for Madelung calculations, and R. C. Haddon, K. Raghavachari, R. E. Walstedt, and M. Schluter for valuable suggestions.

15 July 1991; accepted 26 July 1991

Reprinted with permission from Nature
Vol. 352, pp. 787–788, 27 August 1991
© 1991 Macmillan Magazines Limited

Relation of structure and superconducting transition temperatures in A_3C_{60}

R. M. Fleming, A. P. Ramirez, M. J. Rosseinsky,
D. W. Murphy, R. C. Haddon, S. M. Zahurak
& A. V. Makhija

AT&T Bell Laboratories, Murray Hill, New Jersey 07974, USA

THE discovery of conductivity[1] in A_xC_{60} (where A represents an alkali metal) and superconductivity[2] in K_xC_{60} has been followed by reports of superconductivity in other alkali-metal-doped fullerides with transition temperatures as high as 33 K (ref. 3). Elucidation of phase diagrams and understanding the relationship between structure and superconducting properties is essential to a detailed understanding of superconductivity in these systems. So far, structural data have been reported only for the non-conducting, intercalated body-centred cubic (b.c.c.) structures A_6C_{60} (where A is K or Cs; ref. 4), a body-centred tetragonal structure for A_4C_{60} (where A is K, Rb, Cs; ref. 5) and the superconducting, intercalated face-centred cubic (f.c.c.) material K_3C_{60} (ref. 6). Here we report the preparation of a series of single-phase, isostructural f.c.c. superconductors with composition A_3C_{60} (where A is K, Rb, Cs or a mixture of these), and show that T_c increases monotonically with the size of the unit cell. Extended Hückel band-structure calculations also show a monotonic increase in the density of states at

the Fermi level, $N(E_F)$, with lattice parameter. The primary implication of these results is that all A_3C_{60} superconductors have the same structure and that changes in T_c can be accounted for by changes in $N(E_F)$.

A_xC_{60} samples were synthesized by reaction of C_{60} with stoichiometric quantities of alkali metals. The quantity of alkali metals was controlled volumetrically by cutting lengths of alkali-metal-filled capillary tubes in a dry box. The reagents were sealed in evacuated pyrex tubes and heated for two days at 225 °C, after which no alkali metal was visible in the tube. Annealing from 2 to 25 days while gradually increasing the temperature was necessary to attain equilibrium. Typical final annealing temperatures were 350 °C for samples containing potassium and 400 °C for other samples. Pelletizing the Rb_3C_{60} and Cs_3C_{60} enhanced the rate of the solid-state reaction, indicating that intergrain diffusion is important at 400 °C. The progress of the reactions was monitored by measuring the superconducting volume fraction (that is, the diamagnetic d.c. shielding) and the amount of f.c.c. phase present (by X-ray powder diffraction).

Representative Cu Kα X-ray powder patterns for A_xC_{60} are shown in Fig. 1. The pattern for K_3C_{60} is in good agreement with that of Stephens et al.[6]. The other A_3C_{60} patterns are also readily indexed with isostructural f.c.c. cells. The relative intensities of the diffraction peaks are in agreement with those calculated on the basis of a spherical shell of charge for the C_{60} and occupation by the appropriate alkali-metal ions of the octahedral and tetrahedral sites. All A_3C_{60} samples shown in Fig. 1 are essentially single phase, with only minor inclusions of a second phase which we have identified as A_4C_{60} (ref. 5). We have not attempted to investigate possible ordering between octahedral and tetrahedral sites in the mixed alkali-metal systems, although one expects a preference of the larger A atom for the octahedral sites. In Table 1 we summarize the superconducting volume fraction, transition temperature T_c, and lattice parameter for the single-phase superconducting compositions studied, as well as a calculated density of states at the Fermi level, $N(E_F)$. The latter is discussed in more detail below. Of note is the absence both of superconductivity and of any diffraction evidence for an intercalated f.c.c. Cs_xC_{60}. We infer

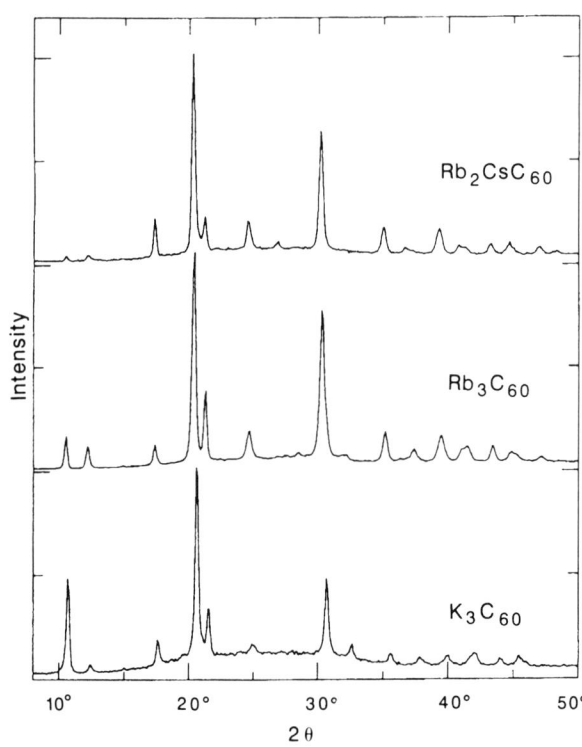

FIG. 1 X-ray powder patterns for K_3C_{60}, Rb_3C_{60} and Rb_2CsC_{60}.

from this that there is steric hindrance of incorporation of Cs, into the relatively small tetrahedral sites of the f.c.c. structure.

The superconducting transition temperatures of the A_3C_{60} compositions increase monotonically as the unit-cell size increases (Fig. 2). (Because of the small range of lattice parameters of the superconducting compositions, there is little change in the shape of the curve if the data are plotted as a function of volume rather than lattice parameter.) The range of T_cs observed for A_3C_{60} as a function of cell size is considerable, but not unusually large if we take $(\Delta T_c/T_c)/(\Delta V/V)$ as a relevant scaling factor. For A_3C_{60} this factor has a value of about 8. For comparison, both the organic β-$(ET)_2X$ compounds[7] and the inorganic $Ba_{1-x}K_xBiO_3$ system[8] have values near 22. The scaling factor is of the order of 1,000 for copper oxide superconductors[9].

If the A_3C_{60} compounds have the same compressibility as C_{60} (refs 10, 11), these results predict a pressure dependence of T_c on the order of 2 K kbar^{-1}. The experimental values[12,13] are about a third of this, perhaps reflecting a stiffening of the structure on alkali-metal intercalation. If we assume that correlation effects are negligible and that the T_c variation both in the pressure studies and in our volume studies arises from the same microscopic origin, then it is possible to estimate a compressibility for A_3C_{60}. Taking $\partial T_c/\partial P = -0.63$ to -0.78 K kbar^{-1} (refs 12, 13) and $\partial T_c/\partial V = 0.088$ K Å^{-3}, we obtain an isothermal compressibility of $(1/V) \times [\partial V/\partial P]_T \approx 2.5 \times 10^{-3}$ to 3.1×10^{-3} kbar^{-1}, compared with 5.5×10^{-1} kbar^{-1} for pristine C_{60} (ref. 11).

We speculated previously[14] that superconductivity in K_3C_{60} and Rb_3C_{60} resulted from electron pairing mediated by high-frequency intramolecular phonons of the C_{60} molecule, with the change in T_c between K_3C_{60} and Rb_3C_{60} explained by a change of the density of states at the Fermi level by about 10%. In this weak-coupling BCS model, the variation of T_c is given by

$$T_c = \omega_{ph}\, e^{-1/VN(E_F)}$$

where ω_{ph} is the energy of the relevant phonons (here 1,000–2,000 K)[1] and V is the electron–phonon coupling strength. We

TABLE 1 Superconducting and structural parameters for A_3C_{60}

A_3*	f.c.c. a_0 (Å)	$N(E_F)$ (states per eV per C_{60})	T_c (K)	Flux exclusion (d.c. %)
Rb_2Cs	14.493 (2)	23.9	31.30	48.0
Rb_3	14.436 (2)	22.9	29.40	55.0
Rb_2K (#2)	14.364 (5)	21.7	26.40	32.0
Rb_2K (#1)	14.336 (1)	21.7	24.40	33.5
$Rb_{1.5}K_{1.5}$	14.341 (1)	21.3	25.1	47.0
K_2Rb	14.299 (2)	20.7	21.80	22.0
K_3	14.253 (3)	20.1	19.28	28.6

* Nominal alkali metal composition. Estimated uncertainty $\pm 10\%$.

have estimated $N(E_F)$ using extended Hückel band-structure calculations. (A similar calculation for C_{60} was reported earlier[15].) We used experimental lattice constants and calculated the density of states of each structure by using the two special points[16] of the f.c.c. Bravais lattice. This approximation overestimates $N(E_F)$ by $\sim 50\%$, but should be adequate to allow comparisons among the isostructural A_3C_{60} compounds. The results are included in Table 1. The 10% variation of $N(E_F)$ required to produce the change in T_c is easily accounted for by the calculated values. This is in contrast to organic superconductors such as β-$(ET)_2X$, for which similar calculations show that $N(E_F)$ is essentially constant[17]. Our results imply a negligible variation in the phonon frequency and electron–phonon coupling among the various A_3C_{60} compounds. This would be consistent with the idea of C_{60} intramolecular phonon modes being important in A_3C_{60} superconductivity, or with intermolecular modes that are weakly affected by the observed changes in lattice parameter. In either case, the density of states seems to be controlling changes in T_c.

Note added in proof. We have found that samples of nominal composition $Cs_{1+x}Rb_{2-x}C_{60}$ for $0 < x < 1.25$ have T_cs of 32.5 K, in agreement with ref. 3. In addition, X-ray diffraction shows these compositions to be multiphase with the f.c.c. fraction having a lattice parameter $a = 14.506$ Å, a value consistent with other members of this isostructural series.

Direct reaction of Cs and C_{60} has not resulted in superconductivity or the formation of an intercalated f.c.c. phase in our experiments. A recent report[18] of superconductivity at 30 K in Cs_xC_{60} synthesized by reaction of C_{60} with CsM_2 alloys (where M is Hg, Tl or Bi) indicates a T_c lower than we would predict for f.c.c. Cs_3C_{60}. These materials probably adopt a different structure or contain a ternary intercalant. □

Received 23 July; accepted 2 August 1991.

1. Haddon, R. C. et al. Nature **350**, 320–322 (1991).
2. Hebard, A. F. et al. Nature **350**, 600–601 (1991).
3. Tanigaki, K. et al. Nature **352**, 222–223 (1991).
4. Zhou, O. et al. Nature **351**, 462–464 (1991).
5. Fleming, R. M. et al. Nature (this issue).
6. Stephens, P. W. et al. Nature **351**, 632–634 (1991).
7. Williams, J. M. et al. Physica B**135**, 520 (1985).
8. Hinks, D. G. et al. Nature **333**, 836–838 (1988).
9. Whangbo, M.-H., Kang, D. B. & Torardi, C. C. Physica C **158**, 371–376 (1989).
10. Fischer, J. E. et al. Science **252**, 1288–1290 (1991).
11. Duclos, S. J. et al. Nature **351**, 380–382 (1991).
12. Schirber, J. E. et al. Physica C **178**, 137–139 (1991).
13. Sparn, G. et al. Science **252**, 1829–1831 (1991).
14. Rosseinsky, M. J. et al. Phys. Rev. Lett. **66**, 2830–2832 (1991).
15. Haddon, R. C. et al. in Carbon Clusters (American Chemical Society, Washington DC, in the press).
16. Chadi, D. J. & Cohen, M. L. Phys. Rev. B **8**, 5747–5753 (1973).
17. Whangbo, M.-H. et al. J. Am. chem. Soc. **109**, 90–94 (1987).
18. Kelty, S. P., Chen, C.-C. & Lieber, C. M. Nature **352**, 223–225 (1991).

ACKNOWLEDGEMENTS. We acknowledge helpful discussions with G. Dabbagh, S. J. Duclos, A. F. Hebard, S. H. Glarum, A. J. Muller and R. Tycko.

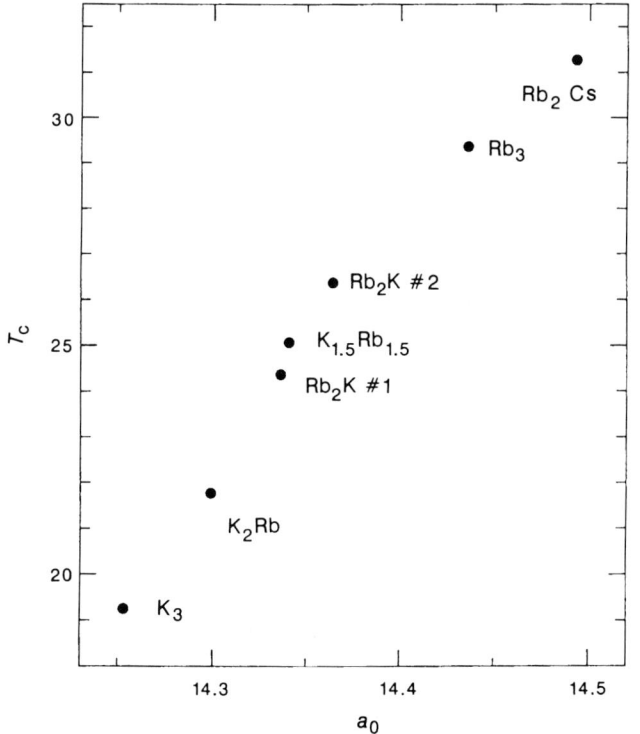

FIG. 2 Variation of the superconducting transition temperature T_c (K) with lattice parameter a_0 (Å) for various compositions of A_3C_{60}.

Reprinted with permission from Science
Vol. 254, pp. 989–992, 15 November 1991
© 1991 The American Association for The Advancement of Science

Superconductivity in the Fullerenes

C. M. Varma, J. Zaanen, K. Raghavachari

Intramolecular vibrations strongly scatter electrons near the Fermi-surface in doped fullerenes. A simple expression for the electron-phonon coupling parameters for this case is derived and evaluated by quantum-chemical calculations. The observed superconducting transition temperatures and their variation with lattice constants can be understood on this basis. To test the ideas and calculations presented here, we predict that high frequency H_g modes acquire a width of about 20% of their frequency in superconductive fullerenes, and soften by about 5% compared to the insulating fullerenes.

THE EXCITING DISCOVERY OF SUPER-conductivity in metallic fullerenes (1) leads us to inquire whether the classic mechanism for superconductivity, namely, effective electron-electron attraction via the interaction of electrons with vibrations of the ions, is applicable here as well. Associated with this is the question of whether the direct electron-electron repulsion in Fullerenes can suppress conventional singlet pairing. In this paper we exploit the special nature of cluster compounds to derive a particularly simple expression for electron-vibrational coupling from which parameters of the superconducting state of fullerenes are easily calculated. Further, we present arguments why the effective repulsions in fullerenes are no different than in conventional metals.

The lattice vibrations couple to the electronic states of metallic fullerenes in two ways: by causing fluctuations in the hopping rate of electrons from one molecule to the other and by causing fluctuations in the electronic structure of a single molecule. The covalent interactions that split the molecular states, which form the bands in the metallic state, are over an order of magnitude larger than the inter-molecular covalency. This is reflected in the intra-molecular splitting $W_{intra} \simeq 20$ eV (2, 3) compared to the width of the t_{1u} bands which is $W_{inter} \simeq 0.6$ eV (4, 5). The electron-vibration coupling is known to be proportional to such covalent splittings (6). Therefore, in the problem of the fullerenes, one needs to consider only the intra-molecular vibration coupling. The same argument rationalizes why the electron-vibrational coupling may be much larger in doped fullerenes than in doped graphite. In the latter, the orbitals near the Fermi-energy are π bonded. The Fullerenes have a significantly larger relevant bandwidth, because of σ admixture due to the non-planar local geometry, and therefore a stronger

AT&T Bell Laboratories, Murray Hill, NJ 07974.

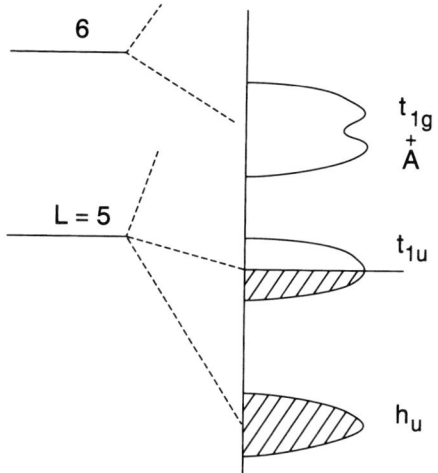

Fig. 1. Artists conception of the electronic structure of the fullerenes, as inferred from photoemission spectroscopies and electronic structure calculations. A denotes the alkali s level.

electron-phonon coupling (5).

Photoemission studies (7), in combination with electronic structure calculations (2–4), have revealed a clear picture of the electronic structure of the fullerenes. The deviation from spherical symmetry of the icosahedra affects the 11 $L = 5$ spherical harmonic states (3), so that they split into a fivefold degenerate h_u and two threefold degenerate t_{1u} and t_{2u} manifolds, respectively. As shown in Fig. 1, the h_u states lie ~2 eV below the t_{1u} states and are occupied. The $L = 6$–derived t_{1g} states lie ~1 eV above the t_{1u} states, and the latter are occupied on doping. The unoccupied alkali s states are roughly degenerate with the t_{1g} states (7). So one may confine attention simply to the threefold degenerate t_{1u} states. These transform as x, y, and z, and so their degeneracy is split by any quadrupolar deformation that makes the cartesian axes inequivalent. With respect to superconductivity these quadrupolar Jahn-Teller modes, having H_g symmetry, are the only relevant ones. This follows from the fact that in icosahedral symmetry

$$t_{1u} \times t_{1u} = A_g + T_{1g} + H_g \quad (1)$$

The T_{1g} mode is asymmetric so that it cannot couple, and the A_g modes do not lift the degeneracy (although they change the local energy level).

The vibrational modes of the C_{60} molecules have been calculated. There are eight such H_g modes, each of which is fivefold degenerate. So of the 174 vibrational modes of C_{60}, 40 can, in a static fashion (Jahn-Teller effect) or, more importantly, in a dynamic fashion (dynamic Jahn-Teller effect) affect the t_{1u} electronic states. Let us now consider the two limiting cases. The intramolecular processes can either lead to a pairing energy much larger than the intermolecular transfer integrals, or the reverse is true. The latter is found more appropriate for the fullerenes, but the former is interesting to discuss first because of its conceptual simplicity. Consider two molecules with average charge $\bar{n}$. The Hamiltonian for the coupled problem is a 3×3 matrix with elements

$$H_{ij} = \bar{E}\delta_{ij} + \sum_{m,\mu} h_{ij}(m,\mu)Q_{m,\mu} + H_{vib} \quad (2)$$

where $i = 1, 2, 3$ labels the degenerate states of t_{1u} symmetry, $Q_{m,\mu}$ are the normal coordinates of the m-th H_g mode with degeneracy $\mu = 1...5$. H_{vib} is the Hamiltonian of the vibrational modes with frequencies ω_m. This Jahn-Teller problem, involving a threefold degenerate electronic state interacting with a fivefold degenerate mode, has been worked out some time ago (8) and the coupling matrix is found to be

$$\frac{1}{2}g_m \begin{pmatrix} Q_{m,5} - \sqrt{3}Q_{m,4} & -\sqrt{3}Q_{m,1} & -\sqrt{3}Q_{m,2} \\ -\sqrt{3}Q_{m,1} & Q_{m,5} + \sqrt{3}Q_{m,4} & -\sqrt{3}Q_{m,3} \\ -\sqrt{3}Q_{m,2} & -\sqrt{3}Q_{m,3} & -2Q_{m,5} \end{pmatrix} \quad (3)$$

where g_m is the characteristic energy per unit displacement of the m-th mode, to be evaluated by microscopic calculations.

In the strong-coupling limit, the t_{1u} levels will split as indicated in Fig. 2 left for one phase of the vibration and Fig. 2 right for the other phase (8). Now if we calculate the energy reduction in this limit (neglecting the zero-point motion) and deduce the effective electron-electron interaction through

$$U_{\bar{n}} = E_{\bar{n}+1} + E_{\bar{n}-1} - 2E_{\bar{n}} \quad (4)$$

where $E_{\bar{n}}$ is the ground state energy for

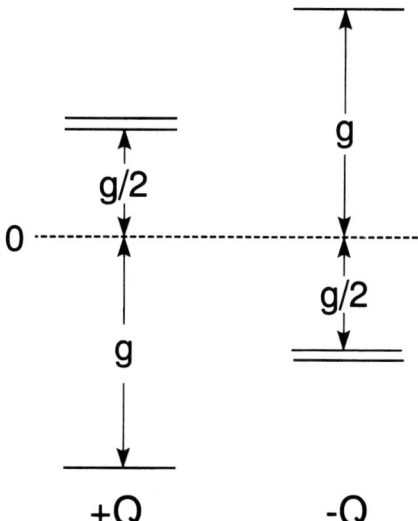

Fig. 2. The splitting pattern of the t_{1u} level by the H_g distortion for the two different phases of the radial displacement Q.

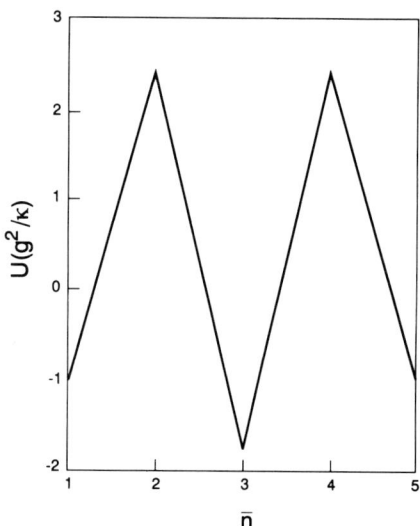

Fig. 3. The local electron-electron interaction (U) as a function of particle number ($\bar{n}$), induced by the Jahn-Teller interactions (κ is the spring constant). A negative U is found for an average occupation ($\bar{n}$) of 3 electrons, and a positive U for 2 or 4 electrons.

average charge n, we find the behavior as in Fig. 3 (9). The actual relevant electron-electron interaction parameter is in this limit, of course $\tilde{U}_{\bar{n}} = U_{\bar{n}} + U_C$, where $U_C \simeq e^2/\varepsilon R \simeq 0.5$ to 1 eV ($\varepsilon \sim 4$) comes from direct electron-electron interactions. $U_{\bar{n}}$ and U_C have in general different high frequency cutoffs.

This picture, which could yield an effective attractive interaction, would be a good starting point for further discussion if $-\tilde{U}_{\bar{n}} >> W$, the conduction electron bandwidth (and the phonon frequency). One would then have local pairing (10) and a superconducting transition temperature of $O(W^2/|\tilde{U}|)$, provided a static Jahn-Teller distortion with a possible phase difference from one molecule to the other, although a simple staggered charge density wave is frustrated in the fcc-lattice (11) were not to occur. For $|U_{\bar{n}}| \leq W$, as we find below, the above adiabatic picture is not valid—an electron runs away from a molecule before it develops pairing correlation with another electron on the same molecule. Then the physics we described above should be incorporated into the traditional way of considering electron-phonon scattering in metallic bands (6).

The 40 intramolecular H_g modes are expected to be nearly dispersionless in the solid state. For this case, the electron-phonon Hamiltonian is particularly simple, and may be written as

$$H_{e-ph} = \sum_{\kappa k,\kappa'k',\sigma,m,\mu} h_{\kappa k,\kappa'k'}(m,\mu)c^+_{\kappa k\sigma}c_{\kappa'k'\sigma}$$
$$\times (Q_{m\mu,k-k'} + Q^+_{m\mu,-k}) \quad (5)$$

Here the scattering matrix element

$$h_{\kappa\mathbf{k},\kappa'\mathbf{k}'}(m,\mu) = \sum_{ij} A^*_{\kappa\mathbf{k}i} h_{ij}(m\mu) A_{\kappa'\mathbf{k}'j} \quad (6)$$

where $A_{\kappa\mathbf{k}i}$ are the elements of the linear transformation from molecular levels to band states—$\mathbf{k}$ is the momentum and κ are the band indices.

For s-wave superconductivity, in weak to intermediate coupling, one is interested in time-reversed states near the Fermi-energy. So one may confine attention to intra-band pairing. For the present case, the dimensionless electron-phonon coupling constant λ has a particularly simple form, which is easily derived from the general expressions (12, 13)

$$\lambda = \sum_{m\mu} \frac{N(0)}{M\omega_m^2} I_{m\mu}^2 \quad (7)$$

$$I_{m\mu}^2 = N(0)^{-2} \sum_{\kappa,\kappa'} \int \frac{dS_{\mathbf{k},\kappa}}{v_{\mathbf{k},\kappa}}$$

$$\times \int \frac{dS_{\mathbf{k}',\kappa'}}{v_{\mathbf{k}',\kappa'}} |h_{\kappa\mathbf{k},\kappa'\mathbf{k}'}(m,\mu)|^2 \quad (8)$$

where the integrals are over the Fermi-surface. Using the fact that the degeneracy of the t_{1u} levels is not lifted in a cubic environment (14) we find

$$\lambda = \frac{5}{6} \sum_m \frac{N(0)}{M\omega_m^2} g_m^2 \quad (9)$$

We can now calculate the superconductive transition temperature T_c through the approximate solution of the Eliashberg equation, given the intra-molecular deformation potentials g_m. These, and the vibrational frequencies, were calculated using the quantum-chemical MNDO semi-empirical technique (15). This method has been successfully used previously on a wide variety of

Table 1. Experimental (20) and calculated frequencies ($\omega_m^{\exp}$, ω_m, in cm^{-1}) and deformation potentials (g_m, in eV/Å), and the corresponding coupling constants λ_m for $N(0) = 10$ per spin per eV per C_{60} molecule for the eight H_g modes of a C_{60}^- molecule.

$\omega_m^{\exp}$	273	437	710	744	1099	1250	1428	1575
ω_m	263	447	711	924	1260	1406	1596	1721
g_m	0.1	0.1	0.2	0.0	0.6	0.2	1.8	1.2
λ_m	0.03	0.01	0.01	0.0	0.06	0.0	0.34	0.11

carbon compounds containing five- and six-membered rings and is known to be reliable for structures, relative energies and vibrational frequencies (16). For example, the MNDO bond-lengths in C_{60} 1.47 and 1.40 Å (17) are in excellent agreement with the best theoretical (18) and experimental (19) estimates. We evaluated the complete matrix of force constants and calculated the associated normal coordinates for all 174 vibrations in pure C_{60}. In Table 1 the results for the eight Raman active H_g mode frequencies are shown which have a mean deviation of only 10% from experiment (20) (Table 1). We calculated the deformation potentials g_m by a frozen-phonon technique. The energy of a C_{60}^- molecule will depend on the amplitude of a frozen-in $Q_{m,5}$ photon as $E = -g_m Q_{m,5} + \kappa_m Q_{m,5}^2/2$ (κ_m is the spring constant in Eqs. 2 and 3). For each of the eight H_g phonons we selected this component ($\sim 3z^2 - r^2$), and we distorted the negatively charged C_{60} molecule along these normal coordinates. The initial slope of the energy as a function of the distortion amplitude yields then the deformation potentials.

In Table 1 the results for the deformation potentials are summarized, together with the coupling constants for the individual modes λ_m, as calculated from Eqs. 7 and 9 with the observed phonon frequencies in C_{60} and the phonon renormalization calculated below. We find that the two highest modes near 1428 and 1575 cm^{-1} are the most strongly coupled. The reason is that these high-lying modes involve bond-stretching, compared to the bond-bending characterizing the lower frequency modes. The former leads to the maximum change in energy for a given (normalized) distortion.

Assuming a square density of states and a bandwidth, which is ~0.6 eV according to band structure calculations, $N(0)$ would be ~5 orbital states per electron volts per C_{60} molecule. However, the band structure calculations (21) give figures for $N(0)$ which are more than twice as large and experimentally $N_0 \sim 10$ to 20 [H_{c1}, H_{c2}, normal state susceptibility (22)]. From Table 1 we find that the overall $\lambda = \Sigma_m \lambda_m$ is in the range of ~ 0.3 to 0.9, and together with the large range in frequency for which the interactions are attractive, high T_c's are to be expected. The large electron-phonon cou-

pling also leads to a strong decrease in the phonon frequencies and a corresponding increase in their linewidth $= \gamma_m \omega_m$. It is known that (24)

$$\frac{\Delta\omega_m}{\omega_m} \simeq -\frac{\lambda_m}{5}; \frac{\gamma_m}{\omega_m} \simeq \pi N(0)|\Delta\omega_m| \quad (10)$$

So we predict a diminution of about 5% in the frequency of the high H_g modes and a linewidth increase of order 300 cm^{-1}, compared to pure C_{60}.

We now present an approximate calculation of T_c. For $\lambda < 1$, McMillan (13) has given a very good (23) approximate solution to the Eliashberg equations (12)

$$T_c = \frac{\omega_{av}}{1.2} \exp\left[-\frac{1.04(1+\lambda)}{\lambda - \mu^*(1 + 0.62\lambda)}\right] \quad (11)$$

Most of the coupling strength is in the two highest lying modes, so the usual average of the phonon frequencies (23) ($\omega_{\log}$) is not appropriate. We find

$$\omega_{av} = \exp\left\{\sum_m \frac{\lambda_m \ln[\omega_m(1-\lambda_m)]}{\lambda}\right\} \quad (12)$$

a better approximation. Eq. 11 includes the Coulomb pseudo-potential parameter μ^*. Because the approximations such as those due to Migdal (25) do not work for electron-electron interactions, it is impossible to estimate μ^* (in contrast to λ). Traditionally μ^* is used as a fitting parameter in comparing T_c, $\Delta(0)$ and the tunneling spectra to theory. For instance, for Pb $\mu^* = 0.12$ (23). μ^* for a Fermi-level in a well-isolated band is smaller than the screened repulsion parameter μ by a factor $[1 + \mu \ln (W/\omega_{av})]$ (26), where W is the smaller of E_F and ω_p, the plasma frequency. E_F/ω_{av} in the Fullerenes is smaller than, say in Pb, by $\sim10^2$. However, if one notes that the actual electronic structure of metallic C_{60} is a ladder of bands of width ~1 eV, spread out over 20 eV, and separated from each other by energies also of order 1 eV, and considers the calculation of μ^* in this situation, one concludes that μ^* is close in value to that of a wide band metal.

We present in Fig. 4 T_c versus $N(0)$ for various values of μ^* (27). T_c has a particularly simple relation to $N(0)$ and therefore to the nearest-neighbor C_{60} distance d in the Fullerenes, because the other factors are

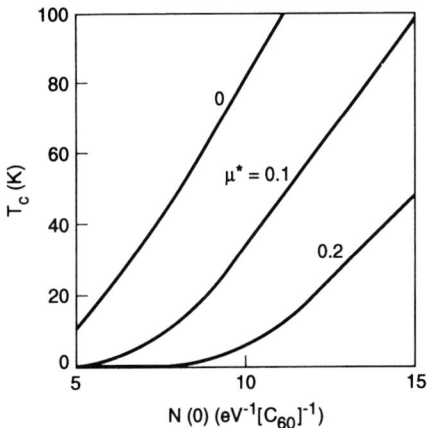

Fig. 4. T_c (in kelvin) as a function of physically likely electronic densities of orbital states $N(0)$ and Coulomb pseudopotentials (μ^*).

intramolecular and do not depend on d. Figure 4 shows that $T_c \sim N(0)$, in agreement with a recent compilation of lattice constants, calculated densities of states and T_c's (21). Given the physical fact we used that most of the coupling is intramolecular, our estimate of λ from the H_g modes should be as good as the determination of vibration frequencies, that is, good to about 10%. One worry is that our calculation of g_m is based on the deformations of a C_{60}^- molecule, whereas the more appropriate calculation would have a neutralizing background. The Migdal approximation for determining T_c is only good in our case to $(\omega_{av}/E_F) \sim 1/5$.

For low density of states obtainable by small doping, we expect the Coulomb interactions to dominate. In that case the intramolecular Hund's rule coupling (owing to orbital degeneracy) plus the almost empty band usually favors ferromagnetism (28). This may be the simple reason for the recent observation of ferromagnetism in the compound $TDAE_1C_{60}$ (29).

Note added in proof: In recent Raman measurements Duclos et al. (30) find that the two highest frequency H_g phonons, which we find couple most strongly to the electrons (Table 1), are clearly seen in C_{60} and K_6C_{60} but disappear in the superconducting compound K_3C_{60}. This is consistent with our prediction based on Eq. 10 for their linewidth.

REFERENCES AND NOTES

1. A. F. Hebard et al., *Nature* **350**, 600 (1991); M. J. Rosseinsky et al., *Phys. Rev. Lett.* **66**, 2830 (1991).
2. R. C. Haddon, L. E. Brus, K. Raghavachari, *Chem. Phys. Lett.* **125**, 459 (1986); M. Ozaki and A. Takahashi, *ibid.* **127**, 242 (1986).
3. S. Satpathi, *ibid.* **130**, 545 (1986).
4. S. Saito and A. Oshiyama, *Phys. Rev. Lett.* **66**, 2637 (1991).
5. J. L. Martins, N. Trouillier, M. Schnabel, preprint (University of Minnepolis, Minnesota, 1991).
6. S. Barisic, J. Labbé, J. Friedel, *Phys. Rev. Lett.* **25**, 419 (1970); C. M. Varma, E. I. Blount, P. Vashista, W. Weber, *Phys. Rev. B* **19**, 6130 (1979); C. M. Varma and W. Weber, *ibid.*, p. 6142.
7. C.-T. Chen et al., *Nature*, in press; J. H. Weaver et al., *Phys. Rev. Lett.* **66**, 1741 (1991); M. B. Jost et al., *Phys. Rev. B*, in press.
8. M. C. O'Brien, *J. Phys. C: Solid State Phys.* **4**, 2524 (1971).
9. In fact, in strong coupling one should consider the true n-particle states instead of simple product states as assumed in the text. This leads to an increase of phonon phase space, which however affects the 2, 3, and 4 particle states in similar ways. Note that in this case, the low spin states would be considered, and Jahn-Teller interactions thus give rise to a negative Hund's rule coupling.
10. C. M. Varma, *Phys. Rev. Lett.* **61**, 2713 (1988); R. Micnas, J. Ranninger, S. Robaskiewicz, *Rev. Mod. Phys.* **62**, 113 (1990).
11. F. C. Zhang, M. Ogata, T. M. Rice, preprint (University of Cincinnati, Cincinnati, OH, 1991). This paper discusses the possibility that the alkali-ion vibrations may lead to an intramolecular electron-electron attraction. We find this unlikely because they screen their repulsive interactions without screening the attractive interactions. Besides, such vibrations have a frequency

of $O(200 \text{ cm}^{-1})$, much smaller than the molecular vibrations we consider and thus are unlikely to be of importance in determining T_c.
12. G. M. Eliashberg, *Zh. Eksp. Teor. Fiz* **38**, 966 (1960); *ibid.* **39**, 1437 (1960) *Sov. Phys.-JETP* **11**, 696 (1960); *ibid.* **12**, 1000 (1961).
13. W. L. McMillan, *Phys. Rev.* **167**, 331 (1968).
14. In the derivation of Eq. 9 we missed the factor 5/6 which was kindly pointed out to us by M. Lannoo.
15. M. J. S. Dewar and W. Thiel, *J. Am. Chem. Soc.* **99**, 4899 (1977); *ibid.*, p. 4907.
16. M. D. Newton and R. E. Stanton, *ibid.* **108**, 2469 (1986).
17. R. E. Stanton and M. D. Newton, *J. Phys. Chem.* **92**, 2141 (1988).
18. M. Haesen, J. Almlof, G. E. Scuseria, *Chem. Phys. Lett.* **181**, 497 (1991); K. Raghavachari and C. M. Rohlfing, *J. Phys. Chem.*, in press.
19. C. S. Yannoni et al., paper presented at the Materials Research Society Meeting, Boston, 1990.
20. D. S. Bethune, *Chem. Phys. Lett.* **179**, 181 (1991).
21. R. M. Fleming, *Nature* **352**, 787 (1991).
22. A. P. Ramirez, M. J. Rosseinsky, D. W. Murphy, R. C. Haddon, in preparation.
23. P. B. Allen and R. C. Dynes, *Phys. Rev. B* **12**, 905 (1975).
24. C. O. Rodriguez et al., *Phys. Rev. B* **42**, 2692 (1990).
25. A. B. Migdal, *Zh. Eksperim. i Teor. Fiz.* **34**, 1438 (1958); *Sov. Phys.-JETP* **7**, 996 (1958).
26. P. Morel and P. W. Anderson, *Phys. Rev.* **125**, 1263 (1962); P. W. Anderson, in preparation.
27. An independent calculation of the electron-phonon

coupling in the Fullerenes has been carried out by M. Schluter et al. (in preparation) using a tight-binding method. Our results differ in some ways from theirs. They find the major contribution from the $\sim 400 \text{ cm}^{-1}$ and $\sim 1600 \text{ cm}^{-1}$ H_g phonons, where the latter contribution is about half of the former. In their calculation of T_c they use an expression averaging over all the phonon frequencies irrespective of their coupling constants.
28. We envisage a competition between Hund's rule couplings and phonon-induced attractive interactions. We note the interesting possibility discussed by S. Chakravarty and S. Kivelson (*Europhys. Lett.*, in press) that in a Hubbard model for C_{60} clusters, the effective electron-electron interaction is repulsive for small U; that is, the Hund's rule is obeyed, but for larger U it may change sign due to configuration interactions.
29. P. M. Allemand et al., *Science* **253**, 301 (1991).
30. S. J. Duclos, R. C. Haddon, S. Glarum, A. F. Hebard, K. B. Lyons, in preparation.
31. We thank B. Batlogg, R. C. Haddon, S. J. Duclos, I. I. Mazin, M. F. Needels, T. T. M. Palstra, A. P. Ramirez, and M. Schluter for helpful discussions. We particularly thank M. Lannoo for carefully going through the manuscript and pointing out several compensating numerical errors. J.Z. acknowledges financial support by the Foundation of Fundamental Research on Matter (FOM), which is sponsored by the Netherlands Organization for the Advancement of Pure Research (ZWO).

23 August 1991; accepted 3 October 1991

Reprinted with permission from Physical Review Letters
Vol. 68, No. 4, pp. 526–529, 27 January 1992

Electron-Phonon Coupling and Superconductivity in Alkali-Intercalated C_{60} Solid

M. Schluter, M. Lannoo,[a] M. Needels, and G. A. Baraff

AT&T Bell Laboratories, Murray Hill, New Jersey 07974

D. Tománek

*Department of Physics and Astronomy and Center for Fundamental Materials Research,
Michigan State University, East Lansing, Michigan 48824-1116*
(Received 16 August 1991)

We propose that superconductivity in A_3C_{60} ($A = $ K, Rb) with $T_c \geq 30$ K results from a favorable combination of high phonon frequencies and the existence of two different energy scales optimizing the coupling constant $\lambda = NV$. Calculations show that electron scattering V is dominated by particular on-ball Jahn-Teller–type modes on the scale of the large on-ball π-hopping energy, while the density of states N is controlled by the weak interball hopping energy. This factorization has several observed experimental consequences. Crucial differences to intercalated graphite explain the much smaller T_c values in the graphite compounds.

PACS numbers: 74.20.−z, 71.25.−s, 71.38.+i, 71.45.Nt

Recently, superconductivity has been observed in crystalline "electron-doped" A_3C_{60} "fullerite" compounds ($A = $ K, Rb, Cs), with superconducting transition temperature T_c exceeding 30 K [1]. Several theoretical models have been proposed to account for this observation [2–7]. Here we report a detailed study of the electron and phonon states and the coupling between them. In particular, we carried out a comparative study of intercalated fullerite and of graphite intercalation compounds (GIC) which exhibit a significantly lower T_c value ($T_c \approx 1$ K). This comparison helps us to identify possible coupling channels which could be responsible for the high T_c in C_{60} compounds.

In our study we concentrate on C_{60}, neglecting the direct influence of the alkali atoms, an approximation which we will justify. We find that the coupling V in C_{60} is dominated by on-ball modes of H_g and A_g symmetry. The coupling is about evenly distributed over lower frequency, predominantly radial modes and higher frequency, predominantly tangential modes. The strength of the coupling can be evaluated from Jahn-Teller–type considerations and amounts to about 40 meV per C_{60}. The scale of this value of V is set by the large on-ball π-electron hopping matrix elements. In fact, V is enhanced by the finite curvature of C_{60} which allows for even stronger σ-hopping admixture. These results for an isolated C_{60} are only slightly modified ($< 10\%$) when C_{60} is placed into a weakly coupled fullerite lattice. For the electron-phonon coupling strength λ, this energy V is combined with the conduction-electron density of states, the scale of which is set by the weak interball hopping. Using average values for $N(\varepsilon_F) \approx 15$ states/eV-spin-C_{60}, estimated from band-structure calculations for fcc C_{60} and from a variety of experiments (see below), we obtain $\lambda \approx 0.6$, which is well within the range of what is needed for $T_c \approx 30$ K. To strengthen the argument we compare the electron-phonon coupling in fullerite with intercalated graphite. Our study shows two main differences between

these compounds: (i) The graphite modes equivalent to the lower-frequency buckling-type modes of C_{60} do *not* couple at all to π electrons at the Fermi surface of graphite to first order, and (ii) the higher-frequency tangential modes couple less efficiently. This results from geometry: The finite curvature of C_{60} allows for finite σ-π admixtures. As a consequence, λ and T_c are much smaller in the GIC. This leaves us with a rather unique situation in C_{60} where λ is factorized into an intramolecular quantity V and an intermolecular quantity N. Several experimental observations, to be discussed below, support this picture.

We begin our studies with a local-density-approximation– (LDA–) density-functional calculation for fcc C_{60} which essentially confirms what is known about the electronic structure [8]. Of importance here are the following points: (i) The conduction-band states are derived from a threefold-degenerate t_{1u} level of C_{60} which broadens into a ~ 0.5-eV-wide band. (ii) The t_{1u}-derived conduction states are predominantly π states, centered at the carbon atoms and pointing nearly radially outwards. There is some finite (a few percent) s,p_σ admixture due to the finite curvature of C_{60}. We can continue our investigation for pure fcc C_{60}, assuming that the three alkali-metal electrons of A_3C_{60} are donated into the t_{1u} band. Reference to calculations by Martins and Troullier [8] for K_3C_{60} indicates only small hybridization effects, which makes our model essentially correct.

We then proceed with a semiempirical (s, p_x, p_y, p_z) tight-binding (ETB) approach, the parameters of which were fitted to a large LDA data base of carbon molecules and solid structures. Details of this Hamiltonian are published elsewhere [9]. For C_{60} its predictions agree well with LDA results (e.g., the overall bandwidth, the density of states, and the symmetry of states near the gap). With this Hamiltonian we not only calculate band structures, but we also obtain the approximate deformation potentials for electronic states. The additional ingredient here is a d^{-n} scaling of all hopping matrix elements with interatomic distance d. Tests against LDA results indicate $n \approx 2$–3. Typical values for on-ball nearest-neighbor atomic deformation potentials range between ~ 4 eV/Å for $pp\pi$ up to ~ 10 eV/Å for $pp\sigma$, which includes the effects of electronic screening. Interball hopping is only approximately described in our ETB approach. To reproduce the LDA t_{1u}-derived bandwidth the individual interball hopping matrix elements can be appropriately scaled and become about a factor of 5 smaller than the corresponding on-ball hopping matrix elements. The corresponding deformation potentials due to interball vibrations are consequently also reduced by this factor. To calculate the electron-phonon coupling strength V we can, therefore, neglect interball hopping to first order. However, the conduction-band width, the Fermi surface, and the exact shape of the density of states near ε_F do depend on details of interball hopping. Because of uncertainties in the relative rotational arrangements of C_{60} molecules these quantities cannot be calculated reliably at present, and we will therefore consider reasonable ranges of possible values of densities of states (see below).

To determine the vibrational modes of the system we use an extension of the simple Keating model [10]. We start with the isolated C_{60} for which we use two nearest-neighbor on-ball elastic constants α,β with a bond-

stretching to bond-bending ratio ranging from 3:1 to 10:1 that covers the range appropriate for carbon. For comparison we also use the results of independent bond-charge model calculations [11]. Results for modes relevant for electron-phonon coupling will be given in Table I. The on-ball modes are distributed in a somewhat bimodal fashion. Modes with predominantly radial displacements are at the lower end of the spectrum, while the high-frequency modes are characterized by tangential displacements. There is some analogy to graphite, where the optical layer stretching modes are near 1600 cm^{-1}, while the buckling modes occupy the lower end of the spectrum. To test the magnitudes of interball scattering, we did add more empirical spring constants to model the ultralow-frequency librational modes in the 10-cm^{-1} regime, the C_{60} interball optical modes in the 100-cm^{-1} regime, and the alkali optical modes in the 100-cm^{-1} regime. The resulting contributions are small and will be neglected.

The dimensionless electron-phonon coupling constant used in the theory of superconductivity is given by [12]

$$\lambda = N(\varepsilon_F)V = N(\varepsilon_F)\sum_v \frac{\langle\langle I_v^2 \rangle\rangle}{M\omega_v^2}, \quad (1)$$

where the sum runs over all vibrational modes v of the system, and where the double brackets denote a double Fermi-surface average over (k,k') of the quantity

$$I_v^{ij}(k,k') = \sum_{\substack{\tau,l \\ \tau',l'}} c_{\tau l}^i(k)^* c_{\tau',l'}^j(k') \sum_R \mathbf{I}\{\mathbf{u}_\tau(v,q)e^{ik'R} - \mathbf{u}_{\tau'}(v,q)e^{ikR}\}\delta(k-k'-q), \quad (2)$$

where $q = k - k'$ is the phonon wave vector. The sum over $\{\tau,l\},\{\tau',l'\}$ runs over all orbitals (l) and sites (τ) of the electronic Hamiltonian, and $c_{\tau l}^i(k)$ are the eigenvector components for state i at wave vector k. The vibrational eigenvectors $\mathbf{u}_\tau(v,q)$ for moving the atom τ along $\{x,y,z\}$ are multiplied by the intrinsic electron-phonon coupling matrix elements between the individual orbitals

$$\mathbf{I} \equiv \mathbf{I}_{\tau l,\tau,R+\tau'l'} = \mathbf{V}_\tau \langle \phi_l(r-\tau)|H|\phi_{l'}(r-R-\tau')\rangle, \quad (3)$$

where the gradient is taken with regard to atom position τ. The normalization $\sum_\tau^{C_{60}}|\mathbf{u}_\tau|^2 = 1$ requires that the density of states in Eq. (1) be normalized to states per eV, per spin, and per C_{60} formula unit.

Because of the weak coupling between balls, we can carry out the double Fermi-surface integral in Eq. (1) analytically in the limit of $t_{inter}/t_{intra} \to 0$. $\langle\langle I_v^2 \rangle\rangle$ in Eq. (1) then becomes $\frac{1}{9}\sum_{i,j}^3 |I_v^{ij}(0)|^2$. This involves only on-ball electron-phonon coupling and V can, therefore, be related to the Jahn-Teller problem of a negatively charged C_{60} cluster, details of which are given elsewhere [13]. Group theory tells us that only fivefold-degenerate H_g modes and onefold-degenerate A_g modes can couple to the t_{1u} electronic states. For these symmetries, the contributions to V in Eq. (1) are $\frac{10}{6}E_{JT}(H_g)$ and $\frac{2}{3} \times E_{JT}(A_g)$, respectively [13], where E_{JT} is the energy lowering due to distortions of C_{60} induced by one added electron.

Using the ETB wave functions and the different phonon models, we calculate [14] $V \approx 40$ meV per C_{60}. De-

TABLE I. Experimental [15] and calculated H_g and A_g phonon frequencies ω_v (in cm^{-1}). Results are given for three different phonon models (see text). The individual mode-coupling constants V_v (in meV), the total coupling strength V [Eq. (1)], and the weighted (logarithmic) average phonon frequency $\omega_{\log}$ (in cm^{-1}) [Eq. (4)] are also given [14].

	ω_v^{expt}	Bond charge (Ref. [11])		Keating $\beta/\alpha=0.3$		Keating $\beta/\alpha=0.1$	
		ω_v	V_v	ω_v	V_v	ω_v	V_v
$H_g(1)$	273	271	3.0	298	4.0	250	0.0
$H_g(2)$	437	410	2.4	411	4.0	347	2.2
$H_g(3)$	710	718	6.0	621	11.0	444	20.0
$H_g(4)$	744	793	4.8	766	4.6	774	1.6
$H_g(5)$	1099	1157	0.0	1162	0.0	1145	0.0
$H_g(6)$	1250	1218	0.6	1226	0.0	1299	0.0
$H_g(7)$	1428	1452	7.0	1500	4.2	1662	0.8
$H_g(8)$	1575	1691	8.4	1718	10.0	1718	13.0
$A_g(1)$	497	499	0.0	476	0.0	492	0.0
$A_g(2)$	1469	1455	5.0	1452	5.2	1678	4.0
V			37.2		43.0		41.6
$\omega_{\log}$		995		874		797	

tailed couplings to the eight H_g modes and two A_g modes are given in Table I, together with the corresponding phonon frequencies. For comparison, experimental [15] Raman mode frequencies are also given. The coupling values for a given phonon model have also been tested by selective LDA frozen phonon calculations. Details will be given in a later publication. For all phonon models there is appreciable coupling both to the lower-frequency buckling modes as well as to the higher-frequency tangential modes. This is in contrast to the results of Ref. [7], where ~80% of the coupling was attributed to the two highest H_g modes. The $q = k - k'$ dependence of the scattering is generally not important since the strength is given by the relatively dispersionless on-ball vibrations. For the H_g Jahn-Teller modes the scattering is dominated by the interband terms (within t_{1u}) including $q = 0$. For the A_g symmetric modes, $q = 0$ scattering is zero since it corresponds to a coherent overall shift of all electronic levels. For finite wave vector $q \neq 0$, A_g mode scattering (interball) is finite, with its scale again given by the on-ball coupling.

The question of the size of $N(\varepsilon_F)$ in Eq. (1) is largely unsettled at this point. Estimates range from values of 1–2 states/eV-spin-C_{60} derived from photoemission data [16], to values of ~6–20 derived from band-structure calculations [8], up to values of 10–15 and > 20 inferred from susceptibility [17] and NMR [18] data, respectively. If we assume an average value of $N(\varepsilon_F) \approx 15$, we arrive at values of $\lambda \approx 0.6$ for the electron-phonon coupling strength. The average temperature $\hbar\omega_{\log}/k_B$ appearing in McMillan's [19] formula for T_c,

$$T_c = \frac{\hbar\omega_{\log}}{1.2 k_B} \exp\left(\frac{-1.04(1+\lambda)}{\lambda - \mu^* - 0.62\lambda\mu^*}\right), \tag{4}$$

can be evaluated for each of the phonon models (see Table I) and is found to be very high, of the order of 1150 to 1450 K. This, when combined with a $\mu^* \approx 0.1$–0.2 [6,20], yields T_c values (i.e., ranging from 5 to 35 K) within the range of the experimentally observed values of T_c.

In order to establish the validity of our model in view of the typical, sizable uncertainties in λ, we attempt to explain the large and qualitative difference between superconductivity in the intercalated C_{60} solid and graphite intercalation compounds. Published values [21] for $N(\varepsilon_F)$ in GIC are similar to the lower range of values for C_{60} (normalized per atom). The key difference is found in the electron-phonon coupling strength. Since the graphite sheets are flat, the lower-energy buckling-type modes do *not* couple in first order to the π electrons near ε_F. The electron-phonon coupling in graphite is solely caused by the high-frequency optic modes, augmented by weak hybridization with intercalant orbitals and by interlayer effects. We calculate λ using the identical procedure as for C_{60} and find it to be reduced by about a factor of 5 as compared to C_{60}.

We conclude on the basis of these calculations that the observed superconductivity in alkali-intercalated C_{60} compounds can be understood in terms of electron-phonon coupling. The main ingredients are strongly scattering on-ball modes set by the on-ball hopping energy scale, a density of states at ε_F set by the low interball hopping energy scale, and a high average phonon frequency reflecting the light carbon mass combined with stiff on-ball modes. The result is a factorization of $\lambda = NV$ into interball and intraball quantities. This simple picture is beautifully confirmed by several experimental observations. First, for a given compound, T_c decreases drastically with increasing pressure [22], which can be explained by the large compressibility of fullerite, leaving individual C_{60} molecules and therefore $V, \hbar\omega_{\log}$ largely unchanged, but resulting in a decreased density of states N with decreasing interball distance. Second, the observed [23] increase in T_c with increasing alkali-intercalant size again supports the same density-of-states argument. In fact, these and the pressure experiments can be explained rather quantitatively assuming simply that $\lambda = NV$ with $N \sim t_{\text{inter}}^{-1} \sim d^n$ ($n \approx 2$–3) as is commonly done for p-electron overlaps. Then, the same values of (λ, μ^*) needed to explain the absolute value of T_c also describe the variation of T_c between K_3C_{60} and Rb_2CsC_{60}, solely on the basis of interball distance (d) variation [23]. A further confirmation of the picture can be found in the apparent disappearance of on-ball Raman phonon lines with metallic intercalation [24], which was first pointed out by Varma, Zaanen, and Raghavachari [7]. The strong on-ball electron-phonon coupling V yields an increased phonon linewidth for selected modes, calculated by us to be of order 5%–10% of the phonon frequency which should wash out most of their spectral features. Individual widths can be extracted from Table I. For $q = 0$, only H_g modes should be broadened, as is clearly seen in Raman scattering [23]. However, for finite q, A_g modes should also be affected. This should be observable in neutron or two-photon Raman scattering experiments.

Finally, we would like to caution that the scenario developed here does not allow T_c to be increased much further. One is approaching the limit where the electron kinetic energy is so small that it becomes comparable with the average phonon energy $\hbar\omega_{\log}$, and Migdal's approximation for calculating T_c breaks down [25]. Furthermore, as the lattice constant of intercalated fullerite is increased, one is approaching the limit where the decreasing interball bandwidth becomes comparable with on-ball Coulomb interactions. At this point the effective one-electron picture used here breaks down and magnetic instabilities are expected to occur.

We thank W. Zhong and Y. Wang for assistance with numerical calculations, S. Duclos, R. Tycko, R. C. Haddon, and A. P. Ramirez for the discussion of their measurements, P. B. Littlewood, M. Grabow, J. C. Tully, S. Shastri, J. Zaanen, C. M. Varma, K. Raghavachari, and A. J. Millis for useful discussions. We would like to thank in particular G. Onida for providing us with the results of his phonon calculations prior to publication. One of us (D.T.) acknowledges financial support by the National Science Foundation, Grant No. PHY-8920927.

(a)Permanent address: Institut Supérieur d'Electronique du Nord, Lille, France.

[1] M. J. Rosseinsky, A. P. Ramirez, S. H. Glarum, D. W. Murphy, R. C. Haddon, A. F. Hebard, T. T. M. Palstra, A. R. Kortan, S. M. Zahurak, and A. V. Makhija, Phys. Rev. Lett. **66**, 2830 (1991); A. F. Hebard, M. J. Rosseinsky, R. C. Haddon, D. W. Murphy, S. H. Glarum, T. T. M. Palstra, A. P. Ramirez, and A. P. Kortan, Nature (London) **352**, 222 (1991).

[2] S. Chakravarty, M. P. Gelfand, and M. P. Kivelson, Science **254**, 970 (1991).

[3] J. L. Martins, N. Troullier, and M. Schabel (to be published).

[4] G. Baskaran and E. Tosatti, Curr. Sci. **61**, 33 (1991).

[5] T. C. Zhang, M. Ogato, and T. M. Rice, Phys. Rev. Lett. **67**, 3452 (1991).

[6] I. I. Mazin, S. N. Rashkeev, V. P. Antropov, O. Jepsen, A. I. Liechtenstein, and O. K. Anderson (to be published).

[7] C. M. Varma, J. Zaanen, and K. Raghavachari, Science **254**, 989 (1991).

[8] R. C. Haddon, L. E. Brus, and K. Raghavachari, Chem. Phys. Lett. **125**, 459 (1986); J. H. Weaver, J. L. Martins, T. Komeda, Y. Chen, T. R. Ohno, G. H. Kroll, and T. Troullier, Phys. Rev. Lett. **66**, 1741 (1991); Q. Zhang, J. Y. Yi, and J. Berhnolc, Phys. Rev. Lett. **66**, 2633 (1991); S. Saito and A. Oshiyama, Phys. Rev. Lett. **66**, 2637 (1991); B. P. Feuston, W. Andreoni, M. Parinello, and E. Clementi, Phys. Rev. B **44**, 4056 (1991); J. L. Martins and N. Troullier (to be published).

[9] D. Tomanek and M. Schluter, Phys. Rev. Lett. **67**, 2331 (1991); (to be published).

[10] R. M. Martin, Phys. Rev. B **1**, 4005 (1970).

[11] G. Onida and G. Benedeck, Europhys. Lett. (to be published).

[12] C. M. Varma, E. I. Blount, P. Vashista, and W. Weber, Phys. Rev. B **19**, 6130 (1979).

[13] M. Lannoo, G. A. Baraff, M. Schluter, and D. Tomanek, Phys. Rev. B **44**, 1210 (1991).

[14] The values in Table I are obtained assuming an $n = 2$ distance exponent in the hopping matrix elements. Using $n = 3$ instead would increase these values by about a factor of 2.

[15] D. S. Bethune, G. Meijer, W. C. Tang, J. H. Rosen, W. G. Golden, H. Seki, C. A. Brown, and M. S. deVries, Chem. Phys. Lett. **179**, 181 (1991).

[16] C. T. Chen, L. H. Tjeng, P. Rudolf, G. Meigs, J. E. Rowe, J. Chen. J. P. McCauley, A. B. Smith, A. R. McGhie, W. J. Romanow, and E. Plummer, Nature (London) **352**, 603 (1991).

[17] A. P. Ramirez, M. J. Rosseinsky, D. W. Murphy, and R. C. Haddon (to be published).

[18] R. Tycko, G. Dabbagh, M. J. Rosseinsky, D. W. Murphy, R. M. Fleming, A. P. Ramirez, and J. C. Tully, Science **253**, 884 (1991).

[19] W. L. McMillan, Phys. Rev. **167**, 331 (1968).

[20] A detailed calculation of μ^* has not been done. However, one can estimate $\mu = NU_c$ to be of order 0.5–1.0 with $U_c \approx$ few eV for carbon orbitals. The renormalization of μ to $\mu^* = \mu/[1 + \mu \ln(\omega_c/\omega_{ph})]$ is drastic, considering a phonon energy $\omega_{ph} \approx 0.2$ eV and the overall Coulomb scattering energy scale $\omega_c \approx 5$–10 eV of the carbon σ, π bands in fcc C_{60} which involves both on-ball and interball polarizations. Strong σ, π plasmons have been observed recently near 28 and 6 eV, respectively [G. Gensterblum et al., Phys. Rev. Lett. **67**, 2171 (1991)].

[21] J. E. Fisher, in *Intercalated Layer Materials*, edited by F. A. Levy (Reidel, Dordrecht, 1979), p. 481.

[22] G. Sporn, J. D. Thompson, S. M. Huang, R. B. Kaner, F. Diederich, R. L. Whetten, G. Gruner, and K. Holczer, Science **252**, 1829 (1991); J. E. Schirber, D. L. Overmyer, H. H. Wang, J. M. Williams, K. D. Carlson, A. M. Kini, M. J. Pellin, N. Welp, and W. K. Kwok, Physica (Amsterdam) **178C**, 137 (1991); O. Zhou, G. B. M. Vaughan, Q. Zhu, J. E. Fisher, P. A. Heiney, N. Coustel, J. P. McCauley, Jr., and A. B. Smith, III (to be published).

[23] R. M. Fleming, A. P. Ramirez, M. J. Rosseinsky, D. W. Murphy, R. C. Haddon, S. M. Zahurak, and A. V. Makhija, Nature (London) **352**, 787 (1991).

[24] S. Duclos, R. C. Haddon, S. Glarum, A. F. Hebard, and K. B. Lyons, Science (to be published).

[25] M. Grabowski and L. J. Sham, Phys. Rev. B **29**, 6132 (1984).

Reprinted with permission from Science
Vol. 254, pp. 970–974, 15 November 1991

Electronic Correlation Effects and Superconductivity in Doped Fullerenes

Sudip Chakravarty, Martin P. Gelfand, Steven Kivelson

A theory of the electronic properties of doped fullerenes is proposed in which electronic correlation effects within single fullerene molecules play a central role, and qualitative predictions are made which, if verified, would support this hypothesis. Depending on the effective intrafullerene electron-electron repulsion and the interfullerene hopping amplitudes (which should depend on the dopant species, among other things), the calculations indicate the possibilities of singlet superconductivity and ferromagnetism.

Buckminsterfullerene, C_{60}, is a molecule of remarkable symmetry and intrinsic beauty; it would attract serious study for this reason alone (1). The recent discovery that in the solid state it can be readily doped and that it exhibits unexpected and fascinating electronic properties has greatly intensified the interest in this and related materials. In particular, when doped with alkali metals, fullerene crystals exhibit metallic conduction and superconductivity at unprecedentedly high temperatures for a molecular solid (2, 3). Most recently, it has been found that when doped with an organic counterion, the crystals become weak, presumably itinerant, ferromagnets (4).

Undoubtedly, the full elucidation of the properties of these materials will require a wide variety of ideas and methods. Nonetheless, we argue that it is ultimately the special electronic states on the scale of a single molecule that produce the novel physics of C_{60} (5). The many-electron states comprising the 60 π electrons of C_{60} provide a scale intermediate between the microscopic energetics at the level of a single carbon atom and the macroscopic scale of the molecular solid. We believe that these π electrons should not be treated within an effective single particle picture; electronic correlation effects are crucial. An interesting consequence of strong electronic correlation effects is the occurrence of an effective attraction between two added electrons (pair-binding), even though the microscopic Coulomb interactions are repulsive. We show that the phenomenon of pair-binding can give rise to superconductivity in doped C_{60}.

We first study the electronic spectrum of a single C_{60} molecule and then use the resulting low lying many-body states as the basis for an effective Hamiltonian that describes the physics at length scales larger than the size of the molecule. We begin with the simplest possible model of the one-electron spectrum and the simplest possible electron-electron interactions, known as the Hubbard model, but defined on the truncated icosahedral C_{60} lattice. The

The authors are in the Department of Physics, University of California at Los Angeles, Los Angeles, CA 90024–1547.

electron-electron interaction is represented in the Hubbard model as a short-ranged on-site interaction. This approach is reasonable as long as electronic screening is adequate as should be the case for doped C_{60} since it is metallic in the normal phase.

The Hubbard model. Let us consider the Hubbard model on a single C_{60} molecule. The Hamiltonian is

$$H = -t \sum_{\langle i,j\rangle\sigma}{}' (c_{i\sigma}^{\dagger}c_{j\sigma} + \text{h.c.}) - t' \sum_{\langle i,j\rangle\sigma}{}'' (c_{i\sigma}^{\dagger}c_{j\sigma} + \text{h.c.}) + \frac{U}{2} \sum_{i\sigma} n_{i\sigma}n_{i-\sigma} \tag{1}$$

where the primed sum runs over all distinct nearest-neighbor bonds on the pentagons and the double-primed sum runs over all distinct bonds that connect these pentagons (h.c. stands for hermitian conjugate). The fermion operator $c_{i\sigma}^{\dagger}$ creates an electron with spin σ on site i, and $n_{i\sigma} = c_{i\sigma}^{\dagger}c_{i\sigma}$ is the density of electrons of spin σ on site i. We shall assume, merely as a guide, that the nearest-neighbor hopping amplitude $t \sim 2$ to 3 eV and the on-site Coulomb interaction $U \sim 5$ to 10 eV, similar to the values known for polyacetylene (6). It is believed (7) that $1.0 < t'/t < 1.3$.

In the noninteracting limit, that is, $U = 0$, we simply recover Hückel theory and the electronic structure of the molecule is well known (8). The states can be labeled according to the irreducible representations of the icosahedral group. We shall not emphasize the labeling of the states according to icosahedral symmetry, although our calculations are fully consistent with it. Because of the nearly spherical shape of the molecule, it is simpler to label the states according to their approximate transformations under the operations of the full rotation group, that is, by their angular momentum.

The threefold degeneracy of the lowest lying unoccupied orbitals is an important property of the molecule. It can be thought of as though the lowest lying unoccupied orbital has "angular momentum" $L = 1$. In the noninteracting limit, the symmetry of the various states with one or more additional electrons in this orbital can be determined with the rules for addition of angular momentum (which in this case reproduce the composition laws of the representations of the icosahedral group). Because of the existence of a gap in the spectrum, the symmetry assignments must persist in the presence of interactions, at least for a finite range of U. Thus, we shall omit the quotation marks when we refer to the angular momentum L.

The pair-binding energy: the RVB picture. In this section we define the central concept of pair-binding and give an intuitive picture of its origin. This intuitive picture is firmly rooted in the resonating valence bond (RVB) picture of Pauling (9) and Anderson (10) and follows from the phenomenon of spin-charge separation. Spin-charge separation in a strongly correlated system implies that an electron is a composite object made up of two quasiparticles, a spinon which carries spin 1/2 and no charge and an eon which carries charge e but no spin (5, 10, 11).

Let us define Φ_0 to be the ground-state energy of the neutral molecule, and $\Phi_n \equiv \Phi_0 + E_n$ to be the ground-state energy of the charged molecule with n added electrons. Then the pair-binding energy, E_{pair}, is given by $E_{\text{pair}} \equiv 2E_1 - E_2 = 2\Phi_1 - \Phi_2 - \Phi_0$. Note that in our notation a positive E_{pair} implies that a pair is favored and not the converse.

From spin-charge separation, the energy to add one electron to a molecule is $E_1 = \Phi_1 - \Phi_0 = E_s + E_e + V_{es}$, where E_e (>0) is the eon creation energy, E_s (>0) is the spinon creation energy, and V_{es} is the spinon-eon interaction energy, which can be of either sign. If the eon-spinon interaction is repulsive, the system would minimize its energy by keeping them as far apart as possible; as a consequence, $V_{es} \approx 0$. If, on the other hand, the eon-spinon interaction is attractive, the eon and spinon would form a bound state with the same quantum numbers as an electron, although the state would have substantial internal structure (6). In this case, $V_{es} < 0$ and E_1 is the creation energy of a renormalized electron.

The energy to add two electrons in a spin singlet state is the energy to create two eons. For simplicity, we ignore the eon-eon and spinon-spinon interactions, because we imagine that C_{60} is large enough that the quasiparticles are weakly interacting. Then $E_2 = 2E_e$, so the pair-binding energy is $E_{\text{pair}} = 2(E_s + V_{es})$. The energy to add two electrons in the triplet state is the energy to add two eons and two spinons, plus the interactions between them, so the splitting between the singlet and the triplet state is also $\Delta_{\text{FM}} \equiv 2(E_s + V_{es})$. If $E_s + V_{es} < 0$, the pair-binding energy is negative and the triplet state of the doubly charged molecule is favored over the singlet, consistent with Hund's rule. However, when $E_s + V_{es} > 0$, pair-binding occurs and the singlet state has lower energy than the triplet state. Within this simple picture, there should be a single critical value of U/t at which both Δ_{FM} and E_{pair} change sign. Which of these two particular cases is realized depends on the values of E_e, E_s, and V_{es} which in turn depend on the microscopic parameters, such as U/t. A similar RVB picture has also been put forward by Baskaran and Tossatti (12), although there are quite significant differences between their theory and ours.

It is easy to extend this analysis to the case of more than two added electrons. We define generalized pair-binding energies ($i = 1$, 3, and 5) to be $E_{\text{pair}}^{(i)} = 2\Phi_i - \Phi_{i-1} - \Phi_{i+1}$, so that $E_{\text{pair}}^{(1)} \equiv E_{\text{pair}}$. We also define generalized splittings between the minimum and maximum spin states, for a given charge ($n = 2$, 3, and 4), to be $\Delta_{\text{FM}}^{(n)} = E_n(S_{\max}) - E_n(S_{\min})$, so that $\Delta_{\text{FM}}^{(2)} \equiv \Delta_{\text{FM}}$. (The low-energy states for $n = 0$, 1, 5, and 6 have unique values of the total spin.) A consequence of the simple quasiparticle picture presented above is that $\Delta_{\text{FM}}^{(n)} \approx E_{\text{pair}}^{(i)} \approx 2(E_s + V_{es})$, independent of i and n.

Perturbation theory. The Hubbard model defined on the truncated icosahedral lattice of C_{60} is a many-electron problem of considerable complexity; consequently, a direct numerical assault on this problem is likely to be unsuccessful. Faced with this dilemma we adopt the textbook approach of second-order perturbation theory in the Hubbard-U (13).

We first discuss the simple case of E_{pair}. In Fig. 1 we show the calculated E_{pair} for the singlet ($L = 0$, $S = 0$) and the triplet ($L = 1$, $S = 1$) states as a function of U/t. The ($L = 2$, $S = 0$) state is always an excited state for this problem. For $U > U_{\text{pair}} \approx 3t$, $E_{\text{pair}} > 0$ in the singlet state, which implies that it is energetically favorable to have a single molecule with two added electrons rather than two molecules with one electron each. The splitting between the singlet and the triplet states vanish at $U_{\text{FM}} \approx U_{\text{pair}}$. There is, however, a narrow region between U_{FM} and U_{pair} in which $E_{\text{pair}} < 0$, but still the singlet state is energetically preferred.

Because the concentration of donors in the superconducting fullerenes is three per molecule, we have calculated the energies Φ_n, hence generalized pair-binding energies for up to six added electrons. As mentioned earlier, we can classify the states according to the total angular momentun and spin. For six electrons, the unique state corresponds to ($L = 0$, $S = 0$); for five electrons, there are a total of 6 states corresponding to ($L = 1$, $S = 1/2$); for four electrons, there are a total of 15 states corresponding to the multiplets ($L = 2$, $S = 0$) (5 states), ($L = 1$, $S = 1$) (9 states), and ($L = 0$, $S = 0$) (1 state); for three electrons, there are 20 states corresponding to the multiplets ($L = 2$, $S = 1/2$) (10 states), ($L = 1$, $S = 1/2$) (6 states), and ($L = 0$, $S = 3/2$) (4 states); for two electrons there are 15 states which were enumerated above; for one electron there are 6 states corresponding to ($L = 1$, $S = 1/2$); the state of the neutral molecule is unique and corresponds to ($L = 0$, $S = 0$). Note that there is an approximate symmetry about three added electrons (where the shell is half-filled); thus, zero is analogous to six, one to five, and two to four. In Fig. 2 we have plotted the energy, E_n, versus the number of electrons, n, added to the neutral molecule for $U = 4t$ and $t'/t = 1.2$ (14). For zero, one, five, and six added electrons, the unperturbed ground-state is nondegenerate except for the degeneracy dictated by the total angular momentum and the total spin. For two, three, and four electrons there are level crossings at $U_{\text{FM}}^{(2)}$, $U_{\text{FM}}^{(3)}$, and $U_{\text{FM}}^{(4)}$, respectively, where the ground state changes from being in the maximal spin sector to the minimal spin sector upon increasing U.

The nature of pairing can be seen in the nonconvexity of E_n versus n in Fig. 2 and only takes place in the form $1 + 1 \rightarrow 0 + 2$ (meaning that it is more favorable to have two molecules with zero and two electrons each rather than two molecules with one electron each), $3 + 3 \rightarrow 2 + 4$, $5 + 5 \rightarrow 4 + 6$, with respective critical values $U_{\text{pair}}^{(1)}$, $U_{\text{pair}}^{(3)}$, $U_{\text{pair}}^{(5)}$. That there is no tendency for the electrons to bind in other ways, such as $3 + 3 \rightarrow 0 + 6$ or $1 + 1 + 1 + 1 \rightarrow 0 + 4$, can be seen from the convexity of the piecewise linear curve obtained by joining the points $n = 0$, 2, 4, and 6. It is perhaps not obvious in Fig. 2 that $E_{\text{pair}}^{(1)} \approx E_{\text{pair}}^{(3)} \approx E_{\text{pair}}^{(5)}$, so in Fig. 3 we plot all of the $E_{\text{pair}}^{(i)}$ versus U for $t'/t = 1.2$. One sees that $E_{\text{pair}}^{(1)}$ and $E_{\text{pair}}^{(5)}$ are quite close at all U, as expected from the approximate particle-hole symmetry within the triplet of lowest lying unoccupied orbitals of C_{60}. The behavior of $E_{\text{pair}}^{(3)}$ is rather remarkable: after decreasing more strongly

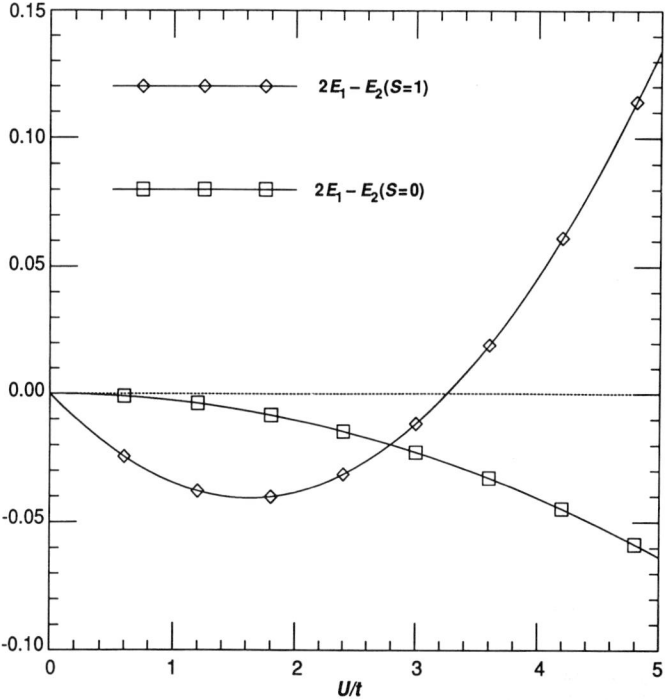

Fig. 1. The C_{60} singlet and triplet E_{pair} (in units of t) as functions of U/t for $t' = t$.

at small U than the other $E_{\text{pair}}^{(i)}$ it rejoins them at $U = U_{\text{FM}}^{(3)}$. We see that all of the $U_{\text{FM}}^{(n)}$ are clustered together and do not lie far from the values of U where the $E_{\text{pair}}^{(i)}$ cross zero. In addition, the typical magnitudes of the $\Delta_{\text{FM}}^{(n)}$ are similar to those of the $E_{\text{pair}}^{(i)}$, hence our RVB arguments above are justified nearly in full. Moreover, the symmetry of Fig. 2 implies that all else being unchanged, the physics should be similar for electron concentrations between two and four electrons per fullerene as it is between zero and two electrons per fullerene.

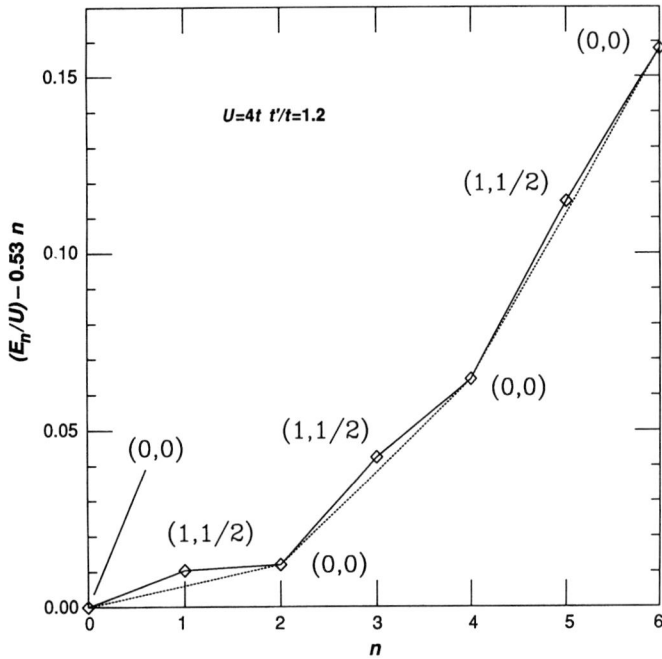

Fig. 2. The energy E_n (in units of t) versus n, for C_{60} at $U = 4t$, $t'/t = 1.2$. Note that a linear piece has been subtracted for clarity. The values of L and S for the lowest energy states at each n are indicated in the form of (L, S).

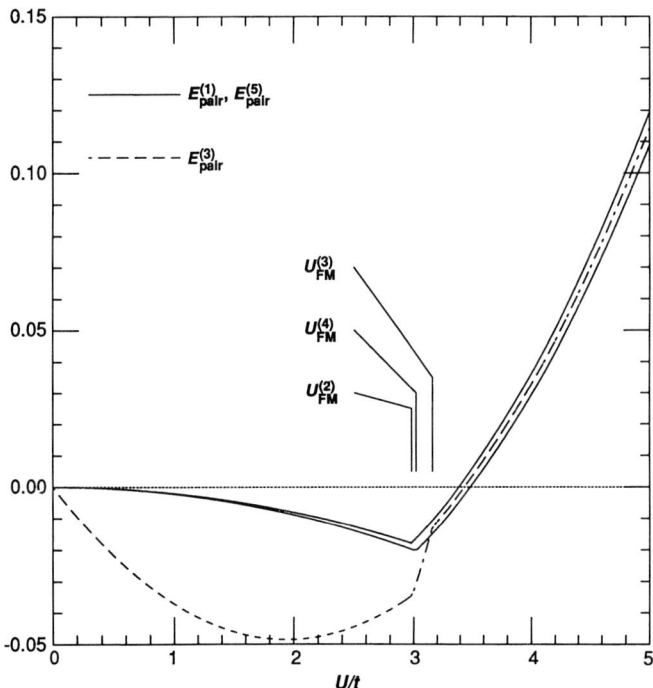

Fig. 3. Values for C_{60} of all of the $E_{\text{pair}}^{(i)}$ (in units of t) versus U/t at $t'/t = 1.2$. $E_{\text{pair}}^{(5)}$ is the lower of the two solid curves. Also indicated are the values of U/t for which the lowest energy states of the two-, three-, and four-charged molecule switch from the maximal to the minimal spin state as U increases.

Positive pair-binding energies, and hence the effective attractive interaction between electrons, are necessarily a core polarization effect. If all of the 60 valence electrons were treated as an inert Fermi sea, the net interaction between two electrons added to a given molecule would necessarily be repulsive; there is no bound-state solution to the Cooper problem with purely repulsive interactions. It is the dynamic interactions with the valence electrons that are crucial in producing overscreening of the purely bare repulsive interaction. Although the first-order term does not involve any virtual excitations, the second-order theory includes important core-polarization effects.

Finally, we note that the reliability of the second-order perturbation theory has been examined in considerable detail elsewhere (15). From exact numerical studies of small Hubbard clusters up to 12 sites we have concluded that indeed the pair-binding can be positive at an intermediate scale, that perturbation theory never produces a misleading sign for the pair-binding energy for relevant values of U smaller than the bandwidth, and that it produces a reasonable magnitude for the pair-binding energy up to values of U of order half the bandwidth. We are therefore quite confident that the perturbative results described above are at least qualitatively correct.

Long-ranged Coulomb interactions and metallic screening. It may appear that the attraction between two electrons that we have obtained is spurious, as we have neglected the effect of the long-range Coulomb repulsion between electrons. If a single molecule is considered in isolation in a dielectric medium, there would be a larger contribution to the energy than those included. In fact, adding electrons to a neutral molecule can be viewed as charging a capacitor. Although, strictly speaking such a macroscopic picture is not correct, it is a reasonable guide for understanding the large electrostatic Hartree contribution. The charging energy is given by $e^2 n^2/2C$, where the capacitance, C, is proportional to the diameter, l, of the molecule. If this contribution is added to the energies calculated above, the pair-binding phenomenon can disappear if the charging energy is sufficiently large.

It is important to note that even in the context of the Hubbard model, this charging energy is already present in a crude form; it is simply the first-order energy in powers of U, that is, the Hartree-Fock contribution to the pair-binding, which in the singlet sector of interest is always strongly repulsive.

The long-range Coulomb contribution to the interaction between electrons must be compared to the repulsion already present in our model, not to the much smaller pair-binding energy which is the difference between a large first-order repulsion and a large second-order attraction between electrons. In other words, we might try obtaining a better estimate of the Hartree charging energy within a model that includes long-range Coulomb interactions, but we must not add this energy to the energies we have already computed from the Hubbard model; rather we should replace the first-order energies with the more realistic estimates.

For instance, for the Hubbard model, the first-order contribution to E_{pair} is $-(1/20)U$ (5), which is ~ -0.5 eV for $U = 4t \sim 10$ eV. Thus, the question to be considered is whether this is a physically reasonable magnitude for the Hartree-Fock contribution to the pair-binding energy. If we compare this estimate with that obtained from the capacitive charging energy, where $C = (\epsilon_\infty l)/2$, $l \approx 8$ Å, and ϵ_∞ is the background dielectric constant (which we estimate to be 2 to 3), then the capacitive estimate of the pair-binding energy is $-(2e^2)/(\epsilon_\infty l)$, which is between -1.2 and -1.8 eV, depending on the value assumed for ϵ_∞. This capacitive estimate is larger than the first-order repulsion present in the Hubbard model calculation for reasonable values of U, but not enormously so.

However, this estimate ignores an important effect, namely, the screening of the repulsion between two electrons on a given molecule due to the rearrangement of the charges on neighboring molecules, that is, metallic screening. This would result in a large decrease in the Hartree-Fock energy.

In order to estimate the effect of metallic screening on the electrons on the same molecule, we imagine embedding a molecule in a spherical metallic cavity. The metallic cavity mimics the effect of the metallic electrons on the other molecules in an effective medium approximation. A simple calculation now leads to the following expression for the charging energy, E_C:

$$E_C = \frac{e^2 n^2}{\epsilon_\infty l}\left(\frac{2d}{l + 2d}\right) \tag{2}$$

where $(l + 2d)$ is the diameter of the cavity. This is reduced from its value in the absence of screening by a factor $[2d/(l + 2d)]$. Here, d should be of the order of the separation between fullerene molecules, that is, $d \approx 3$ Å, so for $\epsilon_\infty = 3$, $(E_C/n^2) \approx 0.27$ eV $\sim (1/10)t$. The corresponding pair-binding energy (approximately -0.5 eV) is roughly the same as the first-order contribution to the pair-binding energy we obtained from the Hubbard model for $U = 4t$. Moreover, if anything, we have overestimated this energy due to the use of a Hartree approximation. Thus, in the presence of metallic screening, the long-range Coulomb interactions are unlikely to destroy the effective attraction between electrons we have derived, at least at low frequencies, where metallic screening is effective.

Superconductivity. The existence of a positive pair-binding energy induced by electron-electron interaction is suggestive of superconductivity. Two different limits of the problem can be addressed, with an expected smooth crossover between them: (i) If the intermolecular hopping matrix element, t_i, is small compared to E_{pair}, we have the "preformed-pair" limit, consisting of a lattice gas of charge $2e$ bosons. In effect, the system would behave as a granular superconductor with each C_{60} molecule playing the role of a superconducting grain coupled through the Josephson mechanism. In this regime we also expect T_c to be an increasing function of t_i and so to increase with pressure. (ii) If t_i is large compared to E_{pair}, then the added electrons should form an extended band of width $W_i \sim 2Zt_i$, where $Z \sim 12$ is the effective coordination number of the fullerene crystal. In this limit E_{pair} would simply play the role of a short-range, weak, instantaneous attraction between electrons. This limit can be treated within the framework of a "BCS-like" mean-field theory, and the gap equation can be solved. Recall, once again, that $E_{pair}^{(n)}$ is approximately independent of n. In this mean-field limit, we expect the superconducting transition to be sharp. Qualitatively, because the superconducting transition temperature $T_c \propto \exp(-W_i/E_{pair})$, T_c should decrease strongly with increasing W_i, hence with pressure. In a conventional electron-phonon superconductor the pressure dependence is usually weak, because the decrease in the density of states is usually compensated by the increased stiffening of the phonons. In contrast, in the present problem the application of pressure would predominantly decrease the density of states at the fermi surface; because E_{pair} is primarily a property of a single molecule, it would not be greatly affected by pressure. Finally, because the effective attractive interaction between the electrons in this limit is mostly nonretarded and spread over the entire bandwidth, we expect that some of the superconducting properties would not be quite "BCS-like." Hence, the phrase "BCS-like" is simply a reminder that mean-field theory of pairing holds, with a pair size large compared to both the separation between electrons and the size of a C_{60} molecule.

Further consequences. *Magnetism.* Because our crude estimate of U places it in the neighborhood of the critical value at which the $E_{pair}^{(i)}$ curves cross 0, it is easy to imagine that if C_{60} is doped with different types of dopants, a regime of parameters can be accessed in which the triplet states have lower energy. Although two electrons

in this regime do not tend to attract on a molecule, if they happen to lie on a single molecule they will be in a triplet state, hence we see the possibility of the existence of itinerant ferromagnetism. The possibility of the formation of a ferromagnet is a unique signature of electron-electron interaction and clearly distinguishes the electronic mechanism for superconductivity from a mechanism based on electron-phonon interaction. In fact, as mentioned earlier, such a ferromagnet has recently been found (4).

Charge density wave. It might be argued that at exact commensurability, for example, when the number of added electrons is three per molecule, the negative U_{eff} ($\propto -E_{pair}$) system would form a charge density wave. However, this is unlikely for two reasons. First, it is easy to see that even a very small departure from commensurability would destroy the charge density wave phase. Secondly, and more importantly, note that the face-centered cubic (fcc) structure (16) of K_3C_{60} is likely to frustrate the formation of the charge density wave in favor of superconductivity. In the small U_{eff} limit this can readily be seen by a simple Hartree-Fock calculation. In the limit $U_{eff} \to -\infty$ this is evident from a second-order degenerate perturbation theory (17).

Doped C_{70}. From an analogous calculation described in the text we found that doped C_{70} should not superconduct unless the relevant value of the intramolecular Hubbard-U is considerably larger. However, because the perturbation theory for large values of U should be suspect, it is an open question whether the pair-binding in C_{70} can ever be positive.

Measurement of E_{pair}. We make some cautionary remarks regarding measurements of pair-binding energy through photoemission. In view of our discussion of the charging energy, it is important that these measurements be carried out in the metallic phase. If the screening is not adequate, the large capacitive charging energy of the doubly charged molecule over the singly charged molecule can swamp the small pair-binding energy.

Novel effects for partially filled shells. Elementary arguments (18) are sufficient to demonstrate that, in the noninteracting limit, E_n versus n should exhibit kinks as the added electrons complete closed shells: this forms the basis for an elementary discussion of the stability of aromatic molecules. What Fig. 2 shows is that similar kinks are found even when a shell (namely the lowest unoccupied level) is only partially filled. This is entirely an electronic correlation effect and signifies a novel mechanism for the stability of certain partially filled shells.

Electron-phonon interactions. Out of 174 intramolecular phonon modes only 2 are symmetry preserving, and these make a positive contribution to E_{pair} (5). The symmetry-breaking Jahn-Teller phonons make a negative, pair-breaking contribution to E_{pair}. The reason is as follows: due to its orbital degeneracy, the singly charged molecule can lower its energy considerably by a Jahn-Teller distortion, whereas for $U > U_{pair}$, the doubly charged molecule is an orbital singlet and so cannot Jahn-Teller distort. The details are discussed elsewhere (19).

Franck-Condon effect. One might ask why have we focused on a subtle electronic correlation effect as the source of the attraction between electrons when there is an obvious source of pair-binding in the form of the Jahn-Teller effect. In the noninteracting limit ($U = 0$), the doubly charged molecule (like the singly charged molecule) is subject to a substantial Jahn-Teller distortion. However, although this is a possible source of attraction between electrons, it is not a possible mechanism for superconductivity. An exponential suppression of the bipolaron bandwidth owing to the Franck-Condon effect results in a nearly infinite bipolaron mass and, consequently, no superconductivity. Rather, a Jahn-Teller distortion would produce self-localized "negative-U" centers (20). An important feature of the present theory is that the correlation effect suppresses the Jahn-Teller distortion and eliminates, to a large degree, the Franck-Condon reduction of the bandwidth.

Experimental consequences. (i) Because pair-binding does not occur for two and four added electrons, we expect that materials such as K_2C_{60} or K_4C_{60} would not superconduct (*21*). (ii) The superconducting transition temperature, T_c, should peak when there are approximately an odd number of electrons per molecule. Of course, some of the long-distance physics not included in this calculation may favor a particular concentration of dopants in the solid (*22*). (iii) With the help of different dopants it is possible to drive the system ferromagnetic. In fact, there is also a narrow range of U, between U_{FM} and U_{pair}, in which the singlet state is favored over the triplet state, and hence in principle it is possible to drive the system antiferromagnetic as well. (iv) The pressure dependences of T_c discussed above should be noted. For the case in which $W_i > E_{pair}$ we predict an approximately linear dependence of $\ln T_c$ on the intermolecular bandwidth (*23*). (v) Because for large enough U, the doubly charged C_{60} should be in an orbital singlet state, the infrared absorption should be quite different from what one would expect from the single particle theory (*19*).

REFERENCES AND NOTES

1. H. W. Kroto *et al.*, *Nature* **318**, 162 (1985); W. Kratschmer *et al.*, *ibid.* **347**, 354 (1990); H. Ajie *et al.*, *J. Phys. Chem.* **94**, 8630 (1990).
2. A. F. Hebard *et al.*, *Nature* **350**, 600 (1991); M. J. Rosseinsky *et al.*, *Phys. Rev. Lett.* **66**, 2830 (1991).
3. K. Holczer *et al.*, *Science* **252**, 1154 (1991); K. Holczer *et al.*, *Phys. Rev. Lett.* **67**, 271 (1991).
4. P.-M. Allemand *et al.*, *Science* **253**, 301 (1991).
5. S. Chakravarty and S. Kivelson, *Europhys. Lett.*, in press.
6. A. J. Heeger *et al.*, *Rev. Mod. Phys.* **60**, 781 (1988).
7. V. Elser and R. C. Haddon, *Nature* **325**, 792 (1987).
8. R. C. Haddon, L. E. Brus, K. Raghavachari, *Chem. Phys. Lett.* **125**, 459 (1986).
9. L. Pauling, *Proc. R. Soc. London Ser. A* **196**, 343 (1949).
10. P. W. Anderson, *Mater. Res. Bull.* **8**, 153 (1973); *Science* **235**, 1196 (1987); in *High Temperature Superconductivity: Proceedings*, K. S. Bedell *et al.*, Eds. (Addison-Wesley, Reading, MA, 1990).
11. S. A. Kivelson, D. S. Rokhsar, J. Sethna, *Phys. Rev. B* **35**, 8865 (1987).
12. G. Baskaran and E. Tossatti, *Curr. Sci.* **61**, 33 (1991).
13. L. Landau and E. M. Lifshitz, *Quantum Mechanics* (Pergamon, New York, ed. 3, 1977), vol. 3.
14. If we define $\epsilon(n) = -100.80069 + 0.321666n$. Then for $t'/t = 1.2$: $\Phi_0(L = 0, S = 0) = \epsilon(0) + 15(U/t) - 0.74785(U/t)^2$, $\Phi_1(L = 1, S = 1/2) = \epsilon(1) + 15.5(U/t) - 0.73772(U/t)^2$, $\Phi_2(L = 1, S = 1) = \epsilon(2) + 16(U/t) - 0.72557(U/t)^2$, $\Phi_2(L = 0, S = 0) = \epsilon(2) + 16.05(U/t) - 0.74237(U/t)^2$, $\Phi_2(L = 2, S = 0) = \epsilon(2) + 16.02(U/t) - 0.73128(U/t)^2$, $\Phi_3(L = 0, S = 3/2) = \epsilon(3) + 16.5(U/t) - 0.71140(U/t)^2$, $\Phi_3(L = 1, S = 1/2) = \epsilon(3) + 16.55(U/t) - 0.72725(U/t)^2$, $\Phi_3(L = 2, S = 1/2) = \epsilon(3) + 16.53(U/t) - 0.71991(U/t)^2$, $\Phi_4(L = 1, S = 1) = \epsilon(4) + 17.05(U/t) - 0.71012(U/t)^2$, $\Phi_4(L = 0, S = 0) = \epsilon(4) + 17.10(U/t) - 0.72671(U/t)^2$, $\Phi_4(L = 2, S = 0) = \epsilon(4) + 17.07(U/t) - 0.71575(U/t)^2$, $\Phi_5(L = 1, S = 1/2) = \epsilon(5) + 17.6(U/t) - 0.70661(U/t)^2$, and $\Phi_6(L = 0, S = 0) = \epsilon(6) + 18.15(U/t) - 0.70087(U/t)^2$. All energies are in units of t. We have explored the range $1.0 \leq t'/t \leq 1.35$ and have found that the basic picture described in the text is the same over the entire range.
15. S. R. White, S. Chakravarty, M. P. Gelfand, S. Kivelson, in preparation.
16. P. W. Stephens *et al.*, *Nature* **351**, 632 (1991).
17. Y. Nagaoka, *Prog. Theor. Phys.* **52**, 1716 (1974). Although Nagaoka discusses the degeneracy of charge density wave and superconducting orders for bipartite lattices, the extension to a frustrated FCC lattice is trivial. In this case the superconducting state has considerably lower energy than the charge density wave state.
18. R. P. Feynman, R. B. Leighton, M. Sands, *The Feynman Lectures on Physics* (Addison-Wesley, Reading, 1964), vol. 3, see especially sec. 15-5.
19. M. Salkola, S. Chakravarty, S. Kivelson, in preparation.
20. P. W. Anderson, *Phys. Rev. Lett.* **34**, 953 (1975).
21. Indeed, recent experiments have shown that A_4C_{60} does not superconduct, where A is K, Rb, or mixtures of K, Rb, or Cs. See R. M. Fleming *et al.*, *Nature* **352**, 701 (1991).
22. The structure of K_3C_{60} is now known to be fcc (*16*), where both the octahedral and the tetrahedral sites are occupied by the K^+ ions, with a lattice constant that is virtually unchanged from the pure C_{60} solid. It is therefore reasonable to assume that doping fills the lowest unoccupied states of the molecule, while leaving the electronic structure of a single molecule essentially unchanged, in particular, the threefold degeneracy of the highest unoccupied orbitals. Any other composition is likely to produce significant distortions in the solid.
23. G. Sparn *et al.*, *Science* **252**, 1829 (1991); G. Sparn *et al.*, in preparation.
24. It is a great pleasure to thank P. W. Anderson for his comments and K. Holczer for numerous exciting discussions of the rapidly developing experimental scene. In particular, we would like to thank P. W. Anderson for drawing our attention to the fact that the symmetry-preserving intramolecular phonons are pair enhancing whereas the symmetry breaking phonons are pair breaking. We thank S. R. White and M. Salkola for collaborations to be reported elsewhere (*15, 19*). Thanks are also due to E. Abrahams, A. Auerbach, V. Emery, I. Pimentel, D. Rokhsar, and P. Scharf for particularly helpful discussions. Supported by the National Science Foundation grants no. DMR-89-07664 (S.C.) and no. DMR-90-11803 (S.K.).

22 September 1991; accepted 10 October 1991

Reprinted with permission from Physical Review Letters
Vol. 68, No. 8, pp. 1228–1231, 24 February 1992

Pressure and Field Dependence of Superconductivity in Rb_3C_{60}

G. Sparn and J. D. Thompson

Los Alamos National Laboratory, Los Alamos, New Mexico 87545

R. L. Whetten, S.-M. Huang, R. B. Kaner, and F. Diederich

*Department of Chemistry and Biochemistry and Solid State Science Center, University of California,
Los Angeles, California 90024-1569*

G. Grüner and K. Holczer[a]

Department of Physics and Solid State Science Center, University of California, Los Angeles, California 90024-1569

(Received 20 September 1991; revised manuscript received 19 November 1991)

Direct measurements of the pressure dependence of the superconducting transition temperature T_c for single-phase Rb_3C_{60} provide for the first time substantial evidence that T_c in A_3C_{60} compounds may be a universal function of the lattice constant (inter-C_{60} spacing). Determination of the lower (H_{c1}) and upper (H_{c2}) critical fields, and consequently the superconducting penetration depth and coherence length, is also reported. Compared to K_3C_{60}, the measured quantities are consistent with a 15% larger electronic-state density in Rb_3C_{60} arising from a greater inter-C_{60} spacing in this compound.

PACS numbers: 74.40.+k, 74.30.Ci

The discovery [1–3] that alkali-buckminsterfullerene compounds form a new family of superconductors has stimulated substantial theoretical and experimental interest. Systematic studies [3–7] of potassium compounds have established that superconductivity appears only in the stoichiometric K_3C_{60} composition [3], which forms an fcc structure [6]. The transition temperature, $T_c = 19.3$ K, decreases strongly with increasing pressure $dT_c/dP = -7.8$ K/GPa [5]. Magnetization measurements [4] have shown K_3C_{60} to be an extreme type-II superconductor with a London penetration depth $\lambda_L = 240$ nm and a superconducting coherence length $\xi_0 = 2.6$ nm. The temperature dependence of the lower critical field $H_{c1}(T)$ and of $\lambda_L(T)$ determined [7] by muon-spin rotation strongly suggests a singlet, s wave pairing. Until now, little has been reported regarding the second member of the family, the superconducting Rb_3C_{60} compound, beyond its transition temperature, $T_c = 29.6$ K [2,3].

Samples were prepared from a solid-phase reaction of high-purity C_{60} powder reacted with Rb vapor in a way similar to that reported for K_3C_{60}. As described in detail elsewhere [8], the major difference lies in the kinetics of the solid-gas reaction, which in the case of Rb results in a strong tendency toward segregation of different stoichiometric compounds, independent of the initial ratio of materials used. The sample used in the present work was prepared from a starting composition of $Rb_{2.5}C_{60}$, heated in vacuum at 200°C for 22 h, and subsequently annealed under He atmosphere over several days at the same temperature. This procedure led to a fractional shielding diamagnetism of ca. 25% on powder, determined relative to a Nb bulk reference sample. X-ray diffraction revealed an fcc crystal structure identical to that of K_3C_{60} but with lattice constant 1.4425 nm, compared to 1.4240 nm for K_3C_{60} [8]. The powder was pressed at 4 MPa to make a pellet of 3 mm diam and 0.7 mm thickness, and then sintered at 200°C for 6 h. Pressing also results in

the appearance of bulk shielding (macroscopic supercurrent) at a somewhat lower temperature than T_c, similar to that described earlier in the case of pressed K_3C_{60} [5].

Pressure and $H_{c2}(T)$ measurements were performed on a 2.57-mg portion cut from the Rb_3C_{60} pellet sealed under 0.1 MPa He gas in a Pyrex capillary. For reasons to be discussed, $H_{c1}(T)$ was determined on a powder sample also sealed with He in Pyrex. In making measurements with a Quantum Design SQUID magnetometer, care was taken to ensure that the sample was not exposed to a field gradient larger than 0.03% of the applied field. The pressure dependence of T_c was measured inductively at a frequency of 457 Hz, using a self-clamping pressure cell [9] with Fluorinert FC-75 as the hydrostatic pressure medium [5]. For these measurements the sample was removed from the Pyrex capillary and immersed directly into the pressure medium. Pressure within the cell was determined from the variation in T_c of a high-purity lead reference.

Figure 1 shows results of ac susceptibility measurements at a series of pressures between ambient pressure and $P = 1.9$ GPa. In the inset are shown typical susceptibility curves, each exhibiting in addition to the shielding-diamagnetism onset at T_c, a second break in slope at $T^* < T_c$, where the macroscopic shielding current is established, i.e., the intergranular coherence appears, as will be described in full detail [10]. A strongly negative pressure dependence is found for both T_c and T^*, as displayed in the figure, from which an initial slope $(dT_c/dP)_{P=0} = -9.7$ K/GPa is inferred, an even larger value than that reported for K_3C_{60} (see Table I) [5]. At the highest pressure attempted, T_c is near 13 K, less than half its ambient pressure value. The data points were taken for both increasing and decreasing pressures, demonstrating that the entire variation as a function of pressure is reversible. Additionally, a slight bump is ob-

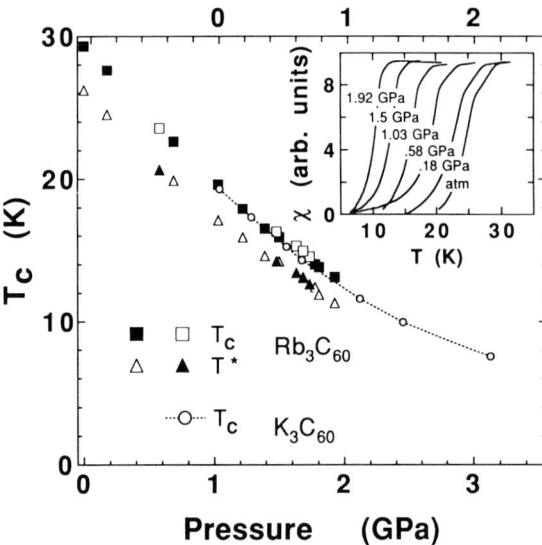

FIG. 1. Inset: Magnetic susceptibility curves $\chi(T;P)$ of Rb$_3$C$_{60}$ obtained under the indicated pressures at nominal zero magnetic field. The transition temperature T_c (inflection point at highest temperature) and a secondary kink (T^*)—not present in unpressed samples and reflective of an intergranular coherence transition—are both clearly visible. The full scale of χ corresponds to 100% (bulk) diamagnetic shielding. The main plot shows the dependence of T_c and T^*, as determined from $\chi(T)$ curves like those in the inset, vs applied pressure. Solid and open symbols correspond to data taken with increasing and decreasing pressure, respectively. For comparison, several points from the $T_c(P)$ curve of K$_3$C$_{60}$ have been plotted on the same scale but with a P scale translated by 1.06 GPa. Both data sets approximately follow $T_c(P) = T_c(0)\exp(-\gamma P)$, where $\gamma = 0.44 \pm 0.03$ GPa^{-1}.

served near 1.5 GPa for both T_c and T^*, which is significant and surely not an instrumental artifact. A similar observation has been reported for V$_3$Si [11], which undergoes a pressure-induced structural transition accompanied by a change in dT_c/dP, associated with a soft phonon mode at ambient pressure. This provides further motivation for structural investigation under pressure.

The determination of the lower critical field H_{c1} is complicated when using a compressed pellet because the applied field H first breaks the intergranular supercurrents, leaving the individual grains acting as a powder sample [10]. Figure 2 shows typical magnetization curves $M(H)$ at two different maximum applied fields (0.02 and 5 T), at both $T=5$ K and ambient pressure. Figure 2(a) shows that the intergranular supercurrents are reduced to near zero for $H=20$ mT; whereas Fig. 2(b) shows even at 5 T substantial hysteresis that we associate with intragranular supercurrents. An analysis of the hysteresis, using a critical-state model described in Ref. [4], gives an intragranular critical-current density $J_c = 1.5 \times 10^6$ A/cm^2 at $H=1$ T and $T=5$ K. This J_c value is about 1 order of magnitude larger than that obtained on K$_3$C$_{60}$ assuming the same 1-μm particle size for

TABLE I. Superconductivity parameters in A_3C$_{60}$ compounds.

Parameter	$A=$Rb	$A=$K
T_c (K)	29.6	19.3
$(dT_c/dP)_{P=0}$ (K/GPa)	−9.7	−7.8
H_{c1} (mT)	12	13
H_{c2} (T)	78	49
H_c (T)	0.44	0.38
J_c (10^5 A/cm^2)$_{T=5\,K}$	15	1.2
ξ_0 (nm)	$2^{+0.3}_{-0.2}$	$2.6^{+0.4}_{-0.2}$
λ_L (nm)	247^{+10}_{-20}	240^{+10}_{-20}
$\kappa = \lambda_L/\xi_0$	124	92

this material.

Because of the possibility that the intrinsic H_{c1} of Rb$_3$C$_{60}$ could be masked by intergranular supercurrents at low fields, dc magnetization measurements were also performed on a powder sample prepared as described above. The inset of Fig. 3(a) shows that a flux penetration field H_a^* can be clearly defined by the deviation from linear M vs H behavior. From values of H_a^* determined at various temperatures, we calculate $H_{c1}=H_a^*/(1-n)$, assuming a demagnetization factor $n=\frac{1}{3}$, appropriate to a sphere, which reflects the geometry of isolated Rb$_3$C$_{60}$ particles. Figure 3(a) gives the temperature dependence of $H_{c1}(T)$ evaluated this way. The zero-temperature extrapolated value of $H_{c1}(0)$ is 12 ± 3 mT.

The upper critical field value H_{c2} has been determined from magnetization curves $M(T)$ taken by zero-field cooling to well below T_c, applying various fields in the range of 0.2 to 5 T, and recording values by warming the sample to well above T_c. Typical magnetization curves for different fields are shown in the inset of Fig. 3(b).

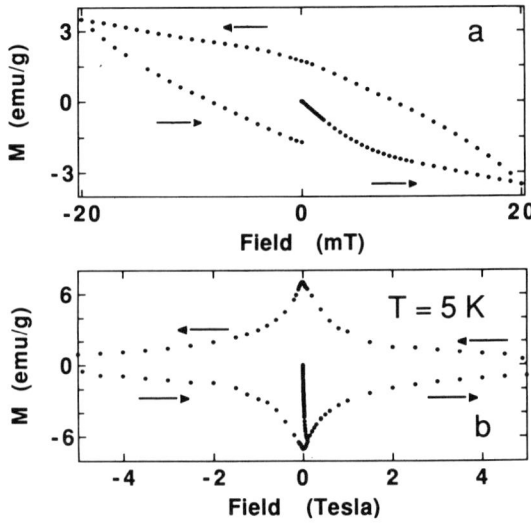

FIG. 2. Representative magnetization curves $M(H)$ recorded at $T=5$ K and ambient pressure, with the field swept as indicated by arrows over two different field ranges: (a) The range -20 to $+20$ mT, where the hysteresis loop in the intergranular regime is evident; (b) $M(H)$ at fields to 5 T.

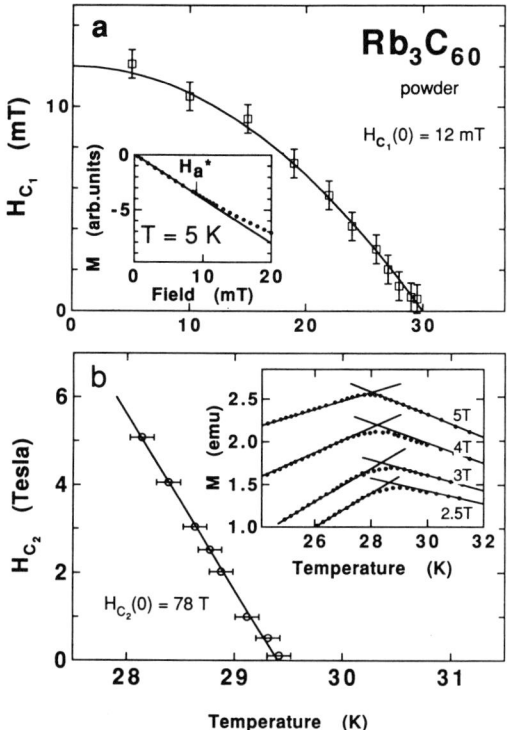

FIG. 3. (a) The temperature dependence of the lower critical field $H_{c1}(T)$ determined on a powder sample as shown in the inset and explained in the text. Error bars represent estimated uncertainty in determining H_a^*. The solid line represents the empirical law $H_{c1}(T)/H_{c1}(0) = 1 - (T/T_c)^2$. (b) The temperature dependence of the upper critical field $H_{c2}(T)$. Inset: The method of determining these values from $M(T)$ curves at fixed field. The error bars in the main $H_{c2}(T)$ plot reflect the rounding of the transition.

Note that the transition is somewhat broader than in the case of the K_3C_{60}, and might be due to Rb deficiency [8]. T_c is determined from the intercept of linear extrapolations from below and above the transition, as shown. Figure 3(b) shows the H_{c2} values obtained in this way as a function of temperature, along with a linear fit resulting in a critical field slope of -3.9 T/K. Using the Werthamer-Helfand-Hohenberg formula [12], a zero-temperature extrapolated value $H_{c2}(0) = 78 \pm 10$ T is obtained. This value exceeds the Pauli weak-coupling limit of 53 T [13].

From $H_{c2}(0)$ and $H_{c1}(0)$, we evaluate the zero-temperature superconducting coherence length ξ_0 and London penetration depth λ_L using the relations [14]

$$H_{c2}(0) = \Phi_0/2\pi\xi_0^2, \quad H_{c1}(0) = (\Phi_0/4\pi\lambda_L^2)\ln(\lambda_L/\xi_0), \quad (1)$$

where Φ_0 is the flux quantum. We find $\xi_0 = 2^{+0.3}_{-0.2}$ nm and $\lambda_L = 247^{+10}_{-20}$ nm. These values are compared in Table I with values obtained for K_3C_{60}. ξ_0 for the Rb compound is somewhat smaller than obtained in the case of K_3C_{60} but still larger than the nearest-neighbor C_{60} distance d (ca. 1 nm). We also estimate the thermodynamic critical field from $H_c^2(0) = H_{c1}(0)H_{c2}(0)/\ln\kappa$ (Ref. [14]) and compare it to K_3C_{60} in Table I.

The outstanding question remains as to why T_c for Rb_3C_{60} is over 10 K higher than for K_3C_{60}. A notable difference between the two alkali metals is the greater spatial extent of the $5s^1$ configuration of Rb relative to the $4s^1$ configuration of K and consequently a lattice parameter for Rb_3C_{60} that is 0.0185 nm greater than in K_3C_{60} (which interestingly corresponds to the difference in ionic radii) [6]. That is, compared to Rb_3C_{60}, K_3C_{60} at ambient pressure is subjected to an effective "chemical" pressure arising from the smaller ionic radius of K. Figure 1 shows that if $T_c(P)$ for K_3C_{60} is shifted rigidly by 1.06 GPa relative to that for Rb_3C_{60}, T_c's of both compounds coincide and have essentially the same pressure dependence. This pressure shift, if driven solely by steric effects, should correspond to the chemical pressure calculated from the lattice-parameter difference and the linear compressibility κ of K_3C_{60}. Unfortunately, κ is not known for K_3C_{60}. However, assuming a linear compressibility $d(\ln a)/dP = 1.21 \times 10^{-2}$ GPa^{-1} for K_3C_{60} gives the required chemical pressure. This assumed value of κ is reasonable given reported values for pure C_{60} that range from 1.8×10^{-2} GPa^{-1} (Ref. [15]) to 2.3×10^{-2} GPa^{-1} (Ref. [16]). We believe this agreement is not fortuitous but provides an important clue to the mechanism of superconductivity, namely, that T_c is a unique function of volume, i.e., lattice spacing. A naive extrapolation of T_c on the lattice constant a_0, using the ionic radius of Cs$^+$, predicts $a_0 = 1.48$ nm and $T_c = 43$ K for Cs_3C_{60}.

The straightforward scaling of T_c with the lattice constant suggests that the mechanism responsible for pair formation is intrinsic to the C_{60} molecule and that the dominant pressure-sensitive parameter is the bandwidth which is governed by the overlap between molecules. The ratio of bandwidths, or equivalently state densities $N(0)$ at the Fermi energy, for Rb_3C_{60} and K_3C_{60} can be estimated from the respective T_c's, coherence lengths, and the relation [14] $\xi_0 \propto v_F/T_c$, where in a free-electron model the Fermi velocity $v_F \propto N(0)$. This relationship gives $N(0)_{Rb_3C_{60}} = 1.18N(0)_{K_3C_{60}}$, if we use values of ξ_0 from Table I, or $N(0)_{Rb_3C_{60}} = 1.24N(0)_{K_3C_{60}}$ assuming ξ_0 is derived from the respective Pauli paramagnetic fields. Further, assuming a BCS expression for $T_c = \omega_D \exp[-1/JN(0)]$ and that the pressure dependence of the cutoff frequency ω_D is negligible, as inferred from T_c scaling with lattice parameter, we find from the ratios of $\partial(\ln T_c)/\partial P$ that $N(0)_{Rb_3C_{60}} = 1.11N(0)_{K_3C_{60}}$. Thus, the pressure dependences of T_c, coherence lengths, and T_c's of K_3C_{60} and Rb_3C_{60} are consistent with a 15% larger state density in Rb_3C_{60} arising from the greater inter-C_{60} spacing in this compound. Further, the slightly larger state density in Rb_3C_{60} is also consistent with a somewhat larger effective electron mass inferred from values of the penetration depth.

This view of superconductivity in fullerenes in which T_c is dominated by a simple volume-dependent electronic state density suggests that models [17] requiring pairing to be mediated by intermolecular vibrational modes that involve the alkali metal are inappropriate and that an alkali-metal isotope effect should be absent. However, models in which pairing requires the unique properties of C_{60} itself and in which the coupling strength does not vary substantially between K_3C_{60} and Rb_3C_{60} are favored. In these latter cases, it is not possible from our experiments to distinguish between models emphasizing intramolecular vibrational modes [18,19] or purely electronic correlations [20] as the pairing mechanism. Observation of a carbon-isotope effect would favor the former interpretation [18,19]. However, what is clear from measurements [21] of the normal- and superconducting-state magnetic properties of K_3C_{60} is that the electronic state can be modeled by Fermi-liquid behavior in a relatively narrow band and that the effective electron mass is not renormalized strongly by Coulomb correlations. Any viable theory must be able to satisfy these experimental constraints.

K.H. and R.L.W. thank Sudip Chakravarty for very useful discussions. Samir J. Anz and Friedericke Ettl produced and separated the molecular carbon samples and O. Klein helped in sample characterization. Work at Los Alamos was performed under the auspices of the U.S. Department of Energy. Funding has been provided by NSF grants (R.L.W. and F.D., and to R.B.K.) and by the David and Lucille Packard Foundation (R.L.W. and R.B.K.).

Note added.—Since this work was completed, Zhou *et al.* [22] measured the compressibility of K_3C_{60} and Rb_3C_{60} and found $d(\ln a)/dP = 1.20\times10^{-2}$ and 1.52×10^{-2} GPa^{-1}, respectively, which validates our assumption. These compressibilities imply logarithmic volume derivatives of T_c of 11 and 7 for K_3C_{60} and Rb_3C_{60}, respectively. Furthermore, Fleming *et al.* have arrived indirectly at conclusions similar to ours on the basis of a structure-T_c relation among $A_xA'_{x-3}C_{60}$ alloy compounds [23].

[a]Permanent address: Central Research Institute for Physics, P.O. Box 49, h1525 Budapest, Hungary.

[1] A. F. Hebard *et al.*, Nature (London) **350**, 600 (1991); R. C. Haddon *et al.*, *ibid.* **350**, 320 (1991).
[2] M. J. Rosseinsky *et al.*, Phys. Rev. Lett. **66**, 2830 (1991).
[3] K. Holczer *et al.*, Science **252**, 1154 (1991).
[4] K. Holczer *et al.*, Phys. Rev. Lett. **67**, 271 (1991).
[5] G. Sparn *et al.*, Science **252**, 1829 (1991).
[6] P. W. Stephens *et al.*, Nature (London) **351**, 632 (1991).
[7] Y. J. Uemura *et al.*, Nature (London) **352**, 605 (1991).
[8] K. Holczer *et al.* (to be published).
[9] J. D. Thompson, Rev. Sci. Instrum. **55**, 231 (1984).
[10] J. D. Thompson *et al.* (to be published). Similar effects were reported for K_3C_{60} in Ref. [5].
[11] G. Fasol, J. S. Schilling, and B. Seeber, Phys. Rev. Lett. **41**, 424 (1978).
[12] N. R. Werthamer, E. Helfand, and P. C. Hohenberg, Phys. Rev. **147**, 295 (1966).
[13] A. M. Clogston, Phys. Rev. Lett. **9**, 266 (1962).
[14] M. Tinkham, *Introduction to Superconductivity* (McGraw-Hill, New York, 1975).
[15] S. J. Duclos *et al.*, Nature (London) **351**, 380 (1991).
[16] J. E. Fischer *et al.*, Science **252**, 1288 (1991).
[17] F. C. Zhang, M. Ogata, and T. M. Rice, Phys. Rev. Lett. **67**, 3452 (1991).
[18] C. M. Varma, J. Zaanen, and K. Raghavachari, Science **254**, 989 (1991).
[19] M. Schluter *et al.*, Phys. Rev. Lett. **68**, 526 (1992).
[20] S. Chakravarty, M. P. Gelfand, and S. Kivelson, Science **254**, 970 (1991).
[21] W. H. Wong *et al.* (to be published).
[22] O. Zhou *et al.* (to be published).
[23] R. M. Fleming *et al.*, Nature (London) **352**, 797 (1991).

Reprinted with permission from Science
Vol. 255, pp. 833–835, 14 February 1992

Compressibility of M_3C_{60} Fullerene Superconductors: Relation Between T_c and Lattice Parameter

Otto Zhou, Gavin B. M. Vaughan, Qing Zhu,
John E. Fischer,* Paul A. Heiney, Nicole Coustel,
John P. McCauley, Jr., Amos B. Smith III

X-ray diffraction and diamond anvil techniques were used to measure the isothermal compressibility of K_3C_{60} and Rb_3C_{60}, the superconducting, binary alkali-metal intercalation compounds of solid buckminsterfullerene. These results, combined with the pressure dependence of the superconducting onset temperature T_c measured by other groups, establish a universal first-order relation between T_c and the lattice parameter a over a broad range, between 13.9 and 14.5 angstroms. A small second-order intercalate-specific effect was observed that appears to rule out the participation of intercalate-fullerene optic modes in the pairing interaction.

A NUMBER OF ISOSTRUCTURAL BInary and pseudobinary alkali metal–C_{60} superconductors have been discovered that have onset temperatures T_c ranging from 18 to 33 K (1). Their general formulas are $M_{3-x}M'_xC_{60}$, and their face-centered-cubic lattice parameters a range from 14.25 to 14.49 Å at atmospheric pressure and 300 K. A monotonic increase of T_c with alkali size is inferred from an empirical linear correlation between T_c and a at constant pressure (1). Moreover, T_c decreases with increasing pressure for the binary compounds with M = K (2, 3) and Rb (4), and the two sets of $T_c(P)$ data can be superposed

by a relative shift of the pressure scales (4). These results both suggest that T_c depends only on the overlap between near-neighbor C_{60} molecules and not explicitly on the

O. Zhou, Q. Zhu, J. E. Fischer, N. Coustel, Laboratory for Research on the Structure of Matter and Department of Materials Science and Engineering, University of Pennsylvania, Philadelphia, PA 19104.
G. B. M. Vaughan and P. A. Heiney, Laboratory for Research on the Structure of Matter and Department of Physics, University of Pennsylvania, Philadelphia, PA 19104.
J. P. McCauley, Jr., and A. B. Smith III, Laboratory for Research on the Structure of Matter and Department of Chemistry, University of Pennsylvania, Philadelphia, PA 19104.

*To whom correspondence should be addressed.

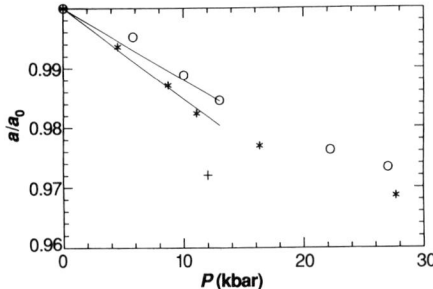

Fig. 1. Fractional reduction of lattice parameter a/a_o versus pressure P: (○), K_3C_{60}; (*), Rb_3C_{60}; and (+), pure C_{60} (10). The experimental errors in a and P are ± 0.002 Å and ± 0.3 kbar, respectively. Solid lines are linear fits to the data points up to 13 kbar.

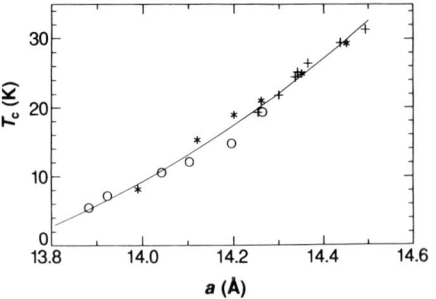

Fig. 2. Superconducting transition temperature versus lattice parameter. The $a(P)$ data from Fig. 1 was converted to $T_c(a)$ by using the raw $T_c(P)$ data of (2) and (4): (○), K_3C_{60}; (*), Rb_3C_{60}; and (+), directly measured points from a number of $M_{3-x}M'_xC_{60}$ compounds at 300 K and 1 bar (1). Solid curve is a quadratic fit.

nature of the intercalate (1, 4). In the context of weak-coupling Bardeen-Cooper-Schrieffer (BCS) superconductivity, both results are interpreted to mean that $N(E_F)$, the density of states at the Fermi energy, increases with increasing intermolecular separation because of reduced orbital overlap and conduction electron bandwidth (assuming that the number of electrons transferred to the conduction band or bands per intercalate is the same for all alkali metals and is independent of P). In the weak-coupling BCS model, $kT_c = \hbar\omega_{ph}e^{-1/VN(E_F)}$, where k is the Boltzmann constant, $\hbar\omega_{ph}$ is a phonon energy, and V is the electron-phonon coupling constant. The dependence of T_c on a is implicit in $N(E_F)$, so the universality of $T_c(a)$ implies that ω_{ph} and V are independent of P, that is, the total differential dT_c/dP is dominated by a single term $[\partial T_c/\partial a][da/dP]$. With the additional assumption of no pressure-induced structural transitions, Fleming et al. (1) predicted a universal compressibility value for all of the isostructural compounds. Similarly, Sparn et al. (4) predicted a compressibility value for Rb_3C_{60} by scaling T_c and a for the two binary compounds.

We measured the pressure dependence of a for K_3C_{60} and Rb_3C_{60} at 300 K using x-ray diffraction. This experiment permits a direct test of the validity of the assumptions noted above. Scaling our $a(P)$ data point-by-point with the measured $T_c(P)$ up to 28 kbar permits us to extend the T_c versus a correlation down to T_c and a values as low as 6 K and 13.9 Å, respectively. The overall behavior is still generally consistent with a universal phenomenon that depends only on intermolecular separation. Not surprisingly, the correlation is no longer linear over the larger range of parameters. Our analysis also reveals a small, second-order, metal-specific effect, the sense of which allows us to rule out pairing mediated by phonons involving the alkali ions. This secondary effect in the $T_c(a)$ correlation is revealed by the observa-

tion that the compressibilities of K_3C_{60} and Rb_3C_{60} are significantly different.

Powder samples of M_3C_{60} were prepared by the dilution technique described in detail elsewhere (5). Equimolar amounts of >99.5% pure C_{60} and saturation-doped M_6C_{60} (6) powders were mixed together in a dry box by grinding and then sealed in evacuated pyrex tubes. These were first heated to 250°C for 24 hours and then annealed in two steps, first at 350°C for 24 hours and then at 400°C for 1 hour. We have previously established by x-ray diffraction that this procedure reproducibly gives single-phase material (5, 7). Diamagnetic shielding measurements of uncompacted powders gave T_c values and shielding fractions of 29.6 K and 56% for Rb_3C_{60} and 19.3 K and 38% for K_3C_{60}, respectively (5).

X-ray diffraction experiments were carried out with synchrotron radiation at the National Synchrotron Light Source X10B beamline. The vertical diffractometer was equipped with a bent Si(111) monochromator crystal for horizontal focusing and a platinum-coated zerodur mirror for vertical focusing and was set for a wavelength λ of 0.947 Å. The resolution $\triangle Q$ was of order 0.0035 Å^{-1} full-width at half-maximum ($Q = 4\pi\sin\theta/\lambda$). Pressures up to 28 kbar were provided by a Merrill-Bassett diamond anvil cell by using mineral oil or pentane-isopentane as hydrostatic media (8). We added CaF_2 powder to the cell and measured pressure from its lattice parameter (determined from three reflections) and the empirical equation of state (9). Four strong Bragg reflections of M_3C_{60} (111, 311, 222, and 420) were recorded at each pressure. No other Bragg reflections were observed in this range of scattering angles up to the maximum pressure. The four measured peaks were fitted with a single lattice parameter, and mixed Gaussian-Lorentzian intensity functions to obtain a at each pressure. A 2θ zero-offset correction was measured at each pressure by recording a CaF_2 reflection with negative scattering angle.

The fractional reduction in lattice parameter a/a_o with increasing pressure for the two M_3C_{60} compounds is shown in Fig. 1, as well as an earlier result for pure C_{60} (10). The contraction is linear up to ~13 kbar within experimental error, from which we derive average linear compressibilities $dlna/dP$ of $1.20 \pm 0.09 \times 10^{-3}$ and $1.52 \pm 0.09 \times 10^{-3}$ kbar^{-1} for K_3C_{60} and Rb_3C_{60}, respectively. These are both less than the value $2.3 \pm 0.2 \times 10^{-3}$ kbar^{-1} for pure C_{60} based on the single data point. The stiffening effect of alkali intercalation is attributed to an increase in lattice energy resulting from the ionic guest-host interaction (1, 2). A similar effect occurs in intercalated graph-

ite (11). K_3C_{60} and Rb_3C_{60} both become stiffer above 13 kbar, as shown by the flattening out of a/a_o versus P at higher pressures. This stiffening also occurs in pure C_{60} (12).

Our measured compressibility for Rb_3C_{60} agrees fairly well with the predicted value 1.7×10^{-3} kbar^{-1} (4). Similarly, the average of our volume compressibilities for K_3C_{60} and Rb_3C_{60}, $4.08 \pm 0.27 \times 10^{-3}$ kbar^{-1}, is consistent with values predicted from the empirical linear relation between T_c and a (1) and the two measurements of dT_c/dP (2, 3), namely 4.7×10^{-3} and 4.0×10^{-3} kbar^{-1}, respectively. These comparisons show that, for a given compound, dT_c/dP is indeed dominated by the term $[\partial T_c/\partial a][da/dP]$. Therefore, within weak-coupling BCS theory and linear approximations for $T_c(P)$ and $a(P)$, any pressure dependences of phonon energies or the coupling constant have negligible effects on T_c below 13 kbar.

The results of converting all of the $a(P)$ data in Fig. 1 point-by-point to $T_c(a)$ are shown in Fig. 2. We interpolated between the $T_c(P)$ data points of Sparn et al. (2, 4) to obtain T_c's at the corresponding P's. This procedure goes beyond the approximation of linear $T_c(P)$ and $a(P)$ at low P but is not rigorously correct because the T_c values were measured near 0 K whereas the P-dependent a values were measured at 300 K. Also plotted in Fig. 2 are Fleming's $T_c(a)$ data at 1 bar for seven different compounds. The $T_c(P)$ and $a(P)$ measurements were performed in two different laboratories, which does not compromise the relative accuracy of the $T_c(a)$ correlations for K- and Rb-doped compounds because both $T_c(P)$ measurements were performed in the same laboratory, as were both $a(P)$ measurements. On the other hand, the absolute relation of the two pressure-dependent data sets to Fleming's 1-bar data is subject to systematic uncertainties. There is good agreement between the two data points for K_3C_{60} (19 K and 14.24 Å) and Rb_3C_{60} (29.6 K and 14.42 Å) at 1 bar. The absolute uncertainties (± 0.5 K %) are smaller than the plot symbols.

The solid curve in Fig. 2 indicates that all three data sets are represented reasonably well by a single quadratic function. We conclude that the proposed universal correlation between T_c and a is correct to first order. The positive sign of the quadratic term may serve as a detailed test of competing models, for example, from the dependence on a of the parameters in the Mac-Millan formula.

Our data also reveal a small second-order intercalate-specific contribution to $T_c(a)$. From Fig. 1, the compressibilities of K_3C_{60} and Rb_3C_{60} are not identical, as would be required by a strictly universal correlation. The three data sets in Fig. 2 can also be reasonably well described by individual linear segments. The "universal" slope of Fleming's directly measured values is 50 K Å^{-1}, which is quite close to our value of 45 ± 1 K Å^{-1} for Rb_3C_{60} but significantly larger than our value of 33 ± 2 K Å^{-1} for K_3C_{60}. The intercalate-specific effect is only significant at small a, that is, at large intermolecular overlap.

The sense of this effect is inconsistent with a proposal that metal-C_{60} optic modes are responsible for the pairing interaction (13). In this hypothesis, superconductivity arises from strong coupling to relatively low-frequency modes rather than from weak coupling to the intramolecular vibrations, which lie at higher frequencies (14, 15). If we assume a linear contribution to $T_c(a)$ from the $e^{1/VN(E_F)}$ factor independent of chemical composition, and that the important coupling phonons are zone-boundary optic modes involving planes of M_3 and of C_{60} (13), there would be an additional scaling of T_c as $\sqrt{(1/m)}$, where m is the reduced mass. This scaling would be in the ratio 14:10 for K_3C_{60} and Rb_3C_{60}, the inverse of the observed 33:45 ratio of slopes.

There are several possible origins for the small intercalate-specific contribution. The first is suggested by the complex Fermi surface of K_3C_{60} (16). In addition to bandwidth variations, which depend directly on the lattice parameter, the volume of the Fermi surface may be subtly different for different intercalates, or may vary with lattice parameter or pressure, or both effects could occur. Another possibility is a pressure-induced phase transition involving molecular orientations. In pure C_{60}, free molecular rotations freeze out at $T_{mo} = 249$ K at 1 bar, locking into specific orientations with respect to the crystal axes (17), and

T_{mo} decreases rapidly with increasing P (18, 19). An x-ray study of K_3C_{60} indicates two equally populated molecular orientations at 300 K and 1 bar (20), and a nuclear magnetic resonance study of Rb_3C_{60} gives evidence for a transition near 300 K (21). It is therefore likely that M_3C_{60} transforms to an orientationally ordered phase at 300 K within the P range of the present data, which could affect our analysis by introducing discontinuities in $a(P)$ or by slightly modifying the Fermi surface. The diffraction signature of this transition is quite subtle in C_{60} (17) and would have been undetectable in the present experiments. More data at the extremes of a would help clarify the situation, for example dT_c/dP and compressibility measurements on Rb_2CsC_{60} (1) and extension of the $T_c(P)$ experiments to higher pressure.

REFERENCES AND NOTES

1. R. M. Fleming et al., Nature **352**, 787 (1991).
2. G. Sparn et al., Science **252**, 1829 (1991).
3. J. E. Schirber et al., Physica C **178**, 137 (1991).
4. G. Sparn et al., unpublished results.
5. J. P. McCauley, Jr., et al., J. Am. Chem. Soc. **113**, 8537 (1991).
6. O. Zhou et al., Nature **351**, 632 (1991).
7. Q. Zhu et al., Science **254**, 545 (1991).
8. We first checked that no degradation in x-ray profiles occurred after immersing M_3C_{60} powder in these liquids for several days. Mineral oil was used only below 8 kbar because of its low glass-transition pressure. The disadvantage of pentane-isopentane is its high vapor pressure, which makes it more difficult to seal the cell without trapping voids in the gasket hole.
9. J. Jayaraman, Rev. Mod. Phys. **55**, 1 (1983); L. Finger and H. Hazen, Comparative Chemistry (Wiley, New York, 1982).
10. J. E. Fischer et al., Science **252**, 1288 (1991).
11. H. Zabel, in Graphite Intercalation Compounds, Topics of Current Physics, H. Zabel and S. A. Solin, Eds. (Springer-Verlag, Berlin, 1990), pp. 101–156.
12. S. J. Duclos et al., Nature **351**, 380 (1991).
13. F. C. Zhang, M. Ogata, T. M. Rice, Phys. Rev. Lett. **67**, 3452 (1991).
14. M. Schluter, M. Lannoo, M. Needels, G. Baraff, D. Tomanek, unpublished results.
15. C. M. Varma, J. Zaanen, K. Raghavachari, Science **254**, 989 (1991).
16. S. C. Erwin and W. E. Pickett, ibid., p. 842.
17. P. A. Heiney et al., Phys. Rev. Lett. **66**, 2911 (1991).
18. G. A. Samara et al., ibid. **67**, 3136 (1991).
19. G. Kriza et al., J. Phys. I (France) **1**, 1361 (1991).
20. P. W. Stephens et al., Nature **351**, 632 (1991).
21. R. Tycko et al., Science **253**, 884 (1991).
22. We thank G. Sparn for providing his raw $T_c(P)$ data, M. Y. Jiang for technical assistance, and also K. S. Liang for the use of the X10B beamline. We acknowledge helpful discussion with S. C. Erwin, A. B. Harris, and E. J. Mele. Supported by NSF grants no. DMR-88-19885 and no. DMR-89-01219 and by the Department of Energy, grant no. DE-FC02-86ER45254.

18 November 1991; accepted 2 January 1992

Reprinted with permission from Physical Review Letters
Vol. 68, No. 12, pp. 1912–1915, 23 March 1992

Electronic Properties of Normal and Superconducting Alkali Fullerides Probed by ^{13}C Nuclear Magnetic Resonance

R. Tycko, G. Dabbagh, M. J. Rosseinsky, D. W. Murphy, A. P. Ramirez, and R. M. Fleming

AT&T Bell Laboratories, 600 Mountain Avenue, Murray Hill, New Jersey 07974

(Received 6 November 1991)

We report the results of ^{13}C NMR measurements on K_3C_{60} and Rb_3C_{60} in the normal and superconducting states. Electronic densities of states at the Fermi energy in the normal state and energy gaps in the superconducting state are estimated from spin-lattice relaxation data. Implications of the relaxation and spectral data for the electronic properties of these materials are discussed.

PACS numbers: 74.30.Gn, 74.70.Ya, 76.60.−k

Current interest in the conducting [1] and superconducting [2-5] alkali fullerides A_3C_{60} (A = K, Rb, Cs) stems from the fact that their superconducting transition temperatures (T_c), at least as high as 32 K in the case of Rb_2CsC_{60} [6,7], are currently the highest known aside from those of the "high-T_c" ceramics, and from the fact that the alkali fullerides are three-dimensional, ionic, molecular solids and therefore appear qualitatively different in structure from other known superconductors. We and others have shown that ^{13}C nuclear magnetic resonance (NMR) measurements provide a wealth of information about the structure [8], molecular dynamics [9-12], phase diagrams [11,12], and electronic properties [12] of both pure C_{60} and the alkali fullerides. In this Letter, we present the results of ^{13}C NMR measurements on *both normal and superconducting K_3C_{60} and Rb_3C_{60}.* We obtain estimates of the electronic energy gap 2Δ in the superconducting state and of the density of electronic states at the Fermi energy $N(E_F)$ in the normal state. We discuss possible implications of the NMR data for the electronic properties in the normal and superconducting states.

Samples of Rb_3C_{60} and K_3C_{60} powder, roughly 60 mg each, were prepared as described previously [3,7,13,14] and shown to be single phase by x-ray diffraction and ^{13}C NMR. All natural-abundance ^{13}C nuclei in the samples contributed to the observed NMR signals, to within the accuracy of our calibration of the signal amplitudes (±10%). No significant attenuation of the signals was observed upon cooling below T_c. The magnetic shielding fractions were 40% and 13% and the zero-field T_c values were 29.4 and 19.4 K for the Rb_3C_{60} and K_3C_{60} samples, respectively, as determined from magnetization measurements. Samples were sealed in 5-mm Pyrex tubes with 0.5 atm He(g). NMR measurements were carried out in 9.39- and 2.37-T fields (100.5- and 25.4-MHz ^{13}C NMR frequencies) using a Chemagnetics CMX spectrometer. Low-temperature measurements were made using an NMR probe based on a modified Janis SuperTran-B cryostat [15]. For good thermal contact, the rf coil and sample were contained inside a sealed copper box that was filled with 1 atm He(g) and mounted on the cold finger. Low rf power levels (5 W peak, < 0.1 mW average) were

used at low temperatures. We estimate our temperature measurements to be accurate to within ± 0.2 K.

^{13}C nuclear spin relaxation times T_1 were measured with the saturation-recovery technique, using a Carr-Purcell echo train to improve the sensitivity [16]. Because the recovery curves were slightly nonexponential, they were fitted to the "stretched exponential" form $S(\tau) = S_a - S_b \exp[-(\tau/T_1)^\beta]$, where τ is the recovery time and S is the peak height in the NMR spectrum, with $\beta = 0.82$ for Rb_3C_{60} and $\beta = 0.80$ for K_3C_{60}. The stretched exponential recovery curves reflect a distribution of spin-lattice relaxation rates [17], with a relative width that appears independent of the temperature. This distribution may arise from a dependence of the rate on the crystallite orientation, which would be possible if dipolar or orbital couplings contributed significantly to the relaxation rate [18], or from the fact that inequivalent carbon nuclei on a C_{60} molecule in the cubic A_3C_{60} structure may have different hyperfine couplings.

The relaxation data in Fig. 1, taken at 9.39 T, display three significant features. First, the product T_1T is nearly constant in the normal states of both K_3C_{60} and Rb_3C_{60}, as expected for a metal [18], with only a gradual increase with decreasing temperature from above 200 K to T_c. This increase may be a consequence of thermal contraction [19], which is expected to reduce $N(E_F)$. At the T_c determined from magnetization measurements, T_1T is 165 K s in K_3C_{60} and 100 K s in Rb_3C_{60}. Using the expression

$$\frac{1}{T_1T} = \frac{\pi k}{\hbar} A^2 N(E_F)^2 ,$$

which applies to relaxation due to a contact hyperfine coupling between nuclear spins and noninteracting conduction electrons, and our previous estimate [12] of the hyperfine coupling constant $A = 1.1 \times 10^{-20}$ erg (0.6 G for electron spins), we obtain $N^K(E_F) = 17$ eV^{-1} and $N^{Rb}(E_F) = 22$ eV^{-1} per C_{60} molecule (per spin state) just above T_c. These values are approximations, since our estimate of A may easily be in error by a factor of 2 and since there may be a distribution of hyperfine couplings. Similar values of $N(E_F)$ have been derived from static magnetic susceptibility [20] and critical field

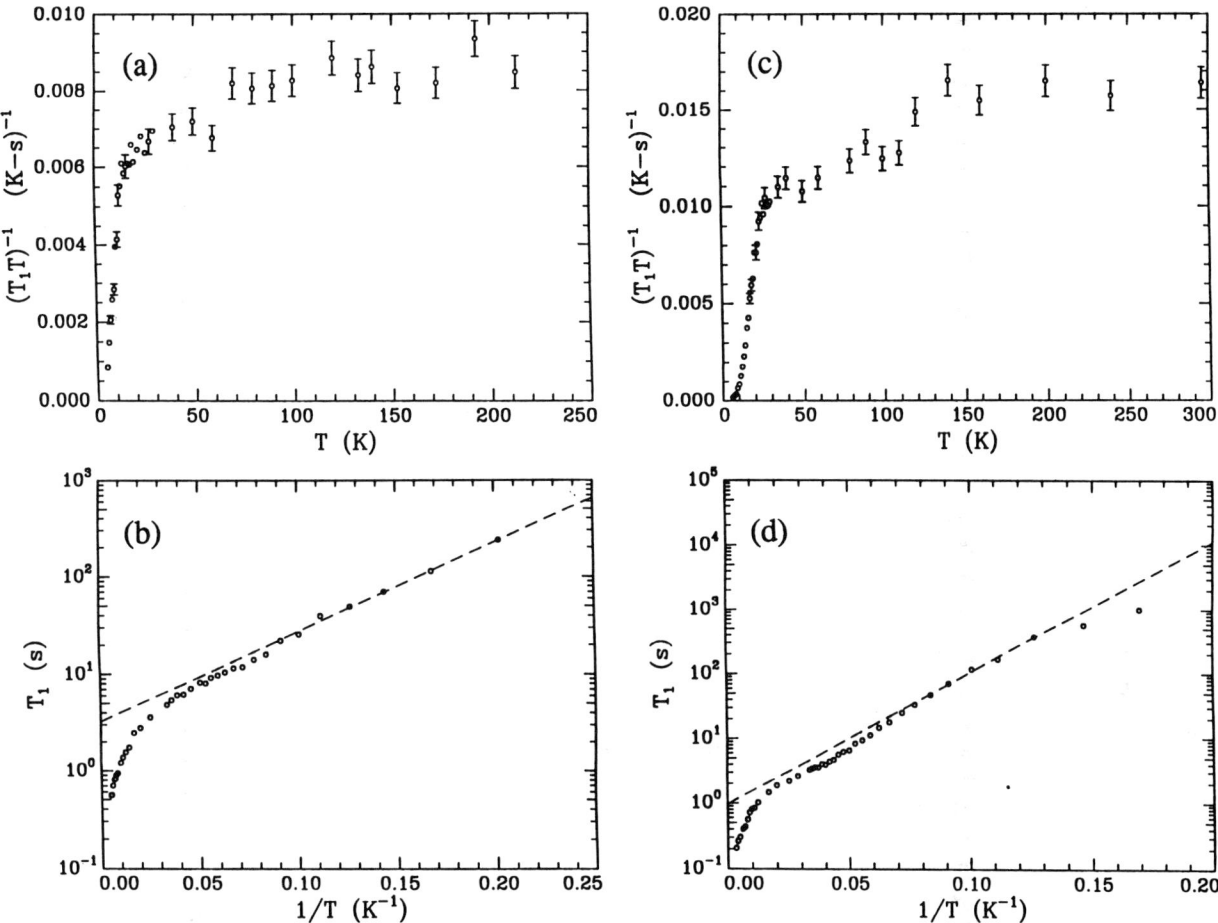

FIG. 1. Temperature dependence of the ^{13}C spin-lattice relaxation time T_1 in (a),(b) K_3C_{60} and (c),(d) Rb_3C_{60}, plotted as $(T_1T)^{-1}$ vs T [(a),(c)] and as $\log T_1$ vs T^{-1} [(b),(d)]. Error bars indicate 1 standard deviation. Dashed lines are fits of the data below 9 K [(b)] and from 8 to 12 K [(d)] by Arrhenius laws, leading to energy gaps $2\Delta^K \approx 42$ K and $2\Delta^{Rb} \approx 94$ K.

[20,21] measurements, while considerably smaller values have been obtained from electron spin resonance [22] and photoemission [23] measurements.

Second, $(T_1T)^{-1}$ decreases dramatically below an apparent superconducting transition temperature T_c^{NMR}, which is 14 ± 1 K for K_3C_{60} and 23 ± 1 K for Rb_3C_{60}. Well below T_c^{NMR}, the data can be fitted by the Arrhenius law $T_1^{-1} = We^{-\Delta/T}$, with $\Delta = 21.3$ K and $W = 0.31$ s^{-1} for K_3C_{60} and $\Delta = 46.8$ K and $W = 1.04$ s^{-1} for Rb_3C_{60}. Such a temperature dependence is expected if there is an energy gap 2Δ for electronic excitations in the superconducting state [24,25]. The values $2\Delta/T_c^{NMR} = 3.0$ for K_3C_{60} and $2\Delta/T_c^{NMR} = 4.1$ for Rb_3C_{60} are to be compared with the weak-coupling BCS result $2\Delta/T_c = 3.5$. Our result for Rb_3C_{60} differs from a recent determination by scanning tunneling microscopy [26] ($\Delta = 77$ K), possibly because NMR relaxation probes the minimum quasiparticle excitation energy, while tunneling probes the maximum in the quasiparticle density of states, or because of differences between surface and bulk properties. Our NMR relaxation data for Rb_3C_{60} clearly deviate from an Arrhenius law below 8 K. At these temperatures, ^{13}C spin-lattice relaxation is very slow (T_1 ~1000 s), so that any weak, extraneous relaxation mechanism may affect the observed T_1 values significantly. One such mechanism is provided by spin diffusion between ^{13}C nuclei in the bulk and the more rapidly relaxing ^{13}C nuclei in normal cores associated with a magnetic flux vortex lattice [27]. We estimate the spin diffusion constant for ^{13}C nuclei at natural abundance to be 1–10 Å^{2}s^{-1} and the distance between vortices to be 150 Å. Spin diffusion will then affect the observed relaxation rates strongly when $T_1 > 500$–5000 s. Vortex motion on the 1000-s time scale would also reduce the observed T_1 values.

Third, within the precision of our measurements, we see no Hebel-Slichter peak [24], which would appear as an *increase* in $(T_1T)^{-1}$ with decreasing temperature immediately below T_c. Our relaxation measurements on K_3C_{60} at 2.37 T (not shown) also do not show a Hebel-Slichter peak. The Hebel-Slichter peak has been observed experimentally in NMR measurements on some superconductors but not others [27,28]. The apparent suppression of T_c^{NMR} in both K_3C_{60} and Rb_3C_{60} by about

3 K below the values expected in a 9.39-T field ($dT_c/dH \approx -0.5$ K/T) may be the vestige of a broadened Hebel-Slichter peak. In other systems, the absence of a Hebel-Slichter peak has been attributed to pair-breaking interactions [27,29–31], gap anisotropy [27], or d-wave pairing [32].

Figure 2 shows the temperature dependence of the center of gravity of the ^{13}C NMR line for K_3C_{60} and Rb_3C_{60}, at 9.39 T, plotted as the frequency shift relative to a tetramethysilane signal (the standard reference for ^{13}C NMR shifts). Pure C_{60} resonates at 143 ppm. The difference between the ^{13}C NMR frequency in pure C_{60} and that in an alkali fulleride will have both chemical shift and Knight shift contributions. The chemical shift contribution will depend on the charge of the C_{60} anion. The Knight shift will only be present in conducting compounds, will be proportional to $N(E_F)$ in the normal state, and will vanish well below T_c. In K_3C_{60}, the shift is essentially independent of temperature in the normal state, lying between 185 and 190 ppm. In Rb_3C_{60}, the shift is more strongly temperature dependent, increasing from 171 ppm at 296 K to 192 ppm at 40 K. In both materials, there is a pronounced decrease in the shift below T_c, by 30–40 ppm. It is tempting to conclude that the Knight shifts in K_3C_{60} and Rb_3C_{60} are therefore positive (to higher frequency) and 30–40 ppm in magnitude. However, one must also take the diamagnetism of the

sample below T_c into account [16]. An extrapolation of magnetization measurements on Rb_3C_{60} taken at fields up to 5 T indicates a magnetization $M \approx -2.0$ G at 9.39 T and 5 K. From the relation $B = H_0 + 4\pi M(1-n)$, we see that the average internal field B is reduced by roughly 20 G (200 ppm) from the applied field H_0, assuming a demagnetizing factor $n = \frac{1}{3}$. Thus, the observed decrease in the ^{13}C NMR shifts below T_c may easily result from the bulk diamagnetism alone. We have also measured shifts of 181 ± 2 ppm for Rb_4C_{60} and 154 ± 5 ppm for Rb_6C_{60} at 296 K. These compounds are neither conductors nor superconductors (T_1T is strongly temperature dependent for Rb_4C_{60}). The precise magnitudes and signs of the ^{13}C Knight shifts in the A_3C_{60} compounds are therefore uncertain, but the Knight shifts appear to be less than 100 ppm in magnitude. The Korringa relation [18] predicts Knights shifts of 160–260 ppm if T_1T is 60–165 K s. The discrepancy may be due to a significant contribution to the relaxation rates from orbital or dipolar couplings, although this seems unlikely because the nonplanarity of the C_{60} molecule leads to a substantial carbon $2s$ orbital component in the conduction band wave functions [33]. An alternative is that the relaxation rates are enhanced relative to the Knight shifts by antiferromagnetic spin fluctuations or a short electron mean free path [34].

Figure 3 shows examples of the spectra from which the data in Fig. 2 are taken. The line shapes in Fig. 3 are superpositions of shift anisotropy powder patterns from the crystallographically inequivalent sites. Above 270 K, the spectrum of K_3C_{60} is narrowed by large-amplitude reorientations of C_{60}^{3-} ions [9–12]. From just above to well below T_c, the lines broaden due to the inhomogeneity of B in the vortex lattice, by roughly 60 ppm in K_3C_{60} and 100 ppm in Rb_3C_{60} (FWHM, assuming Gaussian broadening). These values correspond [35,36] to magnetic penetration lengths $\lambda^K = 6000$ Å and $\lambda^{Rb} = 4600$ Å, in reasonable agreement with the value $\lambda^K = 4800$ Å deter-

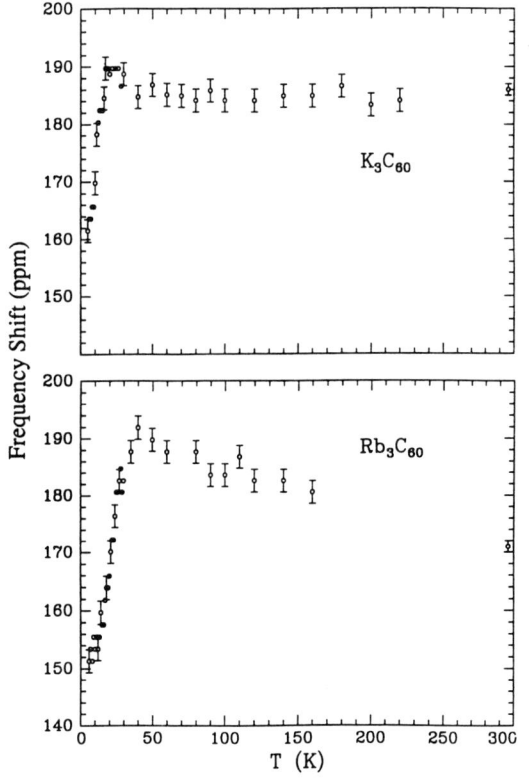

FIG. 2. Temperature dependence of the ^{13}C NMR frequency in K_3C_{60} and Rb_3C_{60}, reported as the shift relative to a tetramethylsilane reference signal.

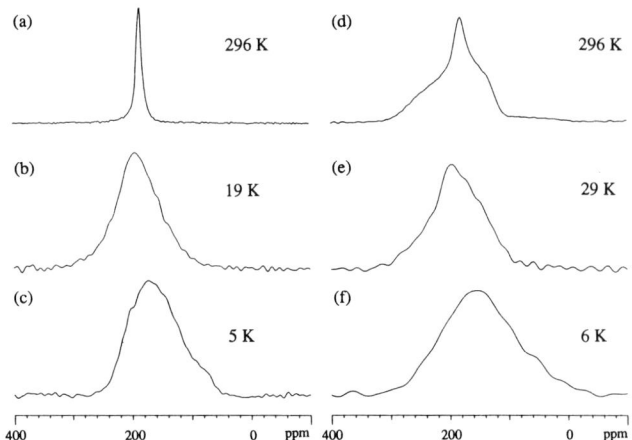

FIG. 3. ^{13}C NMR spectra of (a)–(c) K_3C_{60} and (d)–(f) Rb_3C_{60} at the indicated temperatures.

mined from muon spin relaxation measurements [36].

Our relaxation measurements in the normal state have possible implications for the mechanism of superconductivity. The ratio $R = N^{Rb}(E_F)/N^K(E_F)$ derived from the relaxation data should be more accurate than our estimates of $N(E_F)$, since A is expected to be the same in both materials. Using the BCS expression $T_c = \Omega \exp[-1/VN(E_F)]$, and assuming that the phonon frequency Ω and electron-phonon coupling constant V are the same in K_3C_{60} and Rb_3C_{60}, we obtain

$$\Omega = T_c^K \exp\left[\frac{\ln T_c^{Rb} - \ln T_c^K}{1 - R^{-1}}\right].$$

Using $1.28 < R < 1.40$, the range dictated by the temperature dependence of the T_1T values in Fig. 1, this expression implies $60 \text{ cm}^{-1} < \Omega < 94 \text{ cm}^{-1}$ [and $VN^K(E_F) \approx 0.6$]. Such a small value of Ω suggests that coupling between electrons and low-frequency phonons, i.e., "intermolecular" acoustic, alkali-C_{60} optic [37], or radial "intramolecular" phonons [38], rather than the high-frequency tangential intramolecular phonons [38–40], is important for superconductivity in A_3C_{60}. The same qualitative conclusion follows from the McMillan formula [41], which leads to $150 \text{ cm}^{-1} < \Omega < 230 \text{ cm}^{-1}$. Strong electron-electron interactions [34,42] would make this argument invalid, since then T_1T would not be proportional to $N(E_F)^{-2}$.

We thank B. S. Shastry and P. B. Littlewood for illuminating discussions, R. C. Haddon for the C_{60} starting material, and P. L. Gammel for experimental assistance.

---

[1] R. C. Haddon et al., Nature (London) **350**, 320 (1991).
[2] A. F. Hebard et al., Nature (London) **350**, 600 (1991).
[3] M. J. Rosseinsky et al., Phys. Rev. Lett. **66**, 2830 (1991).
[4] K. Holczer et al., Science **252**, 1154 (1991).
[5] S. P. Kelty, C. C. Chen, and C. M. Lieber, Nature (London) **352**, 223 (1991).
[6] K. Tanigaki et al., Nature (London) **352**, 222 (1991).
[7] R. M. Fleming et al., Nature (London) **352**, 787 (1991).
[8] C. S. Yannoni et al., J. Am. Chem. Soc. **113**, 3190 (1991).
[9] C. S. Yannoni et al., J. Phys. Chem. **95**, 9 (1991).
[10] R. Tycko et al., J. Phys. Chem. **95**, 518 (1991).
[11] R. Tycko et al., Phys. Rev. Lett. **67**, 1886 (1991).
[12] R. Tycko et al., Science **253**, 884 (1991).
[13] W. Kratschmer et al., Nature (London) **347**, 354 (1990).
[14] J. P. McCauley et al., J. Am. Chem. Soc. **113**, 8537 (1991).
[15] K. Carduner, M. Villa, and D. White, Rev. Sci. Instrum. **55**, 68 (1984).
[16] S. E. Barrett et al., Phys. Rev. B **41**, 6283 (1991).
[17] E. Helfand, J. Chem. Phys. **78**, 1931 (1983).
[18] J. Winter, *Magnetic Resonance in Metals* (Oxford Univ. Press, New York, 1971).
[19] P. W. Stephens (private communication). In K_3C_{60}, the lattice parameter determined by x-ray diffraction changes from 14.265 Å at 300 K to 14.168 Å at 11 K, corresponding to a reduction in the nearest-neighbor carbon-carbon distance of 0.07 Å.
[20] A. P. Ramirez et al. (to be published).
[21] K. Holczer et al., Phys. Rev. Lett. **67**, 271 (1991).
[22] S. H. Glarum, S. J. Duclos, and R. C. Haddon (to be published).
[23] C. T. Chen et al., Nature (London) **352**, 603 (1991).
[24] L. C. Hebel and C. P. Slichter, Phys. Rev. **113**, 1504 (1959).
[25] Y. Masuda and A. G. Redfield, Phys. Rev. **125**, 159 (1962).
[26] Z. Zhang et al., Nature (London) **353**, 333 (1991).
[27] D. E. MacLaughlin, in *Solid State Physics*, edited by H. Ehrenreich, F. Seitz, and D. Turnbull (Academic, New York, 1976), Vol. 31, and references therein.
[28] P. C. Hammel et al., Phys. Rev. Lett. **63**, 1992 (1989).
[29] L. Coffey, Phys. Rev. Lett. **64**, 1071 (1990).
[30] P. B. Allen and D. Rainer, Nature (London) **349**, 396 (1991).
[31] Y. Kuroda and C. M. Varma, Phys. Rev. B **42**, 8619 (1990).
[32] H. Monien and D. Pines, Phys. Rev. B **41**, 6297 (1990).
[33] R. C. Haddon, J. Am. Chem. Soc. **109**, 1676 (1987).
[34] B. S. Shastry (to be published).
[35] P. Pincus et al., Phys. Lett. **13**, 21 (1964).
[36] Y. J. Uemura et al., Nature (London) **352**, 605 (1991).
[37] F. C. Zhang, M. Ogata, and T. M. Rice, Phys. Rev. Lett. **67**, 3452 (1991).
[38] D. E. Weeks and W. G. Harter, J. Chem. Phys. **90**, 4744 (1989).
[39] C. M. Varma, J. Zaanen, and K. Raghavachari, Science **254**, 989 (1991).
[40] M. Schluter et al., Phys. Rev. Lett. **68**, 526 (1992).
[41] W. L. McMillan, Phys. Rev. **167**, 331 (1968).
[42] T. Moriya, J. Phys. Soc. Jpn. **18**, 516 (1963).

Reprinted with permission from Science
Vol. 254, pp. 1625–1627, 13 December 1991
© 1991 The American Association for The Advancement of Science

Raman Studies of Alkali-Metal Doped A_xC_{60} Films (A = Na, K, Rb, and Cs; x = 0, 3, and 6)

S. J. Duclos,* R. C. Haddon, S. Glarum, A. F. Hebard,
K. B. Lyons

The room temperature Raman spectra of the intramolecular modes between 100 cm^{-1} and 2000 cm^{-1} are reported for alkali-metal doped A_xC_{60} films. For A = K, Rb, and Cs, phase separation is observed with the spectra of C_{60}, K_3C_{60}, K_6C_{60}, Rb_3C_{60}, Rb_6C_{60}, and Cs_6C_{60} phases reported. The x = 3 phases show only three Raman active modes: two of A_g symmetry and only the lowest frequency H_g mode. The other H_g modes regain intensity in the x = 6 films, with several mode splittings observed. For A = Na, such phase separation is not clearly observed, and reduced mode shifts are interpreted as due to incomplete charge transfer in these films.

CONDUCTING (1) AND SUPERCONducting (2) compounds of A_xC_{60} (0 ≤ x ≤ 6) have recently been prepared by intercalating alkali metals (A) into both powders and films of C_{60} buckminsterfullerene (3–4). Raman spectroscopy is an optical probe of the normal mode frequencies of the C_{60} cage, which are influenced by C–C bond strengths, as well as the electron-phonon coupling important to superconductivity. This report presents Raman spectra of the C_{60} intramolecular modes for all dopant phases observed in films. The spectra show a significant dependence on dopant phase, and represent a convenient optical probe of phase separation in films. Shifts in frequency of these modes as x is increased from 0 to 6 can be attributed to changes in the C–C bonds as the anti-bonding states of the C_{60} molecule become occupied. The relatively high superconducting transition temperatures (18 K to 33 K) (2, 5–8) have stimulated interest in the strength of the electron-phonon coupling in the A_3C_{60} superconductors. The intramolecular phonon modes observed in these Raman spectra are of particular interest owing to their calculated strong scattering of electrons at the Fermi-surface (9–10). Thus the Raman spectra presented here lead to insights on the binary phase diagram of doped C_{60} films, changes in the intramolecular bonding with doping, and the superconducting mechanism of the A_3C_{60} compounds.

The unpolarized Raman spectra were excited by 50 W cm^{-2} (15 mW incident power) of the 514.5 nm Ar ion laser line on approximately 100 μm by 300 μm spots (Fig. 1). A monochromator was used to eliminate laser plasma lines from the spectra. Light was collected in a 45° back scattering geometry, and dispersed by a 3/4-m Spex double spectrometer with 7 cm^{-1} spectral

AT&T Bell Laboratories, Murray Hill, New Jersey 07974.

*To whom correspondence should be addressed.

resolution. Light was detected with a Hamamatsu R585 photomultiplier tube (PMT) and photon counting electronics. Scanning speeds of 10 cm^{-1} per minute were used, and the spectra presented are an average of three scans.

The C_{60} used in the present experiments was first purified by chromatography and reprecipitated in hexane to remove organic contaminants. Films were then grown by sublimation at 425°C for 6 minutes onto the inner surface of evacuated Suprasil tubes of 2-mm inside diameter. X-ray diffraction peak widths indicate a crystallite size of approximately 50 to 60 Å in these films. The Raman spectrum of the Suprasil tube has been measured under the same conditions, and none of the features reported here, unless otherwise stated, can be attributed to it. Doping was accomplished by inserting alkali metal into the tube in a He glove box, sealing in a vacuum of <10^{-3} torr, and heating the entire tube isothermally (11). The superconducting transition temperature, as measured by microwave absorption, of the K- and Rb-doped films was increased by annealing the film portion of the tube after doping. The total doping times and temperatures, and annealing times and temperatures are summarized in Table 1. This resulted in either two (Cs) or three (Na, K, and Rb) visibly sharp bands along the axis of the tube (Fig. 1B). Stability of the doped materials was confirmed by scanning the high frequency A_g mode before and after 24 hours of laser irradiation under the same focus conditions. The Na-doped film showed several anomalous features, and will be discussed separately at the end of this report.

The stoichiometry of the resulting bands has been determined in the following manner. A film that we homogeneously doped with K using the geometry of Haddon et al. (1) was simultaneously probed with both Raman spectroscopy and van der Pauw four-probe resistivity measurements. The spectrum of the material at the resistivity

minimum, identified as K_3C_{60} by previous Rutherford backscattering spectroscopy (RBS) measurements (12), matched that found for the middle band of the K- and Rb-doped films grown in Suprasil tubes. Upon further doping, the spectrum acquired at the resistivity maximum, identified as K_6C_{60} by RBS measurements (12), matched that found in the upper band of the K-, Rb-, and Cs-doped films. The position of the superconducting phase, which we infer to be A_3C_{60}, was also measured by low-temperature microwave absorption (11). For the K- and Rb-doped films this was found to be the same band as that identified as A_3C_{60} by the simultaneous Raman and resistivity measurements. The banding and distinct phases observed here are consistent with the phase separation observed in bulk samples by x-ray diffraction and NMR (13). We note that no evidence of the A_4C_{60} (14) or Cs_3C_{60} (7) phases has been observed in these films.

A spectrum of pristine C_{60} sublimed in a tube and unexposed to air, O_2, or alkali metal was taken (Fig. 2). We have previously shown (15) that exposure of vacuum-grown films to either air or oxygen alters the spectrum, especially in the frequency region of the symmetric A_g mode at 1458 cm^{-1}. In previous experiments (15) the Si substrate inhibited the identification of modes near 520 cm^{-1} and 950 cm^{-1} to 1000 cm^{-1}. In the present experiments, the Si substrate has been eliminated and we can identify the mode at 522 cm^{-1} and the broad feature between 950 cm^{-1} and 1000 cm^{-1} as due to C_{60}. The broad feature may result from a two phonon process involving the 490

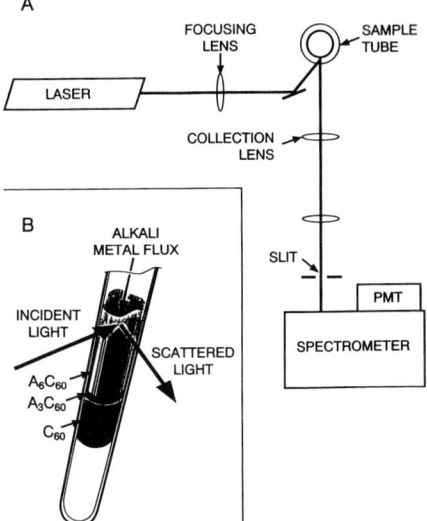

Fig. 1. (**A**) Schematic of the Raman spectrometer used in the present experiments. (**B**) Drawing of the sample tube after alkali metal doping, showing bands of different alkali stoichiometry in the film on the inside surface of the tube.

Table 1. Total doping and annealing times and temperatures for the A_xC_{60} films used in this study. No annealing of the Cs-doped film was done, as it showed no evidence of a Cs_3C_{60} phase.

Alkali metal	Doping		Annealing	
	Temperature (°C)	Time (hours)	Temperature (°C)	Time (hours)
Na	190	4	180	1.5
K	130	1.7	220	32
Rb	100	7.7	120	6
Cs	70	16.4	—	—

cm^{-1} mode. In addition, modes are observed in the 275 cm^{-1} region, which closely corresponds to the calculated frequency of the lowest frequency H_g mode. Splittings in this frequency region, as well as in the 700 cm^{-1} to 775 cm^{-1} range, may be attributable to loss of symmetry of the C_{60} molecule owing to interactions with its neighbors in the solid state. Depolarization ratios·of the modes of the pristine material are consistent with the mode interpretations of Bethune *et al.* (16), who assign the 1458 cm^{-1} peak to the pentagonal-pinch A_g mode, and the 488 cm^{-1} peak to the A_g symmetric breathing mode, and the others as H_g modes, including the low-frequency C_{60} squashing mode. The full width at half maximum (FWHM) of the 1458 cm^{-1} mode is 15 cm^{-1}, and Tolbert *et al.* (17) attribute the anomalous width to rotational-vibrational coupling. The line in our data may be further broadened by the loss of momentum conservation caused by grain boundaries in our microcrystalline films. Incorporation of O_2 into these microcrystalline films results in an intense peak at 1469 cm^{-1} with a resolution-limited FWHM. The Raman spectra of the alkali-doped phases did not depend on

whether or not the starting film had been exposed to air.

We now consider films doped to the A_3C_{60} phase. Raman spectra were obtained for K_3C_{60} (T_c = 18 K) and Rb_3C_{60} (T_c = 24 K) (Fig. 3). The striking feature of these spectra is their simplicity. Only three modes are observed, the two A_g modes and the lowest frequency H_g mode. Further, there is no dependence of the spectra on alkali dopant, indicating no A-C_{60} mode is observed above 100 cm^{-1}. Spectra taken on a variety of K- and Rb-doped samples indicate that the A_g mode frequency in A_3C_{60} is 1445 ± 2 cm^{-1}. The FWHM of this mode is 9 cm^{-1}, which is 6 cm^{-1} less than the value for the pristine material. This reduction is likely due to the stationary orientation of the C_{60} molecules on the time scale of the Raman scattering process, and is similar to the reduction observed in pristine C_{60} as the molecules freeze at low temperatures (17). This result is consistent with recent NMR and x-ray diffraction results that indicate the molecules are jumping between symmetry-equivalent orientations in the K_3C_{60} material at room temperature (13). Finally, we point out that the lowest frequency H_g mode appears to remain, although it is significantly broadened.

We have considered several possible explanations for the disappearance of most of the H_g modes. First, the addition of alkali metal atoms to the lattice will introduce symmetry reduction due to A-C_{60} interactions, which will tend to broaden the H_g modes as their degeneracy is broken. However, on the scale of our resolution these effects should be small and, as discussed below, further doping results in a reappear-

ance of many of these lines, which argues against the intensity reduction solely rising from symmetry breaking. A second explanation involves resonant enhancement of the A_g lines as the electronic structure of the solid is altered during doping, as well as a reduction in overall intensity due to decreased penetration depth of the laser into the metallic material. A factor of 5 decrease in penetration depth would render all of the H_g modes below 1400 cm^{-1} unobservable. Third, recent treatments (9–10) of the electron-phonon coupling in doped C_{60} indicate that the H_g modes couple more strongly than do the A_g modes. As discussed by Allen (18), such coupling leads to a broadening of the affected modes, and for A_3C_{60} will result in a greater broadening for the H_g modes. An assessment of the relative merits of these explanations will be possible only after optical work is performed on the A_3C_{60} materials to determine the penetration depth and the possible effects of resonant enhancement on the observed modes.

Figure 4 shows the Raman spectra of K_6C_{60}, Rb_6C_{60}, and Cs_6C_{60}. The high frequency A_g mode is now at 1430 ± 1 cm^{-1}, and has a resolution-limited FWHM of 7 cm^{-1}. The further narrowing of this mode may be due to continued freezing out of the rotational-vibrational coupling, as NMR (13) and synchrotron x-ray (19) results indicate that the molecules are completely frozen at room temperature in K_6C_{60}. The shift of the A_g mode with doping in the solid appears to be linear at 5 cm^{-1} per electron reduction of C_{60}, and is a convenient probe of reduction state of the molecule in the solid (1).

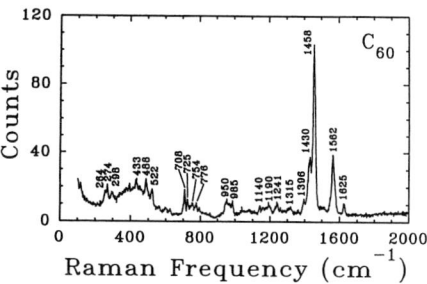

Fig. 2. Room temperature unpolarized Raman spectrum of a pure C_{60} film grown on the inside surface of a Suprasil tube without exposure to air. A broad background between 200 cm^{-1} and 500 cm^{-1} is from the Suprasil tube. The peak at 488 cm^{-1} lies very near a similar weaker feature in SiO_2, but has been observed in samples deposited on Si (15). The laser power density (50 W/cm² at 514.5 nm) is the same as for the spectra in Figs. 3 and 4.

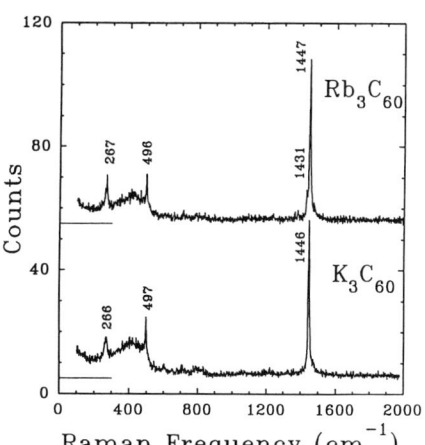

Fig. 3. Room temperature unpolarized Raman spectra of A_3C_{60} (A = Rb and K). Both spectra are on the same scale, but are shifted vertically for clarity. The broad background between 200 cm^{-1} and 500 cm^{-1} is from the Suprasil tube. The shoulder on the low frequency side of the 1447 cm^{-1} peak in Rb_3C_{60} is likely due to a small contamination of Rb_6C_{60}.

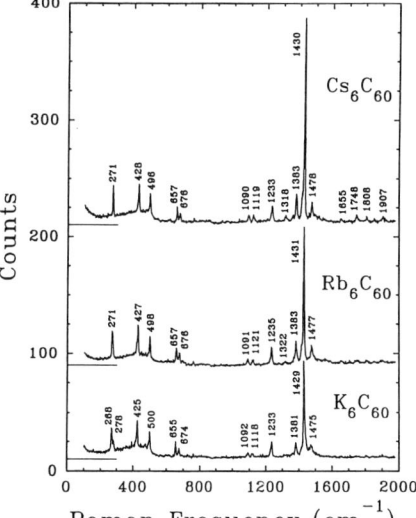

Fig. 4. Room temperature unpolarized Raman spectra of A_6C_{60} (A = Cs, Rb, and K). All of the spectra are on the same scale, but are shifted vertically for clarity. The broad background between 200 cm^{-1} and 500 cm^{-1} is from the Suprasil tube.

The similarity between the A_6C_{60} spectra doped with different alkali metals again indicates that the observed modes are due only to the C_{60} molecule, and that there are no observable A-C_{60} modes above 100 cm^{-1}. The mode at 270 cm^{-1} has a resolution-limited FWHM for Cs_6C_{60}, is broadened in Rb_6C_{60}, and is a doublet in K_6C_{60}. The high-frequency mode of the doublet may result from a two-phonon process involving a low energy–A-C_{60} mode, the frequency of which will increase as the ion mass is reduced. Finally, we also point out that the four modes observed above 1600 cm^{-1} in Cs_6C_{60} (and weakly in Rb_6C_{60} and K_6C_{60}) appear to be sum modes of the two lowest frequency H_g modes (271 cm^{-1} and 428 cm^{-1}) and the two highest frequency H_g modes (1383 cm^{-1} and 1478 cm^{-1}). If this is the case, it is not clear that these modes should be broadened in the same way as the zone center (one-phonon) modes by the electron phonon coupling (9–10).

The Raman spectra of Na_xC_{60} are consistent with the above results with the following exceptions. Although banding was observed in these films upon doping, intermediate values of the A_g mode frequency were observed between 1448 cm^{-1} and 1455 cm^{-1}. Also, a time dependence of this mode frequency has been observed in the laser beam at the power densities used in the studies of the K-, Rb-, and Cs-doped films. These results indicate that Na-doped C_{60} either does not phase separate in Na_xC_{60} with $0 \leq x \leq 3$ or that phases other than $x = 0$ and $x = 3$ are stable. For Na_6C_{60} the A_g mode is at 1434 cm^{-1}, significantly higher than the 1430 $\pm$ 1 cm^{-1} observed in the other A_6C_{60} materials. This is consistent with incomplete electron transfer in Na_xC_{60}, which may be responsible for the lack of superconductivity above 4 K (11). Further work is needed to clarify the phase diagram and electronic structure of the Na_xC_{60} compounds.

REFERENCES AND NOTES

1. R. C. Haddon et al., Nature 350, 320 (1991).
2. A. F. Hebard et al., ibid., p. 600.
3. H. W. Kroto, J. R. Heath, S. C. O'Brien, R. F. Curl, R. E. Smalley, ibid. 318, 162 (1991).
4. W. Krätschmer, L. D. Lamb, K. Fostiropoulos, D. R. Huffman, ibid. 347, 354 (1990).
5. M. J. Rosseinsky et al., Phys. Rev. Lett. 66, 2830 (1991).
6. K. Holczer et al., Science 252, 1154 (1991).
7. S. P. Kelty, C.-C. Chen, C. M. Lieber, Nature 352, 223 (1991).
8. K. Tanigaki et al., ibid., p. 222.
9. C. M. Varma, J. Zaanen, K. Raghavachari, Science 254, 989 (1991).
10. M. A. Schluter, M. Lannoo, M. K. Needels, G. A. Baraff, in preparation.
11. S. H. Glarum, S. J. Duclos, R. C. Haddon, in preparation.
12. G. P. Kochanski, A. F. Hebard, R. C. Haddon, A. T. Fiory, in preparation.
13. R. Tycko et al., Science 253, 884 (1991).
14. R. M. Fleming et al., Nature 352, 701 (1991).

Reprinted with permission from Nature
Vol. 354, pp. 462–463, 12 December 1991
© 1991 Macmillan Magazines Limited

Vibrational spectroscopy of superconducting K₃C₆₀ by inelastic neutron scattering

Kosmas Prassides*, John Tomkinson†,
Christos Christides*, Matthew J. Rosseinsky‡,
D. W. Murphy‡ & Robert C. Haddon‡

* School of Chemistry and Molecular Sciences, University of Sussex,
Brighton BN1 9QJ, UK
† ISIS Science Division, Rutherford Appleton Laboratory,
Didcot, OX11 0QX, UK
‡ AT&T Bell Laboratories, Murray Hill, New Jersey 07974, USA

INTERCALATION of C_{60} (buckminsterfullerene[1,2]) by alkali metals[3] leads to superconducting compounds of stoichiometry A_3C_{60} (refs 4–6) with transition temperatures T_c as high as 33 K (ref. 7). These transition temperatures are considerably higher than those for alkali-metal-intercalated graphite (<0.6 K)[8] and scale with the size of the face-centred-cubic unit cell[9]. Here we present the results of an inelastic neutron scattering study of the vibrational spectrum of the superconducting fulleride K_3C_{60} (T_c = 19.3 K). We find significant changes in the peak positions and intensities principally of the intramolecular H_g vibrational modes, both in the high-energy tangential (130–200 meV) and the low-energy radial (~50 meV) regions, compared with the vibrational spectrum of C_{60} (refs 10, 11). Our results provide strong evidence for the importance of these modes in the pairing mechanism for superconductivity.

The K_3C_{60} sample was prepared by reaction of $K_{6.6}C_{60}$ and C_{60} in a sealed evacuated pyrex tube at 250 °C for 3 days, 350 °C for 24 hours and 250 °C for 8 days[12]. Phase purity was confirmed by X-ray diffraction measurements and Rietveld refinement of high-resolution neutron powder-diffraction data. For the neutron scattering measurements, the 500-mg sample was sealed in a 9-mm-diameter fused silica tube under 0.5 atm helium. Inelastic neutron scattering (INS) measurements were done at 5 K and 30 K with the sample inside a continuous-flow 'orange' cryostat, using the time-focused crystal analyser (TFXA) spectrometer[13] at the ISIS facility, Rutherford Appleton Laboratory; a background run was recorded on an identical empty silica tube at 30 K. TFXA is an inverted-geometry white-beam spectrometer with excellent resolution, $\Delta\omega/\omega \approx 2\%$, over the entire energy transfer range, 2.5–500 meV (1 meV = 8.066 cm⁻¹). Even

for primarily coherent scatterers like carbon, the instrumental characteristics of TFXA permit the use of the incoherent approximation at energy transfers >25 meV. Coherent scattering effects are important only at low-energy transfer.

The scattering law $S(Q, \omega)$ of K_3C_{60} was measured in down-scattering mode between 2.5 and 200 meV. The summed data collected at both 5 and 30 K are shown in Fig. 1. $S(Q, \omega)$ may be written as

$$S(Q, \omega)_\nu \propto (1/3)(Q^2 \, \mathrm{Tr}\mathbf{B}_\nu) \exp(-Q^2 \alpha_\nu)$$

$$\alpha_\nu \approx (1/5)\{\mathrm{Tr}\,\mathbf{A} + 2(\mathbf{B}_\nu : \mathbf{A}/\mathrm{Tr}\,\mathbf{B}_\nu)\}$$

where $\mathbf{A} = \Sigma_\nu \mathbf{B}_\nu$, $\mathbf{B}_\nu$ is the mean square displacement tensor of the scattering atom in mode ν and Q is the momentum transferred during scattering; other terms and conventions are defined elsewhere[14]. Carbon-60 is a molecular solid with negligible dispersion of the intramolecular phonons, and the observed peaks in $S(Q, \omega)$ correspond well to the $Q = 0$ modes observed in infrared and Raman spectroscopy. There are no selection rules in INS and hence all the vibrational modes are observed; in addition, considerations of reduced optical penetration depth in K_3C_{60}, important in Raman and infrared spectroscopy, are not relevant here. The vibrational spectrum of K_3C_{60}, like that of C_{60} (ref. 11), may be divided into two regions: (1) 2.5–25 meV, corresponding to intermolecular and K^+-C_{60}^- modes with a prominent feature at 4.3 meV and a band at 14 meV with full-width at half-maximum of 7 meV; and (2) 25–200 meV, corresponding to the intramolecular modes. The vibrational modes of the molecule may be further subdivided into principally tangential motions, clustering in the high-energy region (110–200 meV) and radial (buckling) motions in the low-energy region (25–110 meV). The symmetry of most of the buckling modes up to 110 meV may be inferred by comparison with C_{60}; there are well-resolved bands at 33 ($H_g^{(1)}$), 44 (T_{2u}, G_u), 49 (H_u), 60 (G_g), 68 (T_{1u}), 71 (T_{1u}), 82 and 87 meV ($H_g^{(3)}$), and a broad band with peaks at 93 and 97 meV ($H_g^{(4)}$). The high-energy region does not show sharp well-defined features and divides roughly into three parts with pronounced cut-offs in intensity: 110–135, 135–170 and 170–200 meV.

Insight into the changes in the intramolecular bonding of C_{60} on intercalation, and the relevance of the various vibrational modes to the mechanism of superconductivity, can be obtained by comparing the low-temperature INS spectra of K_3C_{60} and C_{60}. The striking difference between the two systems is the redistribution of intensity that occurs between the radial and tangential modes. Carbon-60 has intense features up to ~110 meV, followed by many weaker and broader bands up to

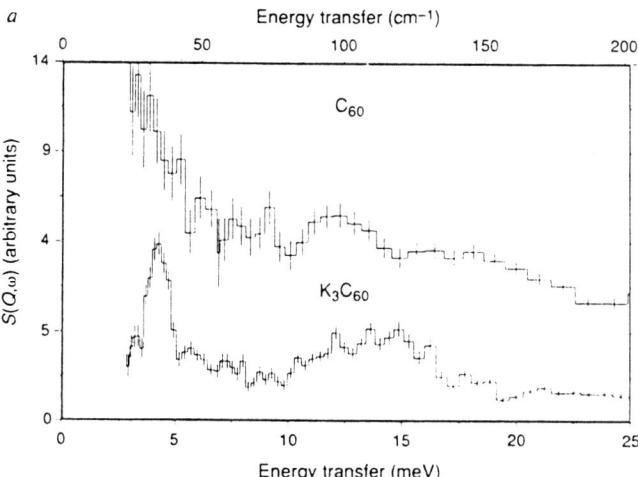

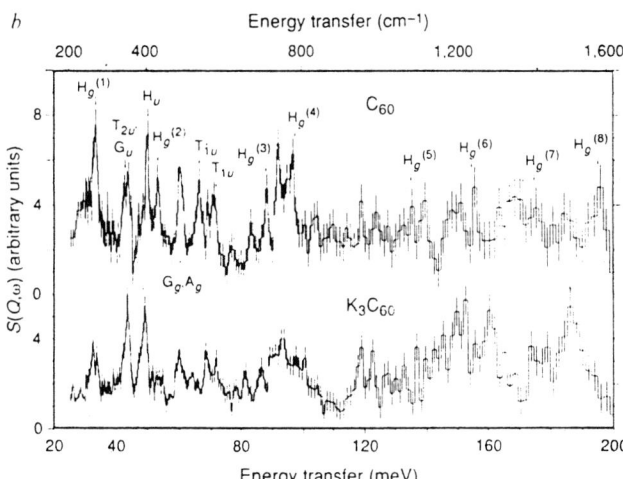

FIG 1 Inelastic neutron scattering spectra in the energy regions (a) 0–25 meV and (b) 25–200 meV of C_{60} at 20 K (top) and K_3C_{60} (bottom); the K_3C_{60} data were collected at 5 K and 30 K and summed. The instrumental resolution is $\Delta\omega/\omega \approx 2\%$, corresponding to decreased resolution at high energy transfers. Some mode assignments in the icosahedral point group (I_h) are shown.

200 meV. In this respect, it resembles graphite which shows the same distribution of intensity between modes involving out-of-plane and in-plane motion[15]. In K_3C_{60}, on the other hand, the high-energy region is intense. The physical reasons behind these changes in intensity are not clear, but they may reflect increased mixing between radial and tangential modes. One possible mechanism involves the strengthening of interaction force-constants. The effect of this would be to switch intensity between the interacting vibrations, without necessarily changing their energy very much.

Theoretical treatments of superconductivity in A_3C_{60} have considered purely electronically induced[16] as well as phonon-induced pairing, involving either low-frequency (~10-20 meV) $K^+C_{60}^{3-}$ optic modes[17] or the high-energy intramolecular modes[18,19]. In addition, the symmetry considerations for the intramolecular coupling models allow only A_g and H_g phonons to couple to the t_{1u} conduction electrons. Schluter et al.[18] find coupling strength distributed over both the low-frequency radial modes and the high-frequency tangential modes, whereas Varma et al.[19] find that the purely tangential $H_g^{(7)}$ and $H_g^{(8)}$ modes at 177 and 196 meV dominate the pairing mechanism. Strong electron-phonon coupling will produce substantial broadening and softening (~10-20%) of the affected modes. Raman spectroscopy is sensitive to the A_g and H_g modes with both A_g and all eight H_g modes evident for C_{60}. Room-temperature Raman spectra[20] of A_3C_{60} thin films show only two A_g modes and the lowest-frequency $H_g^{(1)}$ mode, the absence of the other modes being attributed either to decreased optical penetration depth or lifetime broadening of the spectral function due to electron-phonon coupling.

The INS data confirm the electron-phonon coupling to at least some of the H_g modes. Indeed the most remarkable difference between the K_3C_{60} and C_{60} spectra is the disappearance of the $H_g^{(2)}$ and $H_g^{(8)}$ modes at 54 and 196 meV, consistent with strong electron-phonon coupling. This is an important difference from alkali-intercalated graphite, where the corresponding buckling $H_g^{(2)}$ mode does not couple to the π-electrons because of symmetry considerations. Radial $H_g^{(1)}$, $H_g^{(3)}$ and $H_g^{(4)}$ modes are present in K_3C_{60} at 33, 87 and 97 meV and show only slightly reduced intensities and increased broadening, consistent with minor contributions to the electron-phonon coupling strength. A definitive statement about the effect of intercalation on the $H_g^{(5-7)}$ modes at 136, 155 and 177 meV in C_{60} is difficult because they are weak, but there are significant changes in the spectra of K_3C_{60} in these regions. Although definitive peak assignments must await detailed modelling, it is tempting to assign the broad feature in K_3C_{60} at 186 meV to $H_g^{(8)}$, corresponding to a softening of ~6%. The differences between C_{60} and K_3C_{60} in the 10-20 meV region may be due to K^+-C_{60}^{3-} optic modes.

We have also recorded the INS spectra of K_3C_{60} above (30 K) and below (5 K) the superconducting transition temperature. There is only modest evidence for changes in the intensity of the intramolecular modes in the 135-170 and 170-200 meV regions, as the sample goes through T_c. A decrease in intensity is, however, observed between 30 and 5 K at 4.2 meV; there is also some indication that this is accompanied by an increased broadening. Note that the scattering in this energy range is completely coherent, and we are cutting through a phonon, probably acoustic, branch. These effects may be caused by an as yet undetected structural anomaly at or above T_c, or arise because the phonon energy is close to the superconducting energy gap ($2\Delta \approx 40$ K (ref. 21)). We note that the presence of soft acoustic phonons is well documented in the case of A15-structure superconducting materials[22].

Our data provide strong evidence for the presence of electron-phonon coupling to H_g modes of K_3C_{60}, as predicted by intramolecular phonon theories[18,19]. Higher-resolution data and detailed mode assignments are needed for a more complete picture, particularly the location of the softened H_g modes in

K_3C_{60}, identification of optic modes and the nature of the near-gap scattering law below T_c. □

Received 8 November; accepted 20 November 1991.

1. Kroto, H. W., Heath, J. R., O'Brien, S. C. & Smalley, R. E. Nature 318, 162-164 (1985).
2. Kratschmer, W., Lamb, L. D., Fostiropoulos, K. & Huffman, D. R. Nature 347, 354-358 (1990).
3. Haddon, R. C. et al. Nature 350, 320-322 (1991).
4. Hebard, A. F. et al. Nature 318, 600-601 (1991).
5. Rosseinsky, M. J. et al. Phys. Rev. Lett. 66, 2830-2832 (1991).
6. Holczer, K. et al. Science 252, 1154-1157 (1991).
7. Tanigaki, K. et al. Nature 352, 222-223 (1991).
8. Hannay, N. B. et al. Phys. Rev. Lett. 14, 225-226 (1965).
9. Fleming, R. M. et al. Nature 352, 787-788 (1991).
10. Cappelletti, R. L. et al. Phys. Rev. Lett. 66, 3261-3264 (1991).
11. Prassides, K. et al. Chem. Phys. Lett. (in the press).
12. McCauley, J. P. Jr et al. J. Am. chem. Soc. 113, 8537-8538 (1991).
13. Penfold, J. & Tomkinson, J. Rutherford Appleton Lab. Internal Rep. RAL-86-019 (1986).
14. Tomkinson, J., Warner, M. & Taylor, A. D. Molec. Phys. 51, 381-392 (1984).
15. Hannon, A. C. & Tomkinson, J. Rutherford Appleton Lab. ISIS Ann. Rep. A138 (1989).
16. Chakravarty, S. & Kivelson, S. Europhys. Lett. 16, 751-756 (1991).
17. Zhang, F. C., Ogata, M. & Rice, T. M. Phys. Rev. Lett. (submitted).
18. Schluter, M., Lannoo, M., Needels, M., Baraff, G. A. & Tomanek, D. Phys. Rev. Lett. (submitted).
19. Varma, C. M., Zaanen, J. & Raghavachari, K. Science 254, 989-992 (1991).
20. Duclos, S. J., Haddon, R. C., Glarum, S., Hebard, A. F. & Lyons, K. B. Science (in the press).
21. Tycko, R. et al. Phys. Rev. Lett. (submitted).
22. Shirane, G., Axe, J. D. & Birgeneau, R. J. Solid State Commun. 9, 397-399 (1971).

ACKNOWLEDGEMENTS. We acknowledge J. White and M. Schluter for discussions. We thank R. Fleming for recording the X-ray and R. Ibberson and W. David for the high-resolution neutron diffraction profile. We thank SERC for financial support (K.P.) and for access to ISIS.

Reprinted with permission from the Journal of the American Chemical Society
Vol. 114, pp. 3141–3142, 1992
© 1992 American Chemical Society

Synthesis of Pure $^{13}C_{60}$ and Determination of the Isotope Effect for Fullerene Superconductors

Chia-Chun Chen and Charles M. Lieber*

Department of Chemistry and Division of Applied Sciences, Harvard University Cambridge, Massachusetts 02138

Received January 27, 1992

Herein we report the synthesis of 99% $^{13}C_{60}$ and the measurement of the isotope effect on the superconducting transition temperature (T_c) in potassium-doped materials. To our knowledge, this work represents the first preparation of isotopically pure material and has thus enabled an unambiguous analysis of the isotope shift. Our results demonstrate that T_c is depressed significantly in $K_3{}^{13}C_{60}$ versus $K_3{}^{12}C_{60}$. The depression of T_c for the ^{13}C material indicates that superconductivity arises from phonon-mediated pairing and is not purely electronic in origin.[1,2] The value of the isotope exponent, $\alpha = 0.3$ $(T_c \propto M^{-\alpha})$, is smaller than the theoretical prediction of $\alpha = 0.5$ for a simple phonon mechanism[3] and therefore also suggests that other interactions play a role in determining superconductivity in these cluster-based solids.

Several methods have been used to prepare C_{60} enriched with ^{13}C. One technique involves packing hollow carbon-12 rods with ^{13}C powder and then resistively vaporizing the rods to produce a distribution of $^{13}C/^{12}C$ fullerenes.[4] More recently, Ramirez et al. prepared carbon rods from a mixture of carbon-13 powder and [^{13}C]glucose, but found significant ^{12}C incorporated into the C_{60} product.[5] Pure $^{13}C_{60}$ and not a distribution of $^{13}C/^{12}C$ is

Table I. Experimental and Calculated Infrared Absorption Bands for $^{12}C_{60}$ and $^{13}C_{60}$

	frequency (cm⁻¹)			
$^{12}C_{60}$ (exptl)	1429	1183	576	527
$^{13}C_{60}$ (calcd[a])	1373	1137	553	506
$^{13}C_{60}$ (exptl)	1375	1138	554	506

[a] Calculated from the $^{12}C_{60}$ experimental data using $\nu(^{13}C_{60}) = \nu(^{12}C_{60})[M(^{12}C)/M(^{13}C)]^{1/2}$. IR spectra were recorded on C_{60} thin films using a Nicolet-5PC FT-IR.

needed, however, to determine unambiguously the isotope effect on T_c and other physical properties.

To obtain $^{13}C_{60}$ we prepare isotopically pure carbon-13 rods using a straightforward method that can be carried out with

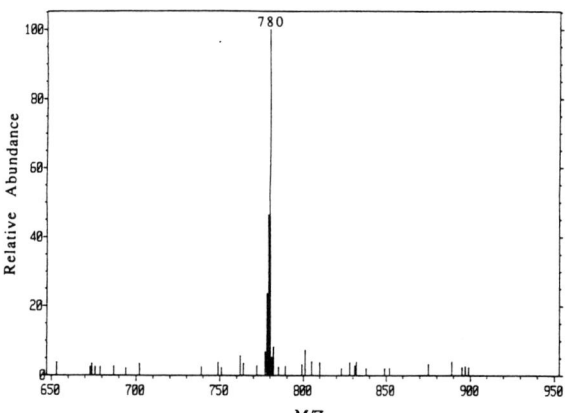

Figure 1. Field desorption (FD) mass spectrum of a purified $^{13}C_{60}$ sample (M^+, 780) recorded using a JEOL AX505H spectrometer. Experimental conditions were as follows: 30 mA FD emitter current; 3 keV ion energy; 9 keV extraction energy. Similar spectra were recorded using fast atom bombardment.

standard equipment. First, ^{13}C powder (99% ^{13}C, Aldrich) was placed in a 0.125-in. i.d. quartz tube between two tantalum rods, and then this assembly was evacuated to ca. 10^{-2} Torr. A constant pressure of 2000 lbs/in² was applied using a laboratory press while current pulses (10 A, 2–3 s) were passed through the sample to drive out air and water. The carbon rods were then formed by sequentially increasing the applied current while the powder was under compression. We found that a sequence of 10-, 50-, and 100-A currents, each applied for 10 s, resulted in the formation of high-quality rods. The rods fabricated in this way were 10–15 mm long and similar in density to commercial ^{12}C rods. The ^{13}C rods were resistively vaporized,[6,7] and pure $^{13}C_{60}$ was isolated from the resulting carbon soot by chromatography, as discussed previously.[8,9]

Figure 1 shows the field desorption mass spectrum of purified $^{13}C_{60}$. The M^+ signal is at m/z 780. The ions at m/z 779 and 778 are isotope peaks resulting from the presence of 1% ^{12}C in the starting material. Since essentially no signal is observed above the background level from m/z 700–777, we can conclude that our material is not contaminated with $^{12}C_{60}$. Furthermore, no signal was detected at m/z 910 (mass $^{13}C_{70} = 910$) in spectra recorded on purified $^{13}C_{60}$ samples. ^{13}C NMR spectra (Brucker AM-500) obtained in C_6D_6 showed the single peak at 143.2 ppm expected for $^{13}C_{60}$;[4] no resonances were detected for $^{13}C_{70}$. In

(1) (a) Chakravarty, S.; Gelfand, M. P.; Kivelson, S. *Science* **1991**, *254*, 970. (b) Chakravarty, S.; Kivelson, S. *Europhys. Lett.* **1991**, *16*, 751.

(2) (a) Varma, C. M.; Zaanen, J.; Raghavachari, K. *Science* **1991**, *254*, 989. (b) Schluter, M. A.; Lannoo, M.; Needels, M.; Baraff, G. A. Submitted for publication.

(3) Bardeen, J.; Cooper, L. N.; Schrieffer, J. R. *Phys. Rev.* **1957**, *108*, 1175.

(4) (a) Johnson, R. D.; Meijer, G.; Salem, J. R.; Bethune, D. S. *J. Am. Chem. Soc.* **1991**, *113*, 3619. (b) Hawkins, J. M.; Loren, S.; Meyer, A.; Nunlist, R. *J. Am. Chem. Soc.* **1991**, *113*, 7770.

(5) Ramirez, A. P.; Kortan, A. R.; Rosseinsky, M. J.; Duclos, S. J.; Mujsce, A. M.; Haddon, R. C.; Murphy, D. W.; Makhija, A. V.; Zahurak, S. M.; Lyons, K. B. *Phys. Rev. Lett.* **1992**, *68*, 1058.

(6) Kratschmer, W.; Lamb, L. D.; Fostiropoulos, K.; Huffman, D. R. *Nature* **1990**, *347*, 354.

(7) Haufler, R. E.; Conceicao, J.; Chibante, L. P. F.; Chai, Y.; Byrne, N. E.; Flanagan, S.; Haley, M. M.; O'Brien, S. C.; Pan, C.; Xiao, Z.; Billups, W. E.; Ciufolini, M. A.; Hauge, R. H.; Margrave, J. L.; Wilson, L. J.; Curl, R. F.; Smally, R. E. *J. Phys. Chem.* **1990**, *94*, 8634.

(8) Diederich, F.; Ettl, R.; Rubin, Y.; Whetten, R. L.; Beck, R.; Alvarez, M.; Anz, S.; Sensharma, D.; Wudl, F.; Khemani, K. C.; Koch, A. *Science* **1991**, *252*, 548.

(9) Chen, C.-C.; Kelty, S. P.; Lieber, C. M. *Science* **1991**, *253*, 886.

3142

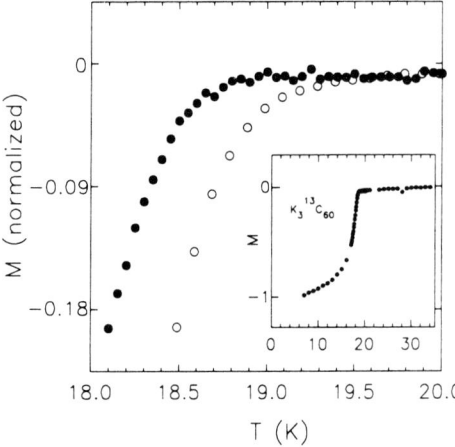

Figure 2. High-resolution temperature-dependent magnetization measurements obtained on $K_3{}^{13}C_{60}$ (●) and $K_3{}^{12}C_{60}$ (○) samples highlighting the depression in T_c for the isotopically substituted material. The samples were initially cooled in zero field to 5 K, and then the curves were recorded on warming in a field of 20 Oe. The curves were normalized to the value of the magnetization at 5 K. The inset shows a full magnetization curve for a $K_3{}^{13}C_{60}$ sample.

addition, we characterized the infrared active modes of $^{13}C_{60}$ and found that the four modes exhibited a classical isotope shift (Table I). These analytical data demonstrate that we prepared isotopically pure $^{13}C_{60}$ and that the purified samples were free from $^{13}C_{70}$ contaminant.

To elucidate the isotope effect on superconductivity, we have studied potassium-doped $^{13}C_{60}$. $K_3{}^{13}C_{60}$ and $K_3{}^{12}C_{60}$ samples were synthesized by heating stoichiometric amounts of solvent-free $^{13}C_{60}$ and $^{12}C_{60}$ with K metal (3:1, K:C_{60}) in sealed quartz tubes (10^{-3} Torr); typically, 1–2 mg of C_{60} was used.[9] The temperature was ramped from 200 to 400 °C over a 1-week period during the reaction.[10,11] $K_3{}^{13}C_{60}$ and $K_3{}^{12}C_{60}$ samples were prepared simultaneously in the same furnace to minimize differences due to the preparative conditions.

Typical temperature-dependent shielding magnetization data obtained from $K_3{}^{13}C_{60}$ and $K_3{}^{12}C_{60}$ samples using a SQUID-based magnetometer (MPMS2, Quantum Design, San Diego, CA) are shown in Figure 2. The full shielding curve for $K_3{}^{13}C_{60}$ (Figure 2, inset) exhibits a sharp transition with an onset at 18.8 K; sharp transition onsets are also observed for the $K_3{}^{12}C_{60}$ samples. The rounding in the low-temperature data is expected since the grain size of these polycrystalline materials is similar to the magnetic

penetration depth; similar broadening is observed in the $K_3{}^{12}C_{60}$ results obtained in this study and reported elsewhere.[5,10] Since the transition onsets are sharp in both the $K_3{}^{13}C_{60}$ and $K_3{}^{12}C_{60}$ curves, we assign T_c as the onset temperature of diamagnetic shielding.[12] The key experimental result of this study shown in Figure 2 is the depression of T_c from 19.2 K for $K_3{}^{12}C_{60}$ to 18.8 K for isotopically pure $K_3{}^{13}C_{60}$. Our analytical data indicate it is unlikely that the decrease in T_c for $K_3{}^{13}C_{60}$ is due to impurities, and thus we attribute the shift in T_c to a true mass effect. Taking into account the uncertainty in all of our data, we find that $\Delta T_c = 0.45 \pm 0.1$ K. Since these data are obtained on isotopically pure compounds, they represent unambiguously the $^{13}C/^{12}C$ isotope effect on T_c.

It is important to compare the observed depression of T_c for isotopically pure $K_3{}^{13}C_{60}$ material with different models of superconductivity. Conventional Bardeen–Cooper–Schrieffer (BCS) theory predicts an isotope effect of $T_c \propto M^{-\alpha}$, where M is the ionic mass and $\alpha = 0.5$.[3] Using the value of ΔT_c determined above we find that $\alpha = 0.3 \pm 0.06$. This experimentally determined value of α places important constraints on the mechanism of superconductivity. First, the observation of the isotope effect strongly suggests that pairing mechanisms involving only electronic interactions are unlikely for these new materials.[1] Secondly, while it is apparent that phonons are important, the value of α indicates that the standard BCS model must be modified to account for superconductivity in the fullerenes. Since a similar value of α is deduced from studies of incompletely ^{13}C-substituted Rb_3C_{60} ($\alpha = 0.37$),[5] we believe that this conclusion is robust. It is known from studies of conventional superconductors that strong coupling can reduce the value of α from the BCS limit of 0.5.[13,14] Interestingly, our recent tunneling measurements on K_3C_{60} and Rb_3C_{60}[15] and theoretical calculations[16] have indicated that these may be strong coupling superconductors.

Acknowledgment. We thank A. N. Tyler for the mass spectroscopy measurements, D. W. Murphy for communication of unpublished results, D. R. Huffman and W. Kratschmer for helpful discussions, and J. L. Huang for technical assistance. C.M.L. acknowledges support of this work by the David and Lucile Packard Foundation and the NSF Presidential Young Investigator Program. The NMR and mass spectroscopy facilities are supported by grants from the NIH and NSF.

(10) Holczer, K.; Klein, O.; Huang, S.-M.; Kaner, R. B.; Fu, K.-J.; Whetten, R. L.; Diederich, F. *Science* **1991**, *252*, 1154.

(11) Fleming, R. M.; Ramirez, A. P.; Rosseinsky, M. J.; Murphy, D. W.; Haddon, R. C.; Zahurak, S. M.; Makhija, A. V. *Nature* **1991**, *352*, 787.

(12) The value of T_c determined this way for $K_3{}^{12}C_{60}$, 19.2 K, agrees well with previous reports.[10,11]

(13) (a) Morel, P.; Anderson, P. W. *Phys. Rev.* **1962**, *125*, 1263. (b) McMillan, W. L. *Phys. Rev.* **1968**, *167*, 331.

(14) The change in lattice constant on isotopic substitution will also affect T_c;[5] we are currently investigating this contribution.

(15) (a) Zhang, Z.; Chen, C.-C.; Lieber, C. M. *Science* **1991**, *254*, 1619. (b) Zhang, Z.; Chen, C.-C.; Kelty, S. P.; Dai, H.; Lieber, C. M. *Nature* **1991**, *353*, 333.

(16) Erwin, S. C.; Pickett, W. E. *Science* **1991**, *254*, 842.

Reprinted with permission from Science
Vol. 254, pp. 1619–1621, 13 December 1991
© 1991 The American Association for The Advancement of Science

Tunneling Spectroscopy of M_3C_{60} Superconductors: The Energy Gap, Strong Coupling, and Superconductivity

Zhe Zhang, Chia-Chun Chen, Charles M. Lieber*

Tunneling spectroscopy has been used to characterize the magnitude and temperature dependence of the superconducting energy gap (Δ) for K_3C_{60} and Rb_3C_{60}. At low temperature the reduced energy gap, $2\Delta/kT_c$ (where T_c is the transition temperature) has a value of 5.3 ± 0.2 and 5.2 ± 0.3 for K_3C_{60} and Rb_3C_{60}, respectively. The magnitude of the reduced gap for these materials is significantly larger than the value of 3.53 predicted by Bardeen-Cooper-Schrieffer theory. Hence, these results show that the pair-coupling interaction is strong in the M_3C_{60} superconductors. In addition, measurements of $\Delta(T)$ for both K_3C_{60} and Rb_3C_{60} exhibit a similar mean-field temperature dependence. The characterization of Δ and $\Delta(T)$ for K_3C_{60} and Rb_3C_{60} provides essential constraints for theories evolving to describe superconductivity in the M_3C_{60} materials.

SUPERCONDUCTIVITY IN ALKALI METal–doped buckminsterfullerene is now well established (1–10), although the mechanism remains an open and intensely investigated question. To date, experimental studies have elucidated several important properties of the superconducting phase, including the face-centered cubic (fcc) structure (5, 6), the coherence length (7), and the penetration depth (7, 8). Investigations of the dependence of the transition temperature (T_c) on the fcc lattice constant (6) and on pressure (9, 10) have further led to the interesting proposal that changes in T_c can be explained through variations in the density of electronic states (DOS) at the Fermi level (E_f). Specifically, using the expression $T_c \propto \hbar\omega\exp(-1/NV)$ from BCS (Bardeen-Cooper-Schrieffer) theory, where $\hbar\omega$ is the excitation energy relevant to electron pairing, N is the DOS at E_f, and V is the electron-phonon coupling strength, it has been suggested that variations in N determine observed changes in T_c, while V and $\hbar\omega$ are essentially constant (2, 6, 9).

Implicit in this analysis is the assumption of weak coupling. Theoretical and experimental studies have also argued, however, that the coupling interaction might be strong (11–13), and hence we have sought to define unambiguously the relative strength of the coupling in the M_3C_{60} superconductors. The superconducting energy gap (Δ) provides a measure of the coupling strength and can therefore address this issue. In particular, weak-coupling BCS theory predicts that there is a universal value for the reduced energy gap, $2\Delta/kT_c$, of 3.53 (14). We have recently reported a preliminary value for the reduced energy gap of Rb_3C_{60} that is significantly larger than this weak-coupling limit (11); however, it is not known whether this large value of the reduced energy gap is universal for the M_3C_{60} superconductors or how Δ depends on temperature. Herein, we describe detailed tunneling spectroscopy studies of the superconducting energy gap of single phase K_3C_{60} and Rb_3C_{60} that answer these questions.

Tunneling spectroscopy is a particularly attractive technique to probe the energy gap since the conductance (dI/dV) determined from current-voltage (I-V) data provides a direct measure of the DOS (15). Because the M_3C_{60} superconductors have short coherence lengths (7), conventional planar junctions prepared on sintered pellets could show extrinsically broadened energy gap features (that is, owing to nonuniform tunneling barriers), and thus we have used a low-temperature scanning tunneling microscope (STM) to make point junctions with a sharpened metal tip. Our data show that the reduced energy gap in the M_3C_{60} materials is independent of M and significantly larger than the weak coupling limit of 3.5. In addition, $\Delta(T)$ exhibits a mean field temperature dependence with the energy gap disappearing at the bulk value of T_c. The implications of these new results to the mechanism of superconductivity in the M_3C_{60} materials are discussed.

Single-phase K_3C_{60} and Rb_3C_{60} materials were prepared by reaction of alkali-metal alloy or alkali-metal with C_{60} as described in detail elsewhere (3, 4). Briefly, a 3:1 mixture of MHg or M (M = K, Rb) and C_{60} were sealed under vacuum in a quartz tube and then heated at 200°C. When the shielding fraction of M_3C_{60} superconducting phase reached about 40%, the tube was opened and the polycrystalline powder was pressed into 3-mm-diameter pellets. The pellets were sintered at 200°C until the shielding fraction approached 100%. Magnetization versus temperature curves typical of the

K_3C_{60} and Rb_3C_{60} samples used in this study are shown in Fig. 1. The transition temperatures of these K- and Rb-doped materials are 19 and 29 K, respectively, and the low temperature shielding fractions are approximately 100% for both samples.

Magnetically characterized M_3C_{60} sintered pellets were mounted on the STM sample holder using silver paint in an inert atmosphere glove box ([O_2] ≈ [H_2O] ≈ 1 ppm). The sample holder was then transferred to the STM which is contained within a vacuum can. The sample is exposed to the atmosphere for a few minutes during the transfer process; however, its superconducting properties do not degrade significantly. After mounting the sample, the evacuated STM assembly was placed in a mechanically and acoustically isolated helium dewar. The metal (tip)-insulator-superconductor (N-I-S) junction was made by mechanically stepping the tip to the sample and then adjusting the junction resistance and position using the tube scanner of the STM. Tunneling measurements were made either through vacuum when the sample surfaces were metallic ($T > T_c$) or in point contact. In the latter case we believe that the partially oxidized sample surface functions as the insulating barrier. The data obtained from these two distinct types of junctions were similar. I-V curves were recorded digitally using custom-built electronics under computer control; the sample temperature was actively maintained for temperatures greater than 4.2 K. Several independent samples of K_3C_{60} and Rb_3C_{60} were examined in these studies, and typically at least 30 I-V curves were recorded at each temperature for each independent sample; the reported data are

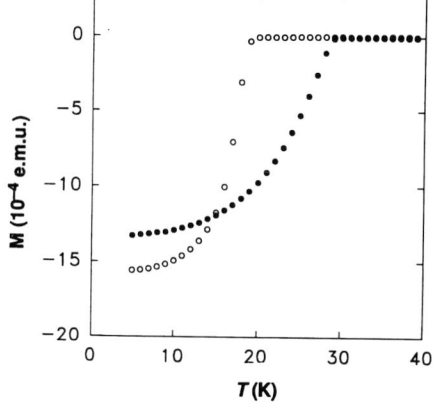

Fig. 1. Temperature (T) dependence of the magnetization obtained for a 4.1-mg K_3C_{60} sintered pellet (open circles) and a 3.8-mg Rb_3C_{60} sintered pellet (filled circles). The T_c's of the K_3C_{60} and Rb_3C_{60} samples are 19.0 and 28.6 K, respectively. The curves were recorded by cooling in zero field to 5 K and subsequent warming in a 10-Oe field. The shielding fractions estimated from these curves are approximately 100%.

Department of Chemistry and Division of Applied Sciences, Harvard University, Cambridge, MA 02138.

*To whom correspondence should be addressed.

representative of this extensive data set.

A series of *I-V* curves recorded at temperatures of 20, 10, and 4.2 K on a K_3C_{60} sample are shown in Fig. 2, A to C, respectively. The curve recorded at 4.2 K exhibits features characteristic of the superconducting energy gap, including (i) a distinct zero-current regime about E_f ($V = 0$), and (ii) conductance onsets at $V \approx \pm 4$ mV. We believe that this structure, which is observed in most of the *I-V* curves recorded at 4.2 K, reflects the modulation in I due to the gap (2Δ) in the DOS of K_3C_{60} probed in the N-I-S tunneling geometry. Similar features are also observed in *I-V* curves recorded on Rb_3C_{60} samples at 4.2 K (Fig. 2D), although the conductance onsets, $V \approx \pm 6$ mV, occur at distinctly larger bias voltage.

Other possible explanations for this structure in the *I-V* curves are a coulomb blockade or superconductor-insulator-superconductor (S-I-S) tunneling. We believe, however, that both of these possibilities are unlikely. First, we do not find that $I \propto V^2$ for small V as predicted and previously observed for Coulomb charging (16, 17). Furthermore, we find that the magnitude of the zero current region around $V = O$ is

reproducible, in K_3C_{60} and systematically larger in Rb_3C_{60} (Fig. 2, C and D), and that for both materials the gap-like structure disappears for $T > T_c$; neither of these observations is consistent with the coulomb blockade. Secondly, for S-I-S tunneling there should be a sharp current jump at $V = 2\Delta$ (15), and not the smooth increase observed in our data. In addition, it is unlikely that the same S-I-S tunneling would be observed for vacuum and point contact tunneling, and thus we believe that the conductance onsets can be assigned with confidence to $\pm \Delta$. Lastly, we note that a small number of *I-V* curves recorded at 4.2 K exhibit large gaps which may be interpreted as S-I-S tunneling; conditions for reproducible observation of this large gap structure are not yet known, and thus we believe it is premature to interpret such data.

An important point evident upon examination of the 4.2 K *I-V* curves is that the gap structure for K_3C_{60} is significantly smaller than for Rb_3C_{60}. These results indicate qualitatively that Δ scales with T_c. To quantitatively assess the magnitude of Δ and the reduced energy gap we have calculated the normalized conductance, $(dI/dV)_S/(dI/dV)_N$ where the subscripts N and S refer to the normal and superconducting states (Fig. 3). Since $(dI/dV)_S/(dI/dV)_N$ is proportional to

the superconducting DOS, N_S, the value of Δ can be determined from a fit of normalized conductance to a model for the DOS. We find that good fits of the experimental data are obtained using the broadened BCS function proposed by Dynes and co-workers, $N_S = \mathrm{Re}[|E-i\Gamma|/((E-i\Gamma)^2-\Delta^2)^{1/2}]$, where E is the energy of the tunneling electron and Γ is a broadening function (18). Dynes *et al.* introduced Γ specifically to account for shortened quasiparticle lifetime, however, here we use Γ as a phenomenological parameter since the mechanism of broadening is not known (for example, it could be due to inelastic scattering or strong coupling effects). The essential result obtained from the fits to the 4.2 K K_3C_{60} data is that the experimental value of Δ, 4.4 meV, is significantly larger than the $T \to 0$ BCS theory prediction of 2.73 meV. Furthermore, the average value of Δ (4.2 K) determined from these experiments yields a reduced energy gap, $2\Delta/kT_c \pm 1$ SD, of 5.3 ± 0.2 for K_3C_{60}. The value of the reduced energy gap at 4.2 K for Rb_3C_{60} reported recently (11) and further refined in this study, 5.2 ± 0.3, is the same within experimental error, and thus Δ clearly scales with T_c in these materials. The large value of $2\Delta/kT_c$ for the M_3C_{60} superconductors shows that the coupling in these materials is strong. Although we are unaware of other experimental studies confirming strong coupling, two recent theoretical calculations have predicted that the coupling in K_3C_{60} will be strong (12, 13).

It is interesting to consider the implications of strong coupling. Within the context of phonon-mediated pairing, theoretical work has shown that large values of Δ arise

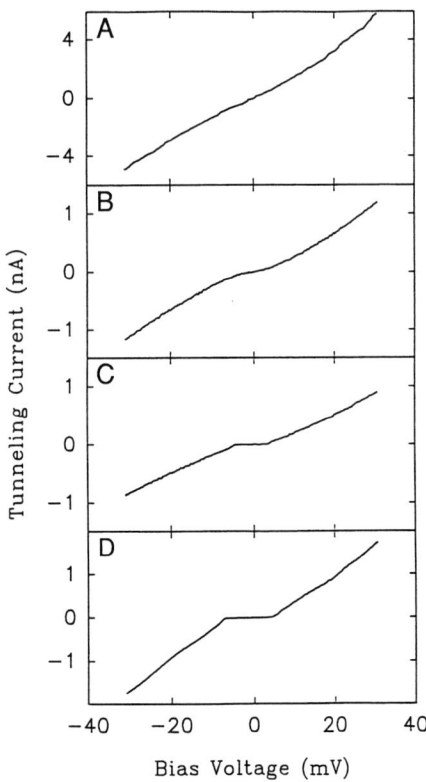

Fig. 2. Current versus voltage (*I-V*) curves recorded on a K_3C_{60} sample at (**A**) 20 K, (**B**) 10 K, and (**C**) 4.2 K, and on a Rb_3C_{60} sample at 4.2 K (**D**). Typically, five sequential *I-V* measurements at a specific sample location were averaged to produce a single curve; these data were not subject to other smoothing routines.

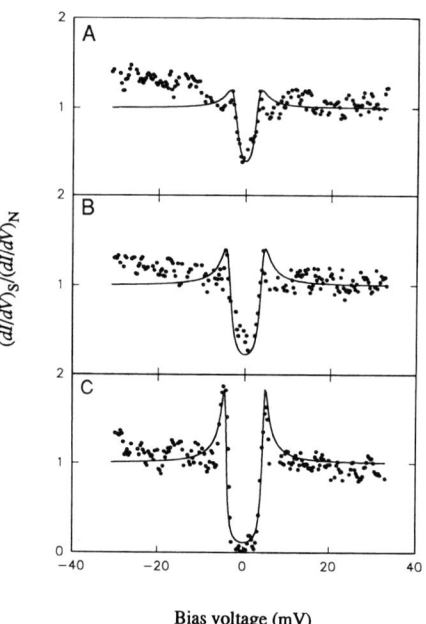

Fig. 3. Normalized conductance versus voltage $[(dI/dV)_S/(dI/dV)_N]$ curves for K_3C_{60} corresponding to temperatures of (**A**) 15 K, (**B**) 10 K, and (**C**) 4.2 K. The experimental data (filled circles) were obtained by numerically differentiating the *I-V* data and then normalizing by dI/dV at 20 K. The solid lines correspond to the best fits of this data to the expression $(dI/dV)_S/(dI/dV)_N = \mathrm{Re}[|E - i\Gamma|/((E - i\Gamma)^2 - \Delta^2)^{1/2}]$. The values of the energy gap Δ and broadening function Γ in meV are (A) $\Delta = 3.0$; $\Gamma = 1.1$, (B) $\Delta = 4.0$; $\Gamma = 0.9$, (C) $\Delta = 4.4$; $\Gamma = 0.5$.

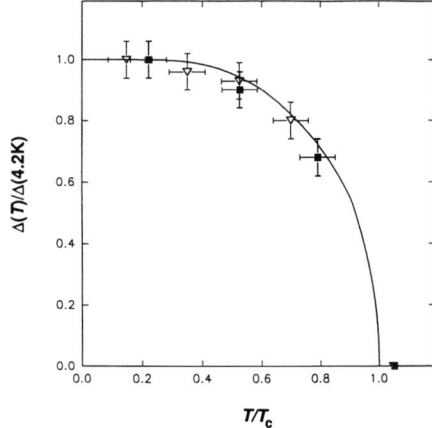

Fig. 4. Summary of the temperature-dependent energy gap results for K_3C_{60} (■) and Rb_3C_{60} (▽). For comparison the values of Δ have been normalized by the respective low temperature (4.2 K) values and the temperature has been scaled by T_c. The solid line corresponds to the temperature dependence of $\Delta(T)/\Delta(0)$ predicted by BCS theory (14).

from strong coupling to low frequency modes (19). In the M_3C_{60} materials, the most obvious low frequency modes are C_{60}-C_{60} intermolecular vibrations or C_{60} rotations. Alternatively, it has been suggested that the M^+ optical phonon could lead to strong coupling (13). High-frequency intramolecular C_{60} modes, which have been implicated in weak-coupling analyses (2, 6), are unlikely to yield the large value of $2\Delta/kT_c$ determined experimentally. Although additional work is needed to define whether the electron-phonon interaction is the operative coupling mechanism and if so, the mode relevant to pairing, our finding of strong coupling should be accounted for in models of superconductivity in these materials.

We have also characterized the temperature dependence of Δ in both K_3C_{60} and Rb_3C_{60} since this can provide additional insight into the mechanism of superconductivity. Representative normalized conductance curves recorded on K_3C_{60} and theoretical fits to these data for 4.2 K $< T < T_c$ are shown in Fig. 3. Qualitatively, we find that Δ decreases as T approaches T_c, and disappears for $T > T_c$. We have summarized the results from these temperature-dependent studies of Δ for K_3C_{60} and for Rb_3C_{60} by plotting $\Delta(T)/\Delta(4.2)$ versus T/T_c (Fig. 4). This figure explicitly shows that the normalized energy gaps of Rb_3C_{60} and K_3C_{60} exhibit a similar temperature dependence, and furthermore, that these data follow the universal temperature dependence predicted by BCS theory. Importantly, our $\Delta(T)$ data indicate that it may be possible to explain superconductivity using a mean-field theory (like BCS) modified for strong coupling. It is also interesting to consider real-space models of superconductivity since the coherence lengths in these materials are so short ($\xi \approx 25$ Å). In particular, it has been suggested that a Bose-Einstein condensation of real-space pairs may explain superconductivity in the short coherence length ($\xi_{ab} \approx 10$ Å) high-T_c copper oxide materials (20, 21). Since $\Delta(T)$ should exhibit a relatively sharp transition near T_c in a Bose-Einstein condensation, we believe that the observed temperature dependence of Δ argues against this interesting possibility.

In conclusion, tunneling spectroscopy has been used to define the energy gap in the M_3C_{60} superconductors. These experimental results have shown that (i) the pair coupling in these materials is strong, (ii) the energy gap scales with T_c, and (iii) the energy gap exhibits a universal temperature dependence. Regardless of the mechanism of pairing in the M_3C_{60} system, we believe that our results will be important constraints for any theoretical explanation of superconductivity in these materials.

REFERENCES AND NOTES

1. A. F. Hebard et al., Nature **350**, 600 (1991).
2. M. J. Rosseinsky et al., Phys. Rev. Lett. **66**, 2830 (1991).
3. K. Holczer et al., Science **252**, 1154 (1991).
4. C.-C. Chen, S. P. Kelty, C. M. Lieber, ibid. **253**, 886 (1991).
5. P. W. Stephens et al., Nature **351**, 632 (1991).
6. R. M. Fleming et al., ibid. **352**, 787 (1991).
7. K. Holczer et al., Phys. Rev. Lett. **67**, 271 (1991).
8. Y. J. Uemura et al., Nature **352**, 605 (1991).
9. G. Sparn et al., Science **252**, 1829 (1991).
10. J. E. Schirber et al., Physica C **178**, 137 (1991).
11. Z. Zhang, C. C. Chen, S. P. Kelty, H. Dai, C. M. Lieber, Nature **353**, 333 (1991).
12. S. C. Erwin and W. E. Pickett, Science, in press.
13. F. C. Zhang, M. Ogata, T. M. Rice, Phys. Rev. Lett., in press.
14. J. Bardeen, L. N. Cooper, J. R. Schrieffer, Phys. Rev. **108**, 1175 (1957).
15. E. L. Wolf, Principles of Tunneling Spectroscopy (Oxford University Press, New York, 1989).
16. D. V. Averin and K. K. Likharev, J. Low Temp. Phys. **62**, 345 (1986).
17. P. J. M. van Bentum, H. van Kempen, L. E. C. van de Leemput, P. A. A. Teunissen, Phys. Rev. Lett. **60**, 369 (1988); R. Berthe and J. Halbritter, Phys. Rev. B **43**, 6880 (1991).
18. R. C. Dynes, V. Narayanamurti, J. P. Garno, Phys. Rev. Lett. **41**, 1509 (1978).
19. B. Mitrović, C. R. Leavens, J. P. Carbotte, Phys. Rev. B **21**, 5048 (1980).
20. R. Friedberg and T. D. Lee, ibid. **40**, 6745 (1989).
21. Z. Schlesinger et al., ibid. **41**, 11237 (1990); B. N. J. Persson and J. E. Demuth, ibid. **42**, 8057 (1990).
22. We thank Z. Schlesinger and J. E. Demuth of IBM for helpful discussions. Supported by the David and Lucile Packard, Alfred P. Sloan, Camille and Henry Dreyfus, and National Science Foundations (C.M.L.).

13 October 1991; accepted 1 November 1991

Reprinted with permission from Science
Vol. 255, pp. 184–186, 10 January 1992
© 1992 The American Association for The Advancement of Science

Electrical Resistivity and Stoichiometry of K_xC_{60} Films

G. P. KOCHANSKI, A. F. HEBARD, R. C. HADDON, A. T. FIORY

Electrical resistances of polycrystalline fullerene (C_{60}) films were monitored while the films were being doped in ultrahigh vacuum with potassium from a molecular-beam effusion source. Temperature- and concentration-dependent resistivities of K_xC_{60} films in equilibrium near room temperature were measured. The resistance changes smoothly from metallic at $x \approx 3$ to activated as $x \to 0$ or $x \to 6$. The minimum resistivity for K_3C_{60} films is 2.2 milliohm-centermeters, near the Mott limit. The resistivities are interpreted in terms of a granular microstructure where K_3C_{60} regions form nonpercolating grains, except perhaps at $x \approx 3$. Stoichiometries at the resistivity extrema were determined by ex situ Rutherford backscattering spectrometry to be $x = 3 \pm 0.05$ at the resistance minimum and $x = 6 \pm 0.05$ at the fully doped resistance maximum.

THE OCCURRENCE OF CONDUCTIVI- ty in alkali fulleride compounds (1) and superconductivity at 18 K in K_xC_{60} (2) has stimulated many investigations of the superconducting properties but little work on transport in the normal state. There has also been considerable effort to determine whether the doped system forms continuous or discrete phases. X-ray scattering shows distinct crystal structures for phases at $x = 0, 3, 4,$ and 6 (3–6), which suggests that there are miscibility gaps that separate the phases. Carbon-13 nuclear magnetic resonance (NMR) shows only two well-defined lines in the range $0 < x < 3$ (7), and photoemission spectra are consistent with linear combinations of $x = 0, 3,$ and 6 phases (8, 9). Shielding diamagnetism curves with a peak near $x = 3$ are interpreted as a single superconducting phase (10). This evidence for discrete phases in K_xC_{60} must be incorporated into any realistic model of the transport properties.

In this report, we present resistivity measurements of K_xC_{60} as a function of temperature and doping. We used ultrahigh-vacuum (UHV) molecular beam deposition to dope polycrystalline C_{60} films with controlled quantities of K while continuously monitoring the time-dependent film resistance near room temperature. The method gives the time, temperature, and concentration dependence of the resistance as well as an estimate for the diffusivity of the K dopant. The stoichiometry of the superconducting phase, which corresponds to a minimum in the

AT&T Bell Laboratories, Murray Hill, NJ 07974.

resistivity at room temperature, has a K concentration given by $x = 3.00 \pm 0.05$, as determined by ex situ Rutherford backscattering (RBS) analyses. We find that the resistivity has a maximum at $x = 6.00 \pm 0.05$. These stoichiometries agree with Raman scattering (1) and photoemission (8, 9) findings.

The results are consistent with a simple model involving a granular microstructure and a miscibility gap between the phases. The conductivity is an activated process with an activation energy derived from the charging energies of individual K_3C_{60} grains. Close to $x = 3$, the grains coalesce, the charging energies become negligible, and the coupled grains give rise to a metallic state with conductivity close to the Mott limit.

The undoped films were grown (1) by vacuum sublimation of C_{60} refined by liquid chromatography from material produced in an electric-arc furnace (11). The amount of deposited C_{60} was monitored with a quartz microbalance calibrated by RBS measurement of the C deposited on clean Si. The thickness was calibrated by an optical interferometric gauge. Potassium areal atomic densities, N_K, were determined by analysis of the RBS K peaks after the film was removed from the UHV chamber. X-ray diffraction and transmission electron microscopy show random polycrystallinity with grain sizes of about 60 Å (12). For electrical measurements, four contact wires were attached with Ag paste to Ag pads that were evaporated at the corners of the 1-cm-square glass or thermally oxidized Si substrate. The C_{60} films were then deposited at ambient temperature. The substrates were mounted by the contact wires on a UHV feedthrough containing a small scanning Auger electron spectroscopy molecular-beam oven (SAES Getters/USA Inc.) and, if necessary, epoxy was used to attach a thermocouple to the back of the substrate.

The UHV chamber was baked at 175°C to outgas the C_{60} film. We deposited K at a base pressure of 10^{-9} torr and monitored the film resistance by a four-wire resistance bridge, using Van der Pauw's method. The two Van der Pauw resistance readings are typically equal within 30%, and the ratio is typically constant to 10% during the exposure, which shows that the K doping is laterally uniform.

We investigated the equilibration of the K in our films several ways. First, the resistivity changes by less than 1% in an hour if deposition is interrupted. Second, in all but the thickest films, changes in film resistance respond promptly to exposure to K. For a film 4450 Å thick, the equilibration time is slow enough to be measured: 400 s for $x < 3$. If this is interpreted as simple Fickian diffusion, then the K diffusivity is on the order of 10^{-12} cm^2 s^{-1}. The ionic conductivity is therefore not important. Third, preliminary thermopower measurements show slow responses, on the order of 3 μV/K over time scales of minutes, which we interpret as a thermally driven redistribution of K. Consequently, we expect the K distribution in

Fig. 1. Time dependence of the resistivity of K_xC_{60} film during exposure to a K molecular beam in UHV at ambient temperature near 74°C. The end point stoichiometry was determined for this particular sample; stoichiometry at the minimum was determined from other samples. The arrows show changes in the resistivity of a similar sample as it was heated from 60° to 134°C (see Fig. 2). The dashed curve is a fit to Eq. 2.

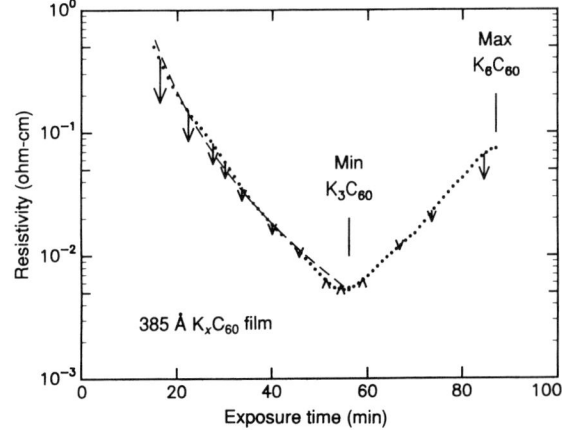

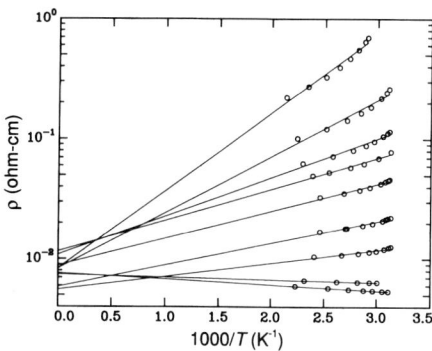

Fig. 2. Plots of the logarithm of resistivity versus inverse temperature for a film with doping in the range $1 < x \leq 3$. These data correspond to the arrows in Fig. 1. The lines are linear least-squares fits to the data and are extrapolated to $T^{-1} = 0$, to show the activated behavior of the resistivity. The doping progresses from top to bottom with the lowest curve corresponding to $x = 3$.

the film to be fully equilibrated on submicrometer length scales.

Figure 1 shows the dependence of resistivity on exposure time to K for a typical film, with arrows denoting the changes for a fixed temperature increment. The stoichiometry $x = 3$ corresponds to the minimum resistance (R_{min}) and $x = 6$ to the final maximum resistance (R_{max}). We established these x values by stopping the K exposure at either a minimum or maximum point, and then measuring the deposited K by RBS. Although the time to reach K_3C_{60} is greater than the interval from K_3C_{60} to K_6C_{60}, we cannot distinguish a stoichiometry-dependent sticking coefficient from possible systematic variations in the K flux. There is no obvious signature for the $x = 4$ phase. Film resistance decreases beyond $x = 6$, possibly because of unreacted metallic K on the surface (1). Data near $x = 0$ were not available for films on oxidized Si substrates because of the nonzero substrate conductivity. In our model, we assume a miscibility gap between the phases, in agreement with the aforementioned NMR, photoemission, and diamagnetism results. The resistivity is approximately activated, with a gap that is nearly symmetrical about $x = 3$, is approximately proportional to $|x - 3|$, and reaches values as high as 0.12 eV near $x = 1$.

In a typical granular metal system, there is a marked change in properties at the percolation threshold. This occurs near a 50% volume fraction of the conducting phase and usually is marked by a sharp change in resistivity and a change in the sign of the temperature coefficient (13). Our data show no signs of dramatic change in either the resistivity or the temperature coefficient (lengths of arrows in Fig. 1) (14) near $x = 1.5$ or $x = 4.5$; instead, the changes occur very near $x = 3$. Thus, a random percolation

model with $x = 0$ or $x = 6$ insulating phase and $x = 3$ conducting phase is inconsistent with the data. A percolation model could explain the data if there were repulsive interactions between grains or if the K_3C_{60} grains are determined by the preexisting grains in the film before doping. If the conducting regions are located in the center of grains, because K_3C_{60} is most energetically favorable there, then the percolation threshold can be pushed arbitrarily close to $x = 3$. Presumably for $x > 3$, the insulating $x = 4$ and $x = 6$ phases would grow in from the grain boundaries and form a similar structure. This explains the symmetry about $x = 3$. Because the energy differences between various phases of K_xC_{60} are not large, on the order of 0.1 eV (5), it is plausible that there are mixed compositions at the grain boundaries where strains are large and disorder is more important.

To explore the exponential doping dependence, we measured the temperature dependence of resistance in the range $46°C < T < 194°C$ (there are irreversible changes in the resistance slightly above 200°C). Over this temperature range, Arrhenius plots of the resistance show a slight positive curvature (Fig. 2), but least-squares fits extrapolate to a compact cluster as $T^{-1} \to 0$. This convergence is the signature of an approximately activated conductivity.

For thicknesses greater than 1000 Å, the data of Fig. 3 show a thickness-independent minimum resistivity $\rho_{min}(K_3C_{60}) = 2.2$ milliohm-cm. The cause of the larger resistivity in films thinner than 1000 Å, which also have the stoichiometry $x = 3$, is not presently understood, although it might be due to the exposure of the C_{60} films to air before doping or to systematic changes in grain size with thickness.

The lattice constant of the $x = 3$ face-centered cubic unit cell is 14.28 Å (4). Accordingly, the carrier density is 4.1×10^{21} cm^{-3}, with four C_{60} molecules and twelve donated electrons per unit cell. This charge density corresponds to a Fermi wave vector $k_F = 0.50$ Å^{-1} which, when substituted into a Boltzmann equation description of the minimum resistivity ρ_{min}, gives $\ell = 2.3$ Å for the electronic mean free path. This unphysically small ℓ implies that, even at $x = 3$, the Boltzmann equation is inadequate for describing a system where intergranular transport may still be limiting the conductivity.

The standard treatment for conduction in granular systems has been reviewed by Abeles (15), who showed that the charging energy of the grains is important to the conduction process. Adding K changes the grain size and therefore the charging energy and conductivity. The simplest model of the

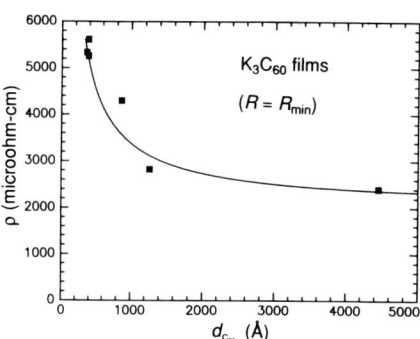

Fig. 3. Dependence of the resistivity of K_3C_{60} films on the thickness of the initial C_{60} film (curve is arbitrary).

medium consistent with our data is a constant number density of metallic grains surrounded by an insulating phase. We assume that the relation between grain spacing, L, and grain size, d, is $d = L(x/3)^{1/3}$ (16). The gap between grains is $s = L - d$. Following Abeles (15), we calculate the energy required to move an electron from one grain to another distant grain

$$E_c^0 = \frac{e^2}{2\pi\epsilon d} \frac{2d}{d + 2s} \qquad (1)$$

where ϵ is the dielectric constant of the insulating phase. The resistivity as a function of x and T is then given by the expression

$$\rho = \frac{h}{e^2} \frac{(d + s)}{\gamma} \exp\left(\frac{E_c^0}{2kT}\right) \exp(2\chi s) \qquad (2)$$

where h is Planck's constant, k is the Boltzmann constant, T is temperature, χ is the decay constant of the wave function between the grains, and γ is a factor of order unity describing the spatial distribution of the tunneling. We obtain a good fit to the data with $L = 83$ Å, $\gamma = 4.1$, and $\chi = 0$, as shown in Fig. 1.

Equation 2 is derived for weakly coupled

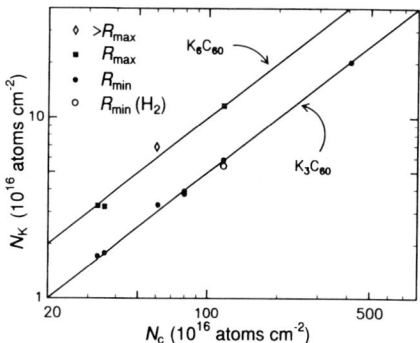

Fig. 4. Areal atomic densities in K_xC_{60} films. Filled symbols correspond to the turning points (min or max) in plots of resistance versus K doping (for example, Fig. 1). The open circle represents a film doped in the presence of 1 torr H_2. The diamond indicates a film doped beyond the R_{max} turning point.

grains, and $E_c^0 >> kT$. As $x \rightarrow 3$, these assumptions fail as the arguments of the exponentials approach zero. We cannot say how tightly coupled the grains become in the real system; it is possible that disorder on the grain boundaries may not allow the metallic regions to completely merge.

We have made an ad hoc improvement to the theory that has the proper behavior as $E_c^0 \rightarrow 0$ and can represent granular metal conductivities and percolation well:

$$\rho = \frac{2h}{e^2} \frac{(d + s)}{\gamma} \sinh\left(\frac{E_c^0}{2kT}\right) \exp(2\chi s) + \rho_g$$
(3)

where ρ_g is a resistivity contribution from intragrain scattering. Our motivation for the replacement $\exp() \rightarrow 2\sinh()$ is that, as $E_c^0 \rightarrow 0$, the barriers to conduction vanish and $\rho \rightarrow 0$ (neglecting intragrain scattering). Fits with this modified theory are similar, with $L = 75$ Å, $\gamma = 6$, $\chi = 0$, and $\rho_g = \rho(x = 3)$. The data cannot be fit well using any reasonable effective medium model unless the conductivity of the percolating phase has a similar activated behavior.

The fitted values of L are in good agreement with the 60 Å grain size for the C_{60} film before doping. The result $\chi = 0$ implies that the energy barriers between grains are comparable to the charging energies (17), and about as many electrons go over the barriers as go through them. The fitted values of γ are close enough to unity to be plausible in the absence of a complete theory (18).

Stoichiometries of films of various thicknesses are shown in Fig. 4. The filled symbols correspond to doping at either the R_{min} or R_{max} points in plots of resistance versus K exposure. These points fall to within an error of ± 0.05 on solid lines drawn respectively for the precise stoichiometries at $x = 3$ or $x = 6$. Although implied by existing work (1), this result demonstrates that the conductivity extrema fall exactly at half and full band filling.

In summary, the diffusion of K into C_{60} films produces a minimum resistivity phase that is barely metallic with $\rho = 2.2$ milliohm-cm at composition $K_{3.00\pm0.05}C_{60}$, and a high resistivity (max) phase at composition $K_{6.00\pm0.05}C_{60}$. Intermediate compositions yield activated conduction with x-dependent activation energy, which can be explained by a simple model with immiscible phases where the metallic phase maximally avoids percolation. The recognition that the normal state has a granular microstructure will also be important in understanding superconductivity in thin films.

REFERENCES AND NOTES

1. R. C. Haddon et al., Nature 350, 320 (1991).
2. A. F. Hebard et al., ibid., p. 600.
3. R. M. Fleming et al., Mater. Res. Soc. Symp. Proc. 206, 691 (1991).
4. P. W. Stephens et al., Nature 351, 632 (1991).
5. R. M. Fleming et al., ibid. 352, 701 (1991).
6. O. Zhou et al., ibid. 351, 462 (1991).
7. R. Tycko et al., Science 253, 884 (1991).
8. D. M. Poirier et al., ibid., p. 646.
9. C. T. Chen et al., Nature 352, 603 (1991).
10. K. Holczer et al., Science 252, 1154 (1991).
11. W. Krätschmer, L. D. Lamb, K. Fostiropoulos, D. R. Huffman, Nature 347, 354 (1990).
12. A. F. Hebard, R. C. Haddon, R. M. Fleming, A. R. Kortan, Appl. Phys. Lett. 59, 2109 (1991).
13. B. Abeles, P. Sheng, M. D. Coutts, Y. Arie, Adv. Phys. 24, 407 (1975).
14. The temperature coefficient of resistance, $R^{-1}dR/dT$ is negative for all films, except for $2.8 < x < 3.2$ where the behavior is metallic with $R^{-1}dR/dT = (7.9 \pm 0.4) \times 10^{-4}$ K^{-1} at $x = 3$. The width of the resistance minimum (and the region of positive temperature coefficient) is comparable to our estimates of doping uniformity (10%), so an ideal experiment might reveal a sharp minimum.
15. B. Abeles, Appl. Solid State Sci. 6, 1 (1976).
16. If the grains were spheres on a simple cubic lattice, they would intersect at $x = \pi/6$; the intersection would correspond to a percolation threshold. Because we see no percolation until $x \approx 3$, the expression we use (corresponding to a lattice of cubes) is the simplest choice.
17. In the derivations in (15), the assumption is made that the barrier is large compared to the charging energies. What is referred to as the "barrier height" is more accurately the barrier height above the charging energy.
18. The calculation of γ in (15) is only valid in the limit $\chi(d + s) >> 1$; its derivation assumes the strong dependence of current on grain separation that is characteristic of a tunneling process.
19. We thank E. E. Chaban, D. J. Eaglesham, R. H. Eick, R. M. Fleming, A. R. Kortan, A. V. Makhija, D. W. Murphy, T. T. M. Palstra, J. C. Phillips, M. J. Rosseinsky, J. E. Rowe, F. C. Unterwald, B. E. Weir, and G. K. Wertheim for their contributions to this work.

19 September 1991; accepted 7 November 1991

Reprinted with permission from Physical Review Letters
Vol. 68, No. 7, pp. 1054–1057, 17 February 1992

Electronic Transport Properties of K_3C_{60} Films

T. T. M. Palstra, R. C. Haddon, A. F. Hebard, and J. Zaanen

AT&T Bell Laboratories, Murray Hill, New Jersey 07974-2070

(Received 23 October 1991)

We report the longitudinal resistivity and Hall-effect data of thin films of K_3C_{60} in the normal and superconducting states in magnetic fields up to 12.5 T. The resistivity is 2.5 mΩ cm at room temperature and near the metal-to-insulator transition. The Hall coefficient is small as expected for a half-filled conduction band, and changes sign at 220 K. The resistivity is interpreted in terms of the granularity of the film which leads to zero-dimensional superconductivity in these systems with a length scale of 70 Å. We find a superconducting coherence length of ~26 Å, which we interpret as a single-grain property. The Pippard coherence length, i.e., in the absence of granularity, is estimated to be ~150 Å.

PACS numbers: 74.70.Mq, 74.70.Kn

The discovery of superconductivity [1,2] in conducting alkali-doped C_{60} phases [3] yielded the second group of materials besides the oxide superconductors [4] to overcome the historic limit of $T_c = 23$ K in intermetallic compounds [5]. This breakthrough has created an intense experimental and theoretical effort to understand the nature of both the superconducting and normal-state properties of these new conductors. We report the dc longitudinal and Hall resistivities of K_3C_{60} in the normal and superconducting states. The absence of reliable transport data on doped C_{60} phases has led to wide discrepancies in estimates of the conduction bandwidth and associated Fermi-level density of states. Early magnetization experiments [6] found an upper critical field of ~45 T, corresponding to a small value of the superconducting coherence length of ~26 Å. This small coherence length was interpreted as arising from a very narrow band with a width of only 500 K. However, it was not verified whether the clean-limit analysis they employed is valid, or whether the coherence length was reduced by a small mean free path. Recent magnetic susceptibility measurements [7] of the compounds $K_x Rb_{1-x} C_{60}$ were interpreted with a density of states arising from a much wider band, and the small coherence length was ascribed to a small mean free path of the order of the size of a C_{60} molecule. Assuming weak-coupling BCS theory, local-density approximation (LDA) band-structure calculations of the Fermi velocity v_F yield a clean-limit coherence length in the range 100–200 Å [8,9].

Our electrical transport data show that our films have a minimum resistivity of ~2.5 mΩ cm, which corresponds to an effective mean free path of ~3 Å and would validate the dirty-limit analysis. However, detailed analysis of our data shows unambiguously that the high resistivity does not arise from microscopic disorder, but comes about from the granular nature of our films [10]. We find a grain size of ~70 Å, and we interpret the experimentally observed short coherence length as a single-grain property, which is significantly reduced from the bulk value by finite-size effects. Using theoretical results pertaining to granular superconductors, we can estimate the clean-limit (Pippard) coherence length, $\xi_0 \approx 150$ Å. The Hall data show that the Hall coefficient is small, as

expected for a half-filled band. However, the charge of the carriers changes sign at 220 K, which can be ascribed to a temperature dependence of the inelastic scattering rate.

The C_{60} thin films were grown by thermal evaporation of pure C_{60} onto a glass substrate of ~10×10 mm^2. Prior to evaporation, four contacts for van der Pauw measurements were made by evaporating Ag, and connecting Pt wires with Ag epoxy. The K doping was performed in a custom UHV stainless-steel reaction cell, with feedthroughs on one end and a metal-to-glass seal at the other end. This reaction cell contained besides the thin film, a SAES K dispenser and a sublimation getter. The cell was evacuated and sealed off after the getter was activated, and the undoped film was annealed at 200°C. The doping was performed at room temperature. As K_3C_{60} is very reactive, the cell was designed to be cooled within the bore of a 14.5-T magnet. The temperature was monitored by a carbon-glass thermometer, and the resistance of the sample was measured using an ac resistance bridge operating at a frequency of 17 Hz.

The undoped film had a resistivity larger than 10^5 Ω cm. The film was doped in discrete steps by resistively heating the K dispenser for one minute and then waiting until the resistivity of the film was constant. The resistivity is affected both by the doping process and by radiative heating of the film due to the hot dispenser. The relaxation time of the doping process is relatively slow, about 200 s for a 1600-Å-thick film, corresponding to a diffusion constant of ~10^{-16} m^2/s [10]. The lowest resistivity obtained on this 1600-Å film was 2.5 mΩ cm, only 10% larger than values obtained under UHV conditions on similar films [10]. Good homogeneity was obtained in this film as the two components of the van der Pauw resistance differed by less than 5%.

Figure 1 shows the temperature dependence of the electrical resistivity between 4 and 300 K. At room temperature the resistivity has a positive temperature coefficient of 0.8×10^{-3} K^{-1}. However, the temperature coefficient changes sign at ~250 K, and the resistivity increases by 30% before reaching the superconducting state at 12.8 K. The inset of fig. 1 shows the superconducting transition in detail. The temperature dependence of the

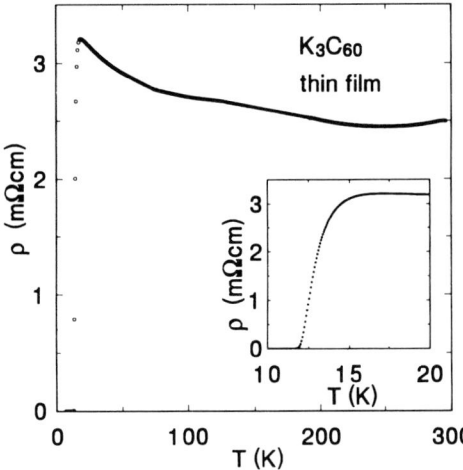

FIG. 1. Temperature dependence of the electrical resistivity of a 1600-Å thin film of K_3C_{60} between 4 and 300 K. The inset shows an expanded scale of the superconducting transition between 10 and 20 K.

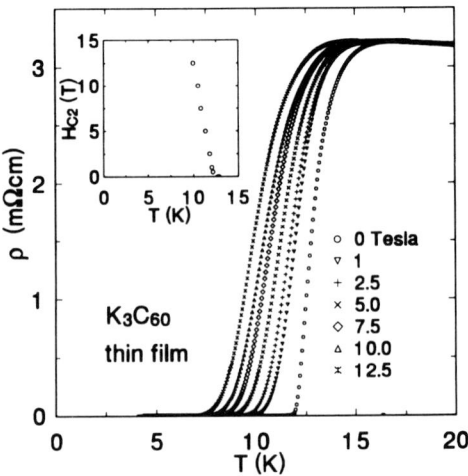

FIG. 2. Temperature dependence of the electrical resistivity of K_3C_{60} in dc magnetic fields of 0, 1, 2.5, 5.0, 7.5, 10.0, and 12.5 T. Inset: Temperature dependence of the upper critical field H_{c2} determined from the midpoints, giving a slope of 5.5 T/K.

normal-state resistivity is negligible on this scale. Therefore, we define the superconducting transition temperature T_c as the midpoint value. The transition is sharp with a 50% to 90% width of only 0.6 K. The onset of the transition is considerably broader, at least part of which is intrinsic to this material (see below).

In Fig. 2 we show the shift of the resistive transition on application of a magnetic field up to 12.5 T. The field was applied perpendicular to the film. Although the transition broadens slightly in a magnetic field, this change is so small that the midpoint is a reasonable estimate of T_c. The inset of Fig. 2 shows the magnetic-field dependence of T_c. We attribute the small upturn at low fields to sample imperfection, and therefore we take the upper-critical-field slope from the high-field data. We obtain a value of 5.5 T/K, which results [11] in a very large upper critical field of ~ 47 T and a Ginzburg-Landau coherence length $\xi_{GL} \approx 26$ Å, in good agreement with magnetic measurements [6,7].

In Fig. 3 we show the Hall coefficient between 30 and 260 K. The Hall coefficient is at low temperatures a factor of 5 smaller than expected from three electrons per C_{60}, which gives $1/ne = -1.5 \times 10^{-9}$ m^3/C. With increasing temperature we observe that the Hall coefficient increases and changes sign at 220 K. Typically one expects electronlike behavior for an almost empty band, and holelike behavior for an almost filled band, based on the energy dispersion in these cases. Near half filling the sign of the charge of the carriers is difficult to predict, and the sign change at 220 K is further evidence that these films are near half filling. This is in agreement with recent band-structure calculations that show both electron and hole orbitals, effectively reducing the Hall coefficient [8]. However, because of the granular microstructure (discussed below), interpretation of the Hall voltage in terms of a carrier density is questionable.

First let us assume that the high resistivity is caused by microscopic disorder. Since the Hall data indicate multiple conduction bands, we assume for simplicity three carriers per C_{60}. This results in an effective scattering time $\tau = m/\rho n e^2 = 3 \times 10^{-16}$ s. Using the theoretical [8] Fermi velocity, $v_F \approx 1.8 \times 10^5$ m/s, we find an effective mean free path l of the order of the interatomic distance. This means that the Ginzburg-Landau coherence length ξ_{GL} is reduced from the Pippard coherence length, $\xi_0 = 0.18 \hbar v_F/k_B T_c \approx 200$ Å, by mean-free-path effects, $\xi_{GL} \approx 0.85 (\xi_0 l)^{1/2} \approx 20$ Å, in good agreement with our experimental value. Since l is of the order of interatomic distances, the system is near the metal-insulator transition. Even though these parameters give a reasonable

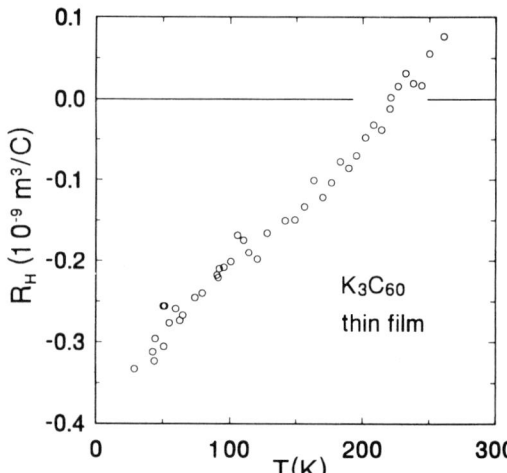

FIG. 3. Temperature dependence of the Hall coefficient between 30 and 260 K. For three electrons per C_{60} the Hall coefficient is -1.5×10^{-9} m^3/C.

description of several normal- and superconducting-state properties, we have compelling evidence that this picture is inappropriate for our films.

We will show that the high resistivity is not caused by microscopic disorder, but by the granular nature of the films. The granularity is evidenced by the observation of zero-dimensional fluctuations in the conductivity, as shown below. Further support of this model is the x-ray linewidth broadening of the pristine films [12], giving an estimated grain size of 60 Å. A granular model has previously been used to model the activated ($x \neq 3$) transport properties above room temperature, using an activation energy derived from electrostatic charging energies of 80-Å-diam grains [10]. We also note that our films have a temperature coefficient of the resistivity that approaches zero at optimum doping [10]. This means that we are near a metal-insulator transition. Imry and Strongin [13] use scaling arguments to propose a critical value at the transition, $\rho_c = (\hbar/e^2)D$, with D the grain size. Using the experimentally observed resistivity of 3 mΩ cm, we can estimate a grain size of ~75 Å. Finally, we can estimate the particle size from the width of the resistive transition [14], $\Delta T \approx T_c(k_B/D^3\Delta C_p)^{1/2}$. Using the experimental width $\Delta T_c \approx 2$ K and the electronic specific-heat term $\gamma = \Delta C_p/1.43 = 70$ mJ/mol K^2 [7], we find that the particle size is ~70 Å.

Usually, superconducting fluctuations [15] in 3D materials are confined to an extremely narrow temperature interval near T_c. However, the small coherence length $\xi_{GL} \approx 26$ Å and the high resistivity in these films makes the fluctuations observable as excess (or para) conductivity in a wide regime above T_c. We have analyzed the zero-field data in more detail in Fig. 4, plotting the excess conductivity as a function of the reduced temperature. The normal-state conductivity can be fitted between 20 and 75 K with the empirical formula, $\sigma_n = 2.1 \times 10^2 + 80 \log_{10}(T)$ (Ω cm)$^{-1}$. Plotting then the excess conductivity versus the reduced temperature $t = (T - T_c)/T_c$, we find power-law behavior for $T_c = 11.75$ K, corresponding to the onset of resistivity. The slope of the data yields a power of -2.2, and is rather insensitive to the value of T_c.

Excess conductivity in superconductors σ' has been extensively studied [15] and depends on the dimensionality d of the system: $\sigma' \propto t^{(d-4)/2}$. For a 3D system the proportionality factor is $\frac{1}{32} e^2/\hbar\xi(0)$ and depends only on the coherence length. Using $\xi(0) = 26$ Å we expect $\sigma' = 3 \times 10^3 t^{-1/2}$, as indicated by the solid line in Fig. 4. It is clear that this functional form cannot account for the data. Moreover, the 3D fluctuation contribution is even larger than the experimental excess conductivity for $t > 0.2$. This means that the experimental curve has a narrower transition than that expected for the 3D fluctuation model and therefore we can exclude the possibility of resistive broadening due to a spread in T_c. The experimentally observed power $(d-4)/2 \approx -2$ suggests that these films exhibit zero-dimensional fluctuations for

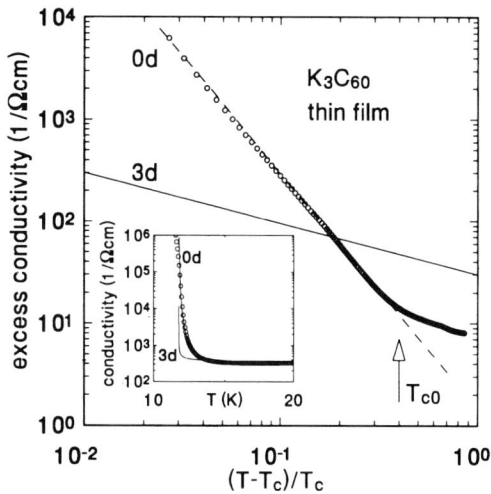

FIG. 4. Excess conductivity vs reduced temperature of K$_3$C$_{60}$. The slope of this curve has been proposed for a zero-dimensional system. A three-dimensional system exhibits fluctuation conductivity indicated by curve "3D," based on the measured coherence length $\xi_{GL}(0) = 26$ Å. Inset: The divergence of the conductivity in unrenormalized quantities. The two lines indicate the theoretical prediction for a 0D and a 3D system, respectively. The arrow indicates the mean-field transition temperature T_{c0}.

$t < 0.4$. For $t > 0.4$ the slope of the curve is $\sim -\frac{1}{2}$, which indicates the transition to 3D behavior at the mean-field transition temperature $T_{c0} = 18$ K.

Zero-dimensional fluctuations can be observed in systems in which the conduction is mediated by superconducting particles that are weakly coupled, i.e., granular materials [16]. Furthermore, the grain size must be smaller than the superconducting coherence length $\xi_{GL}(T)$. Inside each grain the superconducting order parameter is well established, but the long-range phase coherence is destroyed by the weak coupling between the particles. Experimentally, the $\sigma \propto t^{-2}$ behavior has been observed in granular Sn [17] and NbN [18]. As $\xi(T)$ diverges at T_c the particle size can be a few times $\xi_{GL}(0) = 26$ Å. The origin of the granularity [10] may be inherent in the growth of the pristine film or could be related to the phase separation of undoped C$_{60}$, K$_3$C$_{60}$, and the fully doped K$_6$C$_{60}$ [19]. We emphasize that the present film growth procedures consistently give room-temperature resistivities of 2.5 mΩ cm, and the films are completely stable at room temperature.

The reduction of T_c in our films with respect to the bulk value can be ascribed to Coulomb charging effects of the grains. It has been shown that T_c can be substantially reduced, before H_{c2} is affected [20]. The origin of this effect is that for our resistivities larger magnetic fields are required to quench the Josephson coupling compared to the isolated-grain superconductivity [21]. It is well known that for small particles the critical field is a single-grain property, and depends on the size of the grain [22]. Deutscher, Entin-Wohlman, and Shapira

[21] find that in strong magnetic fields and weak coupling the upper critical field is given by $H_{c2} = (5/12\pi^2)^{1/2}\phi_0/\xi_R$, with R the radius of the grain. The appropriate coherence length is that of the single grain [21], $\xi_{grain} = 0.85\xi_0 l[T_c/(T_c - T)]^{1/2}$, and we can take $l \approx R$. Using $H_{c2}(0.9) = 6.5$ T, we find a Pippard coherence length $\xi_0 \approx 150$ Å. (The error bar on this value is mostly related to the determination of the grain size $D \approx 70$ Å.) This coherence length is in good agreement with the LDA calculations [8,9], but, of course, much larger than the single-grain value.

In conclusion, we have characterized thin-film K_3C_{60} to be a granular superconductor, with a typical grain size of 70 Å. The granularity of our films is clearly shown by the presence of zero-dimensional fluctuations above T_c. The grain size of 70 Å leads to a metal-to-insulator transition near a normal-state resistivity of 3 mΩ cm. The upper critical field is strongly affected by the grain size of the film, and using the theory of granular superconductors, we find a Pippard (clean limit) coherence length of ~150 Å, in good agreement with recent theoretical calculations. The sign of the carriers changes with temperature, which shows that in this material at half filling both electron and hole conduction are present.

We gratefully acknowledge stimulating discussions with E. Abrahams, B. Batlogg, S. J. Duclos, D. R. Harshman, G. P. Kochanski, P. B. Littlewood, A. P. Ramirez, S. H. Glarum, D. W. Murphy, M. J. Rosseinsky, and M. Schluter, and technical assistance from A. V. Makhija. J. Zaanen acknowledges financial support from the Foundation for Fundamental Research on Matter (FOM), which is sponsored by the Netherlands Organization for the Advancement of Pure Research (NWO).

[1] A. F. Hebard, M. J. Rosseinsky, R. C. Haddon, D. W. Murphy, S.H. Glarum, T. T. M. Palstra, A. P. Ramirez, and A. R. Kortan, Nature (London) 350, 600–601 (1991).
[2] M. J. Rosseinsky, A. P. Ramirez, S. H. Glarum, D. W. Murphy, R. C. Haddon, A. F. Hebard, T. T. M. Palstra, A. R. Kortan, S. M. Zahurak, and A. V. Makhija, Phys. Rev. Lett. 66, 2830–2832 (1991).
[3] R. C. Haddon, A. F. Hebard, M. J. Rosseinsky, D. W. Murphy, S. J. Duclos, K. B. Lyons, B. Miller, J. M. Rosamilia, R. M. Fleming, A. R. Kortan, S. H. Glarum, A. V. Makhija, A. J. Muller, R. H. Eick, S. M. Zahurak, R. Tycko, G. Dabbagh, and F. A. Thiel, Nature (London) 350, 320–322 (1991).
[4] J. G. Bednorz and K. A. Muller, Z. Phys. B 64, 189 (1986).
[5] L. R. Testardi, J. H. Wernick, and W. A. Royer, Solid State Commun. 15, 1 (1974).
[6] K. Holczer, O. Klein, G. Gruner, J. D. Thompson, F. Diederich, and R. L. Whetten, Phys. Rev. Lett. 67, 271 (1991).
[7] A. P. Ramirez, M. J. Rosseinsky, D. W. Murphy, and R. C. Haddon (to be published).
[8] S. C. Erwin and W. E. Picket, Science 254, 842 (1991).
[9] I. I. Mazin, S. N. Rashkeev, V. P. Antropov, O. Jepsen, A. I. Liechtenstein, and O. K. Andersen (to be published).
[10] G. P. Kochanski, A. F. Hebard, R. C. Haddon, and A. T. Fiory, Science (to be published).
[11] N. R. Werthamer, E. Helfand, and P. C. Hohenberg, Phys. Rev. 147, 295 (1966).
[12] A. F. Hebard, R. C. Haddon, R. M. Fleming, and A. R. Kortan, Appl. Phys. Lett. 59, 2109 (1991).
[13] Y. Imry and M. Strongin, Phys. Rev. B 24, 6353 (1981).
[14] V. V. Shmidt, Pis'ma Zh. Eksp. Teor. Fiz. 3, 141 (1966) [JEPT Lett. 3, 89 (1966)].
[15] W. J. Skocpol and M. Tinkham, Rep. Prog. Phys. 38, 1049 (1975).
[16] G. Deutscher and S. A. Dodds, Phys. Rev. B 16, 3936 (1977); N. A. H. K. Rao, J. C. Garland, and D. B. Tanner, Phys. Rev. B 29, 1214 (1984).
[17] J. Kirtley, Y. Imry, and P. K. Hansma, J. Low Temp. Phys. 17, 247 (1974).
[18] S. Wolf and W. H. Lowrey, Phys. Rev. Lett. 39, 1038 (1977).
[19] R. Tycko, G. Dabbagh, M. J. Rosseinsky, D. W. Murphy, R. M. Fleming, A. P. Ramirez, and J. C. Tully, Science 253 (1991).
[20] T. Chui, P. Lindenfeld, W. L. McLean, and K. Mui, Phys. Rev. B 24, 6728 (1981).
[21] G. Deutscher, O. Entin-Wohlman, and Y. Shapira, Phys. Rev. B 22, 4264 (1980).
[22] P. G. de Gennes and M. Tinkham, Physics (Long Island City, N.Y.) 1, 107 (1964).

Reprinted with permission from Science
Vol. 256, pp. 1190–1191, 22 May 1992
© 1992 The American Association for The Advancement of Science

Synthesis and Electronic Transport of Single Crystal K_3C_{60}

X.-D. Xiang, J. G. Hou, G. Briceño, W. A. Vareka, R. Mostovoy, A. Zettl,* Vincent H. Crespi, Marvin L. Cohen

Sizable single crystals of C_{60} have been synthesized and doped with potassium. Above the superconducting transition temperature T_c, the electrical resistivity $\rho(T)$ displays a classic metal-like temperature dependence. The transition to the superconducting state at $T_c = 19.8$ K is extremely sharp, with a transition width $\Delta T < 200$ mK. In contrast to transport behavior of doped polycrystalline and granular thin films, no anomalous fluctuations are observed near T_c in single crystal specimens.

The discovery of superconductivity in heavy alkali metal–doped C_{60} (1) has generated great theoretical and experimental interest. Although many experimental results of doped fullerenes have begun to shed some light on the underlying physics of these unique materials, nearly all previous measurements have been performed on weakly linked polycrystalline (2) or granular thin films samples (3). Reliable measurements on single crystals are essential for establishing intrinsic properties and determining the superconductivity mechanism.

We report here the synthesis and electronic transport measurements of high-quality single crystals of K_3C_{60}. Measurements of the dc electrical resistivity $\rho(T)$ show an intrinsic metal-like temperature dependence below room temperature, with an extremely sharp transition to the superconducting ground state at $T_c = 19.8$ K; no evidence is found for strong fluctuation effects near T_c. These results are in sharp contrast to the behavior of polycrystalline and thin film samples.

To prepare the undoped crystals, pure C_{60} powder was first extracted from carbon soot via standard liquid chromatography with an alumina column. The powder was baked at 250°C under dynamic vacuum for 24 hours and then sealed in quartz tubes with a few hundred torr of argon gas. Sealed tubes were placed in a gradient furnace with the powder held at 650°C; crystals formed in the tube at about 450°C. With this vapor

Department of Physics, University of California at Berkeley, and Materials Sciences Division, Lawrence Berkeley Laboratory, Berkeley, CA 94720.

*To whom correspondence should be addressed.

transport method, crystals with flat, shiny faces up to a few millimeters across could be obtained in a few days. X-ray diffraction studies confirmed the fcc (face-centered-cubic) crystal structure and lattice constant reported previously (4) for solid C_{60}.

Electrical contacts to the samples were made prior to doping by first evaporating silver onto the crystal surfaces and then attaching gold wires with conducting silver paint. Both Van der Pauw (5) and in line four-probe geometries were employed (with similar results). The mounted samples were then sealed together with fresh potassium metal in a Pyrex glass apparatus with tungsten feedthrough leads. Uniform doping was accomplished using a repetitive high-temperature dope-anneal cycle. First, both the sample and dopant were heated uniformly in a furnace while the sample resistance was continuously monitored. The temperature was raised from room temper-

ature to about 200°C at a rate of 6°C per minute. At about 150°C, the resistance of the sample dropped down to within the measurable range of the ohm meter (20 mΩ); thereafter the resistivity of the sample dropped continually to a few hundred mΩ-cm within a few minutes. The tube was maintained at about 200°C for approximately one-half hour until the resistance of the sample reached a minimum. At this point the potassium end of the tube was cooled to room temperature and the sample alone was annealed at about 200° to 250°C overnight. Then the potassium end was reheated to ≈200°C and the sample was further doped until a lower resistivity minimum was reached. The sample alone was then again annealed for several hours. This doping and annealing process was repeated until the resistance reached an equilibrium state. For transport measurements the sample cell was injected with a helium exchange gas to ensure good thermal conduction, and a diode temperature sensor was mounted in the cell adjacent to the crystal.

The dc electrical resistivity ρ was measured versus temperature for a crystal of K_3C_{60} (Fig. 1). Crystals from different preparation batches yielded similar results. Near room temperature the resistivity is about 5 mΩ-cm, comparable to that obtained for K_3C_{60} films at room temperature (1, 3). However, because of geometrical uncertainties associated with the contact pads, the absolute value of the resistivity should be considered reliable only to within a factor of 2. Below room temperature, $\rho(T)$ falls in a metal-like fashion with distinct curvature. At T_c the resistivity drops abruptly to zero, with a transition width < 200 mK. The inset in Fig. 1 shows $\rho(T)$ near T_c in detail. The temperature has been swept slowly (~50 mK per minute) near the transition temperature showing no difference in T_c between cooling and warming.

The $\rho(T)$ behavior shown in Fig. 1 for the single crystal is in contrast to $\rho(T)$ observed by other groups for potassium-

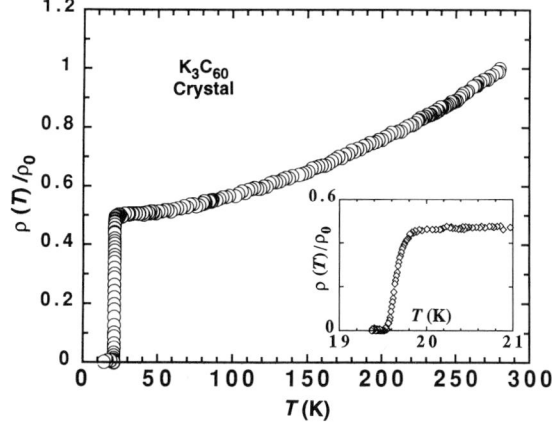

Fig. 1. Normalized dc electrical resistivity $\rho(T)$ for single-crystal K_3C_{60}. The ρ_0 is the resistivity at $T = 280$ K. The inset shows the $\rho(T)$ behavior near the superconducting transition temperature $T_c = 19.8$ K.

doped C_{60} (3, 6). Thin film samples (3) show a semiconductor-like upturn in $\rho(T)$ above T_c and a severely broadened transition to the superconducting state. We conclude that this behavior is not intrinsic but due to imperfections such as grain boundaries (3). In fact, our own transmission electron microscopy studies on thin C_{60} films (7) show that the grain size of freshly deposited films maintained in air at room temperature changes from ~0.5 μm to <200 Å over a period of several days. Hence, conclusions drawn from thin film transport data may be suspect.

The quality of our resistivity data and the sharpness of the transition to the superconducting state allow a test of various fluctuation mechanisms. We have attempted to fit $\rho(T)$ near T_c to three-, two-, one-, and zero-dimensional fluctuation expressions (8). In all cases, the agreement with experimental data is poor. Hence, we conclude that K_3C_{60} displays no substantial fluctuation conductivity near the onset to the superconducting state.

The overall temperature dependence of $\rho(T)$ above T_c places constraints on normal-state transport models. Preliminary analysis indicates that the temperature dependence of the resistivity can be fit to a T^2 functional form, a result consistent with electron-electron scattering, although electron-electron scattering has not been observed at such high temperatures in other systems. The observed temperature dependence can also be accounted for with an electron-phonon scattering mechanism (9) if there is a high-frequency contribution from the intraball phonons and a lower frequency contribution from phonons with frequencies in the range of 10 to 200 K.

REFERENCES AND NOTES

1. R. C. Haddon et al., *Nature* **350**, 320 (1991).
2. K. Holczer et al., *Science* **252**, 1154 (1991).
3. T. T. M. Palstra, R. C. Haddon, A. F. Hebard, J. Zaanen, *Phys. Rev. Lett.* **68**, 1054 (1992).
4. P. A. Heiney et al., *Phys. Rev. Lett.* **66**, 2911 (1991).
5. L. J. Van der Pauw, *Philips Res. Repts.* **13**, 1 (1958).
6. Y. Maruyama et al., *Chem. Lett.* **10**, 1849 (1991).
7. J. G. Hou et al., in preparation.
8. M. Tinkham, *Introduction to Superconductivity* (Krieger, Huntington, NY, 1975), p. 253.
9. J. M. Ziman, *Electrons and Phonons* (Oxford at the Clarendon Press, Oxford, 1979).
10. This research was supported by National Science Foundation grant DMR88-18404 and by the Office of Energy Research, Office of Basic Energy Sciences, Materials Sciences Division of the U.S. Department of Energy under contract DE-AC03-76SF00098. V.H.C. was supported by a National Defense Science and Engineering Graduate Fellowship.

21 February 1992; accepted 16 April 1992

Reprinted with permission from Science
Vol. 253, pp. 301–303, 19 July 1991
© 1991 The American Association for The Advancement of Science

Organic Molecular Soft Ferromagnetism in a Fullerene C_{60}

PIERRE-MARC ALLEMAND, KISHAN C. KHEMANI, ANDREW KOCH,
FRED WUDL, KAROLY HOLCZER, STEVEN DONOVAN,
GEORGE GRÜNER, JOE D. THOMPSON

The properties of an organic molecular ferromagnet [$C_{60}TDAE_{0.86}$; TDAE is tetrakis(dimethylamino)ethylene] with a Curie temperature T_c = 16.1 kelvin are described. The ferromagnetic state shows no remanence, and the temperature dependence of the magnetization below T_c does not follow the behavior expected of a conventional ferromagnet. These results are interpreted as a reflection of a three-dimensional system leading to a soft ferromagnet.

THE QUEST FOR A NONPOLYMERIC organic ferromagnet has intensified during the past 5 years with varying results (1–10). Curie temperatures (T_c's) on the order of 1 to 2 K have been observed (7–12). An organometallic molecular ferromagnet with T_c of ~4 K has been known for sometime (13), and similar systems with T_c of 6.2 K (14) and 8.8 K (15) have recently been reported. Polymeric ferromagnets with T_c's claimed to be greater than 300 K have also been described but their characterization remains incomplete (2, 16), and the possibility of a polaronic, polymeric ferromagnet has appeared (17). We report on the preparation and preliminary characterization of an organic molecular solid with a transition at 16 K to a soft ferromagnet state.

Our interest in the preparation of materials based on the reduction (n-doping) (18, 19) of fullerenes (20–25) prompted us to explore the use of strong organic reducing agents such as tetrakis(dimethylamino) ethylene (TDAE). Addition of a 20 M excess of the donor to a solution of C_{60} in toluene in a dry box afforded a black microcrystalline precipitate of $C_{60}(TDAE)_{0.86}$ [or $(C_{60})_{1.16}TDAE$] (26). The extremely air-sensitive (27) solid was washed with toluene and loaded into a capillary tube prepared from a Pasteur pipette, which had been previously sealed at the narrow end. The wide bore end of the pipette was then connected to a 5-mm vacuum stopcock, and the pipette was removed from the dry box to seal the sample under He in a vacuum line at ~2 cm above the powder fill line. The sample was then cooled in zero field in an ac susceptometer and then allowed to warm in an applied field of ≤0.1 Oe. Surprisingly, the transition observed was apparently to a ferromagnetic state.

The sample was studied in greater detail by dc magnetization (M) measurements taken on a Quantum Design superconducting quantum interference device (SQUID) magnetometer. Care was taken to ensure that the sample was not exposed to a field gradient greater than 0.03% of the applied field. In Fig. 1 we show M as a function of T for the sample cooled and warmed in an applied field H_a ~ 1 Oe. Two significant observations can be made from Fig. 1: (i) although M increases sharply below T_c ~ 16.1 K as expected for a ferromagnet, the temperature dependence of M does not follow that of conventional mean field theory; and (ii) within experimental error, there is no hysteresis between cooling and warming. The structure in $M(T)$ below 10 K is field-dependent; similar measurements in applied fields of 10, 100, and 1000 Oe show that with increasing field the minimum at 8 K

P.-M. Allemand, K. C. Khemani, A. Koch, F. Wudl, Departments of Chemistry and Physics and Institute for Polymers and Organic Solids, University of California, Santa Barbara, CA 93106.
K. Holczer, S. Donovan, G. Grüner, Department of Physics, University of California, Los Angeles, CA 90024.
J. D. Thompson, Los Alamos National Laboratory, Los Alamos, NM 87545.

seen in Fig. 1 evolves into only a change in slope for $H_a = 1000$ Oe. That the 16 K transition is one to a ferromagnetic state also is suggested by susceptibility $\chi \equiv M/H_a$ measurements as a function of temperature with $H_a = 10$ kOe. In Fig. 2, the product χT is plotted as a function of T. Above ~30 K, χT increases approximately linearly with T, reflecting a weakly varying χ at these temperatures. However, below ~30 K, χT increases sharply as expected from the onset of ferromagnetic correlations precursive to a ferromagnetic transition at $T_c = 16.1$ K. The maximum in χT at 10 K and the decrease at lower temperatures is a consequence of moment saturation below T_c. The absence of high-temperature Curie-Weiss susceptibility with a positive θ greater than T_c, though unusual for local moment ferromagnets, may be found in itinerant ferromagnets (see below) (28). Additional evidence supporting a ferromagnetic interpretation for $C_{60}(TDAE)_{0.86}$ comes from measurements of the field dependence of T_c, defined by the intersection of linear extrapolations of $M(T)$ from above and below the rapid rise in $M(T)$. The inset of Fig. 2 shows T_c increasing to 17.2 K for $H_a = 1$ kOe and then increasing approximately linearly with H_a to 24.3 K at 50 kOe. A field-induced enhancement of the Curie temperature is typical for ferromagnets.

Shown in Fig. 3 is M versus H_a measured at 5 K. The "S"-shaped curve is characteristic of ferromagnetism. However, within experimental uncertainty of ± 2 Oe, there is no hysteresis and zero coercivity and remanence. Close inspection of $M(H)$ at various

fixed temperatures below T_c shows that for small H_a, M is linear (nearly vertically) about the origin for positive and negative values of H_a. We define ΔM as the magnetization range over which M is linear in H, that is, ΔM is proportional to the spontaneous magnetization. The temperature variation of ΔM is plotted in the inset of Fig. 3. For reference we note that a value of $\Delta M = 2 \times 10^{-3}$ emu corresponds to a moment of 0.11 μ_B per mole of $C_{60}(TDAE)_{0.86}$. From measurements such as shown in Fig. 3, it is difficult to define unambiguously a saturated moment because of the paramagnetic contribution from the pyrex sample holder, which has not been subtracted from these data. (At the very small fields over which ΔM is defined, the pyrex contribution is negligible.)

An alternative explanation for the data of Fig. 3 is that it arises from superparamagnetism. The field variation of M is similar to that observed in superparamagnetic systems in which H_a is perpendicular to the easy

magnetic axis (29). However, in our experiments the fine $C_{60}(TDAE)_{0.86}$ powder is constrained in the sample holder only by gravity, and it is unlikely that the powder satisfied this condition since it would tend to orient itself with the easy axis parallel to H_a. Further, for superparamagnetism, M should scale as H/T below T_c (30). Experimental $M(H)$ isotherms do not scale with H/T nor with H/T scaled by the saturated moment, which we assume is proportional to ΔM. Consequently, we conclude that $C_{60}(TDAE)_{0.86}$ is a very soft ferromagnet with a remanence of zero or nearly so. The lack of remanence may be attributed to a fully three-dimensional system devoid of domain pinning sites. The exact nature of the ferromagnetism remains an open question. The observation of relatively high ambient temperature conductivity in a compressed pellet (~10^{-2} S cm^{-1}) and preliminary electron spin resonance (ESR) results that show strong narrowing of the

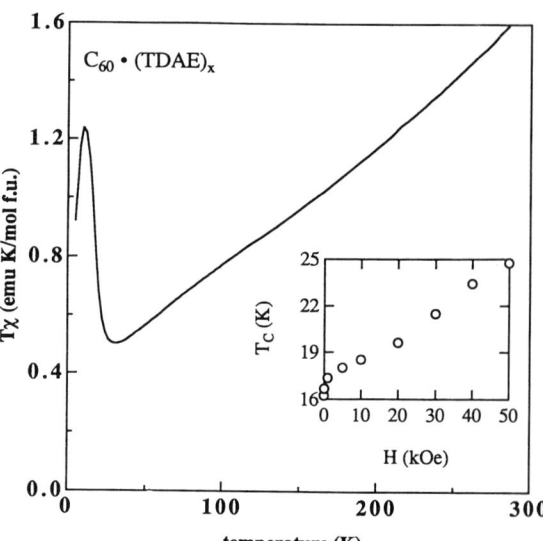

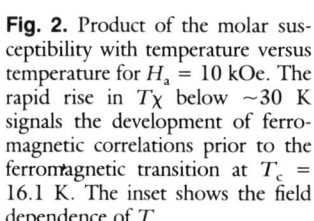

Fig. 2. Product of the molar susceptibility with temperature versus temperature for $H_a = 10$ kOe. The rapid rise in $T\chi$ below ~30 K signals the development of ferromagnetic correlations prior to the ferromagnetic transition at $T_c = 16.1$ K. The inset shows the field dependence of T_c.

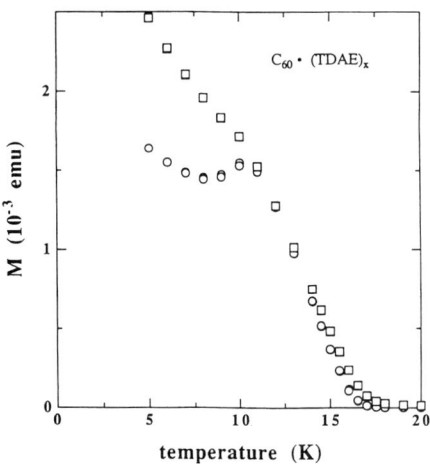

Fig. 1. Magnetization versus temperature for $C_{60}(TDAE)_x$ measured upon cooling and subsequent warming in an applied field of ~1 Oe ($\bigcirc$) and 100 Oe ($\square$). The 1-Oe data were multiplied by 10 to give the same ordinate values as the 100-Oe data. Note disappearance of curvature below 10 K at high field. Data were not corrected for the sample holder or for demagnetizing effects.

Fig. 3. Magnetization as a function of field at 5 K. Within experimental error, there is no hysteresis in these data. The inset shows ΔM, proportional to the spontaneous magnetization, as a function of temperature. See text for details.

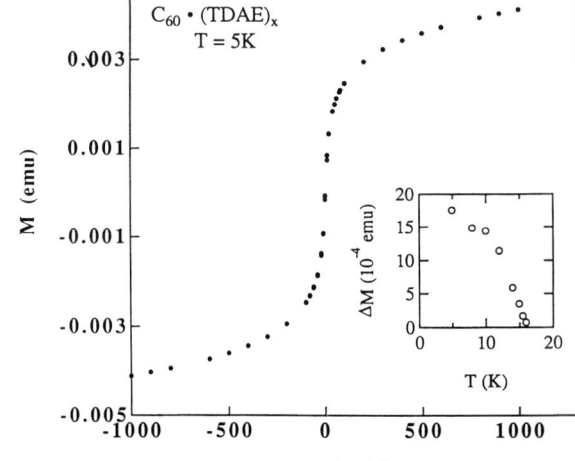

ESR line with decreasing temperature imply that the material might be metallic; if so, it might be an itinerant ferromagnet. Examples of the latter (such as $ZrZn_2$ and Ni_3Ga) show, qualitatively, features that are similar to the ones reported here (31).

$C_{60}(TDAE)_{0.86}$ is a "metallic" organic ferromagnet with a Curie temperature higher than any reported for other molecular ferromagnets based strictly on first-row elements. We note that the same basic molecule, namely C_{60}, supports ferromagnetism, metallic conductivity, and superconductivity (in the forms of K_3C_{60} and Rb_3C_{60}), a rather interesting and unusual occurrence.

REFERENCES AND NOTES

1. R. Breslow, *Mol. Cryst. Liq. Cryst.* **176**, 199 (1989).
2. J. S. Miller and D. A. Dougherty, Eds. *Proceedings of the Symposium on Ferromagnetic and High Spin Molecular Based Materials* [*Mol. Cryst. Liq. Cryst.* **176** (1989)].
3. J. S. Miller, A. J. Epstein, W. A. Reiff, *Chem. Rev.* **88**, 201 (1988).
4. T. LePage and R. Breslow, *J. Am. Chem. Soc.* **109**, 6412 (1987).
5. M. Kinoshita, *Mol. Cryst. Liq. Cryst.* **176**, 163 (1989).
6. H. P. Iwamura, *Pure Appl. Chem.* **58**, 187 (1986).
7. K. Awaga, T. Inabe, U. Nagashima, Y. Maruyama, *J. Chem. Soc. Chem. Commun.* **1989**, 1617 (1989).
8. K. Awaga and Y. Maruyama, *J. Chem. Phys.* **91**, 2743 (1989).
9. _____, *Chem. Phys. Lett.* **158**, 556 (1989).
10. K. Awaga, T. Inabe, U. Nagashima, Y. Maruyama, *J. Chem. Soc. Chem. Commun.* **1990**, 520 (1990).
11. K. Awaga, T. Sugano, M. Kinoshita, *J. Chem. Phys.* **85**, 2211 (1986).
12. P.-M. Allemand, G. Srdanov, F. Wudl, *J. Am. Chem. Soc.* **112**, 9392 (1990).
13. J. S. Miller *et al.*, *J. Chem. Soc. Chem. Commun.* **1986**, 1026 (1986).
14. W. E. Broderick, J. A. Thompson, E. P. Day, B. M. Hoffman, *Science* **249**, 401 (1990).
15. G. T. Yee *et al.*, *Adv. Mater.*, in press.
16. H. Tanaka, K. Tokuyama, T. Sato, T. Ota, *Chem. Lett.* **1990**, 1813 (1990).
17. D. A. Kaisaki, W. Chang, D. A. Dougherty, *J. Am. Chem. Soc.* **113**, 2764 (1991).
18. P.-M. Allemand *et al.*, *ibid.*, p. 1050.
19. P.-M. Allemand *et al.*, *ibid.*, p. 2780.
20. W. Krätschmer, L. D. Lamb, K. Fostiropoulos, D. R. Huffman, *Nature* **347**, 354 (1990).
21. R. E. Haufler *et al.*, *J. Phys. Chem.* **94**, 8634 (1990).
22. R. Taylor, J. P. Hare, A. K. Abdul-Sada, H. W. Kroto, *J. Chem. Soc. Chem. Commun.* **1990**, 1423 (1990).
23. J. M. Hawkins *et al.*, *Science* **252**, 312 (1991).
24. A. F. Hebard *et al.*, *Nature*, in press.
25. R. C. Haddon *et al.*, *ibid.* **350**, 320 (1991).
26. Elemental analysis calculated for $C_{80}H_{24}N_4O_{1.25}$, C, 90.57, H, 2.26, and N, 5.28; found, C, 90.55, H, 2.24, N, 5.31, and O, 1.90 (by difference). Since the material is so reactive, we assume that oxygenation occurred during analysis manipulations and is extrinsic. Therefore throughout the manuscript we omit the oxygen; $C_{80}H_{24}N_4$ corresponds to $C_{60}(TDAE)_{0.86}$ or $(C_{60})_{1.16}TDAE$. The observed mixed valence is in line with material being an electrical conductor.
27. Exposure to the air by accident for a few seconds caused complete loss of magnetic properties, indicating that the phenomenon is not associated with an iron impurity that would exist as magnetic iron oxide.
28. T. Moriya, Ed., *Spin Fluctuations in Itinerant Electron Magnetism* (Springer-Verlag, Berlin, 1985).
29. For example, A. H. Morrish, *The Physical Principles of Magnetism* (Wiley, New York, 1965), p. 332.
30. A. E. Berkowitz and E. Kneller, *Magnetism and Metallurgy* (Academic Press, New York, 1969), p. 393.
31. E. P. Wohlfarth, *Comments Solid State Phys.* **6**, 123 (1975).

32. F.W. thanks the National Science Foundation for support through grants DMR-88-20933 and CHE89-08323. Work at Los Alamos was performed under the auspices of the U.S. Department of Energy.

3 June 1991; accepted 26 June 1991

Reprinted with permission from Nature
Vol. 355, pp. 331–332, 23 January 1992
© 1992 Macmillan Magazines Limited

Lattice structure of the fullerene ferromagnet TDAE–C$_{60}$

Peter W. Stephens*, David Cox†, Joseph W. Lauher‡,
Laszlo Mihaly*, John B. Wiley§, Pierre-Marc Allemand||,
A. Hirsch||, Karoly Holczer§, Q. Li||, Joe D. Thompson¶
& Fred Wudl||

* Department of Physics, State University of New York, Stony Brook,
New York 11794, USA
† Department of Physics, Brookhaven National Laboratory, Upton,
New York 11973, USA
‡ Department of Chemistry, State University of New York, Stony Brook,
New York 11794, USA
§ Departments of Chemistry and Biochemistry and Physics,
and Solid State Science Center, University of California, Los Angeles,
California 90024, USA
|| Departments of Chemistry and Physics and Institute for Polymers
and Organic Solids, University of California, Santa Barbara,
California 93106, USA
¶ Los Alamos National Laboratory, Los Alamos,
New Mexico 87545, USA

IN the brief time since the development of techniques for the purification of condensed phases of C$_{60}$ (refs 1, 2), compounds of C$_{60}$ have been synthesized that exhibit the important solid-state properties of superconductivity[3,4] and ferromagnetism[5]. The fact that the C$_{60}$ molecule participates in these two disparate effects is in itself striking; furthermore, the C$_{60}$-based material has the highest T_c of any molecular organic ferromagnet[6]. An understanding of the physical origin of this ferromagnetism is sure to require a knowledge of the crystal structure. Here we report on an X-ray diffraction study of the ferromagnet TDAE–C$_{60}$, where TDAE is tetrakis(dimethylamino)ethylene, C$_2$N$_4$(CH$_3$)$_8$. We find that the composition is stoichiometric with 1:1 ratio of TDAE to C$_{60}$. The structure has a c-centred monoclinic unit cell suggestive of an anisotropic, low-dimensional band structure. Any theory for the ferromagnetic behaviour will have to take this anisotropy into account.

We prepared samples by allowing an excess of TDAE to react with C$_{60}$. This differs from previous work[5] by the absence of any solvent. The sample used here had a low-temperature saturation moment of $0.33\mu_B$ per formula unit, and a critical temperature $T_c = 16.1$ K. Figure 1 shows the room-temperature diffraction profile measured at beamline X7A of the National Synchrotron Light Source. The pattern consists of a number of strong, sharp peaks from TDAE–C$_{60}$, together with several significantly broader peaks from unreacted face-centred-cubic

(f.c.c.) C$_{60}$ ($a = 14.18$ Å). We have so far been unable to obtain samples free of coexisting unreacted C$_{60}$, apparently because of the low solubility of that molecule. The powder pattern indexing program TREOR[7] found only one good candidate unit cell, with figure of merit $F = 87$: this was a c-centred monoclinic unit cell, with room-temperature dimensions $a = 15.874(4)$ Å, $b = 12.986(2)$ Å, $c = 9.981(3)$ Å, $\beta = 93.31(1)°$ (one sigma error bars in last digit). This cell accounted for all of the diffraction peaks other than those assigned to f.c.c. C$_{60}$. The only systematic absences were found for peaks with $h + k$ odd, suggesting the space group $C2/m$ or the non-centric subgroups Cm and $C2$. If the origin is chosen so that the C$_{60}$ molecules are centred at the $2(a)$ sites (000) and $(\frac{1}{2}\frac{1}{2}0)$, the $2(d)$ sites at $(\frac{1}{2}0\frac{1}{2})$ and $(0\frac{1}{2}\frac{1}{2})$ are obvious locations for the TDAE molecules.

The low intensity of diffraction peaks at high angles makes it impossible to refine all of the internal coordinates of the TDAE molecule, and so we assumed a rigid structure for the latter. An ideal molecular geometry was generated using locally developed molecular mechanics programs[8,9] and is similar to the geometry found for bis(dimethylamino)-methylenemalononitrile[10]. The most interesting feature is a predicted 20° twist about the C=C double bond, caused by the excessive steric bulk of the four dimethylamine substituents. Similarly, the C$_{60}$ molecule was assigned idealized rigid molecule coordinates with bond lengths of 1.37 Å and 1.45 Å, oriented with one of the 2-fold axes along b.

The predicted presence of three 2-fold axes and absence of an inversion centre in the TDAE molecule suggest space group $C2$ as the logical choice. (Other possibilities such as $C2/m$ with disorder in the TDAE orientations must be considered, in analogy to the situation in K$_3$C$_{60}$, where disorder in the C$_{60}$ orientations leads to space group $Fm\overline{3}m$ (ref. 11).) We have performed Rietveld[12] refinements for various possible arrangements of the molecules in the unit cell. A Rietveld fit, excluding the regions of overlap with f.c.c. C$_{60}$ peaks, is shown in Fig. 1, and the structure is in Fig. 2. The best refinement was given with the long axis of the TDAE molecule (in the direction of the ethylene double bond) assumed to be along the monoclinic c-axis. This fit gave temperature factors (assumed isotropic)

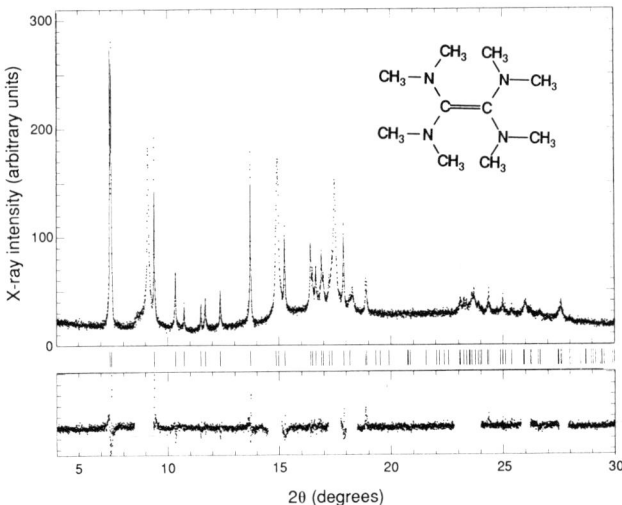

FIG. 1 X-ray diffractogram of TDAE–C$_{60}$ compound. Data were taken at wavelength 1.2995 Å with a channel-cut Ge(111) monochromator and Ge(220) analyser. The smooth curve is a Rietveld fit, as described in text. To define the baseline, we have drawn a piecewise linear curve through regions of the spectrum where there are no allowed diffraction peaks. Inset shows a diagram of the TDAE molecule. The lower panel shows residuals, to the same vertical scale. Regions around untreated f.c.c. C$_{60}$ peaks excluded from the fit are absent from the model curve and residuals.

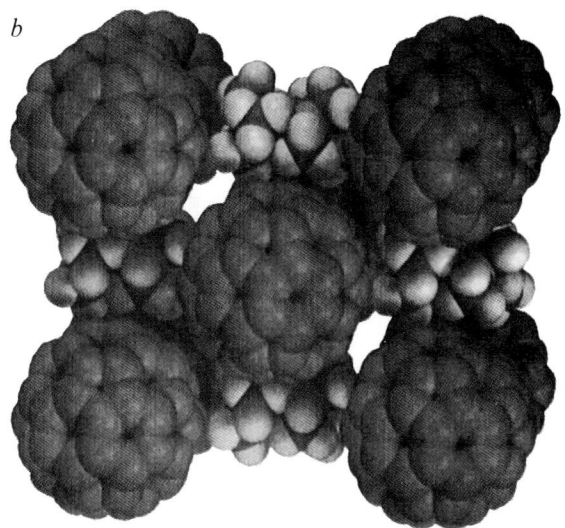

a

b

FIG. 2 *a*, Structure of TDAE-C_{60} determined in the present work. View is along the monoclinic *b*-axis. *b*, View along monoclinic *c*-axis.

$B = 0.9$ Å^2 for C_{60}, 3.8 Å^2 for TDAE, and yielded quality of fit factors $R_I = 15.7\%$, $R_{WD} = 8.6\%$, $\chi^2 = 1.79$.

We have taken a diffraction pattern of a second sample at 11 K. The degree of contamination by unreacted C_{60} is slightly less in that sample. We achieved similar R factors in Rietveld fits. The lattice parameters were $a = 15.807(1)$ Å, $b = 12.785(1)$ Å, $c = 9.859(1)$ Å, $\beta = 94.02(1)°$. We have also studied a sample prepared from C_{60} dissolved in toluene, described in ref. 5, which had a smaller saturation moment of $0.11\mu_B$ per formula unit. It showed a much smaller amount of coexisting f.c.c. C_{60}, but essentially the same monoclinic pattern. This shows that this is a simple two-phase system, and hence confirms the validity of ignoring the f.c.c. C_{60} peaks in the refinement of Fig. 1. Interestingly, the diffraction lines from this sample were not uniformly sharp; in particular, lines with h and l both nonzero were so broad as to be virtually undetectable. Evidently, there were solvent molecules present, which acted both to degrade the lattice perfection and the magnetic properties.

Additional refinements showed that the quality of fit is only weakly sensitive to the C_{60} orientation, and that the TDAE occupancy is $100 \pm 10\%$. Other TDAE molecular orientations differing by 45° and 90° rotations are clearly excluded. Difference Fourier maps do not show any significant structure. The nearest C_{60}-TDAE approach is 2.58 Å, a reasonable distance for a nonbonding C–H distance. Similar refinements in $C2/m$, with the molecules disordered in two sets of mirror-related sites, gave equivalent results.

By crystallographic standards, an R factor of 15.7% is not generally regarded as indicating a satisfactory agreement between data and model. It is equally clear that reliable intramolecular positions are not likely to be obtained from such a small number of reflections. But the quality of the data and of the agreement between data and fit is certainly adequate to determine that the unit cell is correct as specified, to exclude greatly different orientations of the TDAE molecule in the unit

cell, and to prove that the composition is 1:1 TDAE to C_{60}. The initial report[5] that the composition was TDAE$_{0.86}C_{60}$ was based on elemental analysis; in multiphase samples, diffraction experiments are the only accurate means to determine the stoichiometry of the constituent phases. We emphasize that a better refinement will be needed for detailed calculations of band structure.

Our elucidation of the structure helps in the understanding of the ferromagnetism in this material. It was argued previously that TDAE-C_{60} is an itinerant ferromagnet[5]: that is, the unpaired spins are carried by delocalized electrons. We have measured the electron paramagnetic resonance signal in this material, and find a relatively low value for the g-factor, $g = 2.0008$. This implies that the open-shell species reside principally on the C_{60}, and to a lesser extent on the TDAE.

Now that we have determined the unit cell parameters, we see that indeed the C_{60} cores have intermolecular contacts of the same order as observed for the metallic alkali-doped materials. At room temperature, the intermolecular separation is 9.98 Å along the c-axis, and 10.25 Å within the a–b plane. In comparison, in the metallic alkali-doped samples, the observed distances[11,13] range from 10.07 to 10.25 Å. In a tight-binding picture, the electronic bandwidth is proportional to the overlap of electronic wave functions centred on adjacent fullerenes, which decreases exponentially with separation. The observation in this material of the short interfullerene distance, along with the much greater 10.25-Å distance within the a–b plane, leads us to conclude that the TDAE salt should have a very anisotropic band structure. This stands in contrast to the superconducting A_3C_{60} structures, which are all cubic[11,13]. It is well known that low-dimensional bands are subject to a variety of electronic and magnetic instabilities that lead to solid-state phase transitions[14,15]. We therefore conclude that in TDAE-C_{60}, the ferromagnetically ordered state cannot be understood without considering the band anisotropy   □

Received 30 October; accepted 24 December 1991.

1. Krätschmer, W., Lamb, L. D., Fostiropoulis, K. & Huffmann, D. R. *Nature* **347**, 354–358 (1990).
2. Ajie, H. *et al. J. phys. Chem.* **94**, 8630–8633 (1990).
3. Hebard, A. *et al. Nature* **350**, 600–601 (1991).
4. Holczer, K. *et al. Science* **252**, 1154–1157 (1991).
5. Allemand, P. M. *et al. Science* **253**, 301–303 (1991).
6. Kinoshita, M. *et al. Chem. Lett.* 1225–1228 (1991).
7. Werner, P.-E., Eriksson, L. & Westdahl, M. *J. appl. Cryst.* **18**, 367–370 (1985).
8. Lauher, J. W. *J. Am. chem. Soc.* **108**, 1521–1531 (1986).
9. Lauher, J. W. *J. molec. Graphics* **8**, 34–38 (1990).
10. Adhikesavalu, D. & Venkatesan, K. *Acta crystogr.* C**39**, 589–592 (1983).
11. Stephens, P. W. *et al. Nature* **351**, 632–634 (1991).
12. Rietveld, H. M. *J. appl. Cryst.* **2**, 65–71 (1969).
13. Fleming, R. M. *et al. Nature* **352**, 787–788 (1991).
14. Emery, V. J. in *Chemistry and Physics of One Dimensional Metals*, ed. Keller, H. J. (Plenum, New York, 1977).
15. Jacobs, I. S. *et al. Phys. Rev.* B**14**, 3036–3051 (1976).

ACKNOWLEDGEMENTS. Work at Brookhaven was supported by the U.S. Department of Energy, Division of Materials Sciences and at Stony Brook and Santa Barbara by the NSF. Work at Los Alamos was done under the auspices of the U.S. Department of Energy.

V

CHEMICALLY MODIFIED FULLERENES

At the outset, one might have expected that C_{60} would be relatively chemically inert; its compact shape and high symmetry suggest that to attack it chemically may be difficult. However, it reacts readily with appropriate electron-rich reagents, and specific and selective additions are possible.

In this chapter on chemically modified fullerenes, I will discuss only a few representative molecules which incorporate fullerenes, and focus attention on the best-characterized compounds. A comprehensive review of chemical reactions with fullerenes has just been published by Taylor and Walton.[Ta93] It is not appropriate here to attempt a synthesis of the emerging field of the chemical properties of the fullerenes. Covered elsewhere in this book are ionically bonded fullerides, such as the superconductor K_3C_{60} (Chapter IV) and endohedral complexes, in which various atoms are captured *inside* the hollow carbon shell (Chapter VI).

The first chemical reaction to the all-carbon fullerene molecule was made by Haufler *et al.*, who reduced it to a solid of apparent formula $C_{60}H_{36}$, determined by mass spectroscopy.[Ha90] The distribution of H atoms on the surface of the "fuzzyball" was not determined. This reduction was found to be fully reversible.

The first pure fullerene compound to be synthesized and structurally characterized was an osmium derivative, $C_{60}(OsO_4)(4\text{-}tert\text{-butylpyridine})_2$. [Ha91*] The steps and insights leading to this synthesis, as well as further developments, are described in greater detail in a review paper.[Ha92] The single crystal x-ray structure was the first direct proof of the soccer ball structure of C_{60}, inasmuch as the position of each carbon atom could be observed. Recall that a C_{60} molecule has two distinct bonds: thirty "double" bonds at the fusion of two six-membered rings (6:6) and sixty "single" bonds at the fusion of six- and five-membered rings (6:5). The O–Os–O linkage is added across the 6:6 bond. Excluding bonds to the two connecting carbon atoms, the average C–C bond

lengths are 1.386 Å for 6:6 bonds and 1.434 Å for "single" bonds. However, the histograms of the two sets of bond lengths overlap significantly, and so it is clear that the perturbation due to the chemical connection extends across the entire fullerene.

Almost simultaneously, a platinum complex, in which the metal atom is bonded directly to the fullerene cage, was prepared.[Fa91*] The site of attachment is the 6:6 "double" bond, as for the osmium derivative. The carbon bond lengths in the fullerene complexed with Pt are also similar to neat C_{60} and to the Os derivative described above. Further details and other metal attachments are described in a review paper.[Fa92]

A very similar irridium complex is of particular interest because it has been attached both to C_{60} and to C_{70}, and x-ray structures published.[Ba91a, Ba91b] C_{70} has four distinct C–C bonds at the fusion of two six-membered rings, in contrast to the case of C_{60}, where all 6:6 bonds are symmetry-equivalent. The Ir complex attaches only to one of these bonds, apparently the one which can be deformed away from the fullerene with minimal added distortion.

A different family of fullerene-derived molecules has been produced by addition of a carbon "bridge" to a 6:6 bond.[Su91*, Wu92] This bridge is bonded to two phenyl rings, and opens the fullerene 6:6 bond beneath it. Up to six such units can be added. An important feature of this addition is that it apparently leaves the electronic state of the fullerene unchanged, with sixty π-electrons. The ultraviolet-visible spectra and the first few reduction potentials are changed very little as the fulleroid is inflated with up to six diphenyl additions. A number of compounds with various groups attached to, or in place of, the phenyls have been synthesized.[Wu92] This technique has been recently extended to the design and synthesis of a fullerene derivative for the inhibition of HIV.[Si93, Fr93*]

A fullerene oxide, $C_{60}O$, has been prepared by photo-irradiating an oxygenated benzene solution of

C_{60}, followed by appropriate chromatographic separation.[Cr92] Based on NMR data, this molecule is seen to be an epoxide (with the C–C bond under the oxygen intact) rather than an oxidoannulene (broken C–C bond, similar to the diphenyl fulleroids).[Cr92] Solid $C_{60}O$ has an orientational phase transition very similar to that of C_{60}.[Va92] The configuration of fullerene skeletons in the low temperature ($T < 278$ K) simple cubic structure is identical to that of C_{60}, implying that it is stabilized by the electrostatic attraction between bond charges and vacancies on adjacent fullerenes. The external oxygen atoms are apparently distributed randomly among the octahedral and tetrahedral voids.[Va92]

The last chemical addition to fullerenes which we will consider here is bromine. The largest coordination of Br on a single fullerene occurs in the molecule $C_{60}Br_{24}$. This is the largest number of additions that can be made to C_{60} without any two being adjacent. The single crystal x-ray structure of that molecule retains the soccer-ball shape of the fullerene, with either two or three Br atoms attached to each six-membered ring.[Te92] Molecules $C_{60}Br_6$ and $C_{60}Br_8$ have also been synthesized and their structures determined.[Bi92] It may be somewhat surprising to learn that the bromines cluster on one side of the fullerene in these molecules. Indeed, $C_{60}Br_6$ has Br atoms bonded to two adjacent C sites.

So far, we have considered addition reactions to the fullerenes. It is also possible to *substitute* boron for carbon in the fullerene framework. Guo *et al.* accomplished this by mixing BN powder into the graphite target of a laser vaporization supersonic cluster beam apparatus.[Gu91] The mass spectrum around the C_{60} fullerene region showed ions of C_{60}, $C_{59}B$, $C_{58}B_2$, ..., $C_{54}B_6$. It was proved that these clusters were fullerenes by observing that ions with only even numbers of carbon plus boron atoms were produced, and that the ions photofragmented through the loss of C_2 (and presumably BC and B_2). The boron-doped fullerenes were qualitatively as stable to photofragmentation as are pure carbon fullerenes. In a further experiment, it was found that trapped $(C:B)_{60}^+$ clusters exposed to ammonia attached one NH_3 molecule for each B atom within the doped fullerene.[Gu91]

A family of materials closely related to the fullerenes has recently been created by Guo *et al.* [Gu92a, Gu92b] Termed metallo-carbohedrenes, they have the general formula M_8C_{12}, where M represents one of the metals Ti, V, Zr, or Hf. These molecules were first created in a laser-produced plasma above a metal surface, in an atmosphere containing CH_4, C_2H_2, C_2H_4, C_3H_6, or C_6H_6. From a variety of experiments, it is inferred that the structure consists of a dodecahedral network, in which each of the twenty pentagonal faces consists of two metal atoms and three carbons. No fullerene as small as C_{20} has ever been observed, so it is apparent that the metal atoms are important in stabilizing this small cage. More recently, macroscopic quantities of Ti_8C_{12} and V_8C_{12} have been synthesized in an arc discharge, similar to the Krätschmer-Huffman soot technique.[Ca93]

Bibliography

References marked with * are reprinted in this volume.

Ba91a   A. L. Balch, V. J. Catalano and J. W. Lee, "Accumulating Evidence for the Selective Reactivity of the 6-6 Ring Fusion of C_{60}. Preparation and Structure of $(\eta^2$-$C_{60})Ir(CO)Cl(PPH_3)_2 \cdot 5C_6H_6$," *Inorg. Chem.* **30**, 3980–3981 (1991).

Ba91b   A. L. Balch, V. J. Catalano, J. W. Lee, M. M. Olmstead and S. R. Parkin, "$(\eta^2$-$C_{70})Ir(CO)Cl(PPH_3)_2$: The Synthesis and Structure of an Organometallic Derivative of a Higher Fullerene," *J. Am. Chem. Soc.* **113**, 8953–8955 (1991).

Bi92   P. R. Birkett, P. B. Hitchcock, H. W. Kroto, R. Taylor and D. R. M. Walton, "Preparation and Characterization of $C_{60}Br_6$ and $C_{60}Br_8$," *Nature* **357**, 479–481 (1992).

Ca93   S. F. Cartier, Z. Y. Chen, G. J. Walder, C. R. Sleppy and A. W. Castleman, Jr., "Production of Metallo-Carbohedrenes in the Solid State," *Science* **260**, 195–196 (1993).

Cr92   K. M. Creegan *et al.*, "Synthesis and Characterization of $C_{60}O$, the First Fullerene Epoxide," *J. Am. Chem. Soc.* **114**, 1103–1105 (1992).

Fa91*   P. J. Fagan, J. C. Calabrese and B. Malone, "The Chemical Nature of Buckminsterfullerene (C_{60}) and the Characterization of a Platinum Derivative," *Science* **252**, 1160–1161 (1991).

Fa92   P. J. Fagan, J. C. Calabrese and B. Malone, "Metal Complexes of Buckminsterfullerene (C_{60})," *Acc. Chem. Res.* **25**, 134–142 (1992).

Fr93* S. H. Friedman *et al.*, "Inhibition of the HIV-1 Protease by Fullerene Derivatives: Model Building Studies and Experimental Verification," *J. Am. Chem. Soc.* **115**, 6506–6509 (1993).

Gu91 T. Guo, C. Jin and R. E. Smalley, "Doping Bucky: Formation and Properties of Boron-Doped Buckminsterfullerene," *J. Phys. Chem.* **95**, 4948–4950 (1991).

Gu92a B. C. Guo, K. P. Kerns and A. W. Castleman, Jr., "$Ti_8C_{12}^+$ — Metallo-Carbohedrenes: A New Class of Molecular Clusters?" *Science* **255**, 1411–1413 (1992).

Gu92b B. C. Guo, S. Wei, J. Purnell, S. Buzza and A. W. Castleman, Jr., "Metallo-Carbohedrenes $[M_8C_{12}^+$ (M = V, Zr, Hf, and Ti)]: A Class of Stable Molecular Cluster Ions," *Science* **256**, 515–516 (1992).

Ha90 R. E. Haufler *et al.*, "Efficient Production of C_{60} (Buckminsterfullerene), $C_{60}H_{36}$, and the Solvated Buckide Ion," *J. Phys. Chem.* **94**, 8634–8636 (1990).

Ha91* J. M. Hawkins *et al.*, "Crystal Structure of Osmylated C_{60}: Confirmation of the Soccer Ball Framework." *Science* **252**, 312–313 (1991).

Ha92 J. M. Hawkins, "Osmylation of C_{60}: Proof and Characterization of the Soccer-Ball Framework," *Acc. Chem. Res.* **25**, 150–156 (1992).

Si93 R. Sijbesma *et al.*, "Synthesis of a Fullerene Derivative for the Inhibition of HIV Enzymes," *J. Am. Chem. Soc.* **115**, 6510–6512 (1993).

Su91* T. Suzuki, Q. Li, K. C. Khemani, F. Wudl and Ö. Almarsson, "Systematic Inflation of Buckminsterfullerene C_{60}: Synthesis of Diphenyl Fulleroids C_{61} to C_{66}," *Science* **254**, 1186–1188 (1991).

Ta93 R. Taylor and D. R. M. Walton, "The Chemistry of Fullerenes," *Nature* **363**, 685–693 (1993).

Te92 F. N. Tebbe *et al.*, "Synthesis and Single-Crystal X-ray Structure of a Highly Symmetrical C_{60} Derivative, $C_{60}Br_{24}$," *Science* **256**, 822–825 (1992).

Va92 G. B. M. Vaughan *et al.*, "The Orientational Phase Transition in Solid Buckminsterfullerene Epoxide ($C_{60}O$)," *Chemical Physics* **168**, 185–193 (1992).

Wu92 F. Wudl, "The Chemical Properties of Buckminsterfullerene (C_{60}) and the Birth and Infancy of Fulleroids," *Acc. Chem. Res.* **25**, 157–161 (1992).

Reprinted with permission from Science
Vol. 252, pp. 312–313, 12 April 1991
© 1991 The American Association for The Advancement of Science

Crystal Structure of Osmylated C_{60}: Confirmation of the Soccer Ball Framework

JOEL M. HAWKINS,* AXEL MEYER, TIMOTHY A. LEWIS,
STEFAN LOREN, FREDERICK J. HOLLANDER

An x-ray crystal structure that confirms the soccer ball–shaped carbon framework of C_{60} (buckminsterfullerene) is reported. An osmyl unit was added to C_{60} in order to break its pseudospherical symmetry and give an ordered crystal. The crystal structure of this derivative, $C_{60}(OsO_4)(4\text{-}tert\text{-butylpyridine})_2$, reveals atomic positions within the carbon cluster.

IN 1985, KROTO, SMALLEY, AND CO-workers discovered that 60 carbon atoms form a particularly stable cluster in the gas phase. They proposed a simple and beautiful truncated icosahedral structure for C_{60} with a novel carbon framework resembling the seams of a soccer ball, and christened the molecule buckminsterfullerene (1). Late last year, Krätschmer, Huffman, and co-workers (2) reported that C_{60} could be prepared and isolated in macroscopic quantities (3–5). Since then, chemists and physicists have sought to confirm or disprove the soccer ball structure for C_{60}. The infrared (2, 6), Raman (7), ^{13}C NMR (3, 8, 9), and photoelectron spectra (10) are each consistent with icosahedral symmetry and are collectively highly supportive of the originally proposed structure, but they do not strictly prove the soccer ball framework or provide atomic positions. For example, the ^{13}C NMR spectrum (8) does not rule out the possibility of coincident peaks or a fluxional structure. We (11) and others (2, 12) have attempted to obtain a crystal structure of C_{60}, but could not determine specific atomic positions due to extensive disorder in the crystals. While the ball-like molecules pack in an ordered fashion, their nearly spherical symmetry promotes orientational disorder (9). We reasoned that if C_{60} could be derivatized in a way that broke its apparent spherical symmetry, it might crystallize with orientational order. We report here the synthesis of a one-to-one C_{60}–osmium tetroxide adduct and its crystal structure displaying the soccer ball framework of C_{60}.

Our recent report of the osmylation of C_{60} established that heteroatoms can be added to buckminsterfullerene without disrupting its carbon framework (13). Our conditions favored the addition of two osmyl units to C_{60}, giving the two-to-one adduct in 81% yield as a mixture of regioisomers (Scheme I). Chromatographic analysis of the crude reaction mixture revealed six peaks: five peaks corresponding to the precipitate which collectively analyzes with two-to-one stoichiometry, and a single

sharp peak corresponding to toluene-soluble material. Use of one equivalent of OsO_4 increased the yield of the toluene-soluble material to 70%. Osmylation in the absence of pyridine, followed by dimer disruption with pyridine (14), gave the same species in 75% yield. The toluene-soluble material was shown to have one-to-one stoichiometry by converting it to the mixture of two-to-one adducts upon further exposure to the osmylation conditions. Solubility and crystal quality were improved by exchanging the pyridine ligands for 4-tert-butylpyridine.

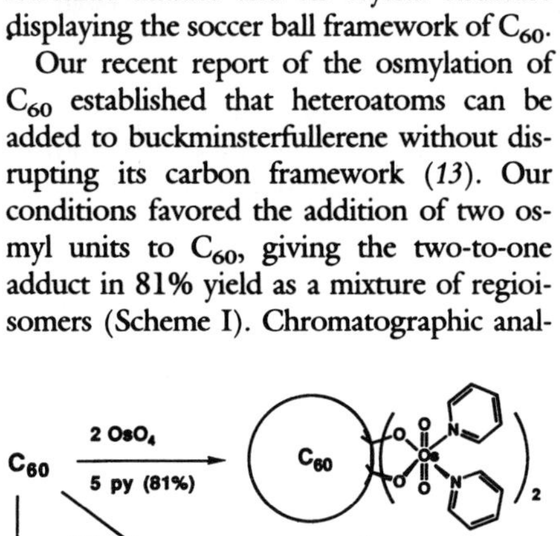

The observation of a single sharp chromatographic peak for the one-to-one adduct suggested that it is a single regioisomer, rather than a mixture of the two regioisomers which are possible from the proposed soccer ball structure for C_{60}. This would be

Scheme I

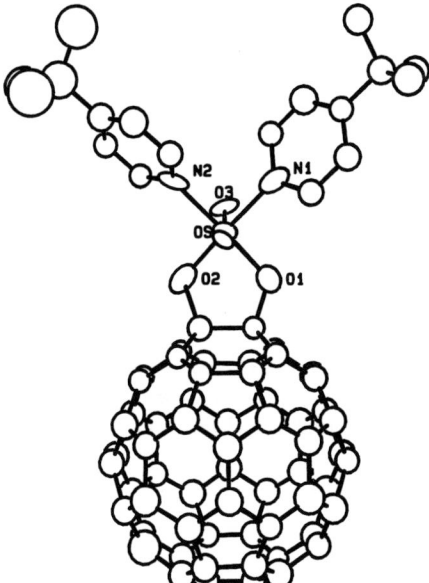

Fig. 1. ORTEP drawing (50% ellipsoids) of the one-to-one C_{60}–osmium tetroxide adduct $C_{60}(OsO_4)(4\text{-}tert\text{-butylpyridine})_2$ showing the relationship of the osmyl unit with the carbon cluster.

Department of Chemistry, University of California, Berkeley, CA 94720.

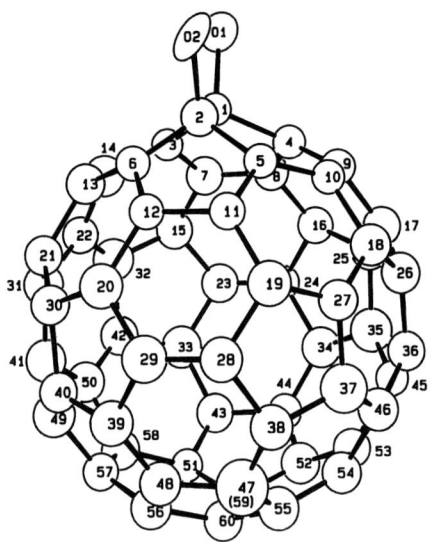

Fig. 2. ORTEP drawing (50% ellipsoids) of the one-to-one C$_{60}$–osmium tetroxide adduct C$_{60}$(OsO$_4$)(4-*tert*-butylpyridine)$_2$ showing the geometry of the C$_{60}$O$_2$ unit and the numbering scheme.

true if the bisoxygenation was strongly favored across one of the two unique bonds in C$_{60}$, the junction of two six-membered rings or the junction of a six- and a five-membered ring. Regioselective osmylation would fix the position of the C$_{60}$ carbon framework relative to the osmyl unit, with the osmyl unit breaking the pseudospherical symmetry of C$_{60}$ as required for an ordered crystal. The one-to-one adduct C$_{60}$(OsO$_4$)(4-*tert*-butylpyridine)$_2$ indeed gave a sufficiently ordered crystal for the determination of atomic positions by x-ray crystallographic analysis (*15*).

The crystal structure (Figs. 1 and 2) confirms the soccer ball–like arrangement of carbon atoms in C$_{60}$ by clearly showing the 32 faces of the carbon cluster composed of 20 six-membered rings fused with 12 five-membered rings. No two five-membered rings are fused together, and each six-membered ring is fused to alternating six- and five-membered rings. The O-Os-O unit has added across a six-six ring fusion, consistent with the regiochemistry predicted by extended Hückel calculations on C$_{60}$ and the principle of least motion or minimum electronic reorganization (*16*).

The tricoordinate carbons C3-C60 all lie within a spherical shell of radius 3.46 to 3.56 Å, with an average distance of 3.512(3) Å from the calculated center of the cluster. The tetracoordinate oxygen-bonded carbons C1 and C2 lie significantly outside of this shell at distances of 3.80(2) and 3.81(3) Å from the center. They have approximately tetrahedral geometry with sums of C-C-C angles equal to 330°, slightly more than 328°, the sum for an ideal tetrahedral atom. The proximate carbons, C3-C6, are the least distorted from planarity within the cluster, with sums of C-C-C angles averaging 353(1)°, compared with 360° for a planar atom. The remaining carbons, C7-C60, have approximately equivalent geometries with sums of C-C-C angles ranging from 344° to 351°. The average sum, 348.0(3)°, equals the value for an ideal junction of two regular hexagons and a regular pentagon. All 60 carbons within the cluster are pyramidalized concave inwards.

The five- and six-membered rings not containing C1 and C2 are planar with deviations from least-squares planes less than 0.05(3) Å. In contrast, tetracoordinate carbons C1 and C2 lie 0.22(2) to 0.30(3) Å outside of the planes defined by the other carbons in the rings which contain them. Excluding bonds to C1 and C2, the average C-C bond lengths are 1.388(9) Å for six-six ring fusions, and 1.432(5) Å for six-five ring fusions. These values are within the range of bond lengths predicted by theory for the two types of bonds in C$_{60}$ (*17*). The C1-C3, C1-C4, C2-C5, and C2-C6 bond lengths average 1.53(3) Å, comparable with normal C(sp^3)-C(sp^2) single bonds. The geometry of the OsO$_2$N$_2$ (diolate) unit is similar to that observed for conventional arene adducts, although our C1-C2 bond length [1.62(4) Å] is longer than the corresponding bonds in the other structures [1.40(4) to 1.54(2) Å] (*18*).

The functionalization of C$_{60}$ via the selective osmylation described here and other organic reactions will allow chemists to go beyond buckminsterfullerene in the pursuit of new and unusual types of organic molecules. We are presently studying the novel cup and band shaped conjugated π-systems of the one-to-one and two-to-one C$_{60}$-OsO$_4$ adducts from chemical, spectroscopic, and theoretical perspectives.

REFERENCES AND NOTES

1. H. W. Kroto, J. R. Heath, S. C. O'Brien, R. F. Curl, R. E. Smalley, *Nature* **318**, 162 (1985).
2. W. Krätschmer, L. D. Lamb, K. Fostiropoulos, D. R. Huffman, *ibid.* **347**, 354 (1990).
3. R. Taylor, J. P. Hare, A. K. Abdul-Sada, H. W. Kroto, *J. Chem. Soc. Chem. Commun.* (1990), p. 1423.
4. R. E. Haufler *et al.*, *J. Phys. Chem.* **94**, 8634 (1990).
5. H. Ajie *et al.*, *ibid.*, p. 8630.
6. W. Krätschmer, K. Fostiropoulos, D. R. Huffman, *Chem. Phys. Lett.* **170**, 167 (1990).
7. D. S. Bethune, G. Meijer, W. C. Tang, H. J. Rosen, *ibid.* **174**, 219 (1990).
8. R. D. Johnson, G. Meijer, D. S. Bethune, *J. Am. Chem. Soc.* **112**, 8983 (1990).
9. R. Tycko *et al.*, *J. Phys. Chem.* **95**, 518 (1991); C. S. Yannoni, R. D. Johnson, G. Meijer, D. S. Bethune, J. R. Salem, *ibid.*, p. 9.
10. D. L. Lichtenberger *et al.*, *Chem. Phys. Lett.* **176**, 203 (1991).
11. J. M. Hawkins *et al.*, *J. Chem. Soc., Chem. Commun.*, in press.
12. R. M. Fleming *et al.*, *Mater. Res. Soc. Symp. Proc.* (Boston), November 1990.
13. J. M. Hawkins *et al.*, *J. Org. Chem.* **55**, 6250 (1990).
14. R. J. Collin, J. Jones, W. P. Griffith, *J. Chem. Soc. Dalton Trans.* (1974), p. 1094.
15. [C$_{60}$(OsO$_4$)(4-*tert*-butylpyridine)$_2$·2.5 toluene]: Tetragonal, space group $I4_1/a$, $a = 30.751(5)$ Å, $c = 24.800(7)$ Å, $V = 23452(14)$ Å^3, $Z = 16$. All atoms were located and all positions were refined. Osmium, oxygen, nitrogen anisotropic, all other isotropic thermal parameters. $R = 10.6\%$, $wR = 10.3\%$, GOF = 1.77; 442 parameters, 3668 observed data.
16. J. R. Dias, personal communication; *J. Chem. Ed.* **66**, 1012 (1989); D. Amic and N. Trinajstic, *J. Chem. Soc. Perkin Trans. 2* (1990), p. 1595. Extended Hückel calculations on a Tektronix CAChe system qualitatively agree with the HMO calculations described in these papers.
17. W. Weltner, Jr., and R. J. Van Zee, *Chem. Rev.* **89**, 1713 (1989).
18. J. M. Wallis and J. K. Kochi, *J. Am. Chem. Soc.* **110**, 8207 (1988).
19. J.M.H. is grateful to the National Science Foundation (Presidential Young Investigator Award, CHE-8857453), the Camille and Henry Dreyfus Foundation (New Faculty Grant), the Merck Sharp & Dohme Research Laboratories (postdoctoral fellowship for A.M.), the Shell Oil Company Foundation (Shell Faculty Fellowship), the Xerox Corporation, and the Monsanto Company for financial support. We thank J. R. Heath, A. F. Moretto, and R. Apodaca for assistance with the synthesis and purification of C$_{60}$, R. E. Smalley and co-workers for a generous gift of soot, and W. D. Shrader for assistance with computer graphics.

Reprinted with permission from Science
Vol. 252, pp. 1160–1161, 24 May 1991
© 1991 The American Association for The Advancement of Science

The Chemical Nature of Buckminsterfullerene (C60) and the Characterization of a Platinum Derivative

Paul J. Fagan,* Joseph C. Calabrese, Brian Malone

Little is known about the chemical nature of the recently isolated carbon clusters (C_{60}, C_{70}, C_{84}, and so forth). One potential application of these materials is as highly dispersed supports for metal catalysts, and therefore the question of how metal atoms bind to C_{60} is of interest. Reaction of C_{60} with organometallic ruthenium and platinum reagents has shown that metals can be attached directly to the carbon framework. The native geometry of C_{60} is almost ideally constructed for dihapto-bonding to a transition metal, and an x-ray diffraction analysis of the platinum complex $[(C_6H_5)_3P]_2Pt(\eta^2\text{-}C_{60})\cdot C_4H_8O$ revealed a structure similar to that known for $[(C_6H_5)_3P]_2Pt(\eta^2\text{-ethylene})$. The reactivity of C_{60} is not like that of relatively electron-rich planar aromatic molecules such as benzene. The carbon-carbon double bonds of C_{60} react like those of very electron-deficient arenes and alkenes.

THE PROPERTIES OF THE RECENTLY isolated carbon clusters (1) have attracted considerable attention with regard to theoretical and physical properties (2–4), but there are few reports concerning the chemistry of these species (3–4). One well-defined derivative has been reported, namely, the osmium tetroxide adduct structurally characterized by Hawkins et al. (3). We investigated organometallic derivatives of C_{60} to ascertain its chemical nature and report the reactions of C_{60} with the reagents $[(C_6H_5)_3P]_2Pt(\eta^2\text{-}C_2H_4)$ (5) and $[Cp^*Ru(CH_3CN)_3]^+O_3SCF_3^-$ $[Cp^* = \eta^5\text{-}C_5(CH_3)_5]$ (6). Zero-valent Pt compounds are well known to react with electron-poor alkenes and arenes bonding in a dihapto-fashion, but are unreactive toward relatively electron-rich aromatic molecules such as benzene (5). In contrast, when $[Cp^*Ru(CH_3CN)_3]^+$-$O_3SCF_3^-$ is reacted with relatively electron-rich planar arenes, the three coordinated acetonitriles are displaced resulting in strong, hexahapto-binding of ruthenium to the six-

Central Research and Development Department, E. I. du Pont de Nemours & Co., Inc., Experimental Station, Wilmington, DE 19880–0328.

*To whom correspondence should be addressed.

membered ring of the arene (6).

Addition of $[(C_6H_5)_3P]_2Pt(\eta^2\text{-}C_2H_4)$ (31 mg) to C_{60} (30 mg) in toluene (2 ml) under a dinitrogen atmosphere resulted in formation of an emerald-green solution from which black microcrystals precipitated over the course of 2 hours. This precipitate was collected by filtration, washed twice with 2-ml portions of toluene and then with 10 ml of hexane, and dried by pulling N_2 through the filter cake. It was recrystallized by first dissolving in tetrahydrofuran (THF), filtering, concentrating, and precipitating with hexane. The isolated yield of this compound was 85% based on the formulation $[(C_6H_5)_3P]_2Pt(\eta^2\text{-}C_{60})$ (THF of crystallization is removed upon drying under vacuum). Elemental analytical data supported this formulation (7). The ^{31}P nuclear magnetic resonance (NMR) spectrum (121.7 MHz, external standard H_3PO_4) of this compound in THF-d_8 displayed a singlet at δ 27.0 ppm with satellites due to coupling of ^{31}P to the spin-1/2 isotope ^{195}Pt (33.8 % abundance) ($J_{P\text{-}Pt}$ = 3936 Hz). For comparison, the shift observed for $[(C_6H_5)_3P]_2Pt(\eta^2\text{-}C_2H_4)$ is δ 34.8 with a coupling constant $J_{P\text{-}Pt}$ = 3738 Hz. Since these coupling constants and chemical shifts were similar, it suggested that the coordination sphere about Pt was nearly identical in both the

ethylene and C_{60} complexes.

In order to substantiate this proposal, we performed a single-crystal x-ray structural analysis of the complex $[(C_6H_5)_3P]_2Pt(\eta^2\text{-}C_{60})\cdot C_4H_8O$. Small multiple needles were grown by slow evaporation from THF. A thin needle was cut in half to obtain a weakly diffracting single crystal from which x-ray data were successfully collected and analyzed (8). Accuracy of the structure was limited because of the small size of the crystal and disorder associated with THF molecules contained in the lattice (Fig. 1). The bonding parameters within the C_{60} framework agree closely with those obtained for the osmium tetroxide derivative of Hawkins et al. (3). In this case, Pt serves to anchor the molecule and reduce disorder problems. Rotation of alkenes about the platinum-alkene bond has a substantial energy barrier (9). The bis(triphenylphosphine)platinum moiety bonds to two carbon atoms of the C_{60} molecule at the junction of two fused six-membered rings rather than at the junction of the five- and six-membered rings. The bonding pattern is reminiscent of other structurally characterized transition metal–alkene complexes (9, 10). Bond distances and bond angles about Pt are shown in Fig. 2. The metrical data agree with those previously established for $[(C_6H_5)_3P]_2Pt(\eta^2\text{-}C_2H_4)$ (10).

It is well known that upon coordination of a transition metal such as Pt to an alkene, the four groups attached to the carbon-carbon double bond splay back away from the metal center (11). One measure of this distortion is to determine the degree to which two groups attached to one end of a double bond bend back relative to remaining planar. This can be defined as the angle between the vector described by the two doubly bonded carbons and the plane defined by one of these carbons and the two groups attached to it. Typical angles in Pt complexes with unconstrained carbon-substituted alkenes range from approximately 22° to 35° (12). In this regard, the natural curvature of C_{60} should permit bonding to a

Fig. 1. View of the $[(C_6H_5)_3P]_2Pt(\eta^2\text{-}C_{60})$ molecule as revealed by the x-ray crystallographic analysis of $[(C_6H_5)_3\text{-}P]_2Pt(\eta^2\text{-}C_{60})\cdot C_4H_8O$. The accuracy of the C–C bonds is limited to ~0.03 Å. The C_{60} distances range from 1.318(29) to 1.534(39) Å. Each of the five-membered and six-membered rings of the C_{60} fragment are planar to within 0.03 to 0.05 Å except for those containing C1 and C2. These atoms are pulled out from the rest of the C_{60} framework. The distances from the centroid of the C_{60} to all of the carbon atoms except C1 and C2 range from 3.48 to 3.60 Å with the average being 3.53; the centroid-C1

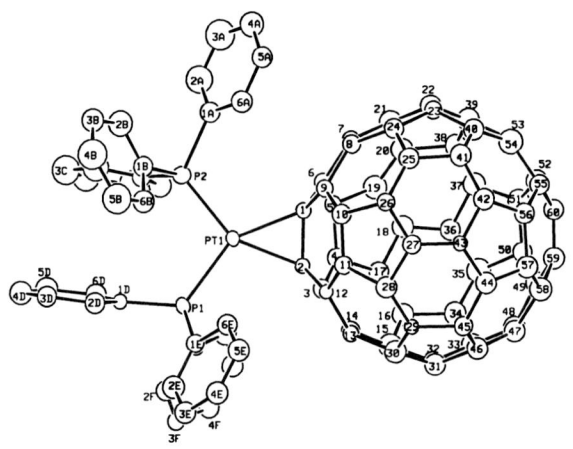

REFERENCES AND NOTES

1. W. Krätschmer, L. D. Lamb, K. Fostiropoulos, D. R. Huffman, *Nature* **347**, 354 (1990); R. Taylor, J. P. Hare, A. K. Abdul-Sada, H. W. Kroto, *J. Chem. Soc. Chem. Commun.* **1990**, 1423 (1990); H. W. Kroto, J. R. Heath, S. C. O. O'Brian, R. F. Curl, R. E. Smalley, *Nature* **318**, 162 (1985); F. J. Stoddart, *Angew. Chem.* **103**, 71 (1991); H. Kroto, *Pure Appl. Chem.* **62**, 407 (1990); H. Kroto, *Science* **242**, 1139 (1988).

2. G. E. Scuseria, *Chem. Phys. Lett.* **176**, 423 (1991), and references therein; R. C. Haddon *et al.*, *Nature* **350**, 46 (1991); C. S. Yannoni, R. D. Johnson, G. Meijer, D. S. Bethune, J. R. Salem, *J. Phys. Chem.* **95**, 9 (1991); P. M. Allemand *et al.*, *J. Am. Chem. Soc.* **113**, 1050 (1991); M. R. Wasielewski, M. P. O'Neil, K. R. Lykke, M. J. Pellin, D. M. Gruen, *ibid.*, p. 2774; P.-M. Allemand *et al.*, *ibid.*, p. 2780; D. M. Cox *et al.*, *ibid.*, p. 2940; C. S. Yannoni, P. P. Bernier, D. S. Bethune, G. Meijer, J. R. Salem, *ibid.*, p. 3190.

3. J. M. Hawkins, A. Meyer, T. A. Lewis, S. Loren, F. J. Hollander, *Science* **252**, 312 (1991); J. M. Hawkins *et al.*, *J. Org. Chem.* **55**, 6250 (1990).

4. R. E. Haufler *et al.*, *J. Phys. Chem.* **94**, 8634 (1990); J. W. Bausch *et al.*, *J. Am. Chem. Soc.* **113**, 3205 (1991).

5. C. D. Cook and G. S. Jauhal, *J. Am. Chem. Soc.* **90**, 1464 (1968); J. Browning, M. Green, B. R. Penfold, J. L. Spencer, F. G. A. Stone, *J. Chem. Commun.* **1973**, 35 (1973).

6. P. J. Fagan, M. D. Ward, J. C. Calabrese, *J. Am. Chem. Soc.* **111**, 1698 (1989).

7. Elemental analysis for $C_{96}H_{30}P_2Pt$: calculated, C, 80.06 and H, 2.10; found, C, 78.88 and H, 1.96.

8. Crystal data: $C_{100}H_{38}OP_2Pt$: black wedge about 0.08 mm by 0.04 mm by 0.40 mm; monoclinic, $P2_1/c$, (no. 14); $a = 22.716(8)$, $b = 14.415(2)$, $c = 19.302(6)$ Å, $\beta = 108.90(1)°$ from 25 reflections; $Z = 4$; $V = 5979.7$ Å³; $T = -70°C$; $FW = 1512.44$; $Dc = 1.680$ g/cm³; Mo Kα radiation, $\mu(Mo) = 24.83$ cm⁻¹; 10,228 data collected, $1.9° \le 2\theta \le 48°$; 2,934 unique reflections with $I \ge 3.\sigma(I)$, corrected for absorption (DIFABS); refinement by full-matrix least squares on F (refined anisotropic, Pt and P; isotropic, C; and fixed atoms, H); 436 parameters, data-to-parameter ratio = 6.72; final $R = 0.073$; $Rw = 0.059$; error of fit = 1.22, maximum $\Delta/\sigma = 0.02$; and largest residual density = 0.96 e/Å³ near C(1C).

9. K. Morokuma and W. T. Borden, *J. Am. Chem. Soc.* **113**, 1912 (1991), and references therein.

10. P. T. Cheng and S. C. Nyburg, *Can. J. Chem.* **50**, 912 (1972).

11. S. D. Ittel and J. A. Ibers, *Adv. Organomet. Chem.* **14**, 33 (1976).

12. R. B. Osborne and J. A. Ibers, *J. Organomet. Chem.* **232**, 371 (1982); J. N. Francis, A. McAdam, J. A. Ibers, *ibid.* **29**, 149 (1971); G. Bombieri, E. Forsellini, C. Panattoni, R. Graziani, G. Bandolini, *J. Chem. Soc. A* **1970**, 1313 (1970).

13. ¹H NMR (CD₃NO₂, 300 MHz): δ 1.82 (broadened singlet, 15 H, Cp*) and 2.62 (broadened singlet, 6 H, CH₃CN). ¹³C{¹H} NMR (CD₃NO₂, 75.6 MHz): δ 4.6 (s, CH₃CN); 9.8 (s, (CH₃)₅C₅); 96.4 (s, (CH₃)₅C₅); 122.6 (q, J_{C-F} = 323 Hz, CF₃SO₃); 130.0 (s, CH₃CN); and 160 to 138 (broad asymmetric envelope, C₆₀). (The broadening of the resonances and the decrease of the linewidths with increasing concentration of the compound suggests a process involving exchange of free and bound acetonitrile is occurring concurrent with migration of ruthenium on the C₆₀ surface.) By preparing an NMR sample with a known weight of $\{[Cp^*Ru(CH_3CN)_2]_x(C_{60})\}^{x+}(O_3SCF_3^-)_x$ and integrating relative to an internal standard of known amount, the value of x was determined to be 3. Elemental analytical data (C, 54.07; H, 3.04; and N, 4.08) was ambiguous but suggests a value for x of either 3 or 4.

14. P. Fowler, *Nature* **350**, 20 (1991); H. W. Kroto, *ibid.* **329**, 529 (1987).

15. This work would not have been possible without the efforts of many of our colleagues here at DuPont. E. Holler purified the C₆₀ by chromatography and W. Marshall assisted in collection of the x-ray crystallographic data. We thank E. Wasserman for helpful discussions and the referees for their comments. Contribution no. 5865.

and centroid-C2 distances are 3.68 and 3.73 Å, respectively. For the bonds not directly associated with platinum binding (that is, excluding carbon bonds, C1–C2, C1–C6, C1–C9, C2–C3, and C2–C12), the average C–C bond length at the junctions of two six-membered rings was 1.388(30) Å. At the junction of fused five- and six-membered rings the average was 1.445(30) Å. These average values for the two types of bonds in C₆₀ are similar to those found in the osmium tetroxide derivative (3).

low-valent transition metal with relatively little deformation of the C_{60} skeleton. For example, examining the geometry about the bond C59–C60 directly opposite where Pt is coordinated, the planes defined by C60–C52–C55 and C59–C49–C58 are tilted away from the C59–C60 axis by 31(2)° and 30(2)°, respectively (numbers in parentheses are the standard error in the last digit or digits). The corresponding angles between

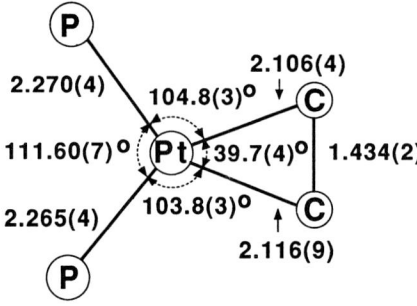

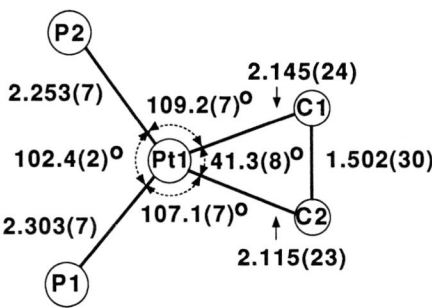

Fig. 2. Comparison of the Pt coordination spheres in $[(C_6H_5)_3P]_2Pt(\eta^2\text{-}C_2H_4)$ (upper) (10) and $[(C_6H_5)_3P]_2Pt(\eta^2\text{-}C_{60})\cdot(C_4H_8O)$ (lower). The atoms P1, P2, Pt1, C1, and C2 are coplanar to within 0.02 Å. The phosphine ligands in the C_{60} complex are bent back more than in the ethylene complex, which may reflect the greater steric bulk of C_{60} versus ethylene.

the C1–C2 axis (attached to Pt) and the planes described by C2–C3–C12 and C1–C6–C9 are 44(2)° and 38(2)°, respectively.

Reaction of C_{60} with a tenfold excess of the reagent $[Cp^*Ru(CH_3CN)_3]^+O_3SCF_3^-$ in CH_2Cl_2 at 25°C for a period of 5 days yielded a brown precipitate, which preliminary data suggests has the formulation $\{[Cp^*Ru(CH_3CN)_2]_x(C_{60})\}^{x+}(O_3SCF_3^-)_x$ (13). Although we were not able to obtain an x-ray crystal structure of this complex, the fact that the metal retains two acetonitrile molecules of coordination leads to the conclusion that each ruthenium bonds to just two carbon atoms of the C_{60} cluster. Whether or not this is the case, this observed chemistry is highly unusual if C_{60} were reacting chemically like a typical planar electron-rich aromatic molecule such as benzene.

Although C_{60} may be perfectly suited for bonding in a dihapto-fashion, this is not so for hexahapto-bonding (and to some extent tetrahapto-bonding), since from above the "plane" of a six-membered ring in the molecule, the carbon p-orbitals are tilted away from the center of the ring. This tilt may weaken the overlap of the highest occupied and lowest unoccupied molecular orbitals of C_{60} with the ruthenium-centered unfilled and filled d-orbitals, respectively. In this case, acetonitrile is apparently a strong enough donor to compete with any such interaction. This is not to say that hexahaptobonding is not possible with a weaker donor than acetonitrile, or with another metal ligand combination.

We note that C_{60} is *not* chemically inert (14) but reacts readily with electron-rich reagents. Its double-bond reactivity resembles that of very electron-poor arenes and alkenes. It follows that much of the reaction chemistry of these types of carbon-carbon double bonds might be successfully applied to C_{60}.

Reprinted with permission from Science
Vol. 254, pp. 1186–1188, 22 November 1991
© 1991 The American Association for The Advancement of Science

Systematic Inflation of Buckminsterfullerene C_{60}: Synthesis of Diphenyl Fulleroids C_{61} to C_{66}

T. SUZUKI, Q. LI, K. C. KHEMANI, F. WUDL,* Ö. ALMARSSON

The synthesis of a new family of spheroidal carbon molecules derived from the fullerenes is described. The fulleroids are produced by incremental addition of a divalent carbon equivalent that has two phenyl (Ph) rings to fullerene C_{60}. The fulleroids Ph_2C_{61}, Ph_4C_{62}, Ph_6C_{63}, Ph_8C_{64}, $Ph_{10}C_{65}$, and $Ph_{12}C_{66}$ have been prepared and characterized.

THE FULLERENES, PARTICULARLY C_{60}, have been well characterized (1–3), but few reports about their chemical reactivity appeared (4–6). It has been proposed that the relatively high electronegativity of this carbon cluster was due to its pyracylene character (2). Further, the inter–five-membered ring bonds are fulvenoid and are potent electrophilic as well as dienophilic and dipolarophilic sites (5, 7).

Although nucleophilic addition reactions to C_{60} produce a myriad of products (7), the electronic character of the products is altered drastically relative to C_{60} to the point that only minor "conjugation" (8) remains (color bleaching to yellow) (7). However, the basic fullerene skeleton could be modified, while retaining the same number of "π" (8) orbitals and attendant unusual electronic character, by a strategy that takes advantage of: (i) the norcaradiene-cycloheptatriene rearrangement; (ii) the aromatic character of the methano[10]annulenes (9); and (iii) the dipolarophilicity of C_{60} (Scheme 1):

A survey of C_{60} reactivity with the dipoles ethyl diazoacetate (10) and phenyl diazomethane (10) indicated that they add

smoothly and, under some conditions, lose N_2. The incomplete pyrolytic nitrogen loss could be attributed to hydrogen migration with concomitant imine formation, as determined by infrared spectroscopy. Also, because the carbon atom bearing the diazo function is asymmetrically substituted, phenyldiazomethane addition gave a large number of isomers upon multiple reaction. In order to prevent hydrogen migration and to reduce the potential number of isomers, we decided to study the reaction of C_{60} with diphenyldiazomethane. We found that this strategy worked and report a method for fullerene inflation (spherical expansion) to produce "fulleroids" (11), expanded fullerenes whose basic electronic properties [ultraviolet-visible (UV-vis) spectroscopy and cyclic voltammetry] are retained.

Reaction of C_{60} with more than one equivalent of diphenyldiazomethane in toluene at room temperature for ~1 hour produced a monoadduct diphenyl fulleroid C_{61} (12). Apparently because compound **1** is similar in molecular shape to $C_{60}Pt(PPh_3)_2$ (4), just as its organometallic analog, it is less soluble in benzene than C_{60} and thus could be obtained in ~40% yield. It was isolated and purified by chromatography on silica gel. Based on its spectroscopic and physico-chemical properties [see (13) and Table 1] we propose structure **1** depicted below and in Fig. 1.

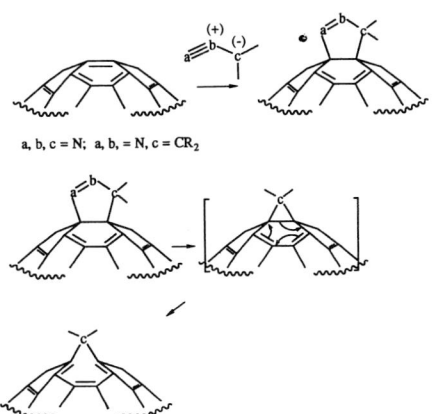

a, b, c = N; a, b, = N, c = CR₂

Scheme 1

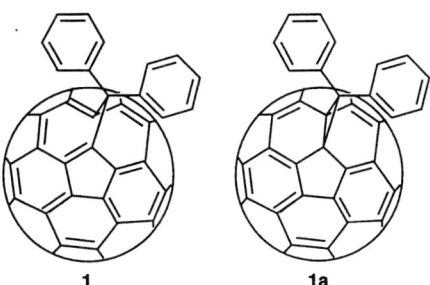

1   **1a**

The salient features that support the proposed structure, and not **1a**, are the similarities of the electronic spectrum (particularly wavelengths of absorption maxima at 212, 258, and 326 nm; see Fig. 1) and cyclic voltammetry of **1** and C_{60} (see Fig. 2 and Table 1). The $(Ph_3P)_2Pt$ analog has struc-

ture **1a** as determined by x-ray crystallography. The ring-closed nature of the derivative is reflected in a major change of its UV-vis spectrum relative to C_{60} as well as a shift of the first reduction wave by 340 mV negative relative to C_{60} (14). Although it is difficult to obtain a ^{13}C nuclear magnet resonance (NMR) spectrum of Ph_2C_{61}, due to insolubility, the ^{13}C NMR results on $Ph_{12}C_{66}$ show only three (broad) resonances in the fullerene region, which is in agreement with octahedral regiochemistry previously observed in the $[(Et_3P)_2Pt]_6C_{60}$ analog whose octahedral structure was determined by x-ray crystallography (5). The 11,11-diphenyl methano[10]annulene could be a reasonable model for ^{13}C chemical shift assignments; however, the compound does not exist. The closest analog is the 11-phenyl methano[10]annulene, whose relevant chemical shifts are shown below (15):

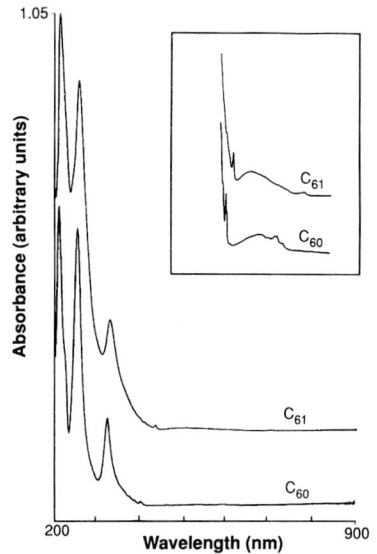

$C_{ortho} = 127.15$ $H_{ortho} = 6.80$
$C_{meta} = 126.38$ $H_{meta} = 6.97$
$C_{para} = 125.62$ $H_{para} = 6.97$

$C_{1,6} = 117.40$

$C_{11} = 46.32$

Since the unsubstituted C_{11} has a chemical shift of 35 ppm and the monophenyl derivative's chemical shift is ~46 ppm, one additional phenyl on the 11-carbon would be

T. Suzuki, Q. Li, K. C. Khemani, F. Wudl, Institute for Polymers and Organic Solids and Departments of Physics and Chemistry, University of California, Santa Barbara, CA 93106.
Ö. Almarsson, Department of Chemistry, University of California, Santa Barbara, CA 93106.

*To whom correspondence should be addressed.

Fig. 1. Comparative ultraviolet-visible spectra of C_{60} and Ph_2C_{61} (labeled "C_{61}"). The inset is the 700- to 350-nm region at 20 times the gain. The absorbance of two spectra has not been normalized.

Table 1. Cyclic voltammetry of the diphenyl fulleroids Ph_2C_{61} to $Ph_{12}C_{66}$. Experimental conditions: Pt working and counterelectrodes; Ag/AgCl/3M NaCl, reference electrode; Ferrocene internal reference (+620 mV); 0.1 M $TBABF_4$ (tetrabutylammonium fluoroborate) in THF (tetrahydrofuran) in a dry box.

Scan rate (mV/s)	Peak position (mV)						
	C_{60}	C_{61}	C_{62}	C_{63}	C_{64}	C_{65}	C_{66}
100	−228	−335	−439	−554	−643	−809	−900
	−826	−920	−974	−1072	−1150	−1278	−1385
	−1418	−1470	−1609	−1632	−1669	−1820	−1915
	−1916	−1929	−1995	−2025	−2120	−2135	
1000	−238	−346	−472	−562	−646	−785	−886
	−838	−924	−988	−1101	−1191	−1350	−1406
	−1418	−1476	−1549	−1654	−1730	−1930	−2001
	−1921	−1953	−1976	−2060	−2150		

Fig. 3. Computer-generated model of Ph_2C_{61}. Note the minor perturbation on the structure of C_{60} caused by the methylene carbon of the diphenylmethylene group.

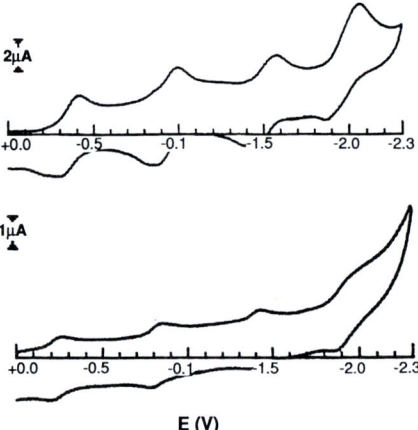

Fig. 2. Comparative cyclic voltammograms of C_{60} (bottom) and Ph_2C_{61} (top) versus Ag/Ag^+ as a reference in tetrahydrofuran with tetrabutylammonium fluoroborate as a supporting electrolyte; a Pt disk and a Pt wire were used as working and counterelectrodes, respectively.

expected to further deshield it to a chemical shift of ~60 ppm. We have not observed any resonances in this region of the ^{13}C NMR spectrum of $Ph_{12}C_{66}$; possibly because of: (i) the expected weakness of the quaternary carbon intensity; (ii) a combination of weak signal and signal broadening observed in most organic fullerene derivatives; or (iii) the strong deshielding effect of the fulleroid electronic structure shifts the resonance to coincide with the aromatic carbon region. An interesting result in the comparison of the fulleroid with methano[10]annulene is the chemical shift of the ortho phenyl hydrogens; in the former they appear at 8.4 ppm (deshielded relative to the meta and para hydrogens), whereas in the later they appear at 6.8 ppm (slightly shielded relative to the meta and para hydrogens).

Whereas the NMR spectra of Ph_4C_{62} are complicated due to restricted rotation around the C–Ph bonds (see Fig. 1), recent results on the similar adduct with 9-diazo-

fluorene (9-Fl), $(9-Fl)_2C_{62}$, in which C–Ph rotation cannot occur, show that only two isomers, "octant" and "polar," were produced (16).

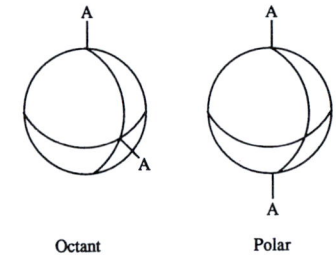

Octant   Polar

Additional corroborating support for the "open" structure type **1** was obtained from results of molecular modeling calculations. The release of strain on going from C_{60} to **1** and from **1a** to **1** was confirmed by a molecular modeling calculation with CHARMm (17). Although the total energy per molecule is not an accurate number, the relative energies may be more reliable; thus, $C_{60} = 1198$, $H_2C_{61} = 1127$, $Ph_2C_{61} = 1165$, $Ph_4C_{62} = 1145$, $Ph_8C_{64} = 1104$, and $Ph_{12}C_{66} = 1059$ kcal/mol, respectively. The "norcaradienoid" form (**1a**), that is, with an unopened cyclopropane, has a minimized energy of 1196 kcal/mol, 31 kcal/mol greater in energy than the proposed structure **1**. The space-filling model structure in Fig. 3 was generated with the Quanta program, above. Although the above offer strong support for the proposed structure, the definitive structure proof requires x-ray crystallographic evidence.

More extended reaction times with different stoichiometries afforded the diphenyl fulleroids C_{62}, C_{63}, C_{64}, C_{65}, and C_{66}. The latter took 8 to 10 days to be formed from a large excess over six equivalents of diphenyl diazomethane. As can be seen in Table 1, with increasing inflation, the fulleroids become more electropositive, which may be a result of a combination of: (i) strain relief; (ii) more π-like overlap due to greater p

character of the 60 p_z orbitals; or (iii) with increasing inflation, closure to the norcaradienoid form may occur. We also note that: (i) the only correct elemental analysis is that of $Ph_{12}C_{66}$ (13) and (ii) the NMR resonance lines are broad (for which we have no simple explanation). After investigating the chemistry of C_{60} for some time (7), we learned that most elemental analyses of C_{60} derivatives give consistently low carbon values. Perhaps as the hydrogen content of the molecule increases, combustion is more complete, which would be the case with $Ph_{12}C_{66}$.

Although one may read that C_{60} will have as profound an influence in organic chemistry as benzene had, for example, in the dye industry, this view is unlikely to be realized because benzene undergoes electrophilic substitution of its hydrogen atoms but C_{60} has no easily replaceable hydrogen atoms. However, with the discovery presented above, one could prepare essentially any functionalized fulleroid. The implication of this discovery is that specifically functionalized, spherically unsaturated molecules can be produced.

REFERENCES AND NOTES

1. W. Krätschmer, L. D. Lamb, K. Fostiropoulos, D. R. Huffman, *Nature* **347**, 354 (1990); R. D. Johnson, G. Meijer, D. S. Bethune, *J. Am. Chem. Soc.* **112**, 8983 (1990); H. Ajie *et al.*, *J. Phys. Chem.* **94**, 8630 (1990); D. L. Lichtenberger, K. W. Nebesny, C. D. Ray, D. R. Huffman, L. D. Lamb, *Chem. Phys. Lett.* **176**, 203 (1991); C. S. Yannoni, R. D. Johnson, G. Meijer, D. S. Bethune, J. R. Salem, *J. Phys. Chem.* **95**, 9 (1991).
2. P.-M. Allemand *et al.*, *J. Am. Chem. Soc.* **113**, 1050 (1991).
3. R. Tycko *et al.*, *J. Phys. Chem.* **95**, 518 (1991); R. M. Fleming *et al.*, *Mater. Res. Soc. Proc.*, in press; R. C. Haddon *et al.*, *Nature* **350**, 320 (1991).
4. P. J. Fagan, J. C. Calabrese, B. Malone, *Science* **252**, 1160 (1991).
5. _____, unpublished results.
6. J. M. Hawkins *et al.*, *J. Org. Chem.* **55**, 6250

(1990); J. M. Hawkins *et al.*, *Science* **252**, 312 (1991); D. Heymann, *Carbon* **29**, 684 (1991); F. Diederich *et al.*, *Science* **252**, 548 (1991); L. S. Sunderlin *et al.*, *J. Am. Chem. Soc.* **113**, 5489 (1991); J. W. Bausch *et al.*, *ibid.*, p. 3205. P. J. Krusic, E. Wasserman, B. A. Parkinson, B. Malone, E. R. J. Holler, *ibid.*, p. 6274.

7. F. Wudl *et al.*, in *Large Carbon Clusters*, G. S. Hammond, Ed. (American Chemical Society, Washington, DC, in press).

8. R. C. Haddon, *Acc. Chem. Res.* **21**, 243 (1988).

9. E. Vogel, in *Aromaticity*, W. D. Ollis, Ed. (The Chemical Society, London, 1967), vol. 21, p. 113.

10. T. Suzuki, unpublished results.

11. We prefer fulleroid over "homo fullerene," as suggested to us by R. Haddon. The naming using the "homo" prefix gets unwieldy as in hexakis homo fullerene.

12. J. M. Wood *et al.* [*J. Am. Chem. Soc.* **113**, 5907 (1991)] have shown by mass spectrometry that species containing $C_{61}O$ to $C_{66}O$ occur in the crude extracts of fullerene preparation. These were not isolated nor characterized further.

13. Properties of the diphenyl fulleroids Ph_2C_{61} to $Ph_{12}C_{66}$. C_{60}: MW (molecular weight) 720; UV-vis [λ_{max} (absorption maximum), cyclohexane] 212, 257, 328, 404, 542, 598, and 620 nm. Ph_2C_{61}: MW 886.882; ^{1}H NMR (CS_2, $CDCl_3$) δ 8.02 (dd, $J = 7.5, 1.2$ Hz, 4H), 7.42 (t, $J = 7.5$ Hz, 4H), and 7.32 (t, $J = 7.5$ Hz, 2H); fast-atom-bombardment (FAB) MS weak 887, stronger 720 (C_{60}); infrared (IR) (KBr) 3020 (w), 1600 (w), 1510 (m), 1495 (m), 1450 (m), 1430 (m), 1385 (w), 1185 (m), 1160 (w), 1080 (w), 1030 (w), 1000 (w), 755 (sh), 740 (m), 700 (s), 575 (w), 550 (w), and 530 (s) cm^{-1}; and UV-vis (λ_{max}, cyclohexane) 212, 258, 326, 429, 495, and 695 nm. Analysis (Anal) for $C_{73}H_{10}$: calculated (calcd) C, 98.86, and H, 1.14; found C, 96.81, H, 1.27, and N, 0.00. Ph_4C_{62}: MW 1053.104; ^{1}H NMR (CD_2Cl_2) δ 8.4 to 6.8 (m, br, 20H); FAB MS 1055 to 1052 (M$^+$); IR (KBr) 3060 (w), 3025 (w), 1600 (m), 1585 (sh), 1495 (m), 1450 (m), 1380 (w), 1185 (w), 1160 (w), 1080 (w), 1030 (w), 1005 (sh), 840 (w), 800 (w), 740 (m), 700 (s), and 520 (br) cm^{-1}; and UV-vis (λ_{max}, cyclohexane) 214, 263, 327, and 510 nm; Anal for $C_{86}H_{20}$: calcd, C, 98.09 and H, 1.91; found C, 95.38, H, 2.48, N, 0.00, and 1.4% of residue. Ph_6C_{63}: MW 1219.326; ^{1}H NMR (CD_2Cl_2) δ 8.4 to 6.8 (m, br, 30H); FAB MS 1220 to 1218 (M$^+$); IR (KBr) 3060 (w), 3025 (w), 1600 (m), 1585 (sh), 1495 (m), 1450 (m), 1380 (w), 1330 (w), 1180 (w), 1155 (w), 1080 (w), 1030 (w), 1000 (w), 835 (w), 740 (m), 700 (s), and 530 (br) cm^{-1}; and UV-vis (λ_{max}, cyclohexane) 260 (sh) and 492 nm. Anal for $C_{99}H_{30}$: calcd C, 97.52 and H, 2.48; found C, 94.29, H, 2.59, N, 0.00, and 2.4% of residue. Ph_8C_{64}: MW 1385.548; ^{1}H NMR (CD_2Cl_2) δ; 8.4 to 6.8 (m, br, 40H); FAB MS 1386 to 1384 (M$^+$); IR (KBr) 3060 (w), 3025 (w), 1600 (m), 1580 (sh), 1495 (m), 1450 (m), 1380 (w), 1320 (w), 1185 (w), 1160 (w), 1080 (w), 1030 (w), 1000 (sh), 740 (m), 700 (s), and 530 (br) cm^{-1}; UV-vis (λ_{max}, cyclohexane) 270 (sh) and 485 nm (sh). Anal for $C_{112}H_{40}$: calcd C, 97.09 and H, 2.91; found C, 92.72, H, 3.26, and N, 0.00. $Ph_{10}C_{65}$: MW 1551.77; ^{1}H NMR (CD_2Cl_2) δ 8.4 to 6.8 (m, br, 50H); FAB MS 1553 to 1551 (M$^+$); IR (KBr) 3060 (w), 3025 (w), 1600 (m), 1585 (sh), 1495 (m), 1450 (m), 1380 (w), 1320 (w), 1185 (w), 1160 (w), 1080 (w), 1030 (w), 1005 (w), 840 (w), 740 (m), 700 (s), and 535 (br) cm^{-1}; UV-vis (λ_{max}, cyclohexane) 280 (sh) and 480 nm (sh). Anal for $C_{125}H_{50}$: calcd C, 96.75 and H, 3.25; found C, 94.97, H, 3.34, and N, 0.00. $Ph_{12}C_{66}$: MW 1717.992; ^{1}H NMR (CD_2Cl_2) δ 8.4 to 6.8 (m, br, 60H); ^{13}C NMR ($CDCl_3$) δ(all broad) 126.4, 127.4, 128.2, 129.1, 130.4, and 132.3 (all due to phenyl) and 140.0, 143.4, and 147.5 (fullerene); FAB MS 1718 to 1716 (M$^+$); IR (KBr) 3060 (w), 3025 (w), 1600 (m), 1585 (sh), 1495 (m), 1450 (m), 1185 (w), 1160 (w), 1080 (w), 1030 (w), 1000 (w), 840 (w), 740 (m), 700 (s), and 530 (br) cm^{-1}; UV-vis (λ_{max}, cyclohexane) 285 (sh) and 490 nm (sh); Anal for $C_{138}H_{60}$: calcd C, 96.48 and H, 3.52; found C, 96.08, H, 3.63, and N, 0.00.

14. P. J. Fagan, S. Lerke, D. Evans, B. Parkinson, personal communication.

15. E. Vogel and H. Wrubel, unpublished results.

16. T. Suzuki, unpublished results.

17. B. R. Brooks *et al.*, *Comput. Chem.* **4**, 187 (1983). Force-field calculations and structure manipulations in the Quanta program (POLYGEN Corp.) were performed on a Silicon Graphics Iris GT/X 220 workstation. Minimizations were performed with a steepest-descent algorithm, followed by adopted basis Newton-Raphson algorithm until the energy change tolerance was less than 10^{-9} kcal/mol. Nonbonded interaction cutoff distance and hydrogen-bonding cutoff distance were chosen to be 11.5 and 7.5 Å, respectively.

18. We thank the National Science Foundation for support through grants DMR-88-20933, DMR-91-11097, and CHE-89-08323, J. M. Hawkins for coordinates of the C_{60} osmylate, and E. Vogel for disclosing data in a personal communication.

ENDOHEDRAL COMPLEXES

Given a hollow molecule such as C_{60}, it is tempting to try to put something inside it. This was accomplished shortly after the observation and hypothesis of C_{60}, by the same group.[He85] They impregnated the graphite target of their laser-vaporization molecular beam apparatus with a lanthanum salt; otherwise, conditions were similar to the experiments which led to the discovery of C_{60}. They observed that the mass spectra of cluster ions contained peaks both for fullerene ions C_n^+ and lanthanum-fullerene ions LaC_n^+. It was hypothesized that the La was inside the carbon shell based on the observation that conditions of multiphoton-induced fragmentation did not detach the La from the carbon cluster.[He85]

Much more direct evidence that metal atoms are encapsulated inside fullerenes was obtained by the same group, using ion cyclotron resonance with magnetically trapped ions.[We88*] In a paper reprinted in this volume, they present several important results. First, trapped ions of KC_{60}^+, CsC_{60}^+, and LaC_{60}^+ do not react with background gas H_2, O_2, NO, and NH_3; a result which is similar to C_{60}^+. This implies that the highly reactive metal atom is somehow protected from the surrounding low-density gas. Furthermore, it was found that when the trapped MC_{60}^+ ions were fragmented by very energetic laser pulses, they did so by loss of C_2 rather than the metal ion. Photofragmentation by C_2 loss indicates that the carbon is a fullerene, because fullerenes consist only of even numbers of carbon atoms.[O'B88] That the metal-fullerene complexes fragment the same way, rather than by losing the metal atom, suggests that the metal atom is bound inside the cage. Most compelling was the result that the ion could be "shrink-wrapped" inside the fullerene, and the number of carbon atoms in the smallest such complex depends on the metal's ionic radius in the expected way. Bare fullerenes can be chipped away to C_{32}^+ by photofragmentation [O'B88]; in contrast, the endpoints for various

metallofullerenes were found to be KC_{44}^+, LaC_{44}^+, and CsC_{48}^+.[We88*]

Macroscopic quantities of endohedral complexes can be produced by either laser or contact arc vaporization of a mixture of graphite with the appropriate metal oxide.[Ch91*, Al91] Starting with La_2O_3 and graphite, Chai et al. used laser ablation to produce a solid film containing a mixture of fullerenes and metallofullerenes, notably C_{60}, C_{70}, LaC_{60}, LaC_{74}, and LaC_{82}, with weaker signals observed for LaC_n for all even n from 70 to 84.[Ch91*] Under "gentle" photoionization conditions, no signal was seen for LaC_{62} to LaC_{68}, as would be expected from the nonexistence of isolated pentagon isomers of C_{62} to C_{68}. It was estimated that a few percent of the fullerenes in a sublimed film contain lanthanum atoms. Another significant result in that paper is that the endohedral complexes could be dissolved in toluene. However, there is a notable difference in the mass spectrum of the toluene extract: the only surviving metallofullerene was LaC_{82}.

The paper by Chai et al. also introduces a useful piece of notation which has rapidly become standard.[Ch91*] The symbol @ is used to indicate a hollow structure, with any atoms to the left of the symbol located inside the cage and any to the right, comprising the cage. In this notation, $La@C_{82}$ represents an encapsulated La inside the C_{82} cage, whereas LaC_{82} presumes no structural information. A fullerene with a boron atom substituted for one carbon atom would be written $@C_{59}B$.

Returning to $La@C_n$, Alvarez et al. showed that the Krätschmer-Huffman contact arc method, with rare-earth oxides mixed with graphite, also produced significant amounts of endohedral complexes.[Al91] They focused their attention on $La_2@C_n$, and found a phenomenon similar to what was seen in $La@C_n$: while the raw soot contains endohedrals with a broad spectrum of n, the toluene extract contains only $La_2@C_{80}$ and $La@C_{82}$. That result was also confirmed by Ross et al., who extended it to endohedral compounds of yttrium. [Ro92] They also created the mixed metallofullerene LaY@C_{80} in pyradine extract.[Ro92]

An important clue for the disappearance of many endohedral species in the process of solvent-extraction is given by Bandow et al.[Ba92] Handling the soot and performing the solvent extraction under anaerobic conditions, they observed LaC_{76} and LaC_{84} in the mass spectra of the solvent extract. Evidently, these species are susceptible to oxidation from the atmosphere and/or dissolved air, which either destroys or immobilizes them. This result calls for a careful reevaluation of their reactivity in an ion trap or similar experiment.

The UCLA group has also studied the resilience of La$_2$@C_{80}^- in collisions with a solid surface, and found that they rebound intact for collision energies up to 200 eV.[Ye92] Unlike the case of C_{60}^-,[Be91*] electron emission is not observed from such collisions.

The charge state of the rare earth atom inside the carbon shell has been studied by two experiments: electron paramagnetic resonance (EPR) and x-ray photoelectron spectroscopy (XPS). The EPR results for La@C_{82} indicate that the unpaired electron spin resides on the carbon cage and the charge state of the encapsulated atom is La^{3+}.[Jo92*] Likewise, the XPS experiments on sublimed films containing unseparated mixtures of La@C_n showed clear evidence of a La^{3+} charge state.[We92] Similar results were obtained for films containing Y@C_n and Y$_2$@C_n.[We92]

A closer look at the EPR spectra of La, Y, and Sc endohedral complexes with C_{82} reveal several extra lines.[Ba92, Su92, Ho92] These are interpreted as resulting from different isomers of the carbon shell, made visible through the hyperfine coupling to ^{13}C nuclei. One anticipates that further work will allow these isomers to be chromatographically separated, and that EPR and NMR experiments will be useful in revealing their structures.

There are noteworthy differences in the results of solvent extractions of dimetallofullerenes among several laboratories. The UCLA and NRL groups see La$_2$@C_{80} from toluene and La$_2$@C_{80} and Y$_2$@C_{80} from pyradine extracts.[Al91, Ye92, Ro92] Several other rare earth metals are also seen as M$_2$@C_{80}; the one UCLA exception is Ho$_2$@C_{82}. [Gi92] On the other hand, the Rice-Minnesota, Mi'e-Meijo, and IBM groups see Y$_2$@C_{82} and higher dimetallofullerenes from toluene and CS_2 extracts. [We92, Sh92b, Ho92] The reason (and significance) of this difference is not clear at present.

There have been a number of hypotheses for the structures of endohedral fullerenes, mostly motivated by the hope of explaining the relative stability of M@C_{82} and M$_2$@C_{80} (or M$_2$@C_{82}). The different carbon atom counts of the most prominent fullerenes vs. endohedral complexes may result from the charge transfer from the encapsulated ion.[Ma92] Yeretzian et al. have observed that there are two possible isomers of C_{80}^{6-} with particularly large HOMO–LUMO gaps, which might stabilize that structure with two trivalent metal ions inside. [Ye92] (We recall that C_{80} has not been isolated, despite the fact that, theoretically, it has six isolated-pentagon isomers with pseudo-closed electronic shells.[Ma91]) However, they point out that no such argument seems to suggest a particular stability for C_{82}^{3-}, which would be the analogous electronic structure for an endohedral complex with a single trivalent ion.

The formation of endohedral complexes is not limited to alkali and rare earth metals. Fe@C_{60} can be created in a contact arc, and is apparently the only metal@C_{60} which is stable to toluene extraction.[Pr92] Scandium endohedrals exist in solvent-extracted soot as a number of species, most notably Sc@C_{82}, Sc$_2$@C_{82}, Sc$_2$@C_{84}, and Sc$_3$@C_{82}.[Sh92a, Ya92] ESR spectra of mixtures of Sc@C_{82} and Sc$_3$@C_{82} have been studied by two groups.[Sh92a, Ya92] (Note that these groups did not perform a chromatographic separation of the Sc, Sc$_2$, and Sc$_3$ endohedrals. The assignments were made on the basis of the simulations of hyperfine spectra, the failure to observe an Sc$_2$ hyperfine multiplet in the EPR spectrum, and the favorable circumstance that the mass spectra of the solvent extracts revealed only C_{82} encapsulating Sc and Sc$_3$.) The Sc$_3$@C_{82} hyperfine spectra indicate that the three scandium ions must be equivalent, implying either fast dynamical averaging or an equilateral triangular geometry. In either case, this provides direct evidence that the Sc are inside the fullerene, and not incorporated within the framework.

Another remarkable endohedral complex which has recently been created is U@C_{28}.[Gu92] This affords a counterexample to the observation that endohedral complexes are typically based on fullerenes larger than C_{60}. Indeed, C_{28} appears to be the smallest fullerene produced in supersonic cluster beams from a laser-vaporization source. Photofragmentation experiments give an additional sense of how strongly the internal U stabilizes C_{28}. Larger, empty fullerenes can be blasted down to C_{32}, but the shrink-wrapping endpoint for higher urano-fullerene ions is actually $U@C_{28}$. Photoemission spectra show that the uranium is in a 4+ valence state. Finally, macroscopic quantities of $U@C_{28}$ can be produced by laser- or arc-vaporization of a UO_2–graphite composite rod, and are stable in a sublimed film, even when exposed to air and water.

Endohedral complexes can also be formed by ion reactions of empty fullerenes with various atoms. The first such experiments collided C_{60}^+ ions with various gases: He, Ne, Ar, and D_2, and confirmed that $He@C_{60}$ etc., and various fragmentation products were formed.[We91, Ro91, Ca91] It is kinematically more favorable to study such reactions in the opposite direction, by colliding the ion to be inserted with neutral C_{60} in the vapor phase.[Wa92a] The paper reporting the insertion of Li^+ and Na^+ into C_{60} is reprinted in this volume.[Wa92b*] The insertion thresholds were 6 and 20 eV, respectively. At higher insertion energies, the fullerenes fragmented. The same group has subsequently found that heavier alkali ions cannot be inserted into an intact fullerene without destroying it, a result which is in accordance with the size of the ion compared with the available opening in the wall of the fullerene.

Due to the presence of helium gas in the Krätschmer-Huffman soot production technique, a fraction of the fullerenes so produced may contain a helium atom. The evolution of He from this stable $He@C_{60}$ has been detected by mass spectroscopy; in commercially prepared C_{60}–C_{70}, approximately one fullerene in a million was actually $He@C_{60}$.[Sa93] It was found that the release of this endohedral helium had an Arrhenius activation energy of 80 kcal mol^{-1} (3.5 eV per molecule), somewhat less than the 6 ± 2 eV threshold determined from C_{60}^+–He collision experiments,[Ca92] and significantly less than the 10 eV predicted barrier for penetration of a He atom through a benzene ring.[Hr92] This suggests that the fullerene network may open and close at elevated temperatures, implying that low concentrations of endohedral complexes of other rare gas atoms could be made simply by heating C_{60} in the appropriate atmosphere. Such an experiment was successful, as evidenced by trace quantities of $^3He@C_{60}$ and $Ne@C_{60}$ produced by heating a sample of C_{60} to 600° C in the presence of ^{3}He and Ne, respectively.[Sa93] The bond-opening model is bolstered by the observation of the same activation energy for thermal insertion of He and Ne.

Bibliography

References marked with * are reprinted in this volume.

Al91   M. M. Alvarez et al., "La_2C_{80}: A Soluble Dimetallofullerene," J. Phys. Chem. **95**, 10561–10563 (1991).

Ba92   S. Bandow et al., "Anaerobic Sampling and Characterization of Lanthanofullerenes: Extraction of LaC_{76} and other LaC_{2n}," J. Phys. Chem. **96**, 9609–9612 (1992).

Be91*   R. D. Beck, P. St. John, M. M. Alvarez, F. Diederich and R. L. Whetten, "Resilience of All-Carbon Molecules C_{60}, C_{70}, and C_{84}: A Surface-Scattering Time-of-Flight Investigation," J. Phys. Chem. **95**, 8402–8409 (1991).

Ca91   K. A. Caldwell, D. E. Giblin, C. S. Hsu, D. Cox and M. L. Gross, "Endohedral Complexes of Fullerene Radical Cations," J. Am. Chem. Soc. **113**, 8519–8521 (1991).

Ca92   E. E. B. Campbell, R. Ehlich, A. Hielscher, J. M. A. Frazao and I. V. Hertel, "Collision Energy Dependence of He and Ne Capture by C_{60}^+," Z. Phys. **D23**, 1–2 (1992).

Ch91*   Y. Chai et al., "Fullerenes with Metals Inside," J. Phys. Chem. **95**, 7564–7568 (1991).

Gi92   E. G. Gillan et al., "Endohedral Rare-Earth Fullerene Complexes," J. Phys. Chem. **96**, 6869–6871 (1992).

Gu92   T. Guo et al., "Uranium Stabilization of C_{28}: A Tetravalent Fullerene," Science **257**, 1661–1664 (1992).

He85   J. R. Heath et al., "Lanthanum Complexes of Spheroidal Carbon Shells," J. Am. Chem. Soc. **107**, 7779–7780 (1985).

Ho92   M. Hoinkis et al., "Multiple Species of $La@C_{82}$ and $Y@C_{82}$. Mass Spectroscopic and Solution EPR Studies," Chem. Phys. Lett. **198**, 461–465 (1992).

Hr92   J. Hrušák, D. K. Böhme, T. Weiske and H. Schwarz, "Ab Initio MO Calculation on the Energy Barrier for the Penetration of a Benzene Ring by a Helium Atom. Model Studies for the Formation of Endohedral $He@C_{60}^+$ Complexes by High-Energy Bimolecular Reactions," Chem. Phys. Lett. **193**, 97–99 (1992).

Jo92*   R. D. Johnson, M. S. de Vries, J. Salem, D. S. Bethune and C. S. Yannoni, "Electron Paramagnetic Resonance Studies of Lanthanum-Containing C_{82}," Nature **355**, 239–240 (1992).

Ma91   D. E. Manolopoulos and P. W. Fowler, "Structural Proposals for Endohedral Metal-Fullerene Complexes," *Chem. Phys. Lett.* **187**, 1–7 (1991).

Ma92   D. E. Manolopoulos, P. W. Fowler and R. P. Ryan, "Hypothetical Isomerisations of LaC_{82}," *J. Chem. Soc. Faraday Trans.* **88**, 1225–1226 (1992).

O'B88   S. C. O'Brien, J. R. Heath, R. F. Curl and R. E. Smalley, "Photophysics of Buckminsterfullerene and Other Carbon Cluster Ions," *J. Chem. Phys.* **88**, 220–230 (1988).

Pr92   T. Pradeep, G. U. Kulkarni, K. R. Kannan, T. N. G. Row and C. N. R. Rao, "A Novel FeC_{60} Adduct in the Solid State," *J. Am. Chem. Soc.* **114**, 2272–2273 (1992).

Ro91   M. M. Ross and J. H. Callahan, "Formation and Characterization of $C_{60}He^+$," *J. Phys. Chem.* **95**, 5720–5723 (1991).

Ro92   M. M. Ross, H. H. Nelson, J. H. Callahan and S. W. McElvany, "Production and Characterization of Metallofullerenes," *J. Phys. Chem.* **96**, 5231–5234 (1992).

Sa93   M. Saunders, H. A. Jiménez-Vázquez, R. J. Cross and R. J. Poreda, "Stable Compounds of Helium and Neon: $He@C_{60}$ and $Ne@C_{60}$," *Science* **259**, 1428–1430 (1993).

Sh92a   H. Shinohara *et al.*, "Encapsulation of a Scandium Trimer in C_{82}," *Nature* **357**, 52–54 (1992).

Sh92b   H. Shinohara, H. Sato, Y. Saito, M. Ohkohchi and Y. Ando, "Mass Spectroscopic and ESR Characterization of Soluble Yttrium-Containing Metallofullerenes YC_{82} and Y_2C_{82}," *J. Phys. Chem.* **96**, 3571–3573 (1992).

Su92   S. Suzuki *et al.*, "Isomers and ^{13}C Hyperfine Structures of Metal-Encapsulated Fullerenes $M@C_{82}$ (M = Sc, Y, La)," *J. Phys. Chem.* **96**, 7159–7161 (1992).

Wa92a   Z. Wan, J. F. Christian and S. L. Anderson, "$Ne^+ + C_{60}$: Collision Energy and Impact Parameter Dependence for Endohedral Complex Formation, Fragmentation, and Charge Transfer," *J. Chem. Phys.* **96**, 3344–3347 (1992).

Wa92b* Z. Wan, J. F. Christian and S. L. Anderson, "Collision of Li^+ and Na^+ with C_{60}: Insertion, Fragmentation, and Thermionic Emission," *Phys. Rev. Lett.* **69**, 1352–1355 (1992).

We88*   F. D. Weiss, J. L. Elkind, S. C. O'Brien, R. F. Curl and R. E. Smalley, "Photophysics of Metal Complexes of Spheroidal Carbon Shells," *J. Am. Chem. Soc.* **110**, 4464–4465 (1988).

We91   T. Weiske, D. K. Böhme, J. Hrušák, W. Krätschmer and H. Schwarz, "Endohedral Cluster Compounds: Inclusion of Helium Within C_{60}^+ and C_{70}^+ Through Collision Experiments," *Angew. Chem. Int. Ed. Engl.* **30**, 884–886 (1991).

We92   J. H. Weaver *et al.*, "XPS Probes of Carbon-Caged Metals," *Chem. Phys. Lett.* **190**, 460–464 (1992).

Ya92   C. S. Yannoni *et al.*, "Scandium Clusters in Fullerene Cages," *Science* **256**, 1191–1192 (1992).

Ye92   C. Yeretzian *et al.*, "Collisional Probes and Possible Structures of La_2C_{80}," *Chem. Phys. Lett.* **196**, 337–342 (1992).

Reprinted with permission from the Journal of the American Chemical Society
Vol. 110, pp. 4464–4465, 1988

Photophysics of Metal Complexes of Spheroidal Carbon Shells[†]

F. D. Weiss,[‡] J. L. Elkind, S. C. O'Brien,[§] R. F. Curl, and
R. E. Smalley*

Rice Quantum Institute and
Department of Chemistry, Rice University
Houston, Texas 77251

Received January 27, 1988

In the fall of 1985 initial evidence was presented for the formation of a new class of organometallic ions consisting of a single metal atom surrounded by a spheroidal cage network of pure carbon.[1] The new species were made by laser vaporization of a $LaCl_3$ impregnated graphite disc in a pulsed nozzle followed by intense laser ionization of the neutral clusters in a supersonic beam. The most prominent of these species was $C_{60}La^+$. Its exceptionally high photophysical stability led to the suggestion that the La atom had been trapped inside the closed shell of the (putatively) icosahedral carbon cage,[2] C_{60}.

To test this rather controversial[3-5] hypothesis, we have performed extensive photophysical and chemical measurements in the magnetic trap of an FT-ICR mass spectrometer. Initial results are presented here with more extensive experiments and discussion to follow.[6] The details of this cluster beam/FT-ICR apparatus have been presented previously[7] as well as several of its early applications to metal[8,9] and semiconductor[10] cluster surface chemistry. Carbon–metal cluster ions were prepared in the supersonic beam by laser-vaporization of a graphite disk impregnated with a salt of the desired metal (La, K, Cs). After injection into the ICR trap and subsequent thermalization with neon gas,[10] the only clusters observed in the 300–1500 amu mass range were the even-numbered bare carbon clusters, C_n^+, and the same clusters with a single metal atom attached, C_nM^+. Ample evidence has been found for the presence of other, more weakly bound metal–carbon clusters in these beams,[1,6] including some containing more than one metal atom (e.g., $C_nK_2^+$).[4,5] However, these weakly

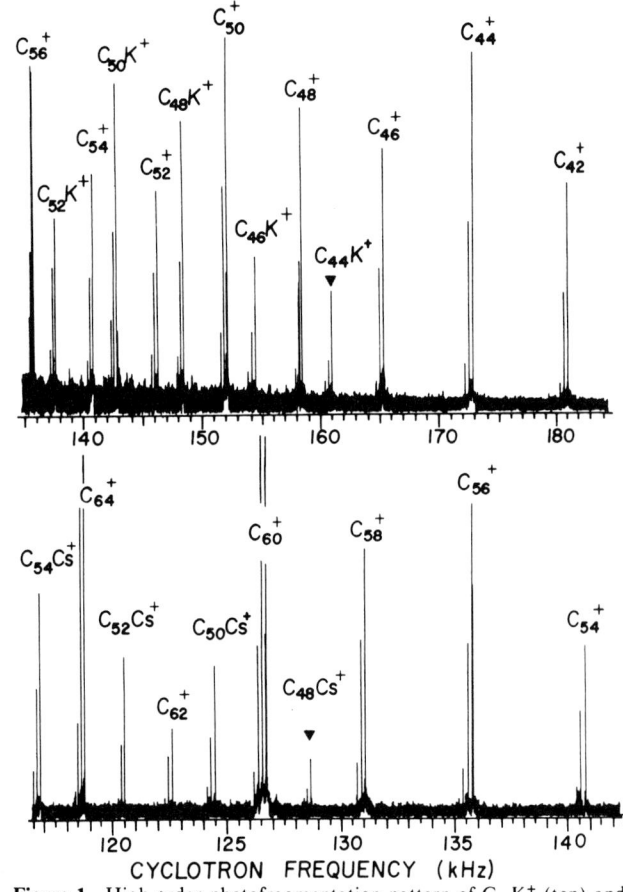

Figure 1. High-order photofragmentation pattern of $C_{60}K^+$ (top) and $C_{60}Cs^+$ (bottom) detected by FT-ICR mass spectrometry. Note the C_nK^+ fragment series breaks off at 44, while the C_nCs^+ series terminates at 48. The bare C_n^+ clusters seen in the top panel are fragments from C_{64}^+ injected into the ICR trap along with $C_{60}K^+$ providing an internal calibration, while in the bottom panel these C_n^+ fragments result from C_{72}^+ parent ions. Conditions (top panel): 200 shots at 10 Hz of ArF excimer laser radiation (193 nm) at 6 mJ cm^{-2} shot^{-1}; (bottom panel): 1600 shots ArF at 3 mJ cm^{-2} shot^{-1}. During laser excitation the pressure in the ICR trap was less than 1×10^{-8} Torr. A few noise spikes have been removed to simplify the figure.

[†] Research supported by the National Science Foundation and the Robert A. Welch Foundation.

[‡] Department of Physics, Rice University.

[§] Nettie S. Autrey Graduate Fellowship.

(1) Heath, J. R.; O'Brien, S. C.; Zhang, Q.; Liu, Y.; Curl, R. F.; Kroto, H. W.; Tittel, F. K.; Smalley, R. E. *J. Am. Chem. Soc.* **1985**, *107*, 7779.

(2) Kroto, H. W.; Heath, J. R.; O'Brien, S. C.; Curl, R. F.; Smalley, R. E. *Nature* (London) **1985**, *318*, 162.

(3) Cox, D. M.; Trevor, D. J.; Reichmann, K. C.; Kaldor, A. *J. Am. Chem. Soc.* **1986**, *108*, 2457.

(4) Kaldor, A.; Cox, D. M.; Trevor, D. J.; Zakin, M. R. *Z. Physik D.* **1986**, *3*, 195.

(5) Cox, D. M.; Reichmann, K. C.; Kaldor, A. *J. Chem. Phys.* **1988**, *88*, 1588.

(6) O'Brien, S. C.; Weiss, F. D.; Alford, M. A.; Elkind, J. L.; Curl, R. F.; Smalley, R. E. *J. Am. Chem. Soc.*, to be submitted.

(7) Alford, J. M.; Williams, P. E.; Trevor, D. J.; Smalley, R. E. *Int. J. Mass Spectrom. Ion Processes* **1986**, *72*, 33.

(8) Alford, J. M.; Weiss, F. D.; Laaksonnen, R. T.; Smalley, R. E. *J. Phys. Chem.* **1986**, *90*, 4480.

(9) Elkind, J. L.; Weiss, F. D.; Alford, J. M.; Laaksonnen, R. T.; Smalley, R. E. *J. Chem. Phys.* **1988**, *88*, 5215.

(10) Elkind, J. L.; Alford, J. M.; Laaksonnen, R. T.; Weiss, F. D.; Smalley, R. E. *J. Chem. Phys.* **1987**, *87*, 2397.

bound species fragmented during the multiple injection/thermalization cycles used to fill the ICR cell,[9] leaving primarily just M^+ atoms as the charged fragment.

In agreement with earlier ICR measurements[11] on C_{60}^+ and the other large even-numbered clusters of carbon, we found no reaction of the bare carbon cluster ions with such species as H_2, O_2, NO, and NH_3. Similar lack of reactivity was found with the $C_{60}La^+$ complex. In agreement with previous photofragmentation studies in a tandem time-of-flight (TTOF) apparatus,[12] ArF excimer laser (6.4 eV) irradiation of these bare clusters in the ICR trap produced fragmentation only as a result of high order multiphoton processes. As expected, the $C_{60}M^+$ clusters were nearly as photoresistant as C_{60}^+ itself.

Extreme photostability of C_{60}^+ and $C_{60}M^+$ was evident from the fact that no measurable dissociation occurred for any of these trapped clusters with less than 0.5 mJ cm^{-2} pulse^{-1} even after over 30-s irradiation with the ArF excimer laser at 50 Hz. Estimates of the absorption cross section of C_{60} at this wavelength[13] indicate that the average such cluster absorbs roughly 300 ArF excimer photons during this period. Cooling by infrared emission between laser shots must therefore be very facile for such clusters.

However, at sufficiently high ArF excimer laser fluence per pulse both C_{60}^+ and the $C_{60}M^+$ complexes did begin to photodissociate. The fluence dependence of this process was roughly the same for the metal-containing complexes as measured previously in the TTOF experiments for the bare clusters.[12] For both, the primary photoprocess was observed to be C_2 loss to form the next smaller even-numbered cluster. Increasing either the laser pulse energy or the length of time the clusters were irradiated in the ICR trap resulted in extensive further dissociation, producing successively smaller even-numbered clusters.

As argued in the earlier TTOF study,[12] loss of C_2 from a carbon cluster is hard to understand *unless* that cluster is a closed, edgeless carbon cage. Otherwise it should lose the much more stable C_3. The linear chain and monocyclic ring clusters in the 2–30 carbon atom range[14] are known[12,15] to lose C_3, and graphitic sheets should also lose C_3. But for closed cages a concerted C_2 loss mechanism

is likely to be the lowest energy process since only then can the next smaller even product form a closed cage.[12] The observation of C_2 rather than M^+ loss from $C_{60}M^+$ suggests that the metal ion is sterically bound, since a bond between K^+ or Cs^+ and carbon should be much weaker than a carbon–carbon bond. These results are then strong evidence that $C_{60}M^+$ clusters are composed of closed carbon cages with the metal ion trapped inside.

The endpoint of the C_2 loss process for bare clusters such as C_{60}^+ is known[12] to be C_{32}^+; it appears to be the smallest viable fragment cage. In this regard the corresponding behavior for the $C_{60}M^+$ complexes presented in Figure 1 is particularly interesting. For potassium–carbon clusters the smallest stable member in the even product ions is clearly seen in Figure 1 to be $C_{44}K^+$. For cesium the smallest fragment cluster is $C_{48}Cs^+$. These results are intriguing since breaks near these cluster sizes are predicted from a simple model which places an M^+ ion in the center of a closed carbon cage, allowing for the known ionic radius of M^+, a 1.65 Å van der Waals radius for each carbon atom, and C–C bond lengths in the range of 1.4–1.5 Å. Similar experiments with $C_{60}La^+$ photodissociation show breakoff to occur at $C_{44}La^+$ (or possibly $C_{42}La^+$), again nicely in accord with expectations for a central La^+ ion completely enclosed in an inert carbon cage.

Registry No. C, 7440-44-0; K, 7440-09-7; Cs, 7440-46-2; La, 7439-91-0.

(11) McElvany, W. N.; Baronovski, A. P.; Watson, C. H.; Eyler, J. R. *Chem. Phys. Lett.* **1987**, *134*, 214.

(12) O'Brien, S. C.; Heath, J. R.; Curl, R. F.; Smalley, R. E. *J. Chem. Phys.* **1988**, *88*, 220.

(13) Heath, J. R.; Curl, R. F.; Smalley, R. E. *J. Chem. Phys.* **1987**, *87*, 4236.

(14) Yang, S.; Taylor, K. J.; Craycraft, M. J.; Conceicao, J.; Pettiette, C. L.; Cheshnovsky, O.; Smalley, R. E. *Chem. Phys. Lett.*, submitted for publication.

(15) Geusic, M. E.; Jarrold, M. E.; McIlrath, T. J.; Freeman, R. R.; Brown, W. L. *J. Chem. Phys.* **1986**, *86*, 1986.

Reprinted with permission from The Journal of Physical Chemistry
Vol. 95, No. 20, pp. 7564–7568, 1991
© 1991 American Chemical Society

Fullerenes with Metals Inside

Yan Chai,[†] Ting Guo, Changming Jin,[†] Robert E. Haufler,[†] L. P. Felipe Chibante,[†] Jan Fure,
Lihong Wang,[†] J. Michael Alford, and Richard E. Smalley*

*Rice Quantum Institute and Departments of Physics and Chemistry, Rice University, Houston, Texas 77251
(Received: August 19, 1991)*

Fullerenes with a single lanthanum atom trapped on the inside of the carbon cage were produced by laser vaporization of a lanthanum oxide/graphite composite rod in a flow of argon gas at 1200 °C. When sublimed with C_{60} and C_{70}, they formed an air-stable film containing principally LaC_{60}, LaC_{70}, LaC_{74}, and LaC_{82}. When dissolved in toluene and exposed to air, LaC_{82} was found to be uniquely stable. Evidence was also obtained for coalescence reactions between these fullerenes at high temperatures to form larger cages with as many as three lanthanum atoms inside. Indications have also been obtained for the successful production of KC_{60}, $C_{59}B$, and $KC_{59}B$ where the boron has substituted for a carbon in the soccerball cage. The use of the @ symbol is advocated for specifying such complex fullerenes as ($K@C_{59}B$).

Introduction

Within a week after the initial 1985 discovery of the special stability of C_{60} and the proposal of its soccerball structure,[1] evidence was obtained in our laboratory that a single lanthanum atom could be trapped inside.[2] Laser vaporization of a graphite target impregnated with $LaCl_3$ produced a supersonic cluster beam that upon intense irradiation with an ArF excimer laser appeared to have one La atom (and only one) attached to each fullerene in a manner such that it could not be knocked off by the laser blast. At the time, the proposal that this tightly bound atom was on the inside triggered some controversy, since a group at Exxon was able to show that two or more atoms could be attached.[3,4] But, in fact, there never was much difference in the results from the two groups. The extra metal atoms in the Exxon experiments were readily blown off by using higher laser power.[4]

Additional very strong confirmation for the geodesic cage structure of fullerenes and the existence of a central metal atom was obtained in a series of "shrink wrap" experiments showing that the carbon cage bursts a point dictated by the ionic radius of the internal atom.[5] Recent experiments by Freiser and co-workers[6,7] have verified that externally attached metal atoms do, in fact, behave in a radically different way: they react readily and are easily knocked off.

Since the discovery by Kratschmer and Huffman last year of a simple method[8,9] for producing macroscopic quantities of C_{60}, C_{70}, and some of the higher empty fullerenes, we have searched for a means of extending this technique to "fill the void" and make macroscopic quantities of internally substituted fullerenes—endohedral fullerene complexes.[10] We report below our first success in this search. It involves a simple revision to the laser-vaporization method originally used in 1985: just do it in an oven at 1200 °C.

Symbolism and Nomenclature

By now the term "fullerene" appears to have achieved uniform acceptance as the general name for hollow carbon structures composed of 12 pentagonal rings and some number of six-membered rings linked together to form a geodesic dome. To discuss the various possible sorts of doped fullerene molecules and materials in a concise fashion, a systematic symbolism and naming convention will need to be worked out. In what follows we need at least to make an initial stab at the first part of this problem.

Ordinarily, a fullerene composed of n carbon atoms with a metal atom, M, will simply be represented as MC_n, regardless of whether the metal atom is inside or outside the cage. However, to facilitate discussion of these and more complicated fullerenes with one or more atoms inside, some attached outside and possibly one or more heteroatoms substituting for carbons in the cage network itself, a more explicit symbolism is essential. We will use a set of

parentheses around the symbol @ to indicate that the atoms listed within the parentheses are grouped to form a fullerene. Within this group all atoms listed to the right of the @ symbol are assumed to be part of the cage network, and all to the left are situated somehow inside the cage. Buckminsterfullerene, under this notation is then ($@C_{60}$), and a C_{60}-caged metal species is written ($M@C_{60}$). A more complex example that will be encountered below is $K_2(K@C_{59}B)$, which denotes a 60-atom fullerene cage with one boron atom substituted for a carbon in the geodesic network, a single potassium trapped inside, and two potassium atoms adhering to the outside.

We have adopted this symbolism since it has the virtue of being concise, while still containing the critical information. It is readily printed and transmitted electronically, and it is compatible with ordinary chemical formula conventions. It also has the virtue of being visually suggestive, and it emphasizes the "superatom" aspects of the fullerenes as new chemical entities.

Experimental Section

The laser-vaporization fullerene generator used in this work has been described briefly in an earlier publication.[9] It is simply a 2.5-cm-diameter 50-cm-long quartz tube mounted in a temperature-controlled tube furnace (Lindberg Model 55035). This tube is O-ring sealed on the front end to an aluminum block to which is attached a fused silica window and some plumbing for gas addition and pressure monitoring. Another aluminum block attached at the rear of the quartz tube connects via an adjustable valve to a mechanical vacuum pump. A rotary vacuum feedthrough (Huntington Model VF-106) mounted in the end of this rear aluminum block is attached inside the quartz tube to a 6-mm-o.d. graphite rod that extends through a graphite centering ring into the heated portion of the tube furnace. The end of this graphite rod is threaded into the rear of 1.25-cm-diameter graphite sample rods of the desired composition to serve as targets for vaporization by the laser. During vaporization, the target rod is rotated at 2 rpm in order to maintain a uniform, reproducible

(1) Kroto, H. W.; Heath, J. R.; O'Brien, S. C.; Curl, R. F.; Smalley, R. E. *Nature* **1985**, *318*, 165.
(2) Heath, J. R.; O'Brien, S. C.; Zhang, Q.; Liu, Y.; Curl, R. F.; Kroto, H. W.; Tittel, F. K.; Smalley, R. E. *J. Am. Chem. Soc.* **1985**, *107*, 7779.
(3) Cox, D. M.; Trevor, D. J.; Reichmann, K. C.; Kaldor, A. *J. Am. Chem. Soc.* **1986**, *108*, 2457.
(4) Cox, D. M.; Reichmann, K. C.; Kaldor, A. *J. Chem. Phys.* **1988**, *88*, 1588–1597.
(5) Weiss, F. D.; O'Brien, S. C.; Elkind, J. L.; Curl, R. F.; Smalley, R. E. *J. Am. Chem. Soc.* **1988**, *110*, 4464.
(6) Roth, L. M.; Huang, Y.; Schwedler, J. T.; Cassady, C. J.; Ben-Amotz, D.; Kahr, B.; Freiser, B. S. *J. Am. Chem. Soc.* **1991**, *113*, 6298.
(7) Huang, Y.; Freiser, B. S. *J. Am. Chem. Soc.*, in press.
(8) Kratschmer, W.; Lamb, L. D.; Fostiropoulos, K.; Huffman, D. R. *Nature* **1990**, *347*, 354.
(9) Haufler, R. E.; Chai, Y.; Chibante, L. P. F.; Conceicao, J.; Jin, Changming; Wang, Lai-Sheng; Maruyama, Shigeo; Smalley, R. E. *Mater. Res. Soc. Symp. Proc.* **1990**, *206*, 627.
(10) Cioslowski, J.; Fleischmann, E. *J. Chem. Phys.* **1991**, *94*, 3730.

*To whom correspondence should be addressed.
[†] Robert A. Welch Foundation Predoctoral Fellow.

surface to the incoming laser beam.

For the experiments described below the vaporization laser was the green second harmonic of a Nd:YAG laser (Quantel) operating at 300 mJ/pulse at 10 pps. The roughly 10-mm diameter nearly Gaussian output beam of this laser was concentrated on the end of the target rod in the furnace using a 50-cm focal length cylindrical lens positioned 60 cm away from the end of the rod, producing an oval vaporization region 1.0 cm high and 0.2 cm wide. As we reported earlier, vaporization of a pure graphite rod in this apparatus produces C_{60}, C_{70}, and a bit of higher fullerenes in excellent yield if the oven is heated to 1200 °C and argon or helium carrier gas at a pressure of several hundred Torr is slowly (0.1–0.2 cm s^{-1}) flowed over the target rod. These fullerenes sublime readily at such temperatures and are carried away in the flowing gas, depositing on cool inner surfaces of the quartz tube at the end of furnace. Argon was used at a pressure of 250 Torr for the vaporization experiments used in this work. With pure graphite targets this resulted in toluene-soluble fullerene yields of well over 10% of the vaporized carbon.

To produce trapped metal fullerenes, composite graphite target rods were made by mixing lanthana powder, La_2O_3 (Aesar AAS grade 99.99%), with graphite powder (Ultra Carbon graphite powder, ultra F purity), together with graphite cement (Dylon Industries, GC grade), and pressing into a 1.25-cm cylindrical die while curing the cement at 100 °C. The resulting rod was then carbonized by slowly heating in argon (1 °C/min to 400 °C, 5 °C/min to 1200 °C) and then baked at 1200 °C for 10 h. In some of the experiments LaB$_6$ was used in place of the La$_2$O$_3$. In others, KCl and/or boron powder was added.

Analysis of the fullerene products was made by Fourier transform ion cyclotron resonance (FT-ICR) study of laser-desorbed sample disks onto which the fullerenes had either been directly sublimed while still in the quartz tube or deposited by evaporating solutions in which the material had been dissolved. The cluster FT-ICR apparatus has been extensively described elsewhere.[11,12] When looking for fullerenes on the target, the vaporization laser (Nd:YAG second harmonic) was kept at low fluence (below 1.0 mJ in a 1-mm-diameter spot on the sample disk). The desorbed clusters were entrained in helium and made into a supersonic beam directed down the central magnetic axis of the FT-ICR device. Ionization was generally accomplished by an ArF excimer laser crossing the neutral cluster beam just before it entered into the analysis magnet. In addition to simply examining their masses, the clusters were probed while levitated in the FT-ICR magnetic trap by pulsing in reactant gases, colliding with argon buffer gas by selective excitation of the cyclotron motion, and irradiating with a XeCl excimer photolysis laser.

Computerization and control of this elaborate cluster FT-ICR facility was recently revamped using a single IBM RISC-6000 Powerstation 320 programmed in C using Xwindows running under the AIX operating system. Copies of the software package and details of the implementation are available from us.

Results and Discussion

Laser vaporization of a 7% (by weight) La$_2$O$_3$ in graphite target rod in the tube furnace produced a black-brown deposit on the cool downstream end of the quartz tube. After the furnace was cooled, a water-cooled copper rod with several copper FT-ICR sample disks was then inserted so that the disks were centered in the quartz tube immediately beneath the black-brown deposit. This was done while maintaining a purge of argon gas so that exposure of the freshly generated fullerenes to air was minimized. After pumping down to vacuum (<20 mTorr), the furnace was moved down the quartz tube so that the black-brown deposit was now well within the heated zone, and the temperature was ramped up to 650 °C over 15 min, subliming the fullerenes onto the sample disks and copper support rod. After cooling, the copper rod was

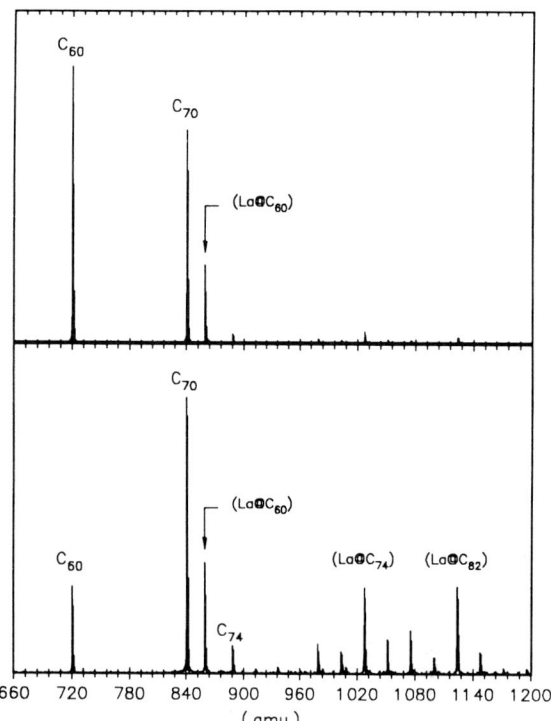

Figure 1. FT-ICR mass spectra of black mirrorlike sublimed fullerene film showing presence of (La@C_n) fullerenes. For the top panel the pulsed decelerator responsible for slowing the clusters down to be trapped in the analysis cell was optimized for the C_{60}–C_{70} mass region. For the bottom panel it was optimized for the region around C_{84}.

removed, and the copper sample disks were found to be covered with a mirrorlike black film. The sublimed film sample was then mounted in the cluster FT-ICR apparatus and pumped down to <10^{-7} Torr.

Figure 1 shows the result of laser-desorption probing of this sublimed film. The top panel FT-ICR mass spectrum was taken with the apparatus optimized to trap clusters in the mass range around C_{70}. Note that both C_{60} and C_{70} are the dominant clusters detected. Under the gentle 0.6-mJ desorption laser conditions used, no signal was detected without the ArF excimer ionization laser. The photon energy of this laser (6.4 eV) is lower than the ionization potentials of C_{60}, C_{70}, and all the other empty fullerenes in this mass range, but they are all very efficiently photoionized by a 1+1 resonant two-photon ionization process with this laser involving the formation of a relatively long-lived triplet state.[13] The absence of signal corresponding to fullerenes less than 60 atoms in size and in the C_{62}–C_{68} range is a good indication that these laser desorption and ArF excimer laser ionization conditions are probing the native composition of the film on the sample disk without fragmentation or fullerene–fullerene reactions.

Note that as labeled in Figure 1 there is a prominent peak corresponding to LaC$_{60}$ present in the film. Careful examination of this feature at high mass resolution showed it had precisely the correct mass and isotope distribution. Exposure to oxygen or ammonia reactant gas in the ICR trap showed no reactivity. Exposure to XeCl excimer light showed it was just as resistant to photofragmentation as C_{60}^+, and when it did fragment, it shrank by successive C_2 losses, just as does C_{60}^+. When this shrinkage was pushed to its limit, it was found to be difficult to fragment past LaC$_{44}^+$ and impossible to go past LaC$_{36}^+$ without bursting the cluster. This is a reasonable result for an internally caged lanthanum atom, since the La$^+$ ionic radius is roughly the same as that of K$^+$, and (K@C_{44})$^+$ is known to be the smallest survivable (K@C_n)$^+$ cluster.[5] In the case of a tightly shrink-wrapped (La@C_n)$^+$ cluster, the ground electronic state may involve higher

(11) Alford, J. M.; Laaksonen, R. T.; Smalley, R. E. *J. Chem. Phys.* **1991**, *94*, 2618.

(12) Maruyama, S.; Anderson, L. R.; Smalley, R. E. *Rev. Sci. Instrum.* **1991**, *61*, 3686.

(13) See, for example: Haufler, R. E.; Wang, L.-S.; Chibante, L. P. F.; Jin, C.; Conceicao, J.; Chai, Y.; Smalley, R. E. *Chem. Phys. Lett.* **1991**, *179*, 449.

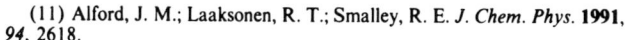

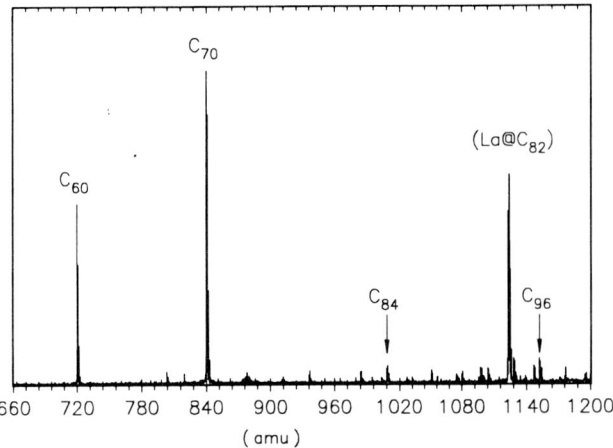

Figure 2. FT-ICR mass spectrum of hot toluene extract of fullerene material produced by laser vaporization of a 10% La_2O_3/graphite composite rod. The sample was exposed to air and moisture.

effective charge on the central metal atom, permitting the "shrink wrapping" to proceed a bit further before the fullerene cage bursts.

We therefore believe that the desired $(La@C_{60})$ clusters had been made in abundance in the quartz tube furnace. They survived the 1200 °C conditions in the furnace during the several minutes necessary to flow to the end where they deposited in the black-brown film. They survived the subsequent sublimation onto the copper sample disk, and when imbedded in a high-quality film made up primarily of C_{60} and C_{70} they survived for several hours exposed to air prior to mounting in the FT-ICR apparatus.

Also evident in the top part of Figure 1 are some small peaks due to higher fullerenes. In the bottom panel, these are seen more clearly after the apparatus had been adjusted to trap them more effectively. Note that, except for C_{74} and a small amount of C_{84}, all this signal is due to $(La@C_n)$ fullerenes, with $(La@C_{74})$ and $(La@C_{82})$ appearing specially abundant. Experiments with reactivity and fragmentation behavior proved all these to have the La atom trapped inside. No larger clusters were found under these conditions of laser desorption. The relative absence of $(La@C_n)$ clusters with $60 < n < 70$ and $n > 84$ is strong evidence that the lasers did not perturb the native $(La@C_n)$ cluster distribution on the sublimed sample film. They too must have been made in the quartz tube furnace and survived sublimation and transport through room air to the FT-ICR apparatus.

The results of Figure 1 give the appearance that a substantial fraction of the sublimed film is made up of $(La@C_n)$-doped fullerenes. However, these species are likely to be directly photoionized by a single 6.4-eV photon of the ArF excimer laser.[3,4] While they are certainly present in the sublimed film in significant quantities, their actual relative abundance in the sublimed film compared to the two-photon-ionized $(@C_n)$ species may be as low as a few percent.

Although this stability of these lanthanum fullerenes is impressive, the fact that they were buried in a good quality C_{60}/C_{70} film means that most of them were never actually exposed to air. When they *were* exposed, a dramatic difference was seen. Figure 2 shows the FT-ICR spectrum of a sample probed in the same, gentle way. Except here the fullerenes were obtained by collecting the initial black-brown deposit in the quartz tube (which is easily brushed off of the tube surface, much like the raw soot from a standard carbon-arc fullerene synthesis). This black powder was then Soxhlet extracted for 2 h in boiling toluene. No attempt was made to exclude air or moisture. The resultant solution was then evaporated, and the extracted material redissolved (or, at least, resuspended) in a small amount of toluene so it could be deposited on a copper FT-ICR sample disk. Note that effectively the only lanthanum-containing fullerene to be extracted by hot toluene and survive exposure to Houston's mid-August weather was $(La@C_{82})$.

This stability and toluene extractability of $(La@C_{82})$ led us to suspect it could be made and recovered as well by the standard carbon arc technique now widely used to produce C_{60}, C_{70}, and

the higher empty fullerenes. Accordingly two 1.25-cm-diameter graphite rods were bored out down the center axis of one end to form a 1.0-cm-diameter, 10-cm-long cavity. They were filled with a mixture of La_2O_3 (22%), graphite powder (45%), and pitch (33%). After being baked to carbonize the pitch, the rods were then used as opposing electrodes in our present scaled-up version of the carbon arc fullerene generator described in an earlier publication from this group.[9] Soot collected from this run was then Soxhlet extracted with toluene, the resulting fullerene solution concentrated down, and a small portion deposited on an FT-ICR sample disk. The resultant FT-ICR mass spectrum was similar to that shown in Figure 2: $(La@C_{82})$ was the only lanthanum fullerene detected. It appeared in substantially lower relative yield that the laser-vaporization sample of Figure 2, but we expect optimization of La_2O_3 loading levels in the graphite electrodes, more thorough carbonization, and adjustment of the operating conditions of the arc will enable $(La@C_{82})$ to be produced at yield levels similar to that of the most stable empty fullerene in this size region, C_{84}.

Such experiments with large amounts of heavy-metal oxides in carbon arcs should be done with caution. The oxides are readily reduced under conditions of the arc to the neutral metal, and there is no guarantee that all of these metal atoms will end up safely on the inside of carbon cages. In addition to their toxicity, finely dispersed lanthanum metal atoms on the *outside* of graphitic carbon structures are highly pyrophoric, as was demonstrated to us quite memorably when we vented the carbon arc apparatus rapidly to air after our first run. We now bleed in a small amount of air while the apparatus is still under vacuum, converting the surface lanthanum into La_2O_3 at a slow, controlled rate.

Why Is $(La@C_{82})$ Special?

This unique stability of a lanthanum-containing fullerene at 82 carbon atoms at first seems strange. For the normal, empty fullerenes the most abundant are usually 60, 70, and 84, and to some extent 76, in decreasing order of apparent stability.[14] When probing sublimed or solvent-deposited C_n fullerene films with this FT-ICR apparatus, this relative abundance pattern is readily seen.

However, with an internal lanthanum atom, there ought to be a change in this pattern of stabilities. Calculations of the electronic structure of $(La@C_{60})$ have been done by Rosen and Wastenburg[15] and more recently by Chang et al.[16] The La is predicted (in the latest calculation) to donate both its 6s electrons into the delocalized aromatic shell system of the carbon cage, leaving behind a single 5d electron. In the case of C_{60} this pair of donated electrons can only go into the t_{1u} LUMO shell, producing a molecule whose reactivity should be like that of the triplet state[17] of normal C_{60}. Although the relevant calculations have not yet been done, it is likely that the same circumstance will apply to the other particularly stable empty fullerenes, C_{70}, C_{84}, and C_{76}, all of which are closed-shell ground-state singlets with substantial HOMO–LUMO gaps.

In view of the results of Figure 1, we suspect there is still a way that one can end up with a closed-shell fullerene with a lanthanum inside. If as in metal clusters such as K_x or Cu_x where the shell model works it is also true in these hollow-shell fullerenes that it is the electron count that matters most, then the stability of C_{84} means that 84 must be a stable shell closing. By dropping down to the C_{82} fullerene and adding two extra electrons from the 6s shell of an internal lanthanum atom, one regains the 84 electron count, and there may be an arrangement of the pentagons and hexagons that preserves this as a shell closing. The resultant

(14) Diederich, F.; Ettl, R.; Rubin, Y.; Whetten, R. L.; Beck, R.; Alvarez, M.; Anz, S.; Sensharma, D.; Wudl, F.; Khemani, K. C.; Koch, A. *Science* **1991**, *252*, 548.

(15) Rosen, A.; Wastenberg, B. *J. Am. Chem. Soc.* **1991**, *110*, 8701.

(16) Chang, A. H. H.; Ermler, W. C.; Pitzer, R. M. *J. Chem. Phys.* **1991**, *94*, 5004.

(17) Arbogast, J. W.; Darmanyan, A. P.; Foote, C. S.; Rubin, Y.; Diederich, F. N.; Alvarez, M. M.; Anz, S. J.; Whetten, R. L. *J. Phys. Chem.* **1991**, *95*, 11.

(18) Wasielewski, M. R.; O'Neil, M. P.; Lykke, K. R.; Pellin, M. J.; Gruen, D. M. *J. Am. Chem. Soc.* **1991**, *113*, 2774.

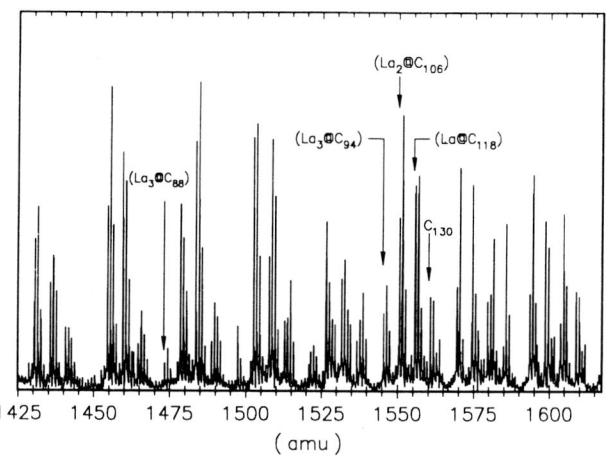

Figure 3. FT-ICR mass spectrum showing the existence of fullerenes with up to three lanthanum atoms trapped inside.

(La@C_{82}) would be a closed-shell singlet with a substantial HOMO–LUMO gap.

This would nicely explain (La@C_{74}) as being the next most stable lanthanum fullerene—it has the same 76-shell electron count as C_{76}. The lower panel of Figure 1 shows that (La@C_{74}) does appear to be nearly as abundant as (La@C_{82}) in the sublimed film. But no such trick can be found at the 60- and 70-electron shell closings since (La@C_{58}) and (La@C_{68}) have no structures that avoid adjacent pentagons.

It will be interesting to see if these simple ideas hold up to detailed calculations. But, regardless of these speculations, it is clear that in reality (La@C_{82}) truly is special.

Metallofullerene Coalescence: More Atoms Inside. In the course of this work we were interested in finding what levels of vaporization laser fluence were capable of producing (La@C_n) fullerenes from the La$_2$O$_3$/graphite composite and whether there were any conditions that would give FT-ICR mass spectra similar to those of Figure 1 even though the metallofullerenes did not originally exist on the FT-ICR sample disk. We found that laser vaporization of a 0.1-cm-thick disk of the same La$_2$O$_3$/graphite rod used to generate the sublimed film sample of Figure 1 did not produce any detectable fullerene signal when probed in the FT-ICR cluster source with desorption laser fluences less than 1 mJ. However, when struck with vaporization laser pulses of >10 mJ, this composite disk did produce readily detectable fullerene and metallofullerene signals in the FT-ICR apparatus. Quite unlike in Figure 1, these cluster distributions were extremely broad, with only slight evidence for special stability of clusters such as C_{60}, (La@C_{60}), and (La@C_{82}).

The most interesting result, however, concerned the time development of the cluster FT-ICR mass spectrum. Initially, as the vaporization laser struck fresh surfaces of the composite sample, the cluster distribution was primarily made up of C_n clusters with $n < 60$, with only a small amount of (La@C_n) lanthanum fullerenes evident. As the target disk was moved around for the second pass, the cluster signal became more intense, extended out to clusters well over 100 atoms in size, and was dominated by (La@C_n). Further laser blasting of the composite disk continued to change the cluster distribution, showing strong signals for clusters of composition La$_2C_n$ and even La$_3C_n$. Figure 3, for example, shows a portion of the FT-ICR cluster mass spectrum obtained after extensive laser-aging of the composite target disk. Signals due to La$_3C_n$ clusters are seen here to grow in importance as a function of cluster size, starting with La$_3C_{88}$.

In a series of XeCl excimer laser photofragmentation and reaction tests with O_2 and ammonia reactants, we were able to prove that all these clusters were, in fact, (La$_x$@C_n) lanthanum fullerenes. As levitated positive ions in the FT-ICR apparatus, they are chemically inert, and they fragment only by successive C_2 losses. Laser "shrink wrapping" experiments showed that the smallest cage that can surround two lanthanum atoms is C_{66}, and the minimum cage that can fit around three is C_{88}.

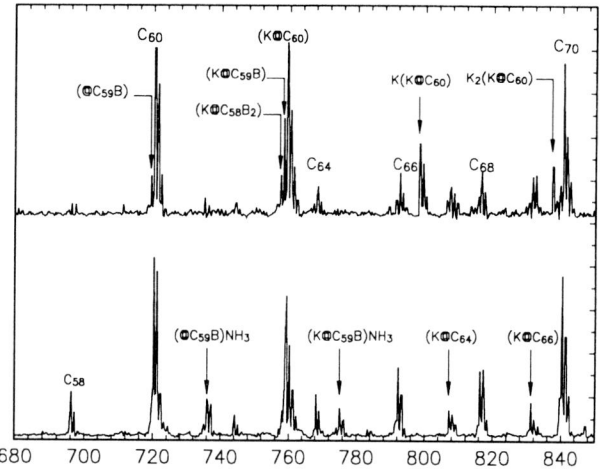

Figure 4. FT-ICR mass spectra showing evidence for the production of boron and boron/potassium doped 60-atom fullerenes. The bottom panel shows the result of reaction with ammonia. Note that the boron-doped clusters have been titrated with ammonia, demonstrating that the boron is substituting for a carbon as part of the fullerene cage. Note also that the clusters in the top panel marked K(K@C_{60})$^+$ and K$_2$(K@C_{60})$^+$ are missing after reaction with ammonia in the bottom panel, demonstrating that the extra potassium atoms were on the outside, unprotected by the fullerene cage.

The mechanism for formation of these clusters is unknown, but from the observed dependence on laser aging the target surface, we suspect they are generated by (La@C_n)–(La@C_n) coalescence reactions occurring either on the target disk surface or in the laser-induced plasma just above this surface. Detailed probes of such laser-zapped metal/graphite surfaces by STM and various types of photoemission may be quite fascinating.

We note in passing that the recent transmission electron microscopy pictures of C_{60} films by Wang and Buseck[19] show evidence of C_{60}–C_{60} coalescence to form cylindrical "bucky tubes" in the solid film, presumably triggered by the 400-keV electron beam. We wonder if metal atom encapsulation events would occur under similar circumstances with metal-doped fullerene films.

Potassium/Boron Doped Fullerenes. In an attempt to produce boron-doped fullerenes[20] using a boron/graphite composite target rod in the 1200 °C tube furnace, we produced a material that when sublimed onto a sample disk gave the FT-ICR mass spectra shown in Figure 4. The top panel shows excellent evidence for the existence in this sublimed fullerene film of not only (@C_{59}B) but also (K@C_{60}), (K@C_{59}B), and (K@$C_{58}B_2$). The potassium had been introduced by accident since at the time the experiment was done the only quartz tube that was available was a dirty one that had been used months before to produce a KCl/graphite composite rod. It still had KCl deposited on the inner surfaces of the tube, baked in so effectively that it could not be removed. The bottom panel shows the resultant FT-ICR mass spectrum after exposure to ammonia reactant gas. The boron-doped fullerenes here are seen largely titrated with ammonia. Note also that the signals labeled in the top panel as K(K@C_{60}) and K$_2$(K@C_{60}) are missing after reaction with ammonia. They are outside complexes, K$_x$(K@C_{60})$^+$, which react with NH$_3$ to produce (we suspect) K$_{x-1}$(K@C60) and KNH$_3^+$. XeCl excimer laser photolysis of the (K@C_{60})$^+$, (K@C_{59}B)$^+$, and (K@$C_{58}B_2$)$^+$ clusters shows only successive C_2 losses, demonstrating that they have the single potassium atom safely doped on the inside.

Conclusion

Experiments such as those discussed above have led us to suspect there will soon be a vast new array of doped fullerene species generated by both laser-vaporization and arc techniques. The method used here involving laser vaporization in a tube furnace,

(19) Wang, S.; Buseck, P. R. *Chem. Phys. Lett.* **1991**, *182*, 1.
(20) Guo, T.; Jin, C.; Smalley, R. E. *J. Phys. Chem.* **1991**, *95*, 4948.

followed by sublimation onto a substrate, has the advantage that even rather reactive open-shell doped fullerenes can be stabilized by embedding in a C_{60}/C_{70} matrix. This matrix is of sufficient quality to permit the film to be exposed to air for substantial periods of time. Subsequent dissolution of these fullerene films in appropriate solvents in air- and water-free environment should permit the more reactive doped fullerenes to be separated and passivated by attachment of protecting groups, or by formation of stable salts or complexes.

Acknowledgment. The delightful suggestion of the use of the @ symbol in formulas for complex fullerenes is due to Ori Cheshnovsky. We thank Lai-Sheng Wang, Jose Conceicao, and Lila Anderson for help and encouragement in the early, fitful beginnings of this work on "dopyballs". This research was supported by the National Science Foundation, the Office of Naval Research, and the Robert A. Welch Foundation, using a cluster FT-ICR apparatus developed with major support from the U.S. Department of Energy, Division of Chemical Sciences.

Reprinted with permission from Physical Review Letters
Vol. 69, No. 9, pp. 1352–1355, 31 August 1992
© 1992 The American Physical Society

Collision of Li $^+$ and Na $^+$ with C$_{60}$: Insertion, Fragmentation, and Thermionic Emission

Zhimin Wan, James F. Christian, and Scott L. Anderson

Department of Chemistry, State University of New York, Stony Brook, New York 11794-3400
(Received 14 May 1992)

Interactions of Li$^+$ and Na$^+$ with C$_{60}$ molecules have been studied over the collision energy range from 0 to 150 eV. We observe insertion of the alkali-metal ions to form the endohedral [LiC$_{60}$]$^+$ and [NaC$_{60}$]$^+$ species, with energy thresholds of 6 and 20 eV, respectively. At higher collision energies, the excited [NaC$_{60}$]$^+$ appears to relax mainly by loss of C$_2$ units from the fullerene cage, yielding [NaC$_{60-2n}$]$^+$. For [LiC$_{60}$]$^+$, escape of the Li$^+$ competes effectively with C$_2$ loss. For both Li$^+$ and Na$^+$, signal for pure carbon fragment ions C$_{60-2n}$$^+$ increases sharply at $\sim$30 eV, which is attributed to thermionic emission.

PACS numbers: 34.50.Lf, 82.20.Pm, 82.30.Fi, 82.40.Dm

Since 1985, gas-phase C$_{60}$ has attracted the interests of many chemists and physicists. Many properties have been investigated experimentally and theoretically, including unimolecular decay [1], electronic structure and UV and visible spectroscopy [2], vibrational Raman and ir spectroscopy [3], and chemical reactivity with atoms and molecules [4]. A review of many aspects of fullerene chemistry and physics has been given recently in a special issue of Accounts of Chemical Research [5]. C$_{60}$ has a stable hollow-cage structure [6], which raises the interesting possibility of preparing compounds where atoms or molecules are trapped inside. Chai *et al.* first prepared metal-doped fullerenes with the metal atom in the endohedral cavity [7]. More recently, Weiske *et al.* [8] reported *insertion* of rare-gas atoms into the endohedral cavity during high-energy collisions of C$_{60}$$^+$ with various atoms and molecules. Independent experiments by Ross and Callahan [9], Caldwell *et al.* [10], and Campbell *et al.* [11] have confirmed this observation. To avoid problems with unknown internal excitation of the C$_{60}$$^+$, and to simplify measurements of collision energetics, we have studied the insertion, fragmentation, and charge-transfer reactions, using collisions of Ne$^+$ with C$_{60}$ [12].

In this paper, we report on collisions of alkali-metal ions M^+ (M = Li and Na) with C$_{60}$ vapor molecules. These systems are interesting for several reasons. Na$^+$ is similar in size and is isoelectronic with neon, allowing direct comparison of the results. Li$^+$ is considerably smaller and lighter (more comparable to helium), and thus may have a very different insertion mechanism. A major difference between the alkali-metal-ion and rare-gas-ion systems is their electronic energies. For the C$_{60}$$^+$ inert-gas-ion system, exoergic charge transfer occurs with a large cross section, releasing over 10 eV, while for the alkali-metal ions, the charge is expected to remain on the alkali atom.

With the exception of the alkali-metal source, the experimental setup for this work is the same as described previously [12,13]. Alkali-metal ions are generated by surface ionization on a hot filament [14]. The nascent M^+ beam is collected by a radio-frequency (rf) octapole ion guide, then accelerated and mass filtered by a magnetic sector. The mass-selected M^+ beam is then de-

celerated, and injected into a second rf octapole ion guide, operated at 4.34 MHz, which guides the reactant ions through a scattering cell. The cell, which is heated to $\sim$340°C, holds a capillary tube containing mixed C$_{60}$/C$_{70}$ powder, which generates C$_{60}$:C$_{70}$$\sim$85:15 vapor. The fullerene pressure is conductance limited to $\sim$10^{-7} Torr, as estimated by the rate of mass loss and cell conductance. This is well below the vapor pressure of C$_{60}$ at 340°C [15], which minimizes condensation on the ion optics.

A small fraction of the reactant ions collide with C$_{60}$ molecules, and product ions scattered into the forward hemisphere in the laboratory frame are collected by the octapole. Product and unscattered reactant ions are mass selected by a double focusing electric and magnetic sector mass spectrometer, and detected by an on-axis Daly detector [16]. In the following we give only relative cross sections, due to uncertainties in the absolute pressure of C$_{60}$, the product collection efficiency, and the alkali-metal ion beam intensities *in the scattering cell*.

The results for Li$^+$+C$_{60}$ are summarized in Fig. 1, which gives the relative cross sections for all product channels as a function of collision energy. The main product ion is LiC$_{60}$$^+$, which appears at 6 eV collision energy. At energies above 20 eV, the LiC$_{60}$$^+$ signal decreases, and a series of LiC$_{60-2n}$$^+$ fragment ions appears. This suggests that C$_2$ loss is an important decay mechanism for excited LiC$_{60}$$^+$. Note that LiC$_{60-2n}$$^+$ fragments for $n > 3$ have negligible intensities—evidently Li loss is favored at the higher energies.

At collision energies above 30 eV, we observe C$_{60}$$^+$ with a cross section about one-fourth as large as for LiC$_{60}$$^+$. As collision energy is increased above 40 eV, the C$_{60}$$^+$ signal decreases and is replaced by a series of C$_2$ loss fragments: C$_{60-2n}$$^+$. The bottom frame of Fig. 1 gives the total cross sections for formation of fullerene product ions, with and without Li. Note that for the C$_n$$^+$ series, the total cross section is roughly energy independent once the appearance energy is surpassed, while for the LiC$_n$$^+$ series, the cross section is nearly zero for collision energies above 70 eV. (Note that the appearance of the C$_n$$^+$ channels coincides with the decrease in the LiC$_n$$^+$ cross section. From the Na$^+$ results, we believe

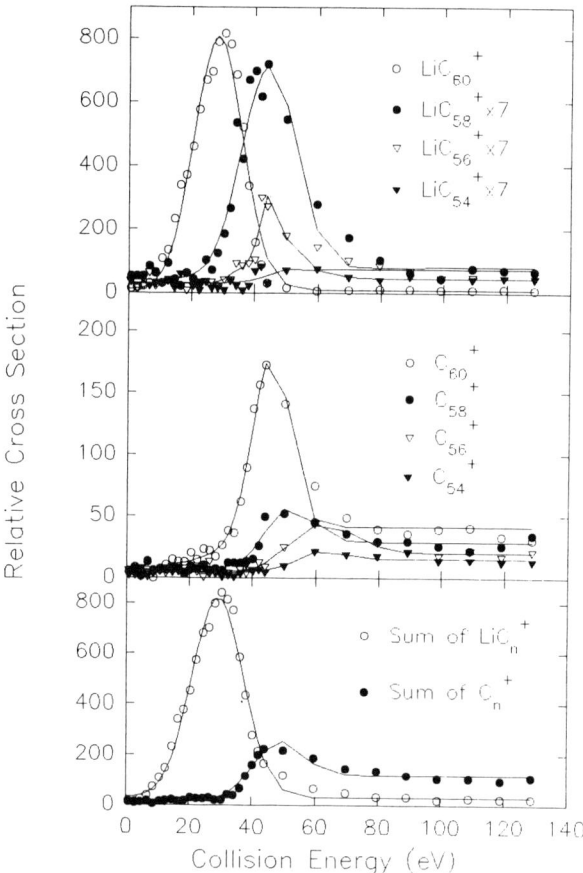

FIG. 1. Relative product cross sections as a function of collision energy for $Li^+ + C_{60}$. The solid lines are simply smooth curves to guide the eye.

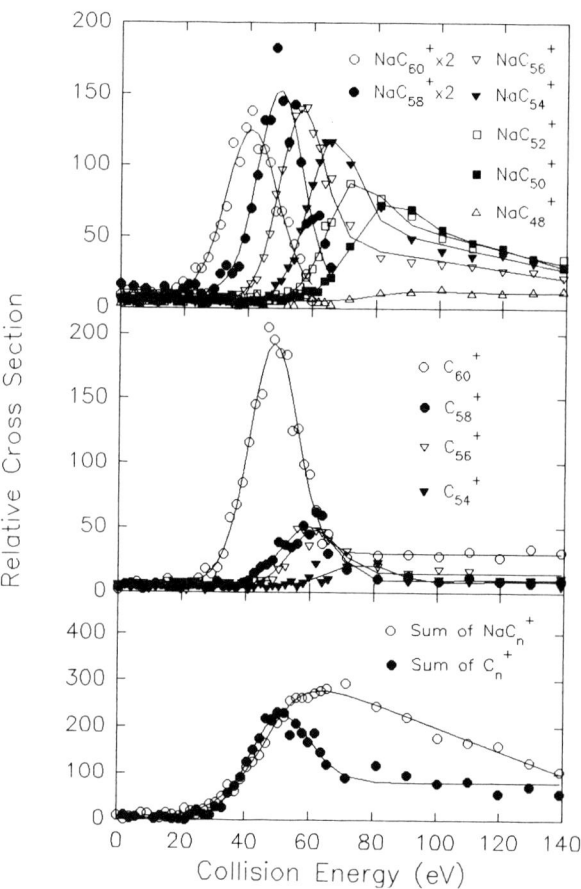

FIG. 2. Relative product cross sections as a function of collision energy for $Na^+ + C_{60}$. The solid lines are simply smooth curves to guide the eye.

this is merely an accident, and not indicative of interchannel competition.)

Figure 2 gives the equivalent data for $Na^+ + C_{60}$. NaC_{60}^+ adduct ions appear at ~20 eV collision energy, and with smaller cross section compared to the lithium case. At energies above 25 eV, we see a series of NaC_{60-2n}^+ fragment ions, predominantly produced by sequential C_2 loss from excited NaC_{60}^+. Note that for Na^+, the intensity of the NaC_{60-2n}^+ sequence continues out to NaC_{50}^+, then drops suddenly for NaC_{48}^+ and smaller fragments. This is in contrast to the LiC_{60-2n}^+ sequence, which effectively terminates at LiC_{54}^+. (Note that the NaC_{60}^+ and NaC_{58}^+ cross sections are not plotted for collision energies above 65 eV. At these high energies, there is background from reaction with the C_{70} impurity in our fullerene sample.)

The middle frame of Fig. 2 gives cross sections for production of bare fullerene fragment ions: C_{60-2n}^+. Note that for $Na^+ + C_{60}$, the appearance energy for C_{60}^+ is 30 eV—nearly identical to the $Li^+ + C_{60}$ case. The bottom frame of the figure gives the total cross sections for product ions with and without Na^+. The NaC_n^+ cross section extends to high collision energies, in sharp contrast to the LiC_n^+ results. On the other hand, the magnitude and energy dependence of the C_n^+ total cross section is

nearly identical for Li^+ and Na^+.

Both the collision energy dependence of the MC_{60}^+ ($M = Li$, Na) signal, and the observed decay by C_2 loss, can best be explained if MC_{60}^+ is an *endohedral* complex, i.e., the alkali atom must be bound inside the fullerene cage. There presumably is an attractive interaction that could bind M^+ on the outside of the cage, but for closed-shell alkali-metal ions, the binding is unlikely to be strong enough to stabilize an exocomplex at the high energies we are probing. In addition, we would expect no activation for this type of binding.

Bakowies and Thiel have estimated, by the modified neglect of differential overlap method, that the energy barriers for Li^+ passage through the six-member and five-member rings of C_{60} are 7.3 and 10.2 eV, respectively [17]. Our appearance energy of 6 eV for LiC_{60}^+ is therefore in reasonable agreement with the calculated barrier. Na^+, which has a hard-sphere radius ~0.3 Å larger than Li^+, must almost certainly break or distort C-C bonds to get into the cage, and our appearance energy for NaC_{60}^+ (~20 eV) is consistent with this picture. Another interesting comparison is with the appearance energy for insertion of Ne atoms into C_{60}^+ (~25 eV) [12], which, as expected, is 3–5 eV higher, since Ne is isoelectronic with and slightly larger than Na^+. (Note that Campbell *et al.* found a threshold of 9 eV for inserting Ne atoms into C_{60}^+ [11]. The reasons for the discrepancy between their threshold and ours are not completely clear, but most likely are related to the fact that the C_{60}^+ in their experiments has tens of electron volts of internal excitation [18], while our C_{60} is thermal. This issue is discussed fully elsewhere [19].)

Once the alkali projectile gets inside, the translational energy of the M^+ should be efficiently converted into internal energy of the MC_{60}^+ adduct, thus trapping the M^+. As the collision energy is increased, the resulting internal energy of the endocomplex also increases, and decomposition becomes likely. Radi *et al.* [1] and Yoo, Ruscic, and Berkowitz [20] have estimated that the binding energy of C_{60}^+ with respect to C_2 loss is 4.6 to 6.5 eV. In our experiment, we observe the first C_2 loss product channel at 25 eV for LiC_{58}^+ and 30 eV for NaC_{58}^+. These are the energies at which the lifetime of the MC_{60}^+ activated complex is short enough to allow measurable fragmentation on the experimental time scale (~ 1 msec).

Our observation that the series of LiC_{60-2n}^+ fragment ions terminates sooner than the corresponding NaC_{60-2n}^+ series is consistent with the endohedral binding picture. Just as Li^+ can insert into C_{60} more easily than Na^+, it should be able to escape more readily. The large drop of ion signals toward bigger n clearly shows that Li^+ loss competes effectively with C_2 loss:

$$LiC_{60}^+ \rightarrow C_2 + LiC_{58}^+ \rightarrow 2C_2 + LiC_{56}^+ \rightarrow \cdots$$
$$\searrow C_{60} + Li^+ \quad \searrow C_{58} + Li^+ \quad \searrow C_{56} + Li^+ .$$

Since the radius of Na^+ is much larger than Li^+ and it is harder to escape from the cage, we continue to observe the C_2 loss series down to NaC_{50}^+ with no large decrease in intensity.

The other interesting product ion is C_{60}^+, which is observed at 30 eV collision energy for both $Li^+ + C_{60}$ and $Na^+ + C_{60}$. The ionization potentials (IPs) of Li and Na are 5.39 and 5.14 eV, respectively, which are more than 2 eV lower than the IP of C_{60} (7.61 eV) [21]. The mechanism for production of fullerene ions can be formally divided into two channels:

$$M^+ (M = Li, Na) + C_{60} \rightarrow M + C_{60}^+ \tag{1}$$
$$\searrow M^+ + C_{60}^* \rightarrow M^+ + C_{60}^+ + e^- . \tag{2}$$

The first channel is formally endoergic charge transfer, which can occur through a variety of mechanisms. For example, in large impact parameter collisions, the time-dependent ion-induced polarization can result in electron hopping from C_{60} to the M^+ ion. This mechanism has been investigated for several atom-cluster ion systems [22], and is only efficient when the following condition is satisfied: $R\Delta IP \sim h\upsilon$, where $\Delta IP = IP(C_{60}) - IP(M)$, υ is the relative velocity of the M^+ ion to the C_{60}, and R is

the interaction range, here about 10 Å. The required collision energy turns out to be ~ 10 keV for Li^+ and ~ 35 keV for Na^+, thus this mechanism is insignificant for our energy range. For small collision energy (~ 100 eV), a more likely mechanism is endoergic charge transfer resulting from small impact parameter, intimate collisions, which may or may not involve formation of an intermediate complex. There is no literature on the energy dependence of this process for such large molecules; however, for small molecular ions the cross sections typically show a quick increase as energy is raised through the thermodynamic threshold. In our results, there is a small C_{60}^+ signal at low energies, which probably comes from this mechanism, but the large increase of ~ 30 eV seems to require a different mechanism.

For large stable molecules like fullerenes, thermionic emission may compete favorably with dissociation as an energy loss mechanism for highly excited collision products. For example, Campbell, Ulmer, and Hertel and Wurz and Lykke have examined delayed ionization of laser-heated C_{60}, and concluded that thermionic emission was the mechanism [23]. Maruyama *et al.* have pointed out that thermionic emission is probably the mechanism for laser-induced multiple ionization of giant fullerenes [24]. In our experiment, when an alkali-metal ion collides with C_{60}, part of the kinetic energy will be transferred to internal energy of the C_{60} molecule. For *high-energy* collisions, we can make a crude estimate of the energy transfer by assuming that the alkali-metal ion simply undergoes an elastic collision with a single carbon atom (i.e., assume that the collision is impulsive with the carbon atom that is impacted). In this limit, the C_{60} internal energy after collision is

$$E_{int} = [4M_A M_C/(M_A + M_C)^2]E_{CE} + (3N-3)kT_0$$
$$= (3N-3)kT_{final}, \tag{3}$$

where M_A and M_C are the alkali and carbon atom masses, E_{CE} is the collision energy, $N = $ sixty carbon atoms, and T_0 is the initial C_{60} temperature (600 K). The second line simply expresses the internal energy as a final temperature, T_{final}. For $E_{CE} = 30$ eV, where we observe a large increase in C_{60}^+ signal, T_{final} is 2430 K for lithium, and 2370 K for sodium. This is certainly in the temperature range where thermionic emission might be expected.

Klots has calculated the thermionic emission rate of small particles [25]. For a C_{60} molecule, this rate becomes

$$k(T) = (4.16 \times 10^{13})T(1 + 1.17T^{1/2} + 0.437T)$$
$$\times \exp(-E_0/kT), \tag{4}$$

where T is the internal temperature in a unit of 1000 K, E_0 (= 7.61 eV) is the ionization potential of C_{60} [21], and the unit of $k(T)$ is 1/sec. The relationship between the rate $k(T)$ of a single C_{60} molecule and the total C_{60}^+ ion intensity is estimated to be $I(C_{60}^+) \sim 100k(T)$ for

1354

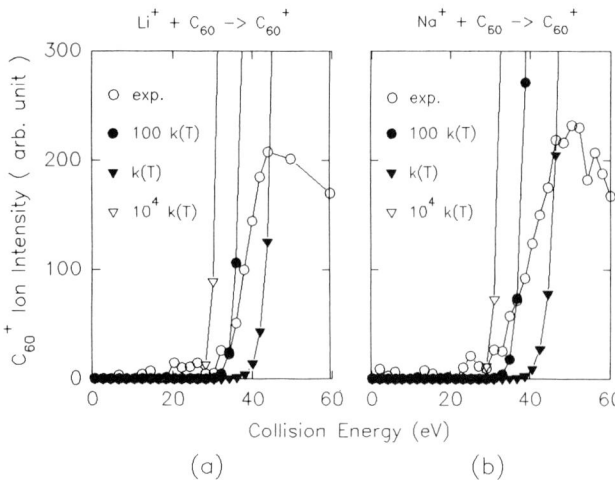

FIG. 3. Thermionic emission model of C_{60}^+ production in $C_{60}+Li^+$ and Na^+ collisions. The experimental data are compared with the theoretical fitting curve $I=A_0k(T)=100k(T)$, where T is calculated from the collision energy by Eq. (3): (a) C_{60}^+ from Li^++C_{60}; (b) C_{60}^+ from Na^++C_{60}. A_0 is a fitting parameter proportional to the C_{60} vapor pressure. $A_0=100$ corresponds to the estimated pressure of 10^{-7} torr. Two other curves, $k(T)$ and $10^4k(T)$, indicate the lower and upper limits of the fitting.

our experimental condition. Using the relationship given above for T_{final}, we plot the thermionic emission rate as a function of collision energy, and compare this with our data in Figs. 3(a) and 3(b). We can see that the experimental data are quite similar to the theoretical curves, which supports our hypothesis that most of our C_{60}^+ is due to thermionic emission.

We recently received a report of a very interesting study of photon-induced thermionic emission by Campbell, Ulmer, and Hertel [26]. They observed delayed ionization when C_{60} molecules absorbed two to four 308-nm (4.03 eV) photons, and fit the results to a thermionic emission model. Although the experiments are quite different, and somewhat difficult to compare quantitatively, their conclusions regarding the temperature (or internal energy) dependence of C_{60} thermionic emission appear to be very similar to ours.

We thank Mark Ross for providing the initial sample of fullerene powder, E. E. B. Campbell and I. V. Hertel for sending us a copy of their work, and C. E. Klots for useful discussion. We are also grateful to Yousef Basir for his effort in preparation of alkali-metal ion sources. This work is sponsored by the Office of Naval Research (N0001492J1202 and URIP N000-14-86-G-0208). S.L.A. is a Camille and Henry Dreyfus Scholar.

[1] P. P. Radi, M. T. Hsu, M. E. Rincon, P. R. Kemper, and M. T. Bowers, Chem. Phys. Lett. **174**, 223 (1990).

[2] J. P. Hare, H. W. Kroto, and R. Taylor, Chem. Phys. Lett. **177**, 394 (1991).

[3] D. S. Bethune, G. Meijer, W. C. Tang, H. J. Rosen, W. G. Golden, H. Seki, C. A. Brown, and M. S. de Vries, Chem. Phys. Lett. **179**, 181 (1991).

[4] L. S. Sunderlin, J. A. Paulino, J. Chow, B. Kahr, D. Ben-Amotz, and R. R. Squires, J. Am. Chem. Soc. **113**, 5489 (1991).

[5] Acc. Chem. Res. **25**, 162 (1992).

[6] J. M. Hawkins, A. Meyer, T. A. Lewis, S. Loren, and F. J. Hollander, Science **252**, 312 (1991).

[7] Y. Chai, T. Guo, C. Jin, R. E. Haufler, L. P. F. Chibante, J. Fure, L. Wang, J. J. Alford, and R. E. Smalley, J. Phys. Chem. **95**, 7564 (1991).

[8] T. Weiske, D. K. Böhme, J. Hrušák, W. Krätschmer, and H. Schwarz, Angew. Chem. Int. Ed. Engl. **30**, 884 (1991).

[9] M. M. Ross and J. H. Callahan, J. Phys. Chem. **95**, 5720 (1991).

[10] K. A. Caldwell, D. E. Giblin, C. S. Hsu, D. Cox, and M. L. Gross, J. Am. Chem. Soc. **113**, 8519 (1991).

[11] E. E. B. Campbell, R. Ehlich, A. Hielscher, J. M. A. Frazao, and I. V. Hertel, Z. Phys. D **23**, 1 (1992).

[12] Z. Wan, J. C. Christian, and S. L. Anderson, J. Chem. Phys. **96**, 3344 (1992).

[13] J. F. Christian, Z. Wan, and S. L. Anderson (to be published).

[14] R. V. Hodges and J. L. Beauchamp, Anal. Chem. **48**, 825 (1976).

[15] J. Milliken, T. M. Keller, A. P. Baronavski, S. W. McElvany, J. H. Callahan, and H. H. Nelson, Chem. Mater. **3**, 386 (1991).

[16] H. M. Gibbs and E. D. Commins, Rev. Sci. Instrum. **37**, 1385 (1966).

[17] D. Bakowies and W. Thiel, J. Am. Chem. Soc. **113**, 3704 (1991).

[18] G. Ulmer, E. E. B. Campbell, R. Kühnle, H.-G. Busmann, and I. V. Hertel, Chem. Phys. Lett. **182**, 114 (1991).

[19] J. F. Christian, Z. Wan, and S. L. Anderson (to be published).

[20] R. K. Yoo, B. Ruscic, and J. Berkowitz, J. Chem. Phys. **96**, 911 (1992).

[21] J. A. Zimmerman, J. R. Eyler, S. B. H. Bach, and S. W. McElvany, J. Chem. Phys. **94**, 3556 (1991); D. L. Lichtenberger, K. W. Nebesny, C. D. Ray, D. R. Huffman, and L. D. Lamb, Chem. Phys. Lett. **176**, 203 (1991).

[22] C. Bréchignac, Ph. Cahuzac, F. Carlier, J. Leygnier, and I. V. Hertel, Z. Phys. D **17**, 61 (1990).

[23] E. E. B. Campbell, G. Ulmer, and I. V. Hertel, Phys. Rev. Lett. **67**, 1986 (1991); P. Wurz and K. R. Lykke, J. Chem. Phys. **95**, 7009 (1991).

[24] S. Maruyama, M. Y. Lee, R. E. Haufler, Y. Chai, and R. E. Smalley, Z. Phys. D **19**, 409 (1991).

[25] C. E. Klots, Chem. Phys. Lett. **186**, 73 (1991); Z. Phys. D **20**, 105 (1991).

[26] E. E. B. Campbell, G. Ulmer, and I. V. Hertel (to be published).

Reprinted with permission from Science
Vol. 355, pp. 239–240, 16 January 1992

Electron paramagnetic resonance studies of lanthanum-containing C_{82}

**Robert D. Johnson, Mattanjah S. de Vries,
Jesse Salem, Donald S. Bethune
& Costantino S. Yannoni**

Almaden Research Center, IBM Research Division, 650 Harry Road,
San Jose, California 95120-6099 USA

THE conjecture that atoms can be trapped inside closed carbon cages such as the fullerenes was first made by Kroto *et al.*[1]. Mass spectroscopic evidence obtained soon after[2] suggested that lanthanum atoms were encapsulated in fullerenes prepared by laser vaporization of a lanthanum-impregnated graphite disk, and these results were later corroborated[3,4]. Recently, helium atoms have been incorporated into fullerenes through collisions in the gas phase[5], and evidence has been obtained for the formation of metal-containing fullerenes during arc burning of composite graphite rods[6]. All of these studies, however, have produced quantities too small for characterization using standard spectroscopic techniques. We report here the preparation of milligram quantities of lanthanum-containing C_{82}, which can be solvent-extracted in yields of about 2% along with empty C_{60} and C_{70} cages. We have

measured the electron paramagnetic resonance spectrum of this mixture, both in solution and in the solid state, which reveals that the lanthanum atom has a formal charge of 3+, and the C_{82} a charge of 3−. This runs contrary to some expectations that the doubly charged fulleride anions would be the most stable species[6,7]; it also reveals that the fullerene cages have the same formal charge as in the superconducting alkali-metal-doped phases[8,9].

Our samples were prepared by using a composite rod made of graphite and La_2O_3. The materials were mixed with a binder (dextrin), partially dried and pressed into a rod. After annealing at 150 °C, the rod was heated to 1,400 °C for 2 h. The soot produced by arc vaporization of the rod under helium at 200 Torr was extracted with toluene, and the extract was washed with diethyl ether and dried. The resulting powder was analysed using mass spectrometry and electron paramagnetic resonance (EPR) spectroscopy.

The time-of-flight mass spectrum of material laser-desorbed from the dry powder (Fig. 1) shows large peaks for C_{60}, C_{70} and $La@C_{82}$. The inset in Fig. 1 shows the fragmentation pattern found for $La@C_{82}$ when the intensity of the ionization laser (wavelength 193 nm) was increased. This pattern is consistent with the pairwise removal of carbons from the skeletal cage, with the La atom remaining inside the resulting smaller fullerene[3]. The two groups of peaks centred about 15 mass units above the C_{60} and C_{70} peaks are due to 25% ^{13}C-enriched C_{60} and C_{70} left over in the arc chamber from an earlier burn and inadvertently vacuumed up along with the material produced for this experiment.

The X-band (9.112 GHz) EPR spectrum of the degassed dry powder shown in Fig. 2a is centred at $g = 2.001$, with a hint of rich structure and an overall width of ∼10 gauss. No other EPR peaks were detected for 2,500 gauss either side of the field corresponding to $g = 2$. The intensity of the EPR signal shows that the $La@C_{82}$ accounts for $2 \pm 1\%$ (by weight) of the sample, close to the 3.5% estimated from X-ray photoelectron spectroscopy on the same material. The powder was then dissolved in 1,1,2,2-tetrachloroethane, degassed several times using a freeze-thaw cycle and sealed in a 5-mm thin-walled pyrex tube. The solution EPR spectrum, shown in Fig. 2b, consists of eight extremely narrow (0.125 gauss) equally spaced (1.25-gauss interval) lines of equal intensity also centred at $g = 2.0010$, with additional intensity in between. The measured g value is similar to those found for fullerene anion radicals[10-13], which range from 1.995 to 2.001. The main eight-line spectrum is unambiguously diagnostic for isotropic hyperfine coupling to a nuclear magnetic moment with spin 7/2 (ref. 14), the value for the ^{139}La nucleus, showing that the unpaired electron has spin

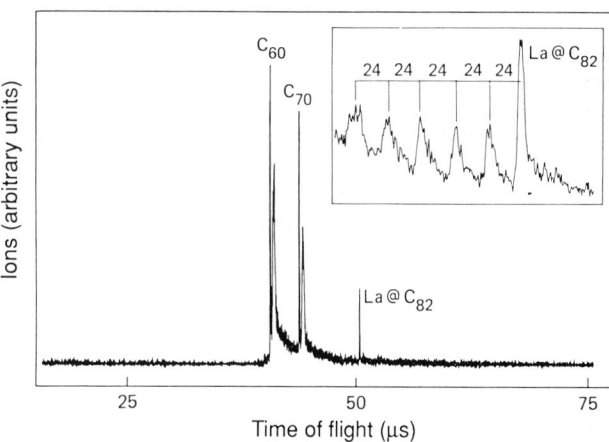

FIG. 1 Mass spectrum of material laser-desorbed from the toluene extract (dried) resulting from arc burning of a composite rod of graphite and La_2O_3 (see text). The inset shows how the peak due to $La@C_{82}$ loses carbons two at a time under strong laser irradiation. The abscissa in the inset is time but the progressive change in mass from peak to peak is shown explicitly.

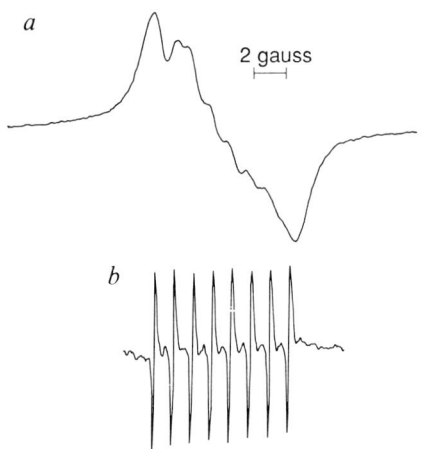

2 gauss

FIG. 2 EPR spectrum (9.112 GHz) at ambient temperature of (a) solid degassed toluene extract (dried) resulting from arc burning of a composite graphite/La_2O_3 rod and (b) a degassed solution of the dried extract in 1,1,2,2-tetrachloroethane.

density on the lanthanum atom. But the ^{139}La hyperfine coupling constant (1.25 gauss) is very small; in contrast, the coupling measured[15] for a La^{2+} defect generated in a CaF_2 lattice (the only experimental value we could find in the literature) is ~50 gauss at 20 K. Furthermore, a calculation of the hyperfine coupling for La^{2+} was made using spin-polarized unrestricted Hartree–Fock wavefunctions and yields a value of 186 gauss. Therefore the small value of the observed hyperfine coupling constant indicates that the lanthanum atom is in the 3+ oxidation state. This is consistent with calculations of the electronic structure of $La@C_{60}$, which found a 3+ charge for the lanthanum atom in that species[16]. Figure 2b shows that there are intensity maxima between all the La lines. As 60% of the $La@C_{82}$ moieties will contain at least one ^{13}C, this intensity is consistent with hyperfine coupling to ^{13}C in natural abundance; the spectrum is expected to be broadened because of the large number of inequivalent carbons in the C_{82} cage[7]. A key point, however, is that observation of coupled electron, ^{139}La and ^{13}C spins provides direct evidence for the association of a La atom with the C_{82} framework.

The EPR spectra reveal a g value consistent with a fullerene anion radical and a La atom in the 3+ oxidation state. We are thus led to a picture of the electronic structure of $La@C_{82}$ in which the La 6s electrons pair in the lowest unoccupied molecular orbital of C_{82} while the third electron occupies the next higher energy orbital. Although this leaves La in a diamagnetic oxidation state, polarization of the atomic orbitals by the paramagnetic π-electron system of C_{82}^{3-} results in unpaired spin density at the site of the ^{139}La nucleus. Therefore, we believe that $La^{3+}@C_{82}^{3-}$ is a ground-state doublet, rather than a closed-shell species as proposed previously[6,7], and furthermore that the paramagnetism originates in an unpaired electron spin in the π-electron system of the carbon framework.

An interesting aspect of this result is that the captive ion (La^{3+} in this case) is relatively unperturbed because the interaction with the cage electrons is small. Thus, we have in this bulk sample a large number of single isolated ions at high concentration (~2% of the molecules), potentially making many spectroscopic studies possible. Efforts to separate the empty fullerenes from metal-containing C_{82} by chromatography are currently in progress. □

Received 24 October; accepted 17 December 1991.

1. Kroto, H. W., Heath, J. R., O'Brien, S. C., Curl, R. F. & Smalley, R. E. Nature 318, 162–163 (1985).
2. Heath, J. R. et al. J. Am. chem. Soc. 352, 7779–7780 (1985).
3. Weiss, F. D., Elkind, J. L., O'Brien, S. C. Curl, R. F. & Smalley, R. E. J. Am. chem. Soc. 110, 4464–4465 (1988).
4. Haufler, R. E. et al. J. phys. Chem. 94, 8634–8636 (1990).
5. Ros, M. M. & Callahan, J. H. J. phys. Chem. 95, 5720–5723 (1991).
6. Chai, Y. et al. J. phys. Chem. 95, 7564–7568 (1991).
7. Manolopoulos, D. E. & Fowler, P. W. Chem. Phys. Lett. (submitted).
8. Hebard, A. F. et al. Nature 350, 600–601 (1991).
9. Holczer, K. et al. Science 252, 1154–1157 (1991).
10. Krusic, P. J., Wasserman, E., Parkinson, B. A., Malone, B. & Holler, E. R. Jr J. Am. chem. Soc. 113, 6274–6275 (1991).
11. Allemand, P.-M. et al. J. Am. chem. Soc. 113, 2780–2781 (1991).
12. Penicaud, A. et al. J. Am. chem. Soc. 113, 6698–6700 (1991).
13. Kukolich, S. G. & Huffman, D. R. Chem. Phys. Lett. 182, 263–265 (1991).
14. Carrington, A. & McLachlan, A. D. Introduction to Magnetic Resonance, Ch. 6 (Harper & Row, New York, 1967).
15. Pilla, O. & Bill, H. J. Phys. C: Solid State Phys. 17, 3263–3267 (1984).
16. Rosen, A. & Waestberg, B. Z. Phys. D: At. Mol. Clusters 12(1–4), 387–390 (1989).

ACKNOWLEDGEMENTS. We thank P. Kasai, R. MacFarlane, H. E. Hunziker, M. Crowder and R. E. Smalley for discussions, P. S. Bagus for the spin-polarized calculation on La^{2+}, D. Hird for XPS data and R. D. Kendrick for help with the g-value measurement.

NANOTUBES

Recently, another new form of carbon has been discovered, associated with the production of fullerenes. Under appropriate conditions, needles with diameters of 4 to 30 nm and lengths up to 1 μm grow on the negative carbon electrode in a dc arc discharge in an atmosphere of 100 Torr of argon.[Ii91*] Electron microscopy reveals that these nanotubes consist of several concentric graphene sheets, separated by a distance which is very close to the basal-plane spacing in graphite. From the observation of the same number of lattice fringes on each side of the nanotube, it is suggested that the structure consists of a number of intact seamless cylinders rather than a single scroll. (However, Dravid *et al.* have found many unterminated graphene sheets in scrolled nanotubes.[Dr93]) The cylinder axis is observed to be different from either the graphite [010] or [110] axes, so that a row of adjacent graphitic hexagons forms a spiral around each cylinder.

In comparison with more familiar graphite fibers (see [Dr88] for a recent comprehensive review of graphite fibers and filaments), the nanotubes are smaller by a factor of some 10^3. Furthermore, the graphene sheets comprising the new nanotubes appear to have a greater degree of structural perfection than they have in "conventional" graphite fibers.

This observation of carbon nanotubes was quickly followed by theoretical work to predict their properties. Mintmire *et al.* performed local-density functional (LDF) band structure calculations of a non-spiral tubule (tube axis parallel to graphite [100] direction), constructed from an infinite chain of C_{10} rings.[Mi92] They found it to be metallic, with estimated carrier densities of the order of good metals. Since the mobility along the axis is expected to be comparable to the high mobility seen in graphite, such a nanotube would be a very good conductor of electricity.

In another set of electronic band structure calculations, the possible helical pitch was considered

explicitly.[Ha92] That group found that the nanotube was either a metal, a narrow-gap semiconductor, or a moderate (0.5–1.0 eV) gap semiconductor, depending upon the exact helical pitch. They gave a simple argument for this result as a consequence of the two-dimensional graphene band structure subjected to periodic boundary conditions in the closed nanotube.

The mechanical properties of an ideal single-shelled nanotube have also been considered theoretically.[Ro92] It was found that the strain energy per atom to roll the graphene sheet into a tube of radius R followed the R^{-2} dependence expected from continuum elasticity, down to the smallest radii. The modulus of elasticity along the nanotube axis was found to be slightly less than bulk graphite for the smallest radii. Carbon nanotubes are also predicted to be exceptionally stiff against bending.[Ov93] However, one must bear in mind that the strength of real materials is usually much less than that of hypothetical, idealized lattice structures. For example, commercial carbon fibers have breaking stress less than 25% of relatively perfect carbon whiskers.[Dr88] Because carbon nanotubes appear to have a high degree of atomic perfection, their strength may approach the theoretical limit.

Further high-resolution electron micrographs lead to interesting results on the topology of the graphitic sheets comprising nanotubes.[Ii92a*] Recall from Chapter I that a network of hexagons requires exactly twelve pentagons to close it with no dangling bonds. Therefore, if the end of a nanotube is to be closed, six pentagons must be incorporated into the network. It was found that tubes were often terminated by cones with a 20° angle. This is exactly what would be expected if the tip contained five pentagons. The sixth pentagon must occur at the point where the cylindrical tube bends inward to form a cone.

One can generalize this topological argument by including seven-sided heptagons. Then, a closed sheet must have twelve more pentagons than heptagons. Indeed, the influence of such heptagons is seen directly in regions of negative curvature, where a tube grows away from the narrow end of a cone.[Ii92a*] The graphene sheets are seen to pass continuously across such an interface, indicating that the network of hexagons is disturbed only by localized $\pm 60°$ disclinations. Further discussion of the growth of nanotubes is given by Iijima et al.[Ii92b].

Conditions for the production of nanotubes have been optimized, and their large-scale synthesis described by Ebbesen and Ajayan.[Eb92] Under appropriate conditions (18 V potential between the graphite rods, 500 Torr atmosphere of helium), some 15% of the starting graphite material is converted to nanotubes.

A different synthesis of carbon nanotubes is described by Ge and Sattler.[Ge93*] They deposited carbon vapor onto a graphite substrate under ultrahigh vacuum conditions, and studied the resulting nanotubes using scanning tunneling microscopy. Tubes with a diameter as small as 10 Å were observed, which would imply that nanotubes can be formed capped by half of a C_{60} molecule. STM images clearly resolve the honeycomb network of the surface graphene sheet. Furthermore, they observe a superpattern that they ascribe to moiré contrast between surface and subsurface graphene sheets having different helicity angles.[Ge93*]

A different set of carbon arc growth conditions is also found to produce very thin nanotubes.[Ii93, Be93] Those groups found that the presence of Fe or Co somehow catalyzed the formation of single-shelled nanotubes. Such samples may be especially useful for comparison to theories of electronic transport and mechanical properties.

Because of their small size and high polarizability, carbon nanotubes should exert a very strong capillary attraction, readily drawing small molecules into the tube. This suggestion was treated in detail by Pederson and Broughton, who used LDF to model the attraction of HF molecules into a hydrogen-terminated open nanotube.[Pe92] Experimentally, Ajayan and Iijima demonstrated that nanotubes which were annealed in the presence of liquid lead lost their caps and became filled with that metal.[Aj93] In most of the nanowires grown in this way, the metal does not appear to be crystalline. A meniscus is clearly seen in some of the samples.

In related work, it has been found that single-crystal particles of LaC_2 in the size range of 20–40 nm are formed inside of carbon shells, under appropriate growth conditions.[Ru93, To93]

It is apparent that carbon nanotubes and nanocapsules are a rapidly growing subfield of fullerene research. As the conditions for their growth become better understood and controlled, we can anticipate that significant contributions to the understanding of mesoscopic systems will emerge. Furthermore, given the great technological importance of graphite fibers, it is likely that carbon nanotubes will have significant application in the production of high strength composite materials.

Bibliography

References marked with * are reprinted in the present volume.

Aj93   P. M. Ajayan and S. Iijima, "Capillarity-Induced Filling of Carbon Nanotubes," *Nature* **361**, 333–334 (1993).

Be93   D. S. Bethune *et al.*, "Cobalt-Catalysed Growth of Carbon Nanotubes with Single-Atomic-Layer Walls," *Nature* **363**, 605–607 (1993).

Dr88   M. S. Dresselhaus, G. Dresselhaus, K. Sugihara, I. L. Spain and H. A. Goldberg, *Graphite Fibers and Filaments* (Springer-Verlag, New York, 1988).

Dr93   V. P. Dravid *et al.*, "Buckytubes and Derivatives: Their Growth and Implications for Buckyball Formation," *Science* **259**, 1601–1604 (1993).

Eb92   T. W. Ebbesen and P. M. Ajayan, "Large-Scale Synthesis of Carbon Nanotubes," *Nature* **358**, 220–222 (1992).

Ge93*   M. Ge and K. Sattler, "Vapor-Condensation Generation and STM Analysis of Fullerene Tubes," *Science* **260**, 515–518 (1993).

Ha92   N. Hamada, S. Sawada and A. Oshiyama, "New One-Dimensional Conductors: Graphitic Microtubules," *Phys. Rev. Lett.* **68**, 1579–1581 (1992).

Ii91*   S. Iijima, "Helical Microtubules of Graphitic Carbon," *Nature* **354**, 56–58 (1991).

Ii92a* S. Iijima, T. Ichihashi and Y. Ando, "Pentagons, Heptagons and Negative Curvature in Graphite Microtubule Growth," *Nature* **356**, 776–778 (1992).

Ii92b   S. Iijima, P. M. Ajayan and T. Ichihashi, "Growth Model for Carbon Nanotubes," *Phys. Rev. Lett.* **69**, 3100–3103 (1992).

Ii93   S. Iijima and T. Ichihashi, "Single-Shell Carbon Nanotubes of 1-nm Diameter," *Nature* **363**, 603–605 (1993).

Mi92   J. W. Mintmire, B. I. Dunlap and C. T. White, "Are Fullerene Tubules Metallic?" *Phys. Rev. Lett.* **68**, 631–634 (1992).

Ov93   G. Overney, W. Zhong and D. Tománek, "Structural Rigidity and Low Frequency Vibrational Modes of Long Carbon Tubules," *Z. Phys.* **D27**, 93–96 (1993).

Pe92   M. R. Pederson and J. Q. Broughton, "Nanocapillarity in Fullerene Tubules," *Phys. Rev. Lett.* **69**, 2689–2692 (1992).

Ro92   D. H. Robertson, D. W. Brenner and J. W. Mintmire, "Energetics of Nanoscale Graphitic Tubules," *Phys. Rev.* **B45**, 12592–12595 (1992).

Ru93   R. S. Ruoff, D. C. Lorents, B. Chan, R. Malhotra and S. Subramoney, "Single Crystal Metals Encapsulated in Carbon Nanoparticles," *Science* **259**, 346–348 (1993).

To93   M. Tomita, Y. Saito and T. Hayashi, "LaC_2 Encapsulated in Graphite Nano-Particle," *Japan. J. of Appl. Phys.* **32**, L280–L282 (1993).

Reprinted with permission from Nature
Vol. 354, No. 6348, pp. 56–58, 7 November 1991
© 1991 Macmillan Magazines Limited

Helical microtubules of graphitic carbon

Sumio Iijima

NEC Corporation, Fundamental Research Laboratories,
34 Miyukigaoka, Tsukuba, Ibaraki 305, Japan

THE synthesis of molecular carbon structures in the form of C_{60} and other fullerenes[1] has stimulated intense interest in the structures accessible to graphitic carbon sheets. Here I report the preparation of a new type of finite carbon structure consisting of needle-like tubes. Produced using an arc-discharge evaporation method similar to that used for fullerene synthesis, the needles grow at the negative end of the electrode used for the arc discharge. Electron microscopy reveals that each needle comprises coaxial tubes of graphitic sheets, ranging in number from 2 up to about 50. On each tube the carbon-atom hexagons are arranged in a helical fashion about the needle axis. The helical pitch varies from needle to needle and from tube to tube within a single needle. It appears that this helical structure may aid the growth process. The formation of these needles, ranging from a few to a few tens of nanometres in diameter, suggests that engineering of carbon structures should be possible on scales considerably greater than those relevant to the fullerenes.

Solids of elemental carbon in the sp^2 bonding state can form a variety of graphitic structures. Graphite filaments can be produced, for instance, when amorphous carbon filaments formed by thermal decomposition of hydrocarbon species are subsequently graphitized by heat treatment[2,3]. Graphite filaments can also grow directly from the vapour-phase deposition of carbon[4,5], which also produces soot and other novel structures such as the C_{60} molecule[6-8].

Graphitic carbon needles, ranging from 4 to 30 nm in diameter and up to 1 μm in length, were grown on the negative end of the carbon electrode used in the d.c. arc-discharge evaporation of carbon in an argon-filled vessel (100 torr). The gas pressure was much lower than that reported for the production of thicker

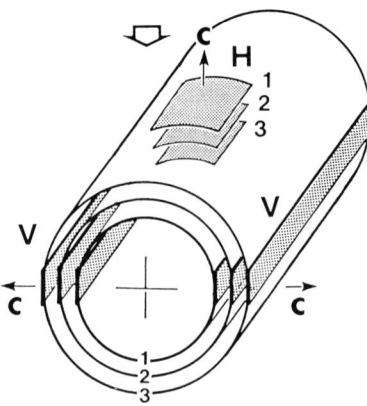

Electron beam

FIG. 2 Clinographic view of a possible structural model for a graphitic tubule. Each cylinder represents a coaxial closed layer of carbon hexagons. The meaning of the labels V and H is explained in the text.

graphite filaments[5]. The apparatus is very similar to that used for mass production of C_{60} (ref. 9). The needles seem to grow plentifully on only certain regions of the electrode. The electrode on which carbon was deposited also contained polyhedral particles with spherical shell structures, which were 5–20 nm in diameter. The needle structures were examined by transmission electron microscopy (electron energies of 200 keV).

High-resolution electron micrographs of typical needles show {002} lattice images of the graphite structure along the needle axes (Fig. 1). The appearance of the same number of lattice fringes from both sides of a needle suggests that it has a seamless and tubular structure. The thinnest needle, consisting of only two carbon-hexagon sheets (Fig. 1b), has an outer and inner tube, separated by a distance of 0.34 nm, which are 5.5 nm and 4.8 nm in diameter. The separation matches that in bulk graphite. Wall thicknesses of the tubules range from 2 to 50 sheets, but thicker tubules tend to be polygonized. This low dimensionality and cylindrical structure are extremely uncommon features in inorganic crystals, although cylindrical crystals such as serpentine[10] do exist naturally.

The smallest tube observed was 2.2 nm in diameter and was the innermost tube in one of the needles (Fig. 1c). The diameter corresponds roughly to a ring of 30 carbon hexagons; this small diameter imposes strain on the planar bonds of the hexagons and this causes two neighbouring hexagons on the ring to meet at an angle of ~6°. For the C_{60} molecule, the bending angle is 42°, which is much larger than for these tubes. The C–C bond energy calculated for the C_{60} molecule is smaller than that of graphite[11], suggesting that bending the hexagons in C_{60} lowers the bond energy. A similar effect of the bending on bonding energies might apply here. One of the key questions about the tubular structure is how the ABAB hexagonal stacking sequence found in graphite is relaxed, as it is impossible to retain this ideal graphite structure for coaxial tubes. There should be a shortage of 8–9 hexagons in going from one circumference of a tube to that inside it. Disordered graphitic stacking is known as turbostratic stacking, but no detailed accounts of stacking patterns in such structures have been reported. The argument here is also applicable to the spherical graphitic particles mentioned earlier[6].

All the electron diffraction patterns (Fig. 3) taken from individual carbon needles are indexed by the {h0l} and {hk0} spots for hexagonal symmetry. The patterns always show strong (00l) spots when the needle axes are perpendicular to the [001] axis, supporting the idea of a coaxial arrangement of graphitic tubes. As shown in Fig. 2, two side portions of each tube (indicated by shading and labelled 'V') will be oriented so that the

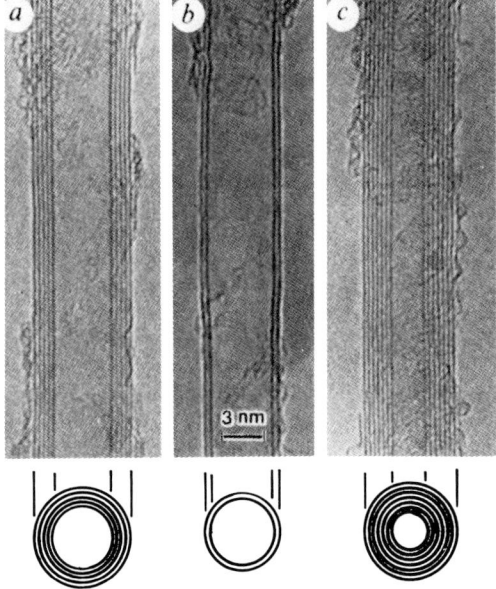

FIG. 1 Electron micrographs of microtubules of graphitic carbon. Parallel dark lines correspond to the (002) lattice images of graphite. A cross-section of each tubule is illustrated. a, Tube consisting of five graphitic sheets, diameter 6.7 nm. b, Two-sheet tube, diameter 5.5 nm. c, Seven-sheet tube, diameter 6.5 nm, which has the smallest hollow diameter (2.2 nm).

(002) planes satisfy the Bragg diffraction condition for the incident electron beam, and thus give (00*l*)-type spots. Individual (002) planes in these portions are directly imaged in Fig. 1.

The {*hk*0} patterns as a whole show *mm*2 mirror symmetry about the needle axis, and consist of multiple sets of {*hk*0} spots. For example, three sets of (*hk*0) spots seem to form ring patterns, only (100)- and (220)-type spots being seen (Fig. 3*a*). The diffraction pattern was obtained from a single tube consisting of seven sheets, confirmed by examining the electron microscope image. Referring again to Fig. 2, the top portion of the outermost tube, labelled 'H', and its counterpart on the bottom of the tube, give independently one set each of {*hk*0} patterns. If these two portions of the cylinder have the same orientation, they produce an identical {*hk*0} pattern. If three hexagon sheets on the tubes 1, 2 and 3 were oriented differently, they would give six different {*hk*0} spot patterns. Such a top–bottom effect is one of the requirements for the mirror symmetry. Another requirement is a helical arrangement of carbon hexagons on individual tubes, described further below.

Consider rotation of individual graphite sheets with respect to the needle axes. To explain the graphitic tube structure, the tube is cut at one side along the needle axis and unrolled. This is illustrated schematically in Fig. 4*a*. Fewer hexagons are drawn, than would form a real tube, but the essential needle geometry correctly represents one of our experimental diffraction patterns. A cylindrical tube can be formed by rolling up the hexagonal sheet about the filament axis (drawn by the heavy line) so as to superimpose hexagons labelled A and B at the top on A' and

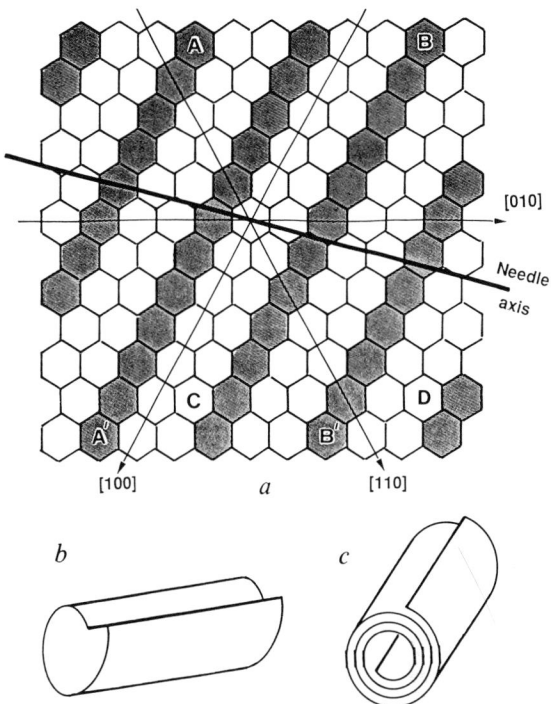

FIG. 4 *a*, Schematic diagram showing a helical arrangement of a graphitic carbon tubule, which is unrolled for the purposes of explanation. The tube axis is indicated by the heavy line and the hexagons labelled A and B, and A' and B', are superimposed to form the tube (for the significance of C and D, see text). *b*, The row of hatched hexagons forms a helix on the tube. The number of hexagons does not represent a real tube size, but the orientation is correct. *c*, A model of a scroll-type filament.

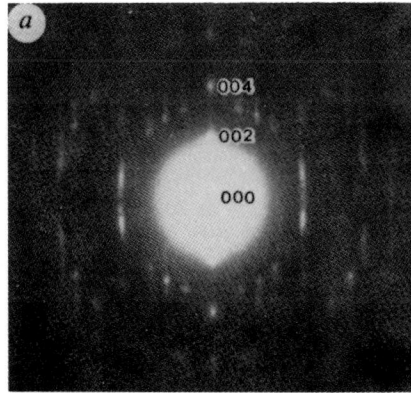

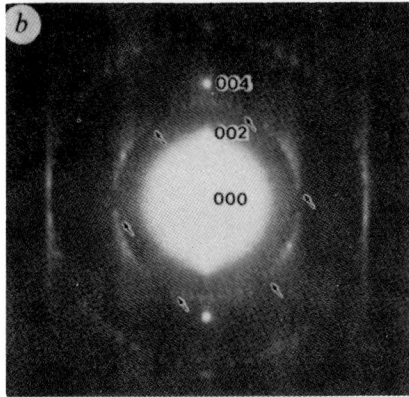

FIG. 3 Electron diffraction patterns from individual microtubules of graphitic carbon. The patterns show *mm*2 symmetry, and are indexed by multiple superpositions of {*h*0*l*}-type reflections and {*hk*0} reflections of graphite crystal. The needle axes are horizontal. *a*, Superposition of three sets of {*hk*0} spots taken from a seven-sheet tubule. *b*, Superposition of four sets of {*hk*0} spots from a nine-sheet tubule.

B' at the bottom. It will be found that the hatched hexagons are aligned perfectly to make a helix around the needle axis (see Fig. 4*b*). The helical arrangement of the hexagons is responsible for the mirror symmetry as observed in the experimental diffraction patterns. One complete spiral rotation leaves a pitch three hexagons in height at the tip of the needle. There are many possible pitches in the helix, depending on the orientation of the sheet with respect to the needle axis. In other words, the orientation of a hexagon sheet can be determined uniquely by referring to the needle axis. If hexagons A and B coincide with those labelled C and D, the needle axis will be along [010] and thus there will be no spiral rows of hexagons. This has, however, rarely been observed.

Because of the helical structure, the {*hk*0} spot patterns should always contain sets of even numbers of spots. Occasionally, sets of odd numbers of spots occur, such as the groups of three shown in Fig. 3*a*. These can be explained by frequent coincidence of the top and bottom sheet orientations. Such an accidental coincidence will be increased for hexagonal symmetry of individual hexagon sheet when it is rolled. Consider the sets of four of {*hk*0} in Fig. 3*b*. Each of the sets of spots, which have equal intensity distributions, is rotated by ~6° about the (000) origin, or the needle axis. We take one set of (*hk*0) spots (indicated by arrows), corresponding to one of the hexagon sheets, as a reference whose [100] axis is rotated by 3° about the needle axis. The other three sets of spots are then rotated by 6°, 12° and 24° from the reference sheet. Considering the fact that the needle consists of only nine sheets and each set of {*hk*0} spots is generated from only three or four sheets, it is reasonable that every three or four sheets are rotated stepwise by 6° about the *c* axis. Any translational shift of the sheets cannot be detected in the electron diffraction patterns. A systematic change in sheet orientations was confirmed by (*h*0*l*)-type lattice image observations. The question of whether the systematic variation in the rotation angles in successive tubules acts to stabilize the structure remains to be answered.

The tips of the needles are usually closed by caps that are curved, polygonal or cone-shaped. The last of these have specific opening angles of about 19° or 40°, which can be rationalized in terms of the way that a perfect, continuous hexagonal network can close on itself.

According to Bacon's scroll model for tubular needle growth, needles could be formed by rolling up single carbon-hexagon sheets to form tubular filaments as illustrated in Fig. 4c. Such filaments should have edge overlaps on their surfaces. But I have observed no overlapping edges for the needles described here. Instead, I have observed concentric atomic steps around the needles (I have not confirmed that these are helical). On the basis of these new experimental findings on needle morphologies, I propose a new growth model for the tubular needles. That is, individual tubes themselves can have spiral growth steps at the tube ends (Fig. 4b). It is worth mentioning that the spiral growth steps, which are determined by individual hexagon sheets, will have a handedness. The growth mechanism seems to follow a screw dislocation model analogous to that developed for conventional crystals, but the helical structure is entirely different from the screw dislocation in the sense that the present crystals have a cylindrical lattice. □

Received 27 August; accepted 21 October 1991.

1. Kroto, H. W., Heath, J. R., O'Brien, S. C., Curl, R. F. & Smalley, R. E. *Nature* **318,** 162–163 (1985).
2. Oberlin, A. & M. Endo *J. Cryst. Growth* **32,** 335–349 (1976).
3. Speck, J. S., Endo, M. & Dresselhaus, M. S. *J. Cryst. Growth* **94,** 834–848 (1989).
4. Tibbetts, G. G. *J. Cryst. Growth* **66,** 632–638 (1984).
5. Bacon, R. *J. appl. Phys.* **31,** 283–290 (1960).
6. Iijima, S. *J. Cryst. Growth* **50,** 675–683 (1980).
7. Iijima, S., *J. phys. Chem.* **91,** 3466–3467 (1987).
8. Kroto, H. W. *Science* **242,** 1139–1145 (1988).
9. Krätschmer, W., Lamb, L. D., Fostiropoulos, K. & Huffman, D. R. *Nature* **347,** 354–358 (1990).
10. Whittaker, E. J. W. *Acta Cryst.* **21,** 461–466 (1966).
11. Saito, S. & Oshiyama, A. *Phys. Rev. Lett.* **66,** 2637–2640 (1991).

ACKNOWLEDGEMENTS. I thank Y. Ando for the carbon specimens.

Reprinted with permission from Nature
Vol. 356, No. 6372, pp. 776–778, 30 April 1992

Pentagons, heptagons and negative curvature in graphite microtubule growth

Sumio Iijima*, Toshinari Ichihashi* & Yoshinori Ando†

* Fundamental Research Laboratories, NEC Corporation, 34 Miyukigaoka, Tsukuba, Ibaraki 305, Japan
† Department of Physics, Meijo University, Nagoya 468, Japan

THE standard carbon-arc synthesis for fullerenes also produces graphitic microtubules with helical structures[1]. In most cases the cylindrical tubes are closed by polyhedral caps, some being first transformed into a conical shape before closure[2]. Here we present images from transmission electron microscopy of a further kind of growth morphology, in which cone-like growth is transformed into cylindrical growth by the incorporation of a defect that induces negative curvature. We suggest that the defect in the hexagonal network responsible for negative curvature may be a single heptagonal ring. Three-dimensional, open negatively curved graphitic structures have been proposed recently by Terrones and Mackay[3]. We discuss more generally the effect of pentagons and heptagons on the growth morphologies of these tubules, and the constraints on the number of pentagonal defects imposed by tube closure.

Graphitic microtubules are formed in the carbon deposit in a carbon arc. We find that the deposit is composed mainly of pyrolytic graphite with no amorphous carbon, as the high reaction temperature (over 2,500 °C) in the arc favours graphitization. The tubules are usually attached to the pyrolytic graphite at one end, the other end being free. This suggests that the tubules grow at their free ends, as is common in whisker growth. The tubules are not always cylindrical; we have seen many variations in shape, especially near the tube tips, and have classified the morphologies into several groups. Polyhedral graphitic particles coexist with the tubules in the carbon deposits. The particles are thought to be formed as a consequence of homogeneous growth in all directions, whereas the tubes grow in one direction because of a helical structure in the carbon hexagon network[1]. The growth nuclei controlling the growth morphologies of the graphitic particles have not yet been understood.

Among many tube-tip shapes, one that we repeatedly observed in the transmission electron microscope (TEM) is represented in Fig. 1. The tubule, hollow at the centre, comprises three portions: on the right, a nanometre-sized tube of five graphitic cylinders with its tube-tip closed by polyhedral surfaces; in the middle a cone with stacked 'ice-cream cone' terminations of inner graphitic shells; and on the left, a wider tube which extends towards the left.

We frequently observed nearly regular cone-like terminations which were more common in thicker tubes. Cone angles seen in TEM images (Fig. 2) are typically ~20°, varying over the range 15–30° (refs 1, 2). The angle corresponds to that of a cone made by connecting two sides of a regular triangle as illustrated in Fig. 3a. The hexagon sheet is perfectly continuous all around the cone surface. To close the apex, five pentagons are required in the positions indicated by open circles; the closed cone is shown in Fig. 3b. In general, the cones observed are 3–5 times larger than this example. A similar cone shape has been observed in graphitic carbons prepared by thermal decomposition of aromatic hydrocarbons (M. Endo and H. Kroto, personal communication). They thought that the cone at the tube-tip was always closed but could grow by breaking some of the carbon bonds.

The (002) lattice fringes appearing near the tube-tips at bottom of Fig. 1 run horizontally through all three sections (the thick tube, cone and thin tube). This means that the graphitic sheets in this side of the tube consist of perfectly continuous hexagon networks. Where the top side of the tube forms a tapered step (middle portion), the lattice fringes run parallel to this step. The number of lattice fringes, and thus the wall thickness of the tube, vary from location to location but are always the same at both sides of the central hollow. This also applies to the cones shown in Fig. 2. The presence of equal numbers of fringes is due to the fact that the individual graphite sheets forming the multi-shell structure are perfectly continuous even around the transition region (across the cone). It is sufficient, therefore, to consider only the outermost shell in the following discussion, as each shell is more or less the same shape.

An immediate question about the tube shape is how the small and large cylinders are connected topologically to the ends of the cone without disturbing the continuity of the graphite sheets.

Considering the characteristic morphologies, we constructed a topological model in which pentagons and heptagons play a key role in the tube-tip shapes. An apex such as that indicated by arrow A in Fig. 1 must incorporate a single pentagon into a hexagon sheet as illustrated in Fig. 4a. The illustration is a projection of the topological surface on the plane of the page. The network, which has 5-fold symmetry, can be obtained by removing the shaded hexagons in the regular triangle shown in Fig. 4b. The structure proposed here is a kind of defect in a topological surface, described as a +60° wedge disclination[4,5]. A surface containing a pentagon is curved positively. The topological surface transforming the tube into a cone can be achieved by superimposing the hexagon edges indicated by heavy lines at both sides of the network in Fig. 4a. The shaded hexagons form the cone shape, and the remaining hexagons are responsible for the tube. In the tube structure, the ideal 5-fold symmetry about the pentagon is lost and mirror symmetry appears about a vertical line. The mirror symmetry, however, will also be

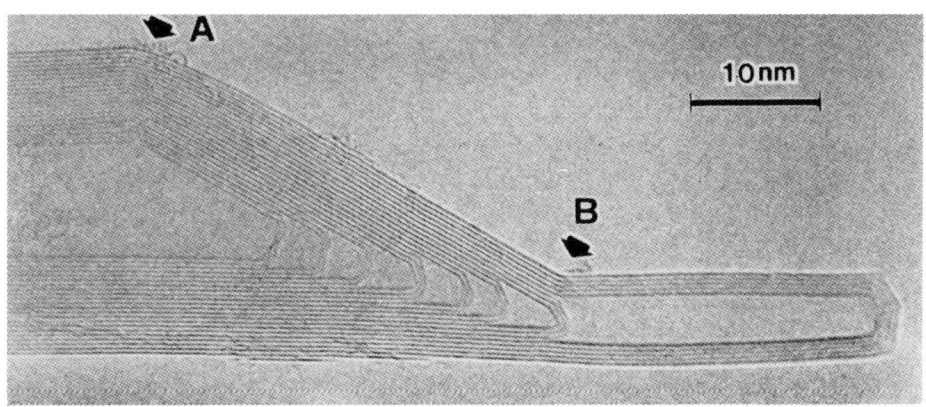

FIG. 1 Electron micrograph showing recurrent terminations of graphitic microtubules consisting of three portions: a thin cylindrical tube (right), a cone (middle) and a thicker cylindrical tube (left). Individual dark lines (lattice images) correspond to graphitic sheets separated by 0.34 nm. The internal cone are often closed in pairs. Regions indicated by A and B contain disclinations.

10 nm

A

B

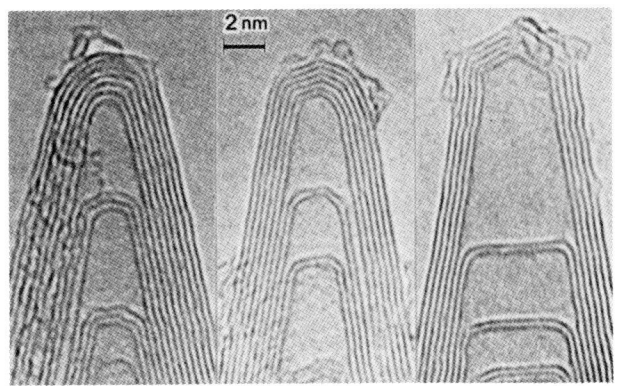

FIG. 2 Terminations with identical conical caps at the tube-tips. Some caps are faceted. The cone angles are roughly 20°, close to that for a cone made by connecting two sides of an equilateral triangle. Shells made up of two or three graphitic sheets are stacked like ice-cream cones. We believe that this is the final form in the cone growth. The caps of individual cones are expected to have five pentagons according to Euler's theorem.

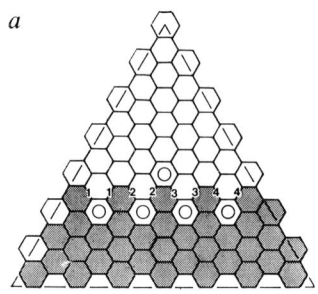

FIG. 3 *a*, The 20° cone shape of Fig. 2 can be reproduced by joining two triangle edges of this carbon hexagon network. *b*, Model for the smallest 20° cone tip, viewed along the cone axis, which has mirror symmetry with five pentagons. The structure is formed by connecting shaded hexagon edges labelled with the same members in *a*. Circles indicate positions where the pentagons are formed.

destroyed in most tubes, as they usually have a helical (and chiral) structure. A helical tube can be formed by joining two sides of the network shown in Fig. 4*b* after sliding one side up or down with respect to the other. Introduction of a pentagon into a hexagon network on the cylinder will result in a strain around the pentagon. This strain effect is noticeable as a slight distortion to the cylinder at A in Fig. 1, indicating that the cross-section of the tube at A is not perfectly circular but oval.

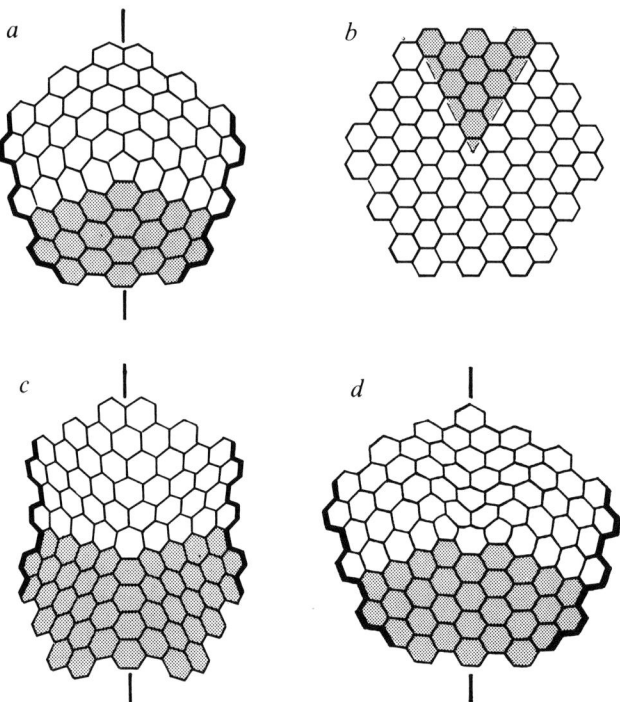

FIG. 4 *a*, A +60° disclination produced by introducing a pentagon into a hexagon network. This takes place at position A in Fig. 1. A three-dimensional network including the portion A can be reproduced by joining the two sides indicated by heavy lines. Shaded hexagons form the 20° cone and the rest form a tube. *b*, The +60° disclination is formed by removing shaded hexagons. *c*, A −60° disclination produced by including a heptagon, to form a saddle-shaped surface (negative curvature) like that at B in Fig. 1. Its three-dimensional structure can be obtained by joining hexagon edges drawn in heavy lines. *d*, Complex disclination, consisting of one heptagon and two pentagons, which is equivalent to the single +60° disclination in *a*.

The opposite situation occurs at the point indicated by arrow B, where the graphitic shell is folded back to lie in the same direction as in the thicker tube. The hexagon network of this region was obtained by introducing a −60° wedge disclination, in which a single heptagon is formed (Fig. 4*c*). Insertion of an extra regular triangle produces a negative surface[4,5]. The network has 7-fold symmetry, but topologically it has a saddle shape with mirror symmetry about a vertical line. Mackay and Terrones[3], and Lenosky and others[6,7], have pointed out the role of seven- or eight-membered rings in forming periodic minimal surfaces from carbon hexagon networks.

The three-dimensional shape corresponding to portion B in Fig. 1 can be obtained by rolling back together the two sides of the hexagon network indicated by heavy lines in Fig. 4*c*. The presence of both a +60° disclination and a −60° disclination on a closed hexagon surface restores the original shape, in this case a cylinder. A combination of two +60° disclinations and one −60° disclination results in the equivalent to a single +60° disclination (Fig. 4*d*), which is another possible candidate for the defect at A in Fig. 1. Euler's theorem leads to a relation $P = S + 12$, where P and S are the numbers of pentagons and heptagons incorporated into a closed hexagonal network. For $S = 1$, we have $P = 13$. In other words, the number of pentagons should always exceed the numbers of heptagons by 12 in any closed surface of the hexagonal network. Such a surface will contain negative curvature. Very complicated morphologies of graphitic spheroidal particles may be caused by these pentagons and heptagons and their combinations[8].

In short, the occurrence of a single pentagon at the open end of a tube leads inevitably to a cone shape, and eventually the cone tip will become closed as illustrated in Fig. 3*b*, unless the formation of a single heptagon on the cone restores a cylinder. We conclude that the tube can continue to grow as long as only hexagons are being formed on the open end of the tube. Formation of a hexagon requires two carbon atoms to be added at each kink site on the periphery, whereas a single atom is required for the pentagon. One possible reason for the formation of pentagons seems to be a drop in carbon supply during carbon-arc evaporation. In the normal evaporation experiment, the arc is fairly unstable and typically lasts for several tens of seconds, after which the carbon-arc electrodes must be realigned for further evaporation. We believe that flux density of carbon atoms decreases towards the end of the arc discharge, and as soon as the arc stops, the tube-tips are closed quickly by forming pentagons. In contrast, a higher flux of carbon atoms may increase the rate of formation of heptagons, which need three carbon atoms at each kink site.

Once a cap is formed on the tip of a cylinder tube, it can

grow no longer. Capping the cylinder tube tips should be completed by clustering six pentagons near the tube tips according to Euler's theorem. The apices of the polyhedral shapes making up the tube tips (Fig. 1), cone tips (Fig. 2) and many spheroidal particles[8,9] are very probably caused by pentagons. To cap a cone-tip, five pentagons will be sufficient because one pentagon has already been used for the $+60°$ disclination. Variations in the polyhedral shapes reflect many possible distributions of the pentagons on the tube tips, unlike the more symmetrical distribution of pentagons in the truncated icosahedron of C_{60}. □

Received 20 February; accepted 18 March 1992.

1. Iijima, S. Nature **354,** 56–58 (1991).
2. Nature **354,** 18 (1991).
3. Mackay, A. L. & Terrones, H. Nature **352,** 762 (1991).
4. Mackay, H. L. Nature **314,** 604–606 (1985).
5. deWit, R. J. appl. Phys. **42,** 3304–3308 (1971).
6. Lenosky, T., Gonze, X., Teter, M. & Elser, V. Nature **355,** 333–335 (1992).
7. Vanderbilt, D. & Tersoff, J. Phys. Rev. Lett. **68,** 511–513 (1992).
8. Iijima, S. J. Cryst. Growth **50,** 675–683 (1980).
9. Kroto, H. Science **242,** 1139–1145 (1988).

ACKNOWLEDGEMENTS. We thank S. Sawada for discussions on tube formation.

Reprinted with permission from Science
Vol. 260, pp. 515–518, 23 April 1993
© 1993 The American Association from The Advancement of Science

Vapor-Condensation Generation and STM Analysis of Fullerene Tubes

Maohui Ge and Klaus Sattler

Fullerene tubular structures can be generated by vapor condensation of carbon on an atomically flat graphite surface. Scanning tunneling microscope (STM) images revealed the presence of tubes with extremely small diameters (from 10 to 70 angstroms), most of which are terminated by hemispherical caps. Atomic resolution images of such structures showed that the tubes have a helical graphitic nature. The formation of the tubes under the quasi-free conditions suggests that the growth to tubular rather than spherical configurations is preferred for "giant fullerenes."

Despite the tremendous interest recently in the physics and chemistry of C_{60} and other fullerenes and their solid forms, little is known about fullerene growth. Neither the onset of nucleation nor the progression toward the fullerene network is understood. The structure of giant fullerenes (containing hundreds of carbon atoms) is also controversial; experimental evidence exists to support two possibilities, spherical molecules versus tubular cages.

Concentric tubular carbon structures were recently found by transmission electron microscopy at the end of carbon rods used for arc discharge [1, 2], and such structures could be produced in macroscop-

Department of Physics and Astronomy, University of Hawaii, Honolulu, HI 96822.

ic quantities by using a similar generation method [3]. The smallest inner tube that was observed [1] had a diameter of 22 Å. Electron diffraction pattern revealed helical arrangement for concentric graphitic networks. The tubes were grown out from a carbon substrate. Therefore a direct comparison to the free fullerene growth could not be made. Although various properties of the tubules were measured, the atomic structure could not be directly determined.

Recent calculations predict that carbon tubules of different diameters and helicities have striking variations in electronic transport, from metallic to semiconducting [4–7]. Also, such tubules are expected to shield guest atoms from external electric and magnetic fields [8]. Besides tubular structures, other low-energy configurations such as negative-curvature fullerene analogs were also suggested (9–11).

We have produced carbon tubules by quasi-free vapor condensation. This method is different from the previously reported generation method of carbon tubes (1, 2). Atomic resolution images obtained with a scanning tunneling microscope (STM) reveal that the detailed structures of the tube surfaces are networks of perfect honeycombs. In addition, we observe a superpattern on the tubes due to an incorporated inner tube with different helicity. The smallest tube imaged in our experiments has a diameter of ~10 Å, which is of the size of C_{60}. We suggest that the tubule growth starts with the formation of a fullerene hemisphere.

Our samples were prepared by vapor deposition of carbon on highly oriented pyrolytic graphite (HOPG) in ultrahigh vacuum (UHV) at a base pressure of ~10^{-8} torr. The use of graphite as a substrate might lead to problems of distinguishing adlayer features from steps, flakes, or other substrate features (12). This possibility, however, can be excluded with our preparation procedure. The graphite (grade-A HOPG) was freshly cleaved in UHV and carefully examined by STM before the deposition. The HOPG surface was atomically flat and defect-free over micrometer dimensions. It was cooled to −30°C during evaporation. The carbon vapor was produced by resistively heating a carbon foil. The deposition rate was monitored by a quartz crystal film-thickness monitor. The average thickness of the carbon adlayers ranged from 20 to 60 Å. After deposition, the samples were transferred to an STM operated at ~10^{-10} torr without breaking vacuum. The STM imaging was performed at room temperature. The microscope was usually operated in the constant-current mode in which the tip-to-sample distance is kept constant by means of an electronic feedback control. The atomic-scale images were obtained in the variable-current mode. Bias voltages of 100 to 500 mV (both positive and negative) and tunneling currents of 0.5 to 3.0 nA were applied. A mechanically shaped Pt-Ir tip was used.

Carbon tubes with diameters between 10 Å and 70 Å and up to 2000 Å in length were found. Three STM images of such tubes are shown in Fig. 1. The cylindrical shape is well displayed. Whereas Fig. 1, A and B, shows parts of much longer tubes, Fig. 1C shows smaller ones. Most of the tubes are terminated by hemispherical caps. Two of them in Fig. 1B are broken at the lower left corner of the image.

In some cases we observed a coaxial arrangement of the outermost and an inner tube (upper right corner, Fig. 1B). The outer tube is terminated in this region, and

the adjacent inner one is imaged simultaneously. We measure an interlayer spacing of 3.4 Å, which is about the graphite interlayer distance (3.35 Å). Such possible coaxial arrangement was confirmed by atomic-scale imaging.

We found that the tubes are placed almost horizontally on the substrate. Irregular nanostructures were also formed as displayed in the images. However, the high occurrence (>50%) of tubes clearly shows that the carbon atoms prefer to condense to tubular structures, as opposed to other nanostructures, under our preparation conditions.

In Fig. 2A we show an atomic resolution image of a carbon tube, 35 Å in diameter. The structure imaged at the upper right corner of the picture comes from another tube. Both of them were ~1000 Å long. A perfect honeycomb surface structure is observed. By taking into account the curvature of the tube surface and the STM imaging profile, we find the same lattice

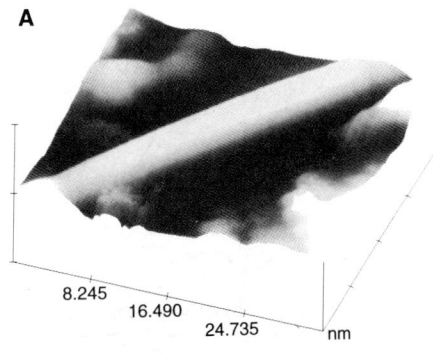

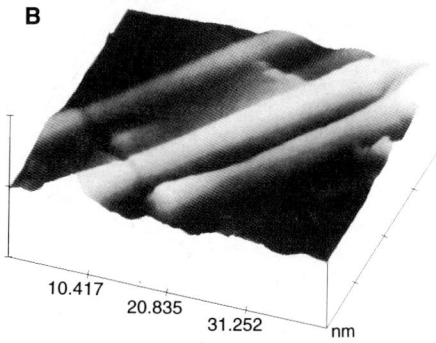

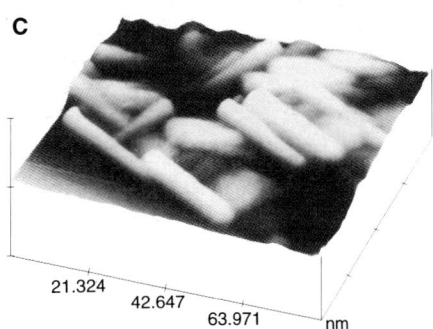

Fig. 1. Three STM images of fullerene tubes on a graphite substrate.

parameter as that of graphite (1.42 Å), demonstrating that the tubular surface is a graphitic network.

One question arises if the observed tubules are scroll-type filaments or perfect cylinders. Scroll-type filaments should have edge overlaps on their surfaces. During our extensive STM survey along the tube surfaces on the atomic scale, we did not observe any edges due to incomplete carbon layers. Therefore we conclude that the tubes are complete graphitic cylinders.

For the tube in Fig. 2A, the honeycomb surface net is arranged in a helical way. A chiral angle of 5.0° ± 0.5° is obtained. The chiral angle is defined as the smallest angle between the tube axis and the C–C bond directions of the honeycomb lattice. We determined the tube axis from a large-scale image.

For the formation of carbon tubes, spherical-type nucleation seeds seem to be required. In a common method for the production of tubular carbon fibers, the growth is initiated by submicrometer-size catalytic metal particles (13). Tube growth out of a graphite rod during arc discharge might also be related to nanoparticle-like seeds present at the substrate (2). Under conditions that resemble those for the production of fullerenes, another type of nucleation seed is possible. In this case, the tubes may grow out of a hollow fullerene cap rather than out of a compact spherical particle. After one-half of a fullerene is formed, the subsequent growth can proceed to a tubular structure rather than to a spherical cage.

Indeed, the smallest tube that we measured had a diameter of 10 Å, which is of the size of C_{60}. It is predicted to be the limiting case of vapor-grown graphitic tubes with monolayer thickness (14). Our observation of hemispherically capped 10 Å tubes suggests that an incomplete C_{60} cluster is the nucleation seed for these tubes. The C_{60}-derived tube could be the core of possible multilayer concentric graphitic tubes. After the fullerene-based tube has been formed, further concentric shells can be added by graphitic cylindrical layer growth.

A "ball-and-stick" model of a C_{60+10j} ($j = 1, 2, 3, \ldots$) tube with C_5 symmetry (Fig. 2B) yields a diameter of 6.83 Å. By considering the charge distribution, the resulting outer diameter is 9.6 Å (14), which is nearly the same as the smallest one observed in our experiment. In the model, one end of the tube is closed by a C_{60} hemisphere. The cap contains six pentagons, and the body part of the tube contains only hexagons. The hexagonal rings are arranged in a helical fashion with a chiral angle of 30° ("armchair configuration"). One obtains this result when the tube axis is the fivefold symmetric axis of the C_{60} cap. Other choices of symmetry axes for C_{60}

caps lead to slightly larger tube diameters and different helicities.

For the structural consideration, the graphitic monolayer tube can be treated as a rolled-up graphite sheet that matches perfectly at the closure line. Choosing the cylinder joint in different directions leads to different helicities. One single helicity gives a set of discrete diameters. In order to obtain the diameter that matches exactly the required interlayer spacing, the tube layers need to adjust their helicities. Therefore, in general, different helicities for different layers in a multilayer tube are expected and were indeed found in our experiment.

In Fig. 2A we observe a zigzag superpattern in addition to the atomic lattice. It appears along the axis at the surface of the tube. The zigzag angle is 120°, and the period is ~16 Å. This superstructure can be considered as part of a honeycomb-type giant lattice. Such giant lattices have been observed for plane graphite when imaged by STM (15–20). They can be explained by misorientation of the top layer relative to the second layer (15–17), which results in a Moire pattern with the same type of lattice structure but a much larger lattice parameter. The angle of misorientation determines the lattice constant. Because of the non-equivalent atomic stacking, the local electron density of states at the Fermi level of the top layer is modulated (16). Therefore the STM can image a giant lattice in addition to the atomic lattice and can give the relative orientation of the first two layers. For multilayer tubes with different helicities between the top and the second layer, one expects such a superpattern to be observed. The helicity of the second layer can be determined from the period of the superstructure.

However, while there is an analogy between two misoriented graphite sheets and two concentric cylindrical graphitic sheets of different helicity, there is also a difference. For planar sheets, the superpattern is extended throughout the whole plane, whereas for cylindrical sheets it should appear only within narrow stripes along the tube axis. This effect occurs because the curvature is different between the two cylinders, which affects the atomic stacking mainly in the direction perpendicular to the tube axis.

In Fig. 2C we illustrate a schematic model of such a superpattern. Two graphite sheets that are rotated relative to each other are superimposed, which results in a giant honeycomb lattice. By choosing an angle of 9° for the misorientation, we obtain a superlattice period of 16 Å.

In general, from such modeling one finds that for small chiral angles of both cylinders the tube axis is approximately along the zigzag direction of the giant honeycomb lattice. The zigzag superpattern along the

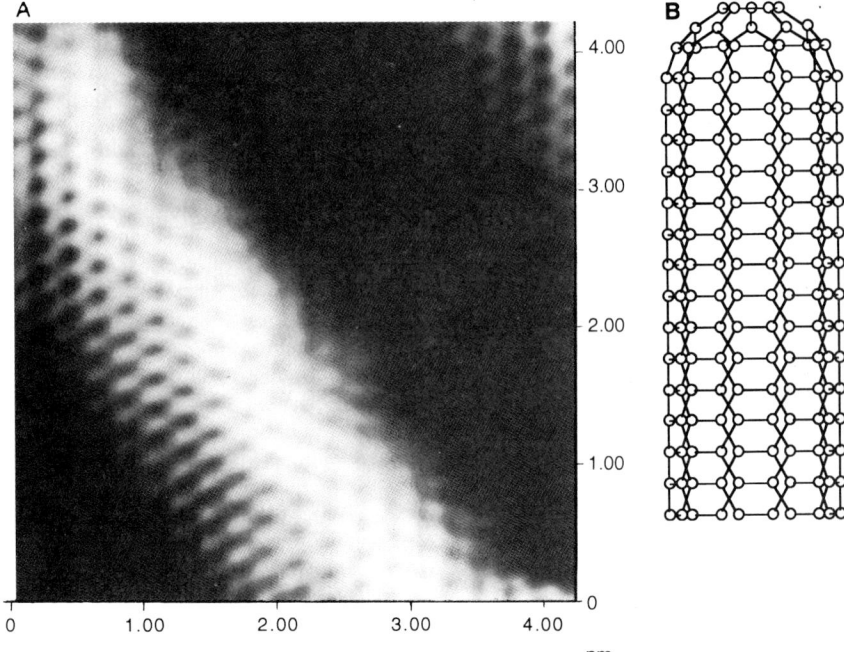

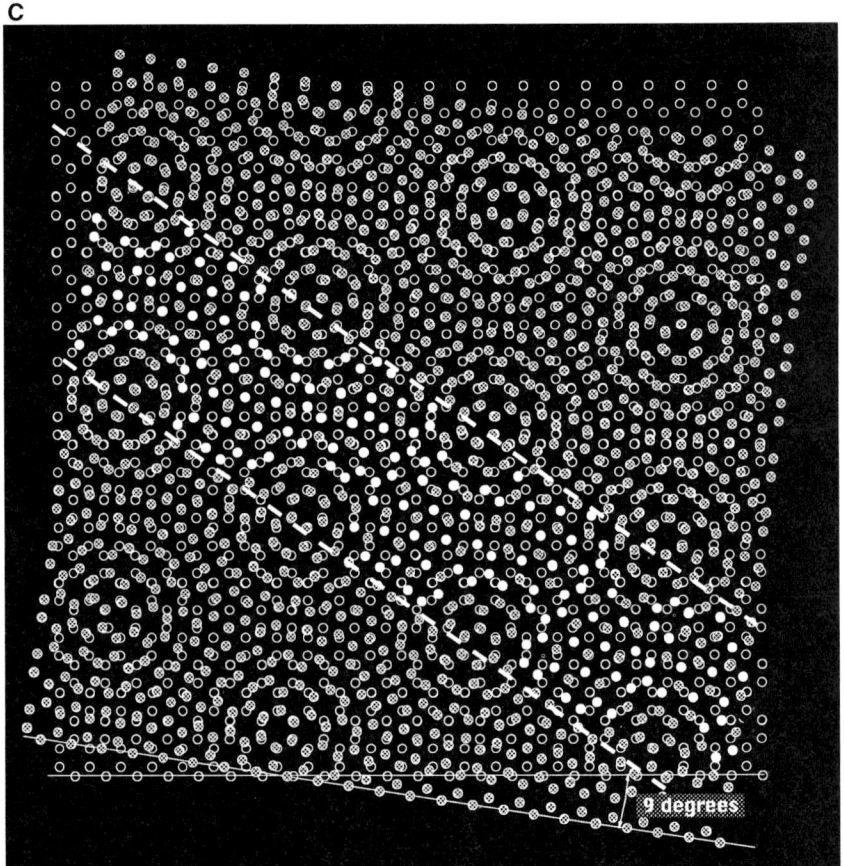

Fig. 2. (A) Atomic resolution STM image of a carbon tube, 35 Å in diameter. In addition to the atomic honeycomb structure, a zigzag superpattern along the tube axis can be seen. **(B)** "Ball-and-stick" structural model of a C_{60}-based carbon tube. The upper part is closed by a C_{60} hemisphere cap. **(C)** Structural model of a giant superpattern produced by two misoriented graphitic sheets. The carbon atoms in the first layer are shaded, and the second layer atoms are open. Between the two dashed lines, we highlight those first layer atoms with white that do not overlap with second layer atoms. Because of their higher local density of states at the Fermi level, these atoms (β-type atoms) appear particularly bright in STM images (16, 21). It results in a zigzag superpattern along the tube axis within the two white dashed lines as indicated.

518

tube axis is directly observed in the STM image of Fig. 2A. The measured period of 16 Å reveals that the second layer of the tube is rotated relative to the first layer by 9°. The first and the second cylindrical layers have chiral angles of 5° and −4°, respectively, showing that the tubes are indeed composed of at least two coaxial graphitic cylinders with different helicities.

In our experiments, hot vapor is suddenly quenched on a cold surface. The substrate is atomically flat and chemically inert. This leads to cooling and supersaturation of the incoming vapor on a nearly noninteracting surface. The resulting quasi-free growth conditions resemble those for vapor-phase fullerene formation. This method gives a new approach to understand the formation of novel hollow graphitic structures. Our observations suggest that the growth to tubular rather than spherical configurations is preferred for "giant fullerenes."

REFERENCES AND NOTES

1. S. Iijima, *Nature* **354**, 56 (1991); _____, T. Ichihashi, Y. Ando, *ibid.* **356**, 776 (1992).
2. S. Iijima, P. M. Ajayan, T. Ichihashi, *Phys. Rev. Lett.* **69**, 3100 (1992).
3. T. W. Ebbensen and P. M. Ajayan, *Nature* **358**, 220 (1992).
4. R. Saito, M. Fujita, G. Dresselhaus, M. S. Dresselhaus, *Appl. Phys. Lett.* **60**, 2204 (1992).
5. N. Hamada, S. Samada, A. Oshiyama, *Phys. Rev. Lett.* **68**, 1579 (1992).
6. J. W. Mintmire *et al.*, *ibid.*, p. 631.
7. K. Harigaya, *Phys. Rev. B* **45**, 12017 (1992).
8. M. Peterson and J. Q. Broughton, *Phys. Rev. Lett.* **69**, 2689 (1992).
9. T. Lenosky *et al.*, *Nature* **355**, 333 (1992).
10. D. Vanderbilt and J. Tersoff, *Phys. Rev. Lett.* **68**, 511 (1992).
11. M. O'Keeffe *et al.*, *ibid.*, p. 2325.
12. C. R. Clemmer and T. P. Beebe, Jr., *Science* **251**, 640 (1991).
13. G. Tibbetts, *J. Cryst. Growth* **66**, 632 (1984).
14. M. S. Dresselhaus, G. Dresselhaus, R. Saito, *Phys. Rev. B* **45**, 6234 (1992).
15. M. Kuwabara, D. R. Clarke, D. A. Smith, *Appl. Phys. Lett.* **56**, 2396 (1990).
16. J. Xhie, K. Sattler, M. Ge, N. Venkatasmaran, *Phys. Rev. B*, in press.
17. P. I. Oden *et al.*, *Surf. Sci. Lett.* **254**, L454 (1991).
18. J. W. Lyding *et al.*, *J. Vac. Sci. Technol.* **A6**, 363 (1988).
19. J. E. Buckley *et al.*, *ibid.* **B9**, 1079 (1991).
20. V. Elings and F. Wudl, *ibid.* **A6**, 412 (1988).
21. D. Tomanek *et al.*, *Phys. Rev. B* **35**, 7790 (1987).
22. We thank N. Venkateswaran and J. Xhie for helpful discussions. Financial support from the National Science Foundation, grant DMR-9106374, is gratefully acknowledged.

VIII

EMERGING APPLICATIONS

The last three years have seen an explosive growth in research on fullerenes. Since it became possible to obtain bulk quantities of C_{60} and related materials, our knowledge of their properties has greatly increased. It is therefore appropriate to consider the question, "Yes, they're fascinating, but what are they good for?"

The last four papers reprinted in this volume describe suggestions for useful applications of fullerenes: the nucleation of diamond films on surfaces, their use as an optical limiter, the incorporation of fullerenes into photoconducting polymers, and the possible use of fullerene derivatives as drugs to combat AIDS. At this time, it is impossible to say if any of these schemes will be of practical importance, but they do point to the utility of some of the properties of fullerenes that have emerged from pure, basic research.

Reprinted with permission from Applied Physics Letters
Vol. 59, No. 26, pp. 3461–3463, 23 December 1991
© 1991 American Institute of Physics

Nucleation of diamond films on surfaces using carbon clusters

R. J. Meilunas and R. P. H. Chang
Department of Materials Science and Engineering, Northwestern University, Evanston, Illinois 60208

Shengzhong Liu and Manfred M. Kappes
Department of Chemistry, Northwestern University, Evanston, Illinois 60208

(Received 18 July 1991; accepted for publication 25 October 1991)

A unique method for nucleating diamond films on surfaces using C clusters is described. The process substitutes the need for diamond polish pretreatment of substrates prior to diamond film growth, as currently practiced in low-pressure (< 1 atm) chemical vapor deposition methods. As an example, the use of C clusters C_{60} and C_{70} as nucleating layers on single-crystal Si surfaces is presented. It is shown that a thin layer (approximately 1000 Å) of pure carbon C_{70} is sufficient for the nucleation and growth of fine grain polycrystalline diamond films. The enhancement of nucleation by the C_{70} layer is nearly ten orders of magnitude over an untreated Si surface. It also follows that C clusters can be used as a one-step lithographic template for growing diamond on selected regions of the substrate. In addition, insight into the mechanism for diamond nucleation from C clusters is given.

Due to the potential applications of diamond films in the protective coatings and electronics industries, a variety of thin-film deposition techniques have been developed. These include chemical vapor deposition (CVD),[1] plasma enhanced chemical vapor deposition (PECVD),[2] and hot filament CVD (HFCVD),[3] as well as gas torch[4] and plasma torch[5] methods. However, for all of the above deposition techniques, a number of basic film growth requirements are necessary which presently limit the full commercial utilization of CVD diamond films. These include high substrate growth temperatures, and the necessity of pretreating nondiamond substrates by polishing them with diamond grit. This pretreatment is essential to obtain a high enough initial nucleation density of diamond crystallites on the substrate to form continuous films. Such a pretreatment is not desirable for coating large or nonplanar surface areas. In this letter, we discuss a nucleation method which substitutes the need for surface abrasion and diamond seeding by using thin layers of C cluster films sublimated onto various nondiamond substrates. The thin C cluster layer produces a large number of nucleation sites suitable for the formation of continuous diamond films. The specificity of the type of cluster is found to aid in elucidating the mechanism for diamond nucleation in low-pressure CVD.

Various studies have described possible diamond forming precursors on which to nucleate diamond. Matsumoto et al.[6] have identified several hydrocarbon cage compounds as possible diamond cluster embryos, or "seeds," which included adamantane, tetracyclopentane, and hexacyclopentane. Angus et al.[7] have proposed various saturated, multiple fused, ring compounds as possible precursors for nucleation. No detailed reports have appeared on the success of using the above model compounds to nucleate diamond. We have used hydrocarbon based oil and adamantane as nucleating agents. But, these hydrocarbon precursors were found to be unstable in H plasma at temperatures for diamond nucleation and growth.

We report here the results of enhancement of diamond nucleation on Si substrates, with the use of C_{60} and C_{70}. C_{60} and C_{70} are obtained at Northwestern from C soot, prepared by graphite evaporation in a He atmosphere. The purified C_{60} and C_{70} clusters are, next, sublimated onto (100) Si substrates at 500 and 600 °C, respectively. The C clusters are sublimated onto the substrates either as continuous films (between 500 and 2000 Å thick) or in the form of 200-μm-diam dots (with the use of a shadow mask).

Diamond films are grown in a microwave plasma reactor operating at 2.45 GHz, as has previously been described.[2] Typical deposition conditions for the diamond samples discussed are pressure of 100 Torr, and a gas composition of 1% methane in H, at a typical flow rate of 100 sccm. The substrates used for diamond deposition are heated to approximately 900 °C by direct interaction with the H/methane plasma operating at 800 W. Since each C atom in either the C_{60} or C_{70} cluster is completely terminated by other C atoms, we provide a pretreatment to help activate the cluster film surface for diamond nucleation. In this pretreatment, a microwave discharge is sustained at a pressure of 15 Torr with a methane concentration of 10% in H. The substrate (either with continuous C_{70} film, or with C_{70} dots) is biased negatively with respect to the plasma, in the voltage range from 200 to 300 V. Following such a pretreatment, which lasts for several minutes, the diamond growth is initiated at the standard processing conditions described earlier.

Figure 1(a) shows a scanning electron micrograph of diamond growth on an array of 1000-Å-thick C_{70} dots. The use of C_{70} dots on the Si surface demonstrates clearly the effect of C_{70} as the nucleating agent. Note the sharp contrast in the density of the diamond growth between the Si surface and the areas which are covered with the C_{70} dots. The massive nucleation and growth occurs only where there is C_{70} present. We would like to point out several important observations here: (1) The pretreatment process (described above) with methane gas does not produce a diamondlike film and/or an amorphous film on the Si surface. This was confirmed by Raman spectroscopy (and surface enhanced Raman spectroscopy). (2) Without (negative)

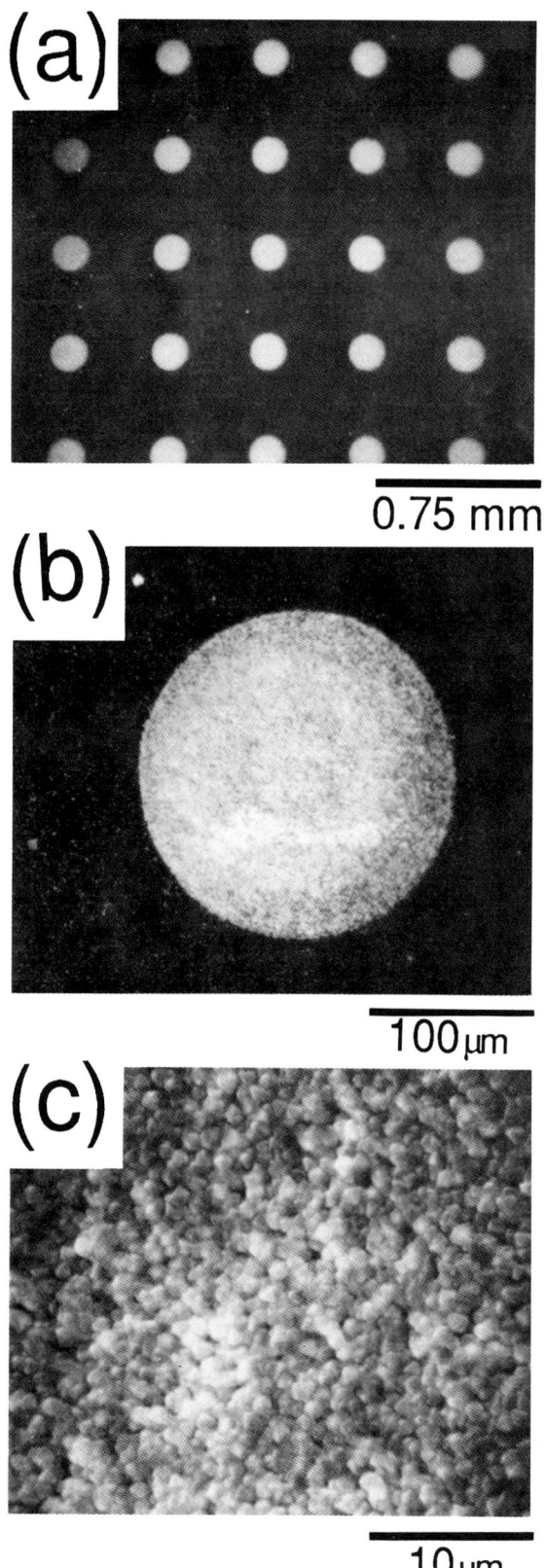

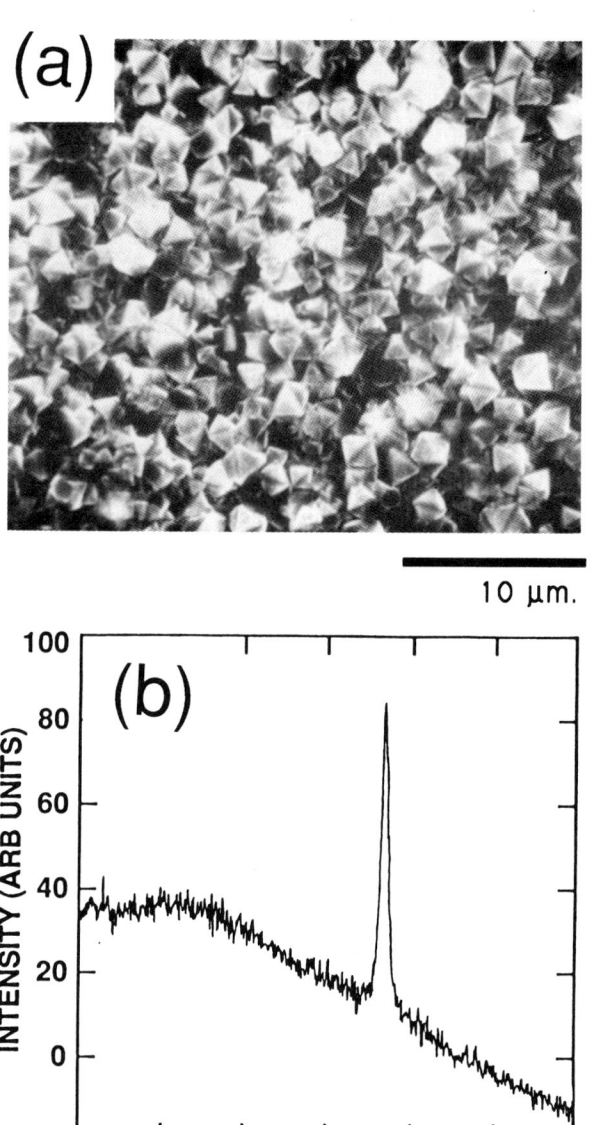

FIG. 2(a) Scanning electron micrograph of a continuous diamond film grown on C_{70} thin film on a Si substrate. (b) Raman spectra of diamond film of Fig. 2(a).

FIG. 1(a) Scanning electron micrographs of diamond growth on C_{70} dot arrays, (b) closer view of a dot, and (c) the underlying polycrystalline diamond morphology.

biasing the substrate with respect to the plasma, there is no nucleation enhancement either on the Si surface, or on the C_{70} dot. By contrast, when a negative bias is applied during

pretreatment only where C_{70} is present, there is massive diamond nucleation or growth. (3) Negative biasing implies ion bombardment by positive plasma charged species on the C_{70} surface, and is necessary to initiate diamond nucleation. Figure 1(b) is a closer view of an individual C_{70} dot, while Fig. 1(c) details the grain size and morphology of the continuous diamond film. The grain size in the film is less than 1 μm, which is comparable to polycrystalline diamond grown on a diamond polished Si surface. The amount of nucleation enhancement due to the presence of C_{70} is estimated to be about ten orders of magnitude, relative to an untreated Si surface. A closer view of a continuous diamond film, under scanning electron microscopy, shows the dominance of (111) faceted crystallites [see Fig. 2(a)]. The presence of diamond has been identified by the characteristic Raman peek at 1332 cm^{-1}, as seen in Fig.

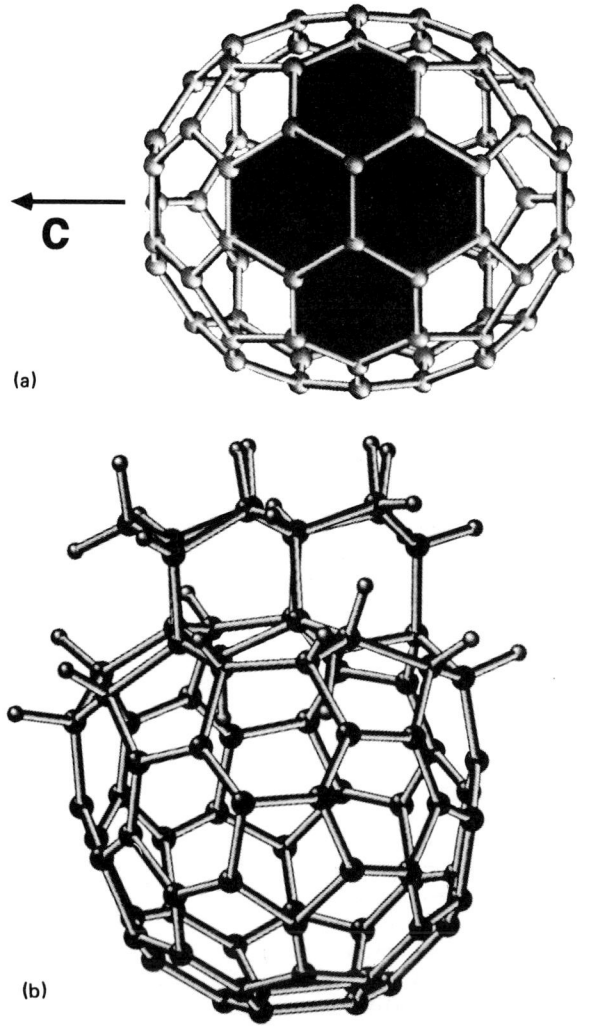

(a)

(b)

FIG. 3(a) Model of C_{70} molecular structure illustrating one ensemble of four fused hexagonal rings as a possible diamond nucleation site. (b) Model of (111) diamond nucleation on C_{70} derivative. The darker and larger atoms represent C, and the lighter and smaller atoms represent H, in this model.

2(b). Other substrates, such as Mo and SiO_2, have also been used, and show similar results as C_{70} in Si. In like manner, a series of experiments were also conducted with C_{60} dots deposited onto Si substrates. Although an enhancement (by a few orders of magnitude) in nucleation is also found for C_{60} films, it never approaches the degree of that of C_{70}. Possible reasons for this are the much higher mobility of C_{60} on the surface of the Si, its lower sublimation temperature, or its different molecular topology.

We propose a simple and speculative model to explain our results. An appropriate model, in this case, must take into account three key points: (1) The stability of the C cluster; (2) the specificity of the C_{70} carbon cluster structure; and (3) the need for surface ion bombardment for activation. The C_{70} film is found to be stable in our H

plasma even at temperatures in the range of 400–500 °C. The stability of C_{70} is necessary to allow time for the diamond to nucleate on its surface. We believe the unique structure of C_{70} molecule, compared to that of C_{60}, for instance, has made the nucleation and epitaxy of diamond on C_{70} possible. By positioning the C_{70} cluster with its c-axis parallel to the horizontal plane, we see five ensembles of four hexagons which, because of their near planar microfaceted structure, are ideal for diamond nucleation, as seen in Fig. 3(a). In this model, energetic cation impact (cations from the plasma) activation results in sp^2 to sp^3 conversion of one of the ensembles, followed by its termination with H or CH_3 growings from the gas-phase species. Such a restructured surface on C_{70} would produce chair-form cyclohexane linkages, which are well known to be the most stable six member ring confrontation free from strain. These linkages are representative of the surface of the (111) plane of diamond [see Fig. 3(b)]. Onto this C_{70} derivatized, hydrogenated/methylated surface, a (111) diamond lattice is then easily propagated. Preliminary surface enhanced Raman results seem to support our simple model. Detailed studies of the nucleation process is currently underway.

In conclusion, we have discussed the use of C clusters as diamond nucleation sites on Si substrates. This nucleation method substitutes the current practice of polishing surfaces with diamond grits. We have also demonstrated how C clusters can be used to selectively grow diamond on Si surfaces. In addition, our process provides a means of better understanding the mechanism of diamond nucleation. From our experiments, we can also speculate the reason why surface pretreatment is not necessary in the case of flame torch diamond deposition methods. We postulate that C clusters formed by the torch are helping to nucleate diamond on surfaces. The use of C clusters for diamond growth on other substrate materials, and the details of diamond nucleation will be reported elsewhere.

This research is sponsored by the Office of Naval Research, and the DOE Basic Energy Science Division (Contract No. DE-FG02-87ER45314). The use of the Materials Research Center facilities, supported by the NSF-MRL grant, is also acknowledged.

[1] S. Koizumi and T. Inuzuka, J. Cryst. Growth **99**, 1188 (1990).
[2] R. Meilunas and R. P. H. Chang, in *Proc. of the 2nd ICEM Conf.*, edited by R. P. H. Chang, Michael Geis, Bernie Meyerson, David Miller, and R. Ramesh (Mater. Res. Soc., Pittsburgh, PA, 1990), p. 609.
[3] A. M. Bonnot, Phys. Rev. B **41**, 6040 (1990).
[4] Y. Tzeng, C. Cutshaw, R. Philips, T. Srivingunon, A. Ibrahim, and B. H. Loo, Appl. Phys. Lett. **56**, 134 (1990).
[5] F. Zhang, Y. Zhang, Y. Yang, G. Chen, and X. Jiang, Appl. Phys. Lett. **57**, 1467 (1990).
[6] S. Matsumoto and Y. Matsui, J. Mater. Sci. **18**, 1785 (1983).
[7] J. C. Angus, M. Sunkara, C. C. Hayman, and F. A. Buch, Carbon **28**, 745 (1990).

Reprinted with permission from Nature
Vol. 356, pp. 225–226, 19 March 1992

Optical limiting performance of C$_{60}$ and C$_{70}$ solutions

Lee W. Tutt & Alan Kost

Hughes Research Laboratories, Malibu, California 90265, USA

OPTICAL sensors used in connection with bright sources such as lasers and arc welders must commonly be protected from damaging light levels by the use of optical limiters[1]. One approach to optical limiting makes use of materials whose optical transmittance decreases at high light levels[2-5]. For most protective applications, the response must be rapid and the saturation threshold low; a lower threshold provides a greater safety margin. Studies of the optical properties of C$_{60}$ have shown that the absorption cross-section of the photoexcited triplet state is greater than that of the ground state[6], suggesting that it may have a nonlinear optical response of the sort useful for optical limiting. Here we report measurements of the optical response of solutions of C$_{60}$ and C$_{70}$ in methylene chloride and toluene, using 8-ns pulses of 532-nm-wavelength laser light. We observed optical limiting behaviour in all cases, with saturation thresholds equal to or lower than those reported for other optical-limiting materials currently in use.

Electro-evaporated soot was purchased from the MER Corporation. The toluene-soluble portion of the soot was separated and purified by the method of ref. 7. The measuring apparatus has previously been described[2]. We use temporally smooth, spatially uniform, optical pulses from a mode-locked, frequency-doubled Nd–YAG laser as the excitation source, and measure the incident and transmitted energy for each pulse. The output from the laser is collimated. The incident fluence is first increased and then decreased to check for evidence of photodegradation.

The optical limiting responses of a 63% and an 80% transmitting toluene solution of C$_{60}$ to 532-nm optical pulses are shown in Fig. 1. At very low fluences the optical response of the solution obeys Beer's law and the transmittance is roughly constant. At an incident intensity of ~100 mJ cm^{-2} the transmittance begins to decrease markedly. The transmitted fluence effectively becomes clamped at ~65 mJ cm^{-2}. The 80% transmitting sample behaved similarly, clamping at 240 mJ out to greater than 3 J cm^{-2}. Solutions of C$_{60}$ in methylene chloride give similar results indicating solvent effects are small.

Figure 2 is a comparison of the performance of C$_{60}$ to that which has been reported for other soluble, absorbing materials

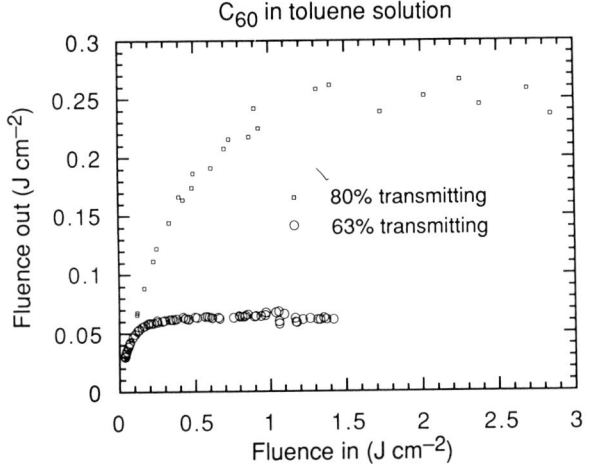

C$_{60}$ in toluene solution

80% transmitting
63% transmitting

FIG. 1 Optical limiting response of 63% and 80% transmitting solutions of C$_{60}$ in toluene to 7 ns, 532 nm optical pulses.

Solutions of C$_{70}$ also show optical limiting action. The output limiting fluence is ~350 mJ cm^{-2} for a 70% C$_{70}$ solution in toluene. The reddish-orange C$_{70}$ solutions have a higher absorption at 532 nm than the magenta C$_{60}$ solutions. It is therefore not surprising that the higher ground-state absorption of C$_{70}$ results in a smaller ratio of excited-state to ground-state absorption and a higher threshold for optical limiting.

For many optical limiting applications, solid films are desirable. Further work is underway to elucidate the response of fullerene films to optical pulses. ☐

1. Wood, G. L., Clark, W. W. III, Miller, M. J., Salamo, G. J. & Sharp, E. J. *SPIE Proc.* **1105**, 154–180 (1980).
2. Tutt, L. W. & McCahon, S. W. *Opt. Lett.* **15**, 700–703 (1990).
3. Hoffman, R. C., Stetyick, K. A., Potember, R. S. & McLean, D. G. *J. opt. Soc. Am.* B**6**, 772–777 (1989).
4. Tutt, L. W., McCahon, S. W. & Klein, M. B. *SPIE Proc.* **1307**, 315–326 (1990)
5. Coulter, D. R. *et al. SPIE Proc.* **1105**, 42–51 (1989).
6. Arbogast, J. W. *et al. J. phys. Chem.* **95**, 11–12 (1991).
7. Ajie, H. *et al. J. phys. Chem.* **94**, 8630–8633 (1990).
8. Lund, D. J., Edsall, P. & Masso, J. D. *SPIE Proc.* **1207**, 193–201 (1990).

ACKNOWLEDGEMENTS. We thank S. W. McCahon and M. B. Klein for valuable discussions.

whose transmittance saturates at a new lower level ('reverse-saturable' materials)[2–5]. We measured the performance of each of these compounds as a 70% transmitting solution at 532 nm. The optical limiter with the lowest threshold for induced absorption is C$_{60}$. It should be noted that chloroaluminum phthalocyanine is a limiter at 532 nm but is a saturable absorber at 650 nm (refs 7, 8).

The optical limiting properties are consistent with the efficient population of a triplet state with a much higher absorption cross-section than the ground state. For pulse lengths less than the lifetime of the triplet state, the triplet state will act as an accumulation site. The limiting action of C$_{60}$ solutions will be most effective for pulses less than 40 μs, the triplet state lifetime. For a longer pulse, a considerable fraction of the excited molecules decay back to the ground state before the end of the pulse. The triplet–triplet absorption spectrum roughly follows the ground-state absorption[6]; we therefore expect the limiting action to be very broadband. Although there may be a nonlinear scattering component to the limiting action, the dominant mechanism is reverse saturable absorption as evidenced by the small solvent dependence.

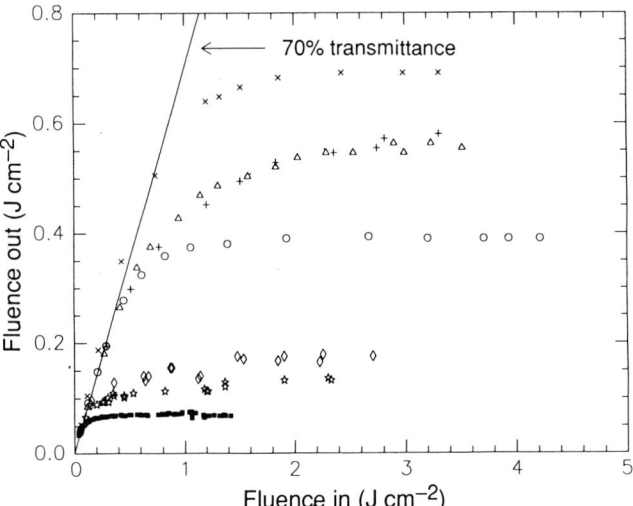

70% transmittance

FIG. 2 Comparison of the optical limiting response of solutions of various reported optical limiters to C$_{60}$ in toluene. All comparison solutions are 70% transmitting at 532 nm, and the solvent was methylene chloride, except with the two exceptions of chloroaluminum phthalocyanine, which was dissoved in methanol, and indanthrone, dissolved in dilute KOH. ○, HFeCo$_3$(CO)$_{10}$ (P(CH$_3$)$_3$)$_2$; △, HFeCo$_3$(CO)$_{12}$; +, (N(C$_2$H$_5$)$_4$)$^+$ (FeCo$_3$(CO)$_{12}$)$^-$; ×, HFeCo$_3$(CO$_{10}$) (P Ph$_3$)$_2$; ◇, indanthrone; ☆, chloroaluminium phthalocyanine; ■, C$_{60}$.

Reprinted with permission from Nature
Vol. 356, pp. 585–587, 16 April 1992
© 1992 Macmillan Magazines Limited

Photoconductivity of fullerene-doped polymers

Y. Wang

Central Research and Development, Du Pont Co.,
Wilmington, Delaware 19880-0356, USA

PHOTOCONDUCTING materials are employed in many technological applications, such as photodetection and electrostatic imaging.[1] We report here the preparation of photoconducting films of polyvinylcarbazole (PVK) doped with fullerenes (a mixture of C_{60} and C_{70}). The performance of this material is comparable with some of the best photoconductors available commercially, such as thiapyrylium dye aggregates[2]. The quantum yield for primary charge separation and the initial ion-pair distance are calculated, within the framework of the Onsager model[3,4], to be 0.9 and 19 Å respectively. The wavelength dependence of photoconductivity is essentially determined by the absorption spectrum of the fullerenes, with the active range extending from about 280 to 680 nm. The isolation of other fullerenes and fullerene derivatives may lead to the development of a large number of fullerene-based polymeric photoconductors in the future.

The fullerenes used were synthesized according to the electric arc method[5], giving a mixture of C_{60} and C_{70} in the ratio ~85:15. Thin films of the fullerene-doped polymer are prepared by spin-coating fullerene and PVK (molecular weight ~5×10^5) from solution in toluene (typically 40 mg of fullerene and 1.5 g of PVK in 12 ml of toluene) onto aluminium substrates. The samples are dried in an oven at 100 °C for several hours before use. The film thickness can be varied from 1 to 30 μm, by changing the concentration of PVK and the spin speed. The films are optically clear and air-stable for at least several months.

Photoconductivity is measured using the standard photo-induced discharge method[6,7]. The sample film, deposited on an electrically grounded aluminium substrate, is first corona-charged (positively or negatively) in the dark. The amount of surface charge is detected by an electrostatic voltmeter (Monroe Electronics, model 244). On exposure to light, the photogenerated electrons and holes migrate to the surface, recombine with surface charges and discharge the surface potential. This photo-induced discharge process is the basis of xerography.

For light of sufficiently low intensity, absorbed within a small fraction of the film thickness, the charge generation efficiency, ϕ, can be obtained from the initial discharge rate of the surface potential, $(dV/dt)_{t=0}$ (refs 6, 7)

$$\phi = -\frac{\varepsilon}{4\pi eLI}\left(\frac{dV}{dt}\right)_{t=0}$$

where ε is the dielectric constant, e the electronic charge, L the film thickness and I the absorbed photon flux. I have determined the wavelength and field dependence of ϕ using a xenon lamp as the light source below 400 nm, and a tungsten lamp for wavelengths >400 nm. The wavelength was selected using a monochromator. A Uniblitz shutter was used to limit the exposure time to 0.1 or 0.3 s. The light intensity used (typically 5×10^{12}–6×10^{13} photons cm^{-2} s^{-1}) was low enough to avoid the space charge effect.

Figure 1 compares the photo-induced discharge curve of pure PVK with that of a fullerene-doped PVK film under identical experimental conditions. Broad band light from a tungsten lamp (50 mW cm^{-2}) was used in this case. This illustrates qualitatively the marked enhancement in the photo-induced discharge rate when PVK is doped to a few per cent with the fullerene mixture. Wavelength-dependent measurements show that the charge generation efficiency is enhanced by a factor of >50 at 500 nm and a factor of ~4 at 340 nm by doping PVK with ~2.7% fullerene (by weight). Figure 1 also shows that the fullerene-doped PVK film possesses low dark conductivity, and a photo-induced discharge that is both fast and complete; three important criteria for electro-photographic applications. I have compared it with polycarbonate film doped with thiapyrylium dye aggregate, one of the best commercial photoconductors[2], and found that both have comparable photo-induced discharge responses over a 40-fold change in the tungsten light intensity.

Figure 2 shows the wavelength dependence of the charge generation efficiency $((\mathrm{d}V/\mathrm{d}t)(1/I)$, also called the gain spectrum) for a fullerene/PVK film with both positive and negative charging. Also shown is the optical absorption spectrum of the film. The gain spectra are similar to, but not identical with, the absorption spectrum, the visible region of the gain spectrum being relatively stronger than that of the absorption spectrum. For wavelengths above 350 nm, efficient photo-induced discharge can be achieved with both positive and negative charging, but in the strongly absorbing region (<350 nm), only positive charging is effective. For the strongly absorbing wavelengths, carriers are generated near the surface and have to migrate across the polymer film; the results show that only holes can achieve this.

Figure 3 shows the field dependence of the charge generation efficiency, measured at 340 nm with a photon flux of 1.62×10^{13} photons cm^{-2} s^{-1}. The Onsager model[3] is recognized as a good starting point for analysing such data, although a modified version may be required for a full interpretation[8]. It provides a means by which the charge-generation efficiency of a range of photoconductors can be compared. The model assumes that the absorption of a photon creates a bound electron–hole pair with a thermalized separation distance of r_0. This bound electron–hole pair either recombines, or separates into a free electron and hole. The fraction of absorbed photons that generate bound electron–hole pairs is the primary quantum yield, ϕ_0, and is assumed to be field-independent. By fitting the experimental data to the Onsager model,[3,4] I obtain values for both r_0 and ϕ_0. Figure 3 shows the best-fit theoretical curve, giving $r_0 = 19$ Å and $\phi_0 = 0.9$. For comparison, I also show curves calculated using $r_0 = 17$ Å and $r_0 = 21$ Å. The thermalization length and primary quantum yield of thiapyrylium dye aggregate films[2] are 44 Å and 0.58 respectively.

The effects of the fullerene on the charge generation and charge transport processes in this new photoconductor still need to be studied in detail. The fullerenes are clearly responsible for the charge generation, as demonstrated by the similarity between the gain spectra and the absorption spectrum (Fig. 2). I have also shown that fullerene can form charge-transfer (CT) complexes with electron-donating aromatic amines such as N,N–diethylaniline[9]. The charge-generating process could thus be due to either the excitation of fullerene (followed by electron hopping from carbazole to fullerene), or the direct excitation of fullerene/carbazole CT complexes. I cannot distinguish between these two possibilities using the gain spectra alone, as

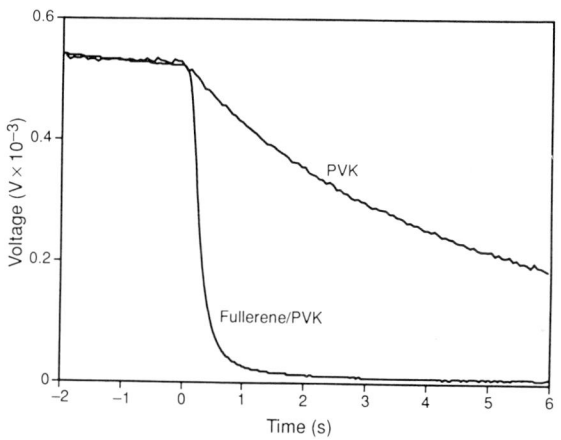

FIG. 1 A qualitative comparison of the photo-induced discharge curves for pure PVK and fullerene-doped PVK under the same experimental conditions. A tungsten lamp (50 mW cm^{-2}) is used as the light source.

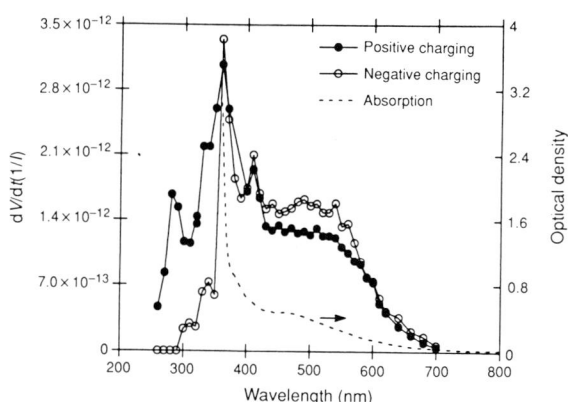

FIG. 2 The wavelength dependence of the charge generation efficiency for a 28 μm fullerene/PVK film, obtained for both positive (●) and negative (○) charging. The dashed line shows the optical absorption spectrum of the film. The applied voltage is ~3×10^5 V cm^{-1}.

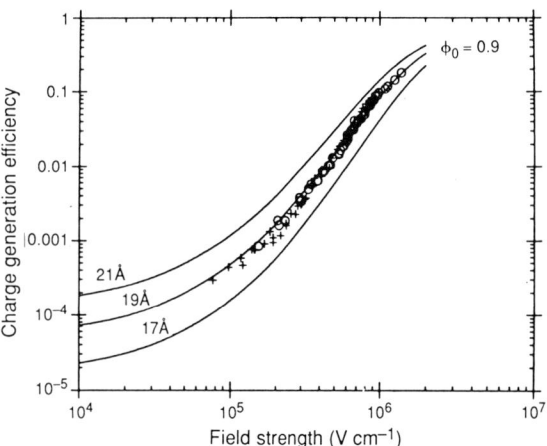

FIG. 3 The field dependence of the charge generation efficiency for 4.5 μm (○) and 9.3 μm (+) fullerene/PVK films, with positive charging and 340 nm irradiation. The solid lines are calculated from the Onsager model. The best-fit curve is obtained with $r_0 = 19$ Å and $\phi_0 = 0.9$.

both the C_{60}/carbazole CT complex and C_{60} have similar absorption spectra: in a separate experiment, I found that the absorption spectrum of the C_{60}/carbazole CT complex differ from that of C_{60} only by an enhanced, broad absorption band in the 500 nm region.

Because the concentration of fullerene used is only a few percent by weight (below the percolation threshold), the transport of carriers has to occur through the PVK polymer matrix. To understand the transport mechanism, it is important to study the effects of dopant level on the charge-carrier mobility. The doping level used in this study is near the maximum imposed by the solubility of fullerene in toluene. To increase the dopant level significantly, a different solvent is required. Increasing the dopant level should further enhance the photoconductive properties through enhanced light absorption and charge-generation efficiency, but any effect on the carrier mobility remains to be determined.

With the fullerene mixtures used in this work (~85% C_{60}), the photoconductor is sensitive only up to wavelengths ~680 nm. It has been reported that some higher fullerenes (for example, C_{76}) have absorption bands in the longer wavelength region[10,11]. It is therefore possible that the range of sensitivity can be extended to the technologically important red and infrared regions using the higher fullerenes or fullerene derivatives. □

---

1. *Photoconductivity and Related Phenomena* (eds Mort, J. & Pai, D. M.) (Elsevier, Amsterdam, 1976).
2. Borsenberger, P. M., Chowdry, A., Hoesterey, D. C. & Mey, W. *J. appl. Phys.* **49**, 5555–5564 (1978).
3. Onsager, L. *Phys. Rev.* **54**, 554–557 (1938).
4. Mozumder, A. *J. chem. Phys.* **60**, 4300–4304 (1974).
5. Kratschmer, W., Lamb, L. D., Fostiropoulos, K. & Huffman, D. R. *Nature* **347**, 354–358 (1990).
6. Regensburger, P. J. *Photochem. Photobiol.* **8**, 429–440 (1968).
7. Chen, I. & Mort, J. *J. appl. Phys.* **43**, 1164–1170 (1972).
8. Braun, C. L. *J. chem. Phys.* **800**, 4157–4161 (1984).
9. Wang, Y. *J. phys. Chem.* **96**, 764–767 (1992).
10. Diederich, F. *et al. Science* **252**, 548–551 (1991).
11. Ettl, R., Chao, I., Diederich, F. & Whetten, R. L. *Nature* **353**, 149–153 (1991).'

ACKNOWLEDGEMENTS. I thank S. C. Freilich for discussions, W. Bindloss and S. C. Freilich for use of the equipment, and S. Harvey for technical assistance.

Reprinted with permission from the Journal of the American Chemical Society
Vol. 115, No. 15, pp. 6506–6509, 1993

Inhibition of the HIV-1 Protease by Fullerene Derivatives: Model Building Studies and Experimental Verification

Simon H. Friedman,[†] Dianne L. DeCamp,[†] Rint P. Sijbesma,[‡] Gordana Srdanov,[‡] Fred Wudl,[‡] and George L. Kenyon[*,†]

Contribution from the Department of Pharmaceutical Chemistry, University of California, San Francisco, California 94143, and Department of Chemistry, University of California, Santa Barbara, California 93106

Received February 16, 1993

Abstract: The ability of C_{60} fullerene ("Bucky Ball") derivatives to interact with the active site of HIV-1 protease (HIVP) has been examined through model building and simple physical chemical analysis. The model complexes generated via the program DOCK3 suggest that C_{60} derivatives will fit snugly in the active site, thereby removing 298 $Å^2$ of primarily nonpolar surface from solvent exposure and driving ligand/protein association. The prediction that these compounds should bind to the active site and thereby act as inhibitors has been borne out by the experimental evidence. Kinetic analysis of HIVP in the presence of a water-soluble C_{60} derivative, bis(phenethylamino–succinate) C_{60}, suggests a competitive mode of inhibition. This is consistent with and supports the predicted binding mode. Diamino C_{60} has been proposed as a "second-generation" C_{60} derivative that will be able to form salt bridges with the catalytic aspartic acids in addition to van der Waals contacts with the nonpolar HIVP surface, thereby improving the binding relative to the tested compound.

Introduction

The protease specific to the human immunodeficiency virus 1 (HIVP) has been shown to be a viable target for antiviral therapy.[1] The active site of this enzyme can be roughly described as an open-ended cylinder which is lined almost exclusively by hydrophobic amino acids (Figure 1a). Notable exceptions to this hydrophobic trend are the two catalytic aspartic acids (Asp 25, Asp 125) which catalyze the attack of water on the scissile peptide bond of the substrate. We hypothesized that since a C_{60} molecule (i.e., fullerene) has approximately the same radius as the cylinder that describes the active site of the HIVP and since C_{60} (and its derivatives) is primarily hydrophobic, an opportunity therefore exists for a strong hydrophobic interaction between the C_{60} derivative and active site surfaces. This interaction should make C_{60} derivatives inhibitors of the HIVP. In this work, we describe model complexes of C_{60} and HIVP generated via the program DOCK3. The surface that is desolvated due to complex formation is shown to be almost exclusively hydrophobic. In addition, kinetic analysis is presented that supports a competitive mode of inhibition of a tested C_{60} derivative, consistent with our intuition and the complexes generated. Finally, we propose and validate the design of an amino-derivatized C_{60} as a reasonable next step in improving the binding energy of C_{60} derivatives to the HIVP.

Results and Discussion

To test the hypothesis regarding the complementarity of the C_{60} with the HIVP active site, a model of C_{60} was created and minimized using the SYBYL package (Version 5.4, Tripos Associates, Inc.). The model produced had a diameter within 0.2 Å of a spectroscopically determined C_{60} structure.[2] This model was fitted into the active site of the so-called "open" (i.e., uncomplexed) form of the HIVP[3] using the program DOCK3.[4] DOCK3 finds optimal orientations of a ligand with its receptor,

scoring on the basis of van der Waals contacts and complementary electrostatics. This procedure produced complexes with the C_{60} squarely in the center of the active site, forming good van der Waals contacts with the active site surface, thereby reinforcing our model. Figures 1b and 1c show the highest scoring complex of C_{60} with HIVP in "front" and "side" views, which show the van der Waals surface contacts.

The change in solvent-exposed surface upon binding was determined in order to approximate the maximum magnitude of hydrophobic interactions. This was accomplished by first determining the total surface area of the active site and C_{60} molecules separately and then subtracting the total surface area of the highest scoring DOCK3 C_{60}/HIVP complex. All surface areas were determined from molecular surfaces generated by the program MS (Michael Connolly, University of California, San Francisco). The calculation indicates that 298 $Å^2$ of primarily hydrophobic surface is removed from solvent exposure. This total desolvated surface was further characterized by summing the individual surface elements according to atom type. The result of this summation (Table I) is that the large majority (273 $Å^2$ or 92%) of the desolvated surface is due to C_{60}-carbon/HIVP-carbon atom contact. The small amount of oxygen desolvation (7%) is due primarily to the partial blockage of the catalytic aspartates. Using the figure of 69.2 cal/(mol·$Å^2$) recently shown to accurately describe the free energy released upon desolvation of hydrophobic molecular surface,[5] we calculated the resultant free energy gain upon binding due to the carbon surface that is desolvated to be 19 kcal/mol. In order to estimate an approximate binding constant of a C_{60} derivative, this value has to be corrected for the free energy cost due to loss of translational/rotational entropy that accompanies binding. This value has been estimated to be on the order of 7–11 kcal/mol.[6] After we took this energetic cost into account, the result is a total ΔG_{bind} of 8–12 kcal/mol. Converting this to K_d values using the expression $\Delta G° = -RT \ln K_d$ results in dissociation constants on the order of 10^{-6}–10^{-9} M. Several factors have been left out of this analysis, for example, rotational entropy persistence of the C_{60} in the active site, conformational energy of the HIVP, and interaction of the

† San Francisco.
‡ Santa Barbara.
(1) Debouck, C. *AIDS Res. Hum. Retroviruses* **1992**, *8*, 153–64.
(2) Liu, S. Z.; Ju, Y. J.; Kappes, M. M.; Ibers, J. A. *Science* **1991**, *254*, 408–410.
(3) Wlodawer, A.; Miller, M.; Jaskolski, M.; Sathyanarayana, B. K.; Baldwin, E.; Weber, I. T.; Selk, L. M.; Clawson, L.; Schneider, J.; Kent, S. B. *Science* **1989**, *245*, 616–21.
(4) Meng, E. C.; Shoichet, B. K.; Kuntz, I. D. *J. Comp. Chem.* **1992**, *13*, 505–524.

(5) Tunon, I.; Silla, E.; Pascual-Ahuir, J. L. *Protein Eng.* **1992**, *5*, 715–716.
(6) Novotny, J.; Bruccoleri, R. E.; Saul, F. A. *Biochemistry* **1989**, *28*, 4735–49.

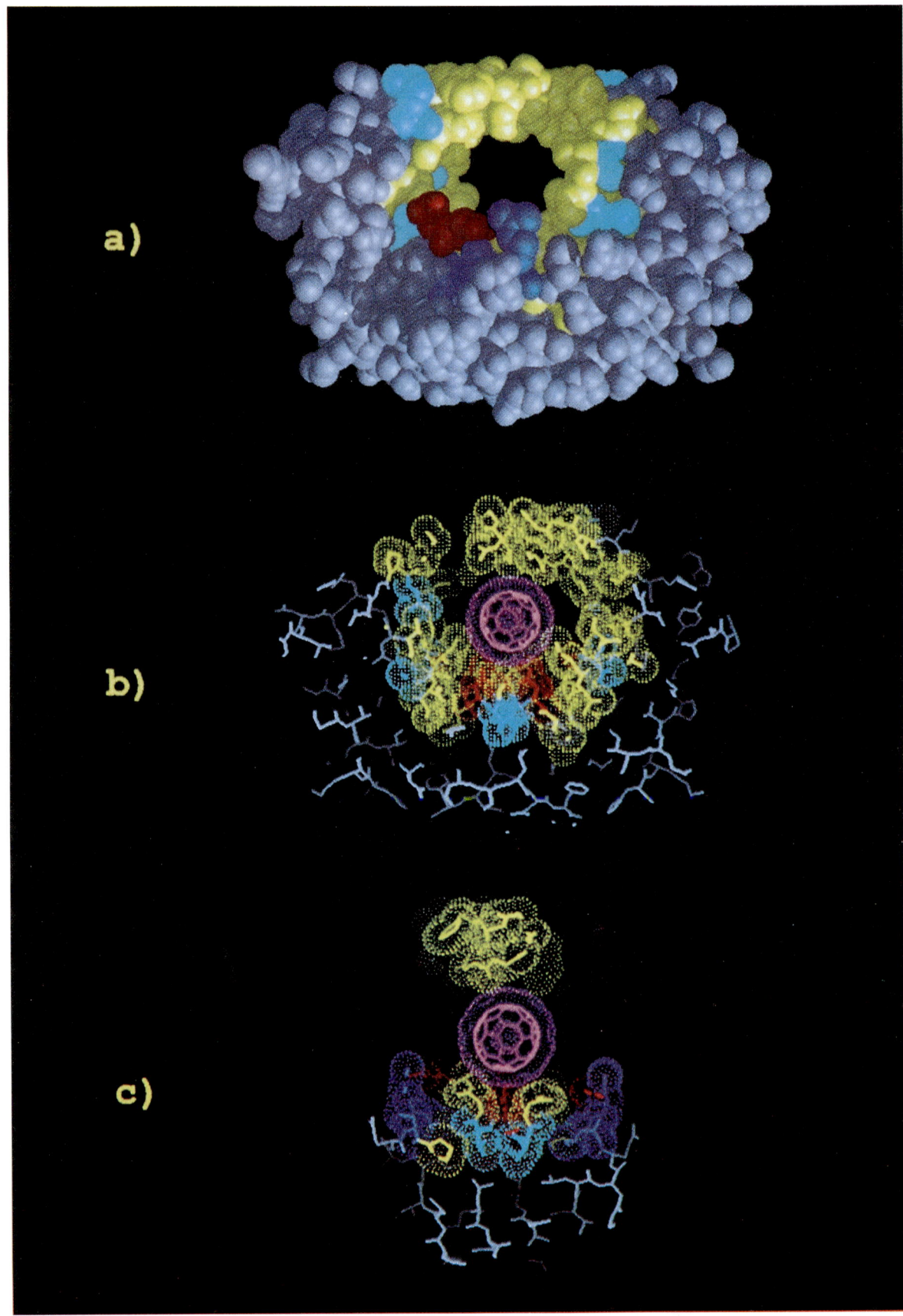

Figure 1. (a) "Front" view of the HIV-1 protease. Color coding is as follows. Yellow: Leu, Ile, Phe, Tyr, Val, Trp, Pro, Gly, Ala. Blue: Lys, Arg. Red: Asp, Glu. Cyan: Thr, Ser, Gln, Asn, Cys, Met, His. Gray: regions greater than 10 Å from the center of the active site. (b) Same view as (a) with the top scoring C_{60} orientation shown. The C_{60} is colored magenta, and the van der Waals surface of the active site and ligand are shown. (c) Same complex as (b) seen at a 90° cross section.

catalytic aspartates with the C_{60} surface. The purpose of this analysis is to account for the factors influencing binding that are reasonably estimated from our understanding of protein–ligand interactions.

Table I. Breakdown of Molecular Surface Changes upon C_{60}/HIVP Complexation According to Atom Type[a]

compd	C	N	O
complex (HIVP + C_{60})	1537.64	109.272	266.456
HIVP	1402.55	112.504	287.898
C_{60}	408.95	0	0
total change	−273.31	−3.232	−21.442

[a] The surface areas of the complex and of HIVP were determined for an identical subset of the total protein structure which contained and flanked the active site.

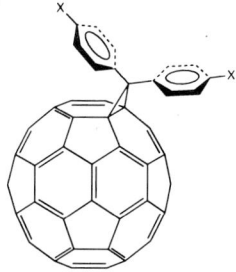

1, X = HOC(O)(CH$_2$)$_2$C(O)NH(CH$_2$)$_2$-

Figure 2. Compound **1**.

A relatively synthetically accessible water-soluble C_{60} derivative, bis(phenethylamino–succinate) C_{60} (compound **1**, Figure 2), was the first test of our hypothesis. The synthesis and characterization of this compound are described in the accompanying manuscript.[7] The highest scoring DOCK3 complex of this compound with the HIVP again positions the core C_{60} in the center of the active site, with the charged side chains extending through the mouth of the active site into solution (Figure 3). The ability of this compound to inhibit the HIVP was assayed with an HPLC method,[8] and its K_i value was found to be 5.3 µM (SE 0.98). The kinetic data fit the pattern of competitive inhibition well (Figure 4). This supports the proposed model complex, as the predicted binding mode of the C_{60} core should preclude any inhibitor binding while substrate is bound. It is of interest to note that compound **1** has been found to inhibit acutely and chronically HIV-1 infected human peripheral blood mononuclear cells (PBMC) with an EC_{50} of 7 µM while showing no cytotoxicity in uninfected PBMC (Raymond F. Shinazi et al., manuscript in preparation).

This introductory example demonstrates the potential for C_{60}-based inhibitors of the HIVP. As a point of comparison, the best peptide-based inhibitors are effective in the subnanomolar range and the best nonpeptide inhibitors are effective in the high nanomolar range.[9–11] The main driving force behind the association of the HIVP and the fullerene derivative examined is presumably hydrophobic interaction between the nonpolar active site surface and the C_{60} surface. In addition, however, there is an opportunity for increasing binding energy by the introduction of specific electrostatic interactions. An obvious possible electrostatic interaction is a salt bridge between the catalytic aspartates on the floor of the active site and a cationic site on the C_{60} surface. It has been found that several dicationic metals are effective inhibitors of the HIVP.[12,13] The authors of this work hypothesized

(7) Sijbesma, R. P.; Srdanov, G.; Wudl, F.; Castoro, J. A.; Wilkins, C.; Friedman, S. H.; DeCamp, D. L.; Kenyon, G. L. *J. Am. Chem. Soc.*, following paper in this issue.

(8) DesJarlais, R. L.; Seibel, G. L.; Kuntz, I. D.; Furth, P. S.; Alvarez, J. C.; Ortiz de Montellano, P. R.; DeCamp, D. L.; Babe, L. M.; Craik, C. S. *Proc. Natl. Acad. Sci. U.S.A.* **1990**, *87*, 6644–6648.

(9) Thaisrivongs, S.; Tomasselli, A. G.; Moon, J. B.; Hui, J.; McQuade, T. J.; Turner, S. R.; Strohbach, J. W.; Howe, W. J.; Tarpley, W. G.; Heinrikson, R. L. *J. Med. Chem.* **1991**, *34*, 2344–2356.

(10) Ikeda, S.; Ashley, J. A.; Wirsching, P.; Janda, K. D. *J. Am. Chem. Soc.* **1992**, *114*, 7604–7606.

(11) DeCamp, D. L.; Babe, L. M.; Salto, R.; Lucich, J. L.; Koo, M. S.; Kahl, S. B.; Craik, C. S. *J. Med. Chem.* **1992**, *35*, 3426–3428.

(12) Woon, T. C.; Brinkworth, R. I.; Fairlie, D. P. *Int. J. Biochem.* **1992**, *24*, 911–914.

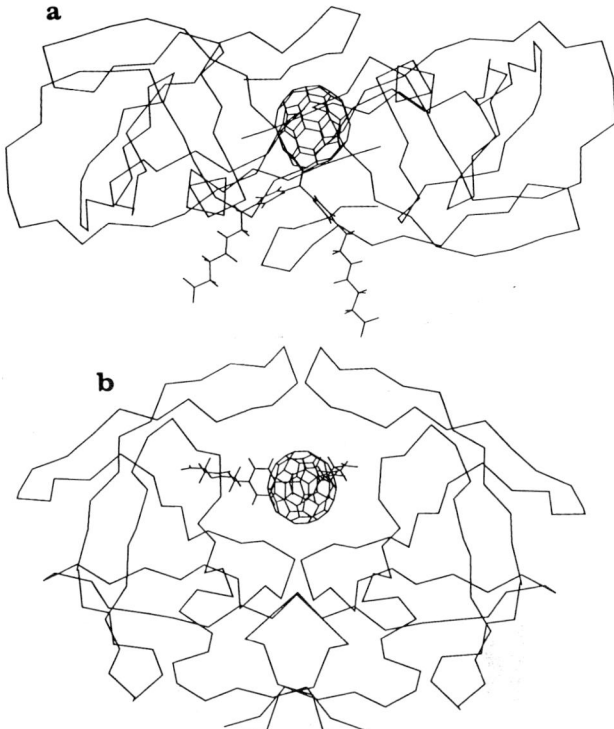

Figure 3. Top DOCK3 complex of compound **1** in (a) "front" and (b) "top" views. For clarity, only the α-carbon chain trace of HIVP is shown.

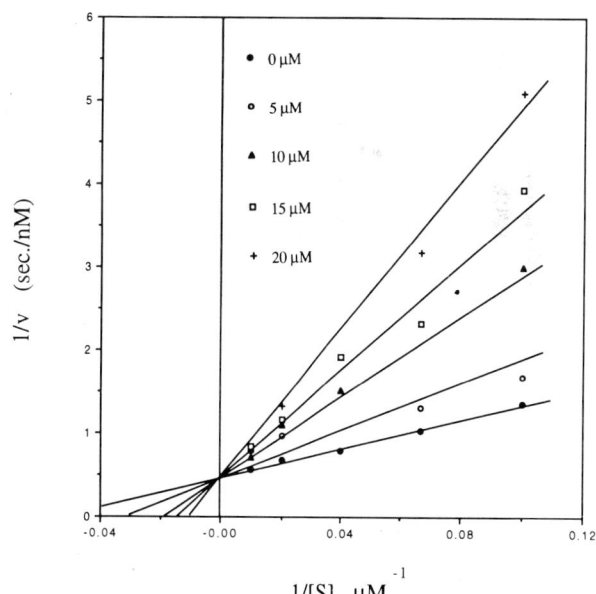

Figure 4. Double reciprocal plot of compound **1** inhibition of HIVP. Assays were performed in buffer containing 50 mM NaAc pH 5.5, 1.0 M NaCl, 5% glycerol, 1% DMSO, and 2 mM EDTA. Kinetic constants were determined by fitting of the data to the equation $v = V_m S/[K_m(1+I/K_i)+S]$ which describes competitive inhibition. K_i: 5.3 µM [0.98]. K_m: 15.9 µM [2.9]. V_m: 1.9 nM/s [0.1]. Standard errors are indicated in brackets.

that a possible mode of inhibition is through a tight electrostatic interaction of the dication with the catalytic aspartates. K_i values for these dications are in the micromolar range, corresponding to ~8 kcal/mol of binding energy, over and above the Gibbs energy loss due to freezing out translational entropy. If even a fraction of the binding energy due to this type of interaction can be added to the existing C_{60} core binding energy, improvements of several orders of magnitude should be realized. It has been shown that the introduction of a single amine/carboxylate salt bridge can increase the binding energy of a ligand to its receptor by ~4 kcal/mol,[14] leading to a 1000-fold improvement in binding.

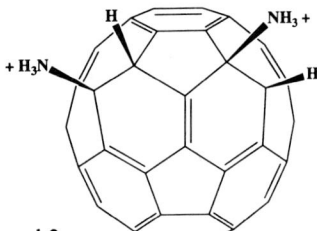

Figure 5. Compound **2**.

With these ideas in mind, we have investigated whether the active site of the HIVP can accommodate a C_{60} that has been derivatized with two amino groups and, in addition, if these groups can be positioned so that they are close enough to the catalytic aspartates to form effective salt bridges. Direct amino adducts can be prepared by reacting a neat amine with the C_{60}, although controlling the stoichiometry of this reaction poses practical problems.[15] Synthetic issues aside, we have modeled compound **2**, 1,4-diamino C_{60}, (Figure 5) and examined its possible interactions with the HIVP active site. DOCK3 is able to orient compound **2** within the active site, placing the core C_{60} in a similar position to that of compound **1**, again allowing extensive nonpolar van der Waals surface interaction. In addition, the two amino groups can effectively bridge the oxygens of the catalytic aspartates, approaching within 2.7 Å and 3.4 Å, respectively (N–O distance), thus making these amino/carboxyl interactions good candidates for improving overall binding (Figure 6).

Conclusions

We have suggested through modeling that C_{60} derivatives should be inhibitors of the HIVP due to their steric and chemical complementarity with the active site. We have demonstrated that this is the case and that this behavior is consistent with a qualitative analysis of hydrophobic surface transfer in model complexes.

Our interest in C_{60} derivatives is twofold. First of all, they represent nonpeptide-based lead compounds that, through careful modeling, may result in effective, tightly binding HIVP inhibitors. Second, they represent a rigid, conformationally restricted scaffold with which we can examine the nature of protein–ligand binding. Because of the steric bulk of C_{60} and its complementarity to the active site surface, there are severe limitations to the orientations it can adopt within the active site. Essentially, the principal degree of freedom of a C_{60} derivative within the active site is rotation around its center. This simplifies the problem of predicting the binding modes of various derivatives. The key to exploiting this system will be the development of the synthetic methodology to facilely and specifically modify the C_{60} surface.

Experimental Section

Modeling. All modeled compounds were generated using the SYBYL 5.4 package. Atomic point charges were calculated using the Gasteiger–Huckel method. For conformationally flexible ligands, torsions were initially set to anticipated low-energy conformers. Minimization to the used model structure was done using the Maximin2 minimizer and Tripos force field and parameters. Docking to the active site of the studied protein was done using the program DOCK3.[4] Grids required by DOCK3 were generated against the dimer formed from the Protein Data Bank file 3hvp, using the standard AMBER united atom charges and van der Waals parameters. Single mode runs of modeled compounds against the active site were performed using the following parameters: dislim = 1.500, nodlim = 5, ratiom = 0.0000, lownod = 4, lbinsz = 0.4000, lovlap = 0.1000, sbinsz = 0.8000, and sovlap = 0.2000. All molecular graphics were produced using the MIDAS Plus system. Molecular surfaces were generated using the program MS, written by Michael Connolly. A probe sphere diameter of 1.4 Å and default values for van der Waals radii were used.

(13) Zhang, Z. Y.; Reardon, I. M.; Hui, J. O.; O'Connell, K. L.; Poorman, R. A.; Tomasselli, A. G.; Heinrikson, R. L. *Biochemistry* **1991**, *30*, 8717–8721.

(14) Santi, D. V.; Pena, V. A. *J. Med. Chem.* **1973**, *16*, 273–280.

(15) Hirsch, A.; Li, Q. Y.; Wudl, F. *Ang. Chem., Int. Ed. Engl.* **1991**, *30*, 1309–1310.

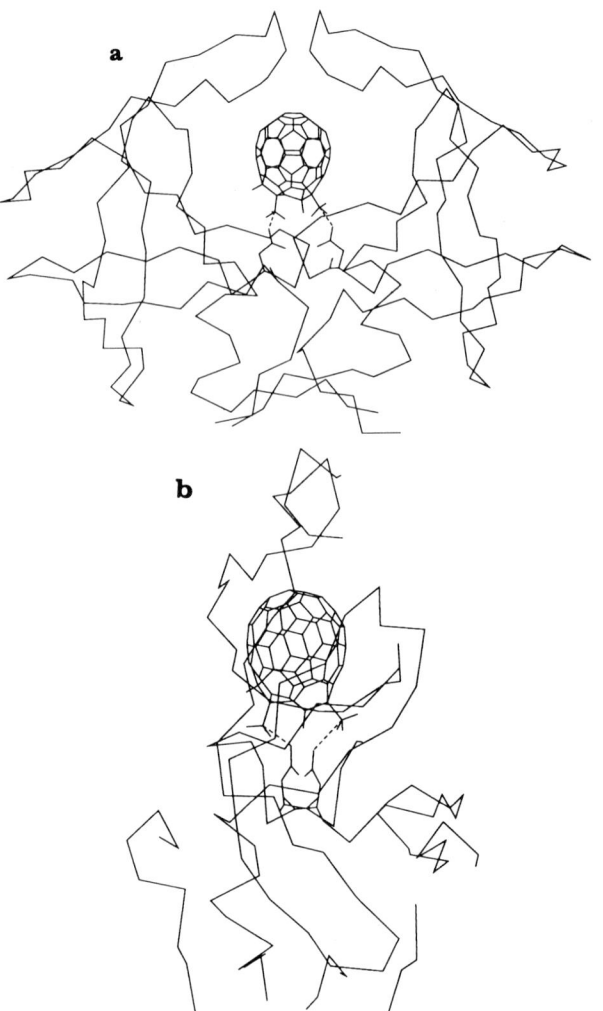

Figure 6. DOCK3 complex of compound **2** with the HIVP in (a) "front" and (b) "side" views. Close approach of compound **2** amino groups and HIVP catalytic aspartate oxygens is highlighted with dashed lines. For clarity, only the catalytic aspartates and the α-carbon chain trace of HIVP are shown.

Enzyme Assays. Compound **1** was assayed against recombinant, affinity-purified HIV-1 protease produced by Bachem Biosciences (0.16 mg protein/mL) at 25 °C. Assays were performed in 100 μL volumes under final conditions of 50 mM NaAc pH 5.5, 1.0 M NaCl, 5% glycerol, 1% DMSO, and 2 mM EDTA. Inhibitor was preincubated with ~0.05 μg of enzyme for 5 min, at which time the reaction was initiated by addition of substrate. The reaction was quenched at <15% product formation by the addition of 15 μL of 10% TFA. The cleavage products of the substrate peptide H-Lys-Ala-Arg-Val-Tyr-*p*-nitro-Phe-Glu-Ala-Ile-NH$_2$ (made by Bachem) were assayed by HPLC using a 10–40% (acetonitrile, 0.1% TFA):(water, 0.1% TFA) gradient over 30 min at 1 mL/min. Product was quantitated by integration of peak areas followed by comparison to product standard curves. Determination of kinetic constants was done with the program KinetAsyst (IntelliKinetics).

Acknowledgment. This work was supported by NIGMS grant GM39552 and NSF grant DMR9111097. The support of an American Chemical Society Medicinal Chemistry Fellowship and an A.F.P.E. Fellowship (S.H.F.) is acknowledged. We thank Diana C. Roe, Elaine C. Meng, and Rafael Salto for valuable discussions and Charles S. Craik for providing recombinant protease.

Note Added in Proof: The parent compound to compound **1**, where X = (CH$_2$)$_2$NH$_2$, was tested and found to have a K_i of ~2 μM. This insensitivity of binding to the nature of the C_{60} side chain supports the predicted binding mode, which positions the side chains away from the active site into full solvent contact.